To John,
With thanks
and best wishes,

FOOD SECURITY AND SOCIOPOLITICAL STABILITY

Food Security and Sociopolitical Stability

EDITED BY

CHRISTOPHER B. BARRETT

UNIVERSITY PRESS

Great Clarendon Street, Oxford, OX2 6DP,
United Kingdom

Oxford University press is a department of the University of Oxford.
It furthers the University's objective of excellence in research, scholarship,
and education by publishing worldwide. Oxford is a registered trade mark of
Oxford University press in the UK and in certain other countries

First Edition published in 2013
Impression: 1

Published in the United States of America by Oxford University Press
198 Madison Avenue, New York, NY 10016, United States of America

British Library Cataloguing in Publication Data
Data available

Library of Congress Control Number: 2013938359

ISBN 978–0–19–967936–2

Printed and bound in Great Britain by
CPI Group (UK) Ltd, Croydon, CR0 4YY

Acknowledgments

After food riots wracked many low- and middle-income families during the global food price spike of 2007–8, and then again in 2010–11, food policy issues suddenly re-emerged on the global stage with a prominence they had lacked since the mid-1970s. An era of agricultural plenty ushered in by the Green Revolution had lulled policymakers and many scientists into a false sense of security that food would always be abundant and food prices low. Reawakened to the challenges of meeting the food needs of a growing, and increasingly well-off and urban, world population facing growing threats of climate change, the global policy community suddenly became keen to learn how food security and sociopolitical stability relate to one another. It also became re-sensitized to the daily preoccupation of a large share of the world's population with their struggles to feed their families.

Thankfully, a cadre of exceptional scholars remained very engaged with these issues even while much of the world's focus drifted elsewhere. The contributors to this volume—each an eminent scholar in his or her field—generously contributed their considerable expertise and goodwill to help pull this volume together. My greatest thanks goes to them, an extraordinary group of brilliant men and women without whom this volume could never have been produced. I learned an enormous amount from our interactions over the year it took from the inception of this project, through our authors' workshop, to a completed book manuscript. Not only are they a remarkable group of sharp minds and cooperative spirits, but they are admirably committed to translating the oftentimes-technical details of their scholarship into clear language accessible to a wide range of prospective readers beyond their core academic disciplines. This volume is intended for a broad audience of scholars, policymakers and their advisers, students, and interested lay readers who want to deepen their understanding of the complex relationship between food security and sociopolitical stability in the twenty-first century.

Such an ambitious project could not have progressed without valuable contributions from many people in addition to the contributing authors. Kathryn Boor, the Ronald Lynch Dean of the College of Agriculture and Life Sciences at Cornell University, and Mike Hoffmann, Director of the Cornell University Agricultural Experiment Station, first brought this opportunity to my attention and were invaluable advocates in getting this project launched. Linda Zall, Rich Engel, and their staffs in the United States intelligence community provided encouragement and a tangible target audience for the contributors. This project was also supported financially by the United States intelligence community. Any opinions, findings, conclusions, or recommendations expressed in this publication are those of the author(s) and do not necessarily reflect the views of the organizations or agencies that provided support for the project.

Cynthia Mathys, Ellen McCullough, Michael Mulford, Marc Rockmore, and Joanna Upton offered invaluable editorial comments on first drafts of the chapters. Art DeGaetano, Harry DeGorter, Drew Harvell, Ron Herring, Steve Kyle, David Lee, Erin Lentz, Phil McMichael, Victor Nee, Rebecca Nelson, Daryl Nydam, David Patel, Ken Roberts, Tony Shelton, Jery Stedinger, Erik Thorbecke, and Lindy Williams each offered insightful comments on first draft chapters at the June 2012 authors' workshop we hosted at Cornell University. The audience at that workshop, especially Jim Baker, Marc Levy, and Mike McElroy, also offered many helpful comments to the contributing authors and me. Charlotte Ambrozek, Alec Kane, and Andrew Pike did a superb job assisting Cynthia Mathys in copy editing and formatting the final chapter manuscripts. Sheri Englund's expert editorial eye and her guidance to the Cornell editorial office made our task considerably easier. Adam Swallow, Jenny Townshend, and Aimee Wright at Oxford University Press had the confidence and patience to take this project from a raw idea to a completed manuscript. My sincere thanks goes to each of these outstanding people.

Cynthia Mathys choreographed all aspects of this project with exceptional professionalism, skill, and good humor. She deserves the lion's share of the credit for the pulling together of the innumerable details of this project so quickly and well.

Finally, I thank my family for their patience and unwavering support as I slogged away at this project. My wife, Clara, and our children, Brendan, Mary Catherine, Joanna, Julia, and Lizzy, are a fabulous support and inspiration. My sincere hope is that the scholarship reflected in this volume and the private and public policy choices it helps inform will help usher in substantial progress in advancing food security and sociopolitical stability worldwide over the course of their lifetimes.

Chris Barrett
Ithaca, NY

December 2012

Contents

List of Figures

List of Tables

List of Abbreviations

ALBA	Bolivarian Alliance for the Americas
AMO	Atlantic Multidecadal Oscillation
AMV	Atlantic Multidecadal Variability
AO	Arctic Oscillation
BACI	before-after-control-impact
BFSP	Brazil's food security policy
BiOS	Biological Innovation for Open Society
BPL	below poverty line
BRIIC	Brazil, Russia, India, Indonesia, and China
CA	conservation agriculture
CAADP	Comprehensive Africa Agriculture Development Program
CAMBIA	Centre for Applications of Molecular Biology in International Agriculture
CBD	Convention on Biological Diversity
CCT	conditional cash transfer
CGIAR	Consultative Group on International Agriculture
CI	contour interval
CIMMYT	International Maize and Wheat Center
COMESA	The Common Market for Eastern and Southern Africa
CONSEA	National Food Security Council (Brazil)
CPI	Consumer price index
CPN–M	Communist Party of Nepal—Maoist
CTE	consumer tax equivalents
DDA	Doha Development Agenda
DPRK	Democratic People's Republic of Korea
DSR	direct seeding of rice
ECCAS	Economic Community of Central African States
ECOWAS	The Economic Community of West African States
ENSO	El Niño–Southern Oscillation
FAO	United Nations Food and Agriculture Organization
FAO-FIVIMS	Food Insecurity and Vulnerability Mapping System
FCI	The Food Corporation of India
FEWS NET	The Famine and Early Warning System Network
GATT	General Agreement on Tariffs and Trade
GCM	General Circulation Models
GEAC	Genetic Engineering Approval Committee
GHI	Global Hunger Index
GIEWS	Global Information Early Warning System

GM	genetically modified
GMO	genetically modified organisms
HDI	Human Development Index
ICISS	International Commission on Intervention and State Sovereignty
IFPRI	International Food Policy Research Institute
IGAD	Intergovernmental Authority on Development
ILRI	International Livestock Research Institute
IMPACT	International Model for Policy Analysis of Agricultural Commodities and Trade
IP	intellectual property
IPR	intellectual property rights
IPO	Interdecadal Pacific Oscillation
IRRI	International Rice Research Institute
ITCZ	Intertropical Convergence Zone
LOC	low-income countries
MAS	marker-assisted selection
MCC	Maoist Communist Center
MDA	Ministry of Agrarian Development
MEP	minimum export prices
MST	Movimento dos Trabalhadores Sem Terra
NAM	Northern Annular Mode
NAO	North Atlantic Oscillation
NARES	National Agricultural Research and Extension Systems
NERICA	New Rice for Africa program
NRA	nominal rates of assistance (to an industry or sector)
OECD	The Organization for Economic Co-operation and Development
OFAC	Office of Foreign Assets Control
OPEC	Organization of the Petroleum Exporting Countries
PAA	Food Acquisition Program (Brazil)
PDO	Pacific Decadal Oscillation
PDV	Pacific Decadal Variability
PIPRA	Public Intellectual Property Resource for Agriculture
PMV	Plan Maroc Vert
PNSAN	National Food and Nutritional Security Plan (Brazil)
PRONAF	National Program for Strengthening Family Farming (Brazil)
PSNP	Productive Safety Net Program
PVP	plant variety protection
PWG	The People's War Group
R2P	responsibility to protect
R&D	research and development
REDD	Reducing Emissions from Deforestation and Forest Degradation
RMB	renminbi
RRA	relative rate of assistance

SADC	The Southern African Development Community
SAM	Southern Annular Mode
SELA	The Latin American and Caribbean economic system
SIDS	small-island developing states
SLP	sea-level pressure
SOCLA	Sociedad Scientifica Latinoamericana de Agroecologia
SOI	Southern Oscillation Index
SRI	system of rice intensification
SST	sea surface temperature
SSTA	sea surface temperature anomaly
s/u	stock-to-use
TAL	transcription activator-like
TAO	Tropical Atmosphere Ocean
TFG	Transitional Federal Government (Somalia)
TFP	total factor productivity
TOGA	Tropical Ocean Global Atmosphere Program
TRIPS	trade-related aspects of intellectual property
UNDP	United Nations Development Program
UNESCO	United Nation Educational, Scientific, and Cultural Organization
UPOV	International Union for the Protection of New Varieties of Plants
USAID	United States Agency for International Development
USDAERS	United States Department of Agriculture Economic Research Service
WFP	World Food Programme
WTO	World Trade Organization

List of Contributors

Arun Agrawal, University of Michigan

Eddie H. Allison, University of East Anglia

Kym Anderson, University of Adelaide, Australian National University, and Center for Economic Policy Research

Christopher B. Barrett, Cornell University

Prapti Bhandary, International Food Policy Research Institute

Mark A. Cane, Lamont-Doherty Earth Observatory of Columbia University

Luc Christiaensen, World Bank

Joshua E. Cinner, James Cook University

Samuel Crowell, Cornell University

Klaus Deininger, World Bank

Dolapo Enahoro, International Livestock Research Institute

Mario Herrero, International Livestock Research Institute

Upmanu Lall, Columbia University

Dong Eun Lee, Lamont-Doherty Earth Observatory of Columbia University

Travis J. Lybbert, University of California, Davis

Timothy R. McClanahan, Wildlife Conservation Society

Susan McCouch, Cornell University

John McDermott, International Food Policy Research Institute

Robert McLeman, Wilfrid Laurier University

Daniel Maxwell, Feinstein International Center, Tufts University

Heather R. Morgan, American University of Cairo

Ryan Nehring, International Policy Centre for Inclusive Growth

Mark W. Rosegrant, International Food Policy Research Institute

Johan Swinnen, LICOS Centre for Institutions and Economic Performance, University of Leuven

C. Peter Timmer, Harvard University and Australian National University

Simla Tokgoz, International Food Policy Research Institute

Joanna B. Upton, Cornell University

Kristine Van Herck, LICOS Centre for Institutions and Economic Performance, University of Leuven

Wendy Wolford, Cornell University

1

Food or Consequences: Food Security and Its Implications for Global Sociopolitical Stability

CHRISTOPHER B. BARRETT

Throughout history, human lives have generally been short, unhealthy, and relatively unproductive, in large part due to insufficient nutrient intake. Conditions changed dramatically, however, after World War II. The world food supply leapt from a level adequate to feed only about two billion people to one that satisfies the full nutritional needs of roughly five billion today, and—with a more equitable distribution—could satisfy all seven-plus billion of us. Dramatic advances in food production, processing, and distribution, coupled with associated real (inflation-adjusted) price declines and income growth among the poor sharply expanded the availability of and access to a nutritionally adequate diet. The reinforcing feedback between nutritional status and productivity then helped to catalyze unprecedentedly rapid and widespread improvements in living standards, manifested in rapid declines in global poverty and advances in various measures of health and nutrition (Barrett 2002; Dasgupta 1993; Fogel 2004).

A major driver of this progress was an unprecedented, concerted effort at agricultural research and development and technology transfer, commonly known as the "Green Revolution." The Green Revolution sharply expanded food production, increasing crop yields by 20 to 25 percent, reducing food prices by 35 to 66 percent, and augmenting per capita caloric intake by 13 to 14 percent relative to a scenario without those scientific advances (Evenson and Gollin 2003). The Green Revolution's achievements were deemed so important that American agronomist Norman Borlaug was awarded the Nobel Peace Prize in 1970 as the "father of the Green Revolution." Nonetheless, the Green Revolution was no panacea: it had its limits and unintended negative environmental and social consequences, although these were more often due, not to the technologies themselves, but rather to the policies used to promote rapid agricultural intensification (Pingali 2012).

Agricultural productivity growth was by no means the only driver of food security improvements. Infrastructure improvements and technological advances in communications and transportation reduced the costs of commerce, enabling newfound surpluses produced in one region to be sold elsewhere inexpensively, reinforcing supply expansion effects on real food prices. These and other advances were abetted by public policy reforms in many countries that reinforced farmers' property rights in land and water and enhanced their discretion in production and marketing decisions, most notably in East and Southeast Asia. Most importantly, the rapid and broad-based income growth and poverty reduction associated with all of these changes then turned increased food availability into tangible food security improvements. Families previously too poor to afford an adequate diet suddenly could.

Unfortunately, success bred complacency, or at least naive optimism. By the mid-1980s, global political and thought leaders seemingly took for granted that the world would enjoy plentiful food in perpetuity. Governments, donors, and scholars turned their attention away from food systems, encouraged by development theories that emphasized industrialization, rather than agricultural development, as the engine of economic growth and poverty reduction. As a consequence, and in spite of widely documented high rates of return on investments in agricultural research and development (R&D), public investment in agriculture from 1991 to 2000 declined by 0.6 percent per year on average in the high-income countries, while the share of foreign aid devoted to agriculture fell even more precipitously, from nearly 17 percent in 1970 to just 4 percent by 2003 (Alston et al. 2000; Pardey et al. 2006; Raitzer and Kelley 2008).

The neglect of agricultural research and related investments quite predictably dampened productivity growth. Annual yield increases in staple cereals—maize, rice, and wheat—fell from around 3 to 5 percent in the late 1970s and early 1980s to just 1 to 2 percent by the early years of the twenty-first century (World Bank 2008). Reversing decades during which production growth steadily outpaced demand increases, by the turn of the millennium, demand had caught up to and soon surpassed supply expansion.

The economics of commodity storage and price dynamics clearly predicts that food price spikes emerge as the natural consequence of demand growth consistently outpacing supply expansion, thereby depleting stocks of storable commodities in competitive markets to the point where small shocks generate outsized price adjustments (Williams and Wright 1991; Deaton and Laroque 1992; Wright 2012). Having grown accustomed to stable, even declining, real food prices, consumers and governments were shocked as food prices spiked repeatedly after 2007, following years of steadily diminishing stocks-to-use ratios. Real food prices reached historic highs in 2008, retreated a bit in 2009 and 2010 as the global economy fell into recession, and then spiked higher still in 2011, with a second, not-quite-as-pronounced spike in summer 2012.

There is little reason to anticipate any substantive change in this fundamental challenge before 2025. Demand-side pressures on national and global food economies—due principally to population and income growth, urbanization, and dietary change—will continue to build. The world's population will grow by roughly one billion persons, with most of that growth occurring in areas of Africa and Asia already struggling to secure adequate food for current populations. Migration will further concentrate consumers in urban and peri-urban areas, multiplying the demands on food marketing channels to cheaply, quickly, and safely move more food from the countryside to cities.

Compounding the demographic effects, today's low- and middle-income country economies are expected to grow at an average annual rate two to three times that of the high-income countries (Conference Board 2012). Because poorer people spend a far greater share of each additional dollar they earn on food, income growth in low-income countries generates five to eight times the added food demand of similar income growth in high-income countries. As a result, by 2025 we should expect food demand in sub-Saharan Africa and Asia to increase at least 50 percent relative to 2012, somewhat less in Latin America and the Middle East and North Africa, and significantly less in Australia, Europe, and North America. Market-mediated demand will increase even faster due to urbanization, placing considerable added stress on marketing infrastructure.

Moreover, the pressure on natural resources such as land and water increases faster still due to dietary transitions that naturally accompany income growth and favor animal-sourced foods. Currently, about one-third of current global cereals production becomes animal feed, which is then converted into eggs, dairy products, and meat for humans at an average protein conversion efficiency of roughly 10 percent (CAST 1999; Smil 2000; Reijinders and Soret 2003). The historically observed dietary transition away from grains and pulses towards animal-sourced foods as income grows and the feed conversion inefficiency of livestock imply that per capita output of grain, oilseed, and pulses will rise disproportionately in spite of less direct consumption by humans of those foods.

The net result is that relatively little can be done to slow food demand growth over the coming decade, since most of it arises due to population and income growth, neither of which can be dampened without distasteful interventions that seriously infringe on individual liberties or suppress the legitimate aspirations of today's poor and middle classes. There is perhaps some modest capacity to slow growth in consumption of meat and other animal-sourced foods through nutritional education and other policy measures. And there is considerable scope to reduce demand for food crops such as maize and cassava for biofuels production through energy policy changes in high-income countries and to reduce net demand by tempering post-harvest losses in the food-marketing channel. But with the main drivers of demand growth effectively locked in, and given the difficulty of rapidly expanding food production, the prospects are slim that food demand growth will decline to the recent pace of supply expansion. Most expert predictions therefore call for continued rising real global food prices over the coming decade. For those poor families that do not fully participate in the expected economic growth of the emerging market economies over the coming decade, demand expansion in the face of slow-growing supply poses a serious threat of increased food insecurity due to higher prices.

This is worrisome for multiple reasons, including ones related to sociopolitical stability. The 2008 price spikes coincided with a surge of political unrest in low- and middle-income countries. Angry consumers took to the streets in at least 48 different nations across Africa, Asia, the Caribbean, and South America (Brinkman and Hendrix 2011). In some places, food riots turned violent, pressuring governments, and in a few cases contributed to their overthrow, as in Haiti and Madagascar. Another burst of unrest came with the 2011 global food price spike, clearly demonstrating a strong correlation between food prices and riots (Lagi et al. 2011; Bellemare 2012). Those with a keen sense of history recalled the turmoil associated with food riots associated with the French Revolution and other moments of political instability (Patel and McMichael 2009; Bellemare 2012). Most dramatically, many popular commentators tied the upheaval in the Middle East and North Africa during the so-called "Arab Spring" of 2011 to food prices, although that seems a gross oversimplification of the complex causality of those events. Seemingly more frequent and severe weather events—droughts, floods, and storms—and growing awareness of climate change fuel further concerns about prospective unrest (Nordås and Gleditsch 2007; Hsiang and Burke 2012).

The specter of sociopolitical instability unleashed by chaotic food markets—themselves the bitter harvest of two decades' neglect of global food systems and potentially aggravated by a changing climate—has caught high-level policymakers' attention. In the past few years, governments and philanthropic foundations began redoubling efforts to resuscitate agricultural R&D and technology transfer, as well as to accelerate the modernization of food value chains to deliver high-quality food inexpensively, faster, and in greater volumes to urban consumers. The media also began publicizing scientists' predictions of impending food riots if food prices could not be tamed (Lagi et al. 2011). But will these efforts be enough?

This chapter, and the broader volume it introduces, explores the complex relationship between food security and sociopolitical stability. We focus on near-term change to roughly 2025, because a key audience is policymakers and those who advise them. There is virtually no chance of anything longer term happening on contemporary policymakers' watch, so the more distant future becomes a low priority, interesting intellectually but not politically imperative to address. Without endorsing this short-sighted perspective, the contributors to this volume accept it as a real constraint on policy choice and tailor analyses accordingly.

The prospective stressors on developing country food security over the coming decade are many and there will be tremendous variation around the globe as to which stressors have greatest salience in sociopolitical stability terms. There are likely to be heightened challenges to managing global commons, such as climate change and ocean fisheries, as well as the production of global public goods such as improved agricultural technologies. Growing scarcity of and competition for critical natural resources, in particular arable land and fresh water, will constrain agricultural production growth, heighten competition for increasingly valuable resources, and place an ever-greater premium on technological change, some of which—such as the use of transgenic crop and livestock varieties—are themselves controversial and a source of intra- and interstate tension. Meanwhile, government policy responses with respect to intellectual property rights, migration, trade, and humanitarian response will not only affect the food security of subject populations, but may also have significant spillover effects on other countries' food security status.

The translation of food security stressors or shocks into sociopolitical instability is not automatic. That link is mediated by a host of factors associated with the preventive and responsive measures taken by governments, firms, civil society organizations, and individuals. Some strategic responses, such as export bans, foreign land investments, expansion into new fisheries, gene grabs, or diversion of fresh water resources, could relieve domestic tensions while undermining stability in other countries. Other responses to food security stress—in particular, accelerated development and release of improved cultivars or disease-resistant livestock breeds, well-timed releases from strategic grain reserves, improved food assistance, and social protection programming—could relieve both domestic and foreign food security stress. A key message of this chapter and volume is that actions taken in an effort to address food security stressors may have consequences for food security, stability, or both—domestically as well as in other countries—that ultimately matter far more than the direct impacts of biophysical drivers such as climate or land or water scarcity.

WHAT DO WE MEAN BY FOOD SECURITY AND SOCIOPOLITICAL STABILITY?

This chapter and volume explore food security[1] and sociopolitical stability (or the converse, unrest or conflict),[2] and their interrelationship. While these terms are commonly used rather loosely, below we briefly clarify what we mean so as to add a bit of useful precision to the ensuing discussion of the complex mechanisms that link them.

What is Food Security?

Among the various definitions currently in use, the prevailing definition, from the 1996 World Food Summit, holds that food security represents "a situation that exists when all people, at all times, have physical, social and economic access to sufficient, safe and nutritious food that meets their dietary needs and food preferences for an active and healthy life." Food insecurity exists when this condition is not met. Food security is commonly conceptualized as resting on three pillars: availability, access, and utilization. Some analysts add a fourth pillar, stability. These concepts are inherently hierarchical, with availability necessary but not sufficient to ensure access, which is in turn necessary but not sufficient for effective utilization, and none of which ensures stability of food security over time (Webb et al. 2006).

Availability reflects the supply side of the food security concept. In order for all people to have "sufficient" food, there must be adequate availability. As Chapter 13 demonstrates, availability remains a key limiting factor in achieving food security in sub-Saharan Africa; the same is true of a few nations elsewhere. But at least since the Green Revolution, and in many food emergencies over the centuries (Sen 1981), food availability is generally not limiting. While adequate availability is necessary, it does not ensure universal access to "sufficient, safe, and nutritious food;" it is by no means a sufficient condition for food security.

Hence the second pillar of food security: access. Access is most closely related to social science concepts of individual or household well-being: what is the range of food choices open to people, given their income, prevailing prices, and formal or informal safety net arrangements through which they can access food? As the Nobel Laureate Amartya Sen (1981: 1, emphasis in original) famously wrote, "Starvation is the characteristic of some people not *having* enough food to eat. It is not the characteristic of there *being* not enough food to eat. While the latter can be a cause of the former, it is but one of many *possible* causes." Access reflects the demand side of food security, as manifest in uneven inter- and intra-household food distribution due to variation in incomes, prices, and formal and informal social safety nets. The access challenge underscores problems of uninsured risk exposure and recourse to safe coping mechanisms to mitigate the effects of adverse shocks such as unemployment spells, price spikes, or the loss of livelihood-producing assets. Through the access lens, food security's close relationship to poverty and to social, economic, and political disenfranchisement comes into clearer focus as a structural explanation for human deprivation. All of the geographically focused chapters in this volume emphasize issues of access, whether through increasingly volatile markets or through social protection programs in low- and middle-income countries that too often prove unreliable, fiscally unsustainable, or both.

Utilization reflects whether individuals and households make good use of the food to which they have access. Do they consume nutritionally essential foods they can afford, or do they choose a nutritionally inferior diet? Are the foods safe and properly prepared, under sanitary conditions, so as to deliver their full nutritional value? Are people's health such that they absorb and metabolize essential nutrients? Utilization considerations typically focus on household-level phenomena somewhat removed from the concerns of this volume.

Stability captures the susceptibility of individuals to food insecurity due to interruptions in availability, access, or utilization. The temporal aspect of stability links to the distinction between chronic and transitory food insecurity. Chronic food insecurity reflects a long-term lack of access to adequate food and is typically associated with structural problems of availability, access, or utilization. Transitory or acute food insecurity, by contrast, is associated with sudden and temporary disruptions in availability, access, or less commonly, utilization. The most common transitory food insecurity is seasonal, recurring quite predictably, especially among rural populations during the period preceding harvest, when grain stocks run low and food prices typically hit annual peaks (Devereux et al. 2008). Some transitory food insecurity is regular but not periodic, as in the case of regular droughts that routinely strike semiarid regions. The most serious episodes of transitory food insecurity are commonly labeled "famines." Famine is an elusive concept typically, but not always, associated with a critical food shortage, mass undernutrition or starvation, and excess mortality (Devereux 1993; Ravallion 1997).

Much progress in food security stems from greater food availability made possible by agricultural technological change associated with plant breeding, improved agronomic practices such as intercropping and crop rotations, irrigation, and the emergence of mechanical implements and chemical fertilizers. Food security has therefore often been oversimplistically equated with food availability, typically measured in terms of satisfaction of dietary energy requirements, such as calories per person per day, with the assumption that food insecurity arises due to insufficient and unstable production. An availability-based view of food security naturally leads policymakers to pursue food self-sufficiency strategies to ensure domestic production will suffice to feed the population.

But although the most severe food insecurity is typically associated with disasters such as drought, floods, war, or earthquakes, most food insecurity is not associated with catastrophes, but rather with chronic poverty, which limits food access. The rapidly growing global population affected by disasters is merely a modest fraction of the undernourished, who in turn represent a minority of those living in poverty at any given time. Many experts, including Sen (1981), therefore eschew food security to focus more broadly on poverty and the satisfaction of human rights.

Food, however, occupies a privileged place in the human experience, so reducing it to just one among a host of substitutable commodities in consumers' budgets or to one among many internationally recognized human rights is to overlook food's distinctive hold on the human psyche. Food is essential instrumentally, as a vehicle for providing the nutrients that maintain cognitive and physical functioning, especially for the weakest members of society: the very young and the elderly. But food also possesses intrinsic value as a deep cultural symbol. In particular, food is a key lubricant for human relationships, as reflected in the etymology of the word "companion" which, like its French and Spanish cousins *copain* and *compañero*, respectively, derives from the Latin *com* for "with" and *panis* for "bread." Food mediates the human interactions that both make markets work and political movements form and mobilize. Disrupt the food economy and one frays the fabric of society. When people struggle to eat as they have grown accustomed, or if they anticipate struggling if current conditions do not change, their distress may be more psychosocial than economic or physiological (Timmer 2012). But either way, it carries political importance. When enough people share such distress, individual grievances morph into societal ones. Food riots can erupt. Moreover, firms, governments, and individuals may take preemptive actions intended to secure their food futures, and in so doing, sow the seeds of future struggles over productive resources. Food security therefore matters far beyond its strong inverse correlation with poverty.

What is Sociopolitical Stability?

The simplest definition of sociopolitical stability is the absence of coordinated human activities that cause widespread disruption of daily life for local populations. Note that this excludes violent personal crimes, such as murder, and natural disasters. But this definition encompasses a continuum of activities that we can array according to the magnitude of their human consequences, from nonviolent riots or large-scale political protests and work stoppages at one end, through violent versions of such organized actions, to guerilla movements and terrorism by state and non-state actors, to outright civil war, and finally to interstate war at the other. Boulding (1978) defined peace as the absence of war and emphasized that peace does not require the resolution of all conflicts within or among nations, merely that such conflict remain nonviolent. As used here and in the rest of this volume, stability is an even more utopian state than mere peace. For example, many of the food riots of the past several years proved extremely disruptive to the populations affected—and threatening to governments—but did not turn violent, at least in the sense of causing deaths. We consider such events moments of instability, even though peace prevailed.

This sort of hierarchical ordering is instructive, as it underscores two fundamental points made directly or indirectly by multiple contributors to this volume. First, not all instability is bad. When peaceful, structured, political, legal, and economic conflict occurs where the probability of large-scale conflict is negligible, mobilization against state policy is not automatically negative. Indeed, nonviolent social protest movements can be important forces for productive change. Social movements often push states to adopt policies that ultimately enhance both food security and sociopolitical stability by offering some redress for longstanding structural grievances that might otherwise lead to violence, even war.

This leads directly to the second fundamental point: the greatest dangers come not from lower-level instability associated with protests, riots, and work stoppages, but rather from violence at scale, especially in the form of organized civil or interstate war. Preserving peace is far more important, in human, economic, and geostrategic terms, than is maintaining stability. Indeed, a certain level of nonviolent instability can help to secure a stable peace if it compels the state to take actions that preempt the intensification and spread of deeper structural grievances—actions it would not choose without pressure. Riots are dangerous to local populations primarily insofar as they enable an opposition to build larger, more durable coalitions for violent political struggle against a regime. State and private actions can defuse more threatening and dangerous guerilla movements, terrorism, and civil or interstate war. Underappreciation of the central place of preventive and responsive action in mediating the relationship between food security and sociopolitical stability is perhaps the greatest deficiency of recent debates, which tend to treat the sociopolitical risks of food insecurity as driven largely by exogenous forcing variables such as climate or global market prices.

THE COMPLEX RELATIONSHIP BETWEEN FOOD SECURITY AND SOCIOPOLITICAL STABILITY

The horrific crescendo of violence—multiple world wars and genocides—that afflicted humankind in the first half of the twentieth century yielded to what has been termed the "Long Peace" over the ensuing 70-plus years (Gaddis 1989; Pinker 2011). The decline in secular violence and conflict over that period in much of the world may be only coincidentally related to unprecedented advances in food security that relaxed competition for food and for the resources needed to produce it. But perhaps there are deeper, causal links between these two broad historical phenomena.

What is the nature of the relationship between food security and sociopolitical stability? What can governments do, both in the low- and middle-income countries where instability occurs most frequently (and violently), and among the global powers that take geostrategic, humanitarian, and economic interest in stability? The food riots of 2008 and 2011 and the loose association between the dramatic upheavals of the Arab Spring and simultaneous global market price spikes in wheat, that region's staple grain, have drawn widespread recent attention and commentary in the media, leading policy think tanks, and intelligence communities around the world.[3] But the complex relationship between food security and sociopolitical stability remains poorly understood and warrants deeper exploration.

Strong correlations between food availability or food access and sociopolitical stability often exist—but certainly not always and everywhere. The linkages are, however, variable across space and time, highly conditional on the responses of multiple private and state actors, and difficult to pin down causally. Sociopolitical unrest, like market shocks, often has strong behavioral foundations that intersect with underlying structural pressures to spark extreme social events. People's expectations regarding their food security can deviate from their sense of entitlements, due either to real changes in availability or access, or due to sociopolitical movements that stir up grievances, based on past experience, others' experiences, or both (Timmer 2012). Governments that have previously employed policies, such as food subsidies, intended to make people feel secure about their access to food are sometimes perceived as reneging on that commitment when food prices rise sharply. This can then be interpreted as a signal that the government intends to renege on other social services as well, leading to unrest that is at once food- and politically-based. Perceptions are important and create difficult-to-observe psychosocial connections between these two central variables.

When not overlooked entirely, the relationship is too often oversimplified. For example, over the past several years, many public commentators have implied—or even stated explicitly—that violence arises due to a food price shock's economic impacts on the poorest and most vulnerable populations. But riots in response to a staple food price spike rarely occur among the most food-insecure and politically marginalized peoples—rural landless and smallholder farmer households—much less among those who are most vulnerable to permanent impairment due to temporary disruptions in access to adequate nutritious food: fetuses, infants, and young children.

Rather, it appears that rioters are disproportionately better-off, predominantly urban populations. Transitory (acute) food insecurity associated with a temporary price spike—or even the threat of acute food insecurity—seems more likely to spark unrest than does the chronic food insecurity associated with long-term deprivation. Both chronic and transitory food insecurity reflect real hardship, to be sure. But the relationship between welfare impacts and propensity to riot seems weak and likely mediated more by grievances related to unmet expectations than to tangible, measurable hardship. As Collier and Hoeffler (2004) point out, beyond the grievance motive, organized violence might also emerge in response to attractive profit opportunities where individuals with a low opportunity cost of time—due to high unemployment and low productivity—become willing to fight when the prospective spoils from rebellion become sufficiently attractive. Regardless of whether greed or grievance accounts for unrest, both should be strongly correlated with exogenous shocks to food prices.

This raises the most fundamental challenge in teasing out the relationship between food insecurity and sociopolitical instability: the two key variables not only affect one another, they also share so many common drivers that it becomes impractical to identify causal mechanisms with statistical rigor. For example, extreme weather events not only disrupt food production and distribution, they often also disrupt energy, health, transport, and water systems, and these disruptions can give rise to grievances unrelated to food. Careful studies routinely find a strong association between random weather events or climate oscillations and outbreaks of human violence, although the strength of that relationship and the causal mechanisms that underpin it remain poorly understood (Miguel et al. 2004; Hsiang et al. 2011; Scheffran et al. 2012; Hsiang and Burke 2012). Likewise, competition for land and water resources, large-scale population movements, and other phenomena can generate a strong correlation that makes it difficult to tease out the causal link from food security to stability. This means that most analysis is necessarily only descriptive or even speculative, and virtually none can convincingly identify causal pathways.

So be it. These nevertheless remain strategically important questions on which policymakers must act—indeed, are acting—and cannot wait for analyses that satisfy the highest scientific standards. So while researchers laudably struggle to pin down these relationships precisely and rigorously, our objective in this volume is far more modest. We aim to offer the best feasible short-to-medium-term analysis given the present, underdeveloped state of the research, to be forthright about the limitations and unanswered questions, and to offer our inevitably flawed expert assessment of how, where, and why food security and sociopolitical stability will evolve—and in turn affect one another—in the particular geographic or thematic context studied. Because the adverse effects of unstable states on food security are rather intuitive (on which, more below), this volume focuses more on the proposition of reverse causality, from food insecurity to sociopolitical instability, which remains far less clear. We aim to identify key stressors, many of which may be associated with policy responses intended to avert food security at national scale, but which have domestic or international spillover effects on food security, stability, or both.

THE IMPACT OF SOCIOPOLITICAL STABILITY ON FOOD SECURITY

The adverse effects of war and less extreme forms of sociopolitical instability on food insecurity are reasonably well-established (Cohen and Pinstrup-Andersen 1999; Collier 1999; Chen et al. 2008; World Bank 2011; Gates et al. 2012). Thus we discuss that direction of the relationship only briefly here.

Violence leads to loss of life, livestock, food stores, and disruption of the input and output marketing systems that regulate food production and distribution. When violence becomes widespread, as in cases of civil and inter-state war, agricultural production and marketing systems often collapse. Predatory and destructive acts deplete the productive capacity of the agricultural economy, sometimes long past the period of conflict if land, livestock, labor, or infrastructure are destroyed or rendered unusable for an extended period (for example, due to land mines and unexploded ordnance in the countryside).

It is not merely the experience of loss from violence that sets back conflict-ravaged societies: it is equally the risk of loss from violence (Rockmore 2012). The precautionary actions of people trying to protect themselves and their families from loss systematically reduce incomes and investment, thereby hampering both food availability—due to productivity losses in conflict-affected agricultural areas—and food access—due to income losses and price increases that result both from adverse effects on local production and from increased costs and risk premia faced by merchants.

Food also gets used as a weapon as warring parties commonly aim to disrupt and destroy food supplies that might support their enemies, often laying siege to civilian populations in the process. Even humanitarian response has been increasingly tied to donors' political and security objectives, such that the degree and nature of instability heavily affects the international community's food assistance to affected populations. For example, nine of the top ten recipient countries of humanitarian assistance in 2011 experienced conflicts that involved at least one group labeled as a foreign terrorist organization by the US government.

While advances in early warning systems and emergency response over the past generation have substantially reduced losses and accelerated recovery in the wake of natural disasters, the most intractable and protracted food emergencies are increasingly in places subject to internal conflict (Barrett et al. 2011). Of the 20 countries that have received the most humanitarian aid since 2000, 18 of them have experienced intra-state violence during that time. The vast majority of such "complex emergencies" have now become protracted, defined as eight of the past ten years. Food security is demonstrably worse in these protracted, complex emergencies. Sociopolitical instability, especially that manifest in violent internal conflict, plainly compromises food security.

HOW FOOD SECURITY AFFECTS SOCIOPOLITICAL STABILITY

The adverse effects of instability on food security are intuitive and undesirable. But is a concern about food security the added factor that will motivate policymakers to pursue peace if they would not otherwise? Probably not. By contrast, the prospective reverse causal relationship, from food security to sociopolitical stability deserves attention in part because if it has merit, it may attract high-level support for a long-neglected food security agenda. Since the complacency about food systems that set in following the Green Revolution's successes bears at least part of the blame for rising real food prices and resulting sociopolitical pressures, those consequences should focus high-level policymaker attention on the impending medium-to-long-term food security challenge.

The main food security stressors vary enormously among countries and regions, but quite a few broad patterns emerge nonetheless. In particular, there exist four main pathways through which food security might impact sociopolitical stability.

Food Price Spikes and Urban Unrest

Several recent papers document a robust effect of international food prices on different indicators of sociopolitical unrest and intra-state conflict (Arezki and Brueckner 2012; Bellemare 2012; Dell et al. 2012). The most apparent channel for such impacts takes the form of food price spikes that spark spontaneous, largely urban unrest. Food riots that suddenly erupted across the globe in 2008 and 2011 drew considerable attention to this particular mechanism. Although the direct welfare impacts of food price spikes might be greatest among the rural poor (Barrett and Bellemare 2011; Ivanic et al. 2012), food riots have historically been an overwhelmingly urban and middle-class phenomenon. Partly this is because urban populations are less dispersed, more educated, and enjoy better communications infrastructure than the rural poor, so they can organize more easily and quickly. But this also underscores that food riots arise at least as much from a price spike's symbolic, subjective, and psychological effects as from its substantive economic or nutritional impacts.

Food price riots are a very real concern, especially where they help bolster a preexisting political opposition ready and willing to attempt to overthrow the state, by violent means if necessary. But the importance of food riots is commonly exaggerated, as is the importance of exogenous factors such as grain price spikes in provoking food riots. This is perhaps most apparent in the simple observation that most countries that experience food price shocks do not suffer riots. Understanding why food riots are actually relatively uncommon responses to global food price spikes sheds significant light on the nature of the mechanisms behind this particular food security–sociopolitical stability link.

There are multiple reasons for the uneven sociopolitical effects of food price spikes. First, global market prices transmit incompletely into and within many developing countries, due to significant transport and other marketing costs, government trade and exchange rate policies, and imperfect substitutability between domestic and imported varieties of the same basic commodities (Barrett 2008; Ivanic et al. 2012). With the important exception of rice, most low-income countries are neither major exporters nor importers of basic grains. In part because of the low value-to-weight of staple food commodities and the relatively high costs of commerce in developing countries, the vast majority of food consumed—consistently over 90 percent in sub-Saharan Africa, for example—is produced domestically. When food markets are partly insulated from global markets by infrastructure, policy, or both, price transmission gets dampened in many places, as happened in the 2010 to 2011 wheat price spike (Ivanic et al. 2012).

Second, social protection policies explicitly aim to buffer the effects of market shocks so that even if the global market price spike does pass through into domestic markets, food assistance, employment guarantee, cash transfer, or other schemes can effectively buffer vulnerable populations and clearly signal the state's solidarity with consumers who might otherwise riot. The rapid rise over the past decade or so of conditional cash transfer and other social protection and food assistance programs to the poor throughout Latin America likely helps explain the limited social unrest in that region in response to staple grains price shocks that might otherwise have sparked significant urban unrest (Wolford and Nehring 2012).

Knowing that policies to mitigate the adverse impacts of food price spikes might preempt unrest, governments will commonly behave strategically, adopting policies that protect not only the vulnerable but also state power. Indeed, Carter and Bates (2012) find a positive effect of food prices on alternative measures of intra-state conflict; but when they condition on government policy responses to mitigate the impact of food price rises on urban consumers, they find the correlation between food prices and instability disappears.

The political economy of food policies, especially food subsidies—like the generous wheat product subsidies typical in much of North Africa prior to the 2011 Arab Spring—can also help explain why food riots occur disproportionately among the urban middle class, who consistently benefit most from such policies (Lipton 1977; Bates 1981; Pinstrup-Andersen 1993). As subsidized food stores' shelves run bare, those with the most to lose are not the most food-insecure rural folks, but rather the urban middle class. Urban food riots in response to price spikes are thus a natural outgrowth of the political economy of food price policies that have routinely subsidized urban consumers in most developing countries over many years.

Third, price spikes are typically short lived (Deaton and Laroque 1992), in the absence of a political opposition ready to foment, organize, and sustain social unrest. Whatever temporary instability might result from food rioters is unlikely to persist and seriously destabilize a regime if the government can effectively intervene for a short period of time through social protection measures, carefully managed strategic grain reserve releases, or variable import duty adjustments. High food prices can instead erupt into violence against ethnic minority groups—in particular, immigrant food traders—perceived to hold and exercise market power. On occasion, states may even abet such ethnic strife as a means of deflecting grievances away from the government.

Fourth, powerful states can effectively use force or the threat of force to discourage aggrieved groups from mobilizing or regular citizens from spontaneously taking to the streets and joining with a broader political opposition to challenge the state. As Wilkinson (2009) emphasizes in his authoritative review of the social science literature on riots, analysts commonly underestimate the central role of state preventive or coercive action in shaping patterns of rioting. Individual and group willingness to riot turns in part on their private assessments of how the state is likely to respond. Food riots are commonly a manifestation not just of high food prices, but perhaps more importantly, of weak states against which the population harbors a broader set of grievances to which food price spikes become a potent rallying point (Tilly 1978). This is surely part of the reason why states with serious and widespread food security problems—for example, countries such as Bangladesh, Ethiopia, or India, with very high rates of children under five suffering from stunting—avoided widespread violent food riots in 2008 and 2011.

Overall, food price spikes appear largely a proximate—not a root—cause of sociopolitical unrest, the proverbial straw that breaks the camel's back when there exist considerable preexisting grievances about other matters that foment collective action against the state (Tilly 1978). High food prices can temporarily unite and mobilize aggrieved subpopulations against the state when people perceive that they are suffering unnecessary and growing structural deprivation, that the state is not providing adequate social protection, and that the state is not strong enough to suppress dissent.

Intensified Competition Leading to Conflict over Rural Resources

Food price spikes arise as the natural consequence of depleted stocks when demand growth outpaces supply growth. Demographic momentum and likely income growth in low- and middle-income countries ensure rapid growth in demand for basic grains and oilseeds. Governments will therefore need and want to work to accelerate supply growth in order to avert price spikes, and profit-seeking agribusinesses and farmers will naturally respond to higher prices by investing in expanding output. One core theme of the chapters in this volume is that, contrary to the focus of recent popular commentary on food price riots, the risk of food insecurity leading to violent unrest—especially the sort that leads to insurgencies, civil war, and government overthrow or interstate warfare—is likely to arise gradually, and outside of cities, from firms' and governments' supply responses to high prices, rather than from the high prices themselves.

Supply expansion comes from one or more of three responses: (i) increased use of productive inputs such as land and water, (ii) improved efficiency in the use of existing inputs and technologies or the development and diffusion of new technologies that increase output, and (iii) policy interventions that (temporarily) reallocate existing supply across space or time through trade or storage policies. These different paths all carry potentially significant—but very different—implications for sociopolitical instability.

Competition over natural resources is ubiquitous and not intrinsically problematic. In stable societies, markets and non-market customary and legal institutions peaceably mediate such competition through allocative rules that are universally, if sometimes begrudgingly, accepted. Problems arise when the traditional mediating institutions cease to function reliably. That might happen because the intensity of the resource competition escalates beyond the management capacity of existing institutional arrangements, or new players enter the mix and operate outside of those institutions. Alternatively, higher stakes might induce key players—such as local or central governments or powerful private actors—to intentionally override or change the institutions that govern resource competition. Under changing conditions, resource competition can readily devolve into instability and violent conflict as the allocation of productive resources becomes perceived as inequitable, unjust, or unreliable.

Insurgencies that lead to violent civil wars commonly emerge in response to the predation of valuable assets—especially immobile ones such as land—and competitors' calculation that they stand more to gain from defying or overturning existing institutional arrangements than in working within them, even after accounting for the risks associated with conflict (Bates 1987; Kalyvas 2006; Blattman and Miguel 2010). As a result, insurgencies typically begin and are fought primarily in the hard-to-police rural countryside and have at least symbolic, and typically substantive, connections to contested real resources.[4]

As rapidly growing global food demand causes markets to tighten, rising prices increase the value of and thus demand for agricultural land, water, and animal and plant genetic material, most of which are reasonably fixed in supply, at least in the short-to-medium run. Following Mark Twain's wry advice during the United States' nineteenth-century land rush to "Buy land, they're not making it anymore," as world food prices began rising in 2006 domestic and multinational agribusinesses—as well as nontraditional investors such as sovereign wealth funds, pensions, and insurance companies—began acquiring vast amounts of farmland in search of new sources of profit, a secure source of future food or biofuel supplies, or both. Conservation-minded organizations, accurately perceiving heightened demand for increasingly scarce natural resources, have likewise rushed to buy up and protect environmentally fragile lands (Fairhead et al. 2012). Both land grabs of the former type and "green" land grabs of the latter sort have ignited controversy.

The geography of land acquisition follows predictable patterns of supply and opportunity cost in alternate uses. The arable land frontier has been effectively exhausted in Asia, Western Europe, and the Middle East and North Africa, and climate change is likely to aggravate land and water availability in the Middle East and North Africa and South Asia, in particular. By contrast, abundant land exists with only modest opportunity cost, given low current agricultural yields, in sub-Saharan Africa and parts of South America, where the overwhelming majority of land investors have focused. For example, land acquisitions in Africa in 2009 alone totaled almost 40 million hectares, an area greater than the whole of the agricultural lands of northwestern Europe: Belgium, Denmark, France, Germany, Luxembourg, the Netherlands, and Switzerland combined! Although most attention on land acquisitions has focused on foreign investors, these pressures arise within national borders as well, as in China where rapid farm consolidation into more efficient, larger-scale, mechanized farms is accelerating rapidly, creating newfound tensions as it displaces peasant families.

The newfound demand for rural resources can generate windfall gains for incumbent operators/owners if new investors can deploy capital or technologies that substantially increase the productivity of the underlying resource, translate that potential into remunerative offers, and the rightful users of those resources reap the benefits of those offers. But those multiple necessary conditions commonly go unmet (White et al. 2012). As the stakes of resource control grow, so do the prospects for conflict in the absence of effectively functioning institutions to reconcile the competing interests of incumbent resource users and those who seek to take control over the resource. This may partly explain why Black (2010) finds a strong negative association between change in arable land supply and the onset of civil war.

In much of the low-income world, the tenurial institutions that govern property rights in fisheries, forests, land, water, and other natural resources operate locally, following customary rules. Such institutions often break down in the face of a sudden influx of migrants, especially ones with economic and political power that may enable them to run roughshod over customary arrangements (Binswanger et al. 1995). Even where formal, transferable titles have been introduced to enshrine legal concepts such as fee simple, their sociopolitical legitimacy among local residents often fails to match the newfound legal status of formalized property rights, leading to disputes and grievances when transfers take place outside the domain of customary resource reallocations.

This is perhaps especially true in places with checkered precolonial and colonial legacies of deprivation and discrimination that sow seeds of distrust. Hence widespread concerns about land acquisitions on less-than-transparent terms in places where local populations lack formal rights to land and water that are leased or sold out from under them (White et al. 2012).

The threat of instability from rural resource grabs was vividly demonstrated when the Madagascar government fell in early 2009 after it surreptitiously arranged a sweetheart deal for Daewoo Logistics, a South Korean conglomerate, to lease 1.3 million hectares, nearly one-third of the nation's arable land. In a nation where rice producers comprise a majority of the population, the threat posed by this extralegal land reallocation quickly merged with preexisting grievances about a heavy-handed and corrupt state to lead to an insurrection that toppled the government. To date Madagascar remains the only case where land grabs have directly led to instability and government overthrow. But the potential for heightened rural resource competition to mutate into instability seems considerable, in part because weak property rights have been an important attribute of lands targeted for acquisition to date (Arezki et al. 2011). Such concerns are perhaps especially pronounced in countries, like Ethiopia and several other countries, where the state formally owns all land and merely grants explicitly revocable use rights to producers.

Across continents and centuries episodes of conflict erupting from competition over agricultural land and water have been commonplace, dating back to the biblical story of Cain and Abel. Some of the most common conflicts have been between transhumant or nomadic herders and crop farmers who expand cultivation into dry season grazing reserves or trekking routes. Customary arbitration mechanisms can defuse conflict quickly (Haro et al. 2005), but as population growth expands the cultivated frontier and intensifies competition for land and water, the competition often begins to exceed the conflict resolution capacity of traditional institutions. The spread of semi-automatic weaponry and the entrance of nontraditional actors—for example, external investors in rangelands or mineral prospectors—along with state weakness in more sparsely populated arid and semiarid lands have made rural resource conflicts more deadly and widespread in areas such as Central Asia, Horn of Africa, and the Sahel. As in Darfur, northern Mali, or Somalia, once resource conflicts take root, they can quickly morph into internecine battles whose resource origins combatants can no longer recall, especially where ethnic, racial, or religious differences become more salient signals of differences.

Land doesn't move, but water does. Users' behaviors in one location impact users downstream, so intensifying competition for water is likely to foster increased tensions within and between states over the coming decade (DIA 2012). Concerns are greatest in regions like the Middle East and North Africa and South Asia that are already suffering more frequent and intense droughts due to climate change. Because agriculture accounts for 93 percent of the global consumption of withdrawn fresh water (DIA 2012), the prospect of growing agricultural demands on increasingly scarce water resources—exacerbated by climate change—lies at the heart of most scenarios for water conflict, whether in the Nile or Niger Basins in Africa, or the Brahmaputra, Ganges, Indus, Mekong, or Tigris and Euphrates basins in Asia.

The possibility of reducing massive inefficiencies in water use that arise from distorted water-pricing policies and poor irrigation infrastructure and dam and reservoir management in many countries offers some prospect for peaceably resolving increased competition for water. One of the major problems of irrigation-intensive Green Revolution technologies has been excessive water use in many Asian countries where governments have given water—or the electricity needed to pump groundwater—away for free, or at prices far below cost. Low-cost technologies, such as drip irrigation and treadle pumps, are widely and inexpensively available to help accommodate prudent but politically difficult water-policy reforms in many developing countries. But absent substantial progress in improved freshwater management, groundwater depletion and degradation will continue, making the competition for agricultural water resources that much more intense.

Historically, disputes over water have led to formal, negotiated agreements over access. But upstream communities—for example, Ethiopia on the Nile, Turkey on the Euphrates and Tigris, or China on the Mekong—can take preemptive actions, such as dam and reservoir construction, to secure their own water access, threatening access by downstream users. These actions can lead directly to unrest when negotiated solutions prove elusive, or to displacement of downstream production and populations, with the resulting migration prospectively destabilizing receiving communities as well.

This raises a different, crucial dimension of the problem of rural resource competition: the accommodation of displaced rural populations. We presently know little about whether agricultural land grabs or water diversions accelerate rural-to-urban migration that adds to the stock of un- or under-employed young adults with prospective grievances who might join urban food riots or provide urban support for rural insurgent movements; but there is building evidence of climate change-induced migration. In China, as many as 350 million people are expected to move from the countryside to the cities over the coming decade or so, prospectively the largest internal migration in human history. This has enormous potential to generate domestic unrest as cities and towns struggle to find housing and jobs for internal migrants, and food value chains struggle to accelerate deliveries. Migration to cities also intensifies demand for peri-urban lands, transmitting the land competition from more remote rural areas to the outskirts of cities.

The pressure to meet rising food demand can spark at least two other sorts of resource grabs that have the potential to sow intra-state, and especially inter-state conflict. As income growth fuels disproportionately rapid increase in demand for animal source foods, including fish, this will fuel added pressures in both freshwater and marine fisheries, as well as further competition for land-based aquaculture. As McClanahan et al. (Chapter 6) report, marine capture fisheries have expanded by one degree of latitude annually, to the point that harvests from two-thirds of the world's continental shelves already equal or exceed maximum sustainable yield. Expanded production of fish must therefore come either from aquaculture that competes with crop agriculture for land and water or from more intensified competition for—and potentially unsustainable use of—maritime fisheries. Weak governance of the seas, compounded by uncontrolled and illegal fishing and other maritime criminal activity, have regularly sparked unrest or "fish wars" over access to and resource declines in fisheries (Pomeroy et al. 2007).

Land and water are not the only natural resources that become more valuable as food demand grows. So does the animal and plant genetic material on which all food production depends. Until the past generation, the development of improved crop varieties and livestock breeds has depended fundamentally on publicly available genetic accessions in collections managed as global public goods. As public funding for international germplasm conservation has fallen over time, however, private breeding enterprises and some governments (most prominently, China) have begun aggressively seizing and asserting intellectual property rights and trying to appropriate entire germplasm collections from organizations that still support free exchange of genetic material. As a result, the uptake of new technologies and the spillover effects of global investments in agricultural R&D have slowed. These widely-overlooked "gene grabs" represent the early stages of escalating competition to control the genetic building blocks of agriculture and the biodiversity that offers an essential hedge against unanticipated abiotic and biotic stresses that might emerge with climate change. Heightening tensions between scientific communities over access to both genetic resources and the technologies used to harness their potential are unlikely to lead to direct sociopolitical unrest, but can make it more difficult for firms and governments to pursue an alternative strategy to increased resource competition as a response to accelerating food demand growth: technological improvement.

Improving Technologies and Technical Efficiency

Although the prospect of increased resource competition induced by accelerating food demand merits concern, it is equally important to bear in mind that for the past half century, the bulk of output increase has come not from increased input use but rather from growth in total factor productivity (TFP)—essentially, the efficiency with which inputs such as labor, land, and water are turned into consumable food. Between 1961 and 2007, total global agricultural land use expanded by only 10 percent, and the share of input growth slowed steadily as the share of output expansion due to TFP growth increased steadily (Fuglie 2010). The proven potential of productivity growth offers an alternate route to food security, one that can mitigate the pressures of increased competition for scarce rural resources.

But productivity growth has been extremely uneven across and within countries, favoring high-potential agroecological zones and countries that have invested heavily in agricultural R&D, such as Brazil and China, or addressed longstanding inefficiencies by relaxing state control over farm management and marketing decisions, as in China and the countries of the former Soviet bloc. Elsewhere, investment in agricultural R&D has slowed, especially in the public sphere where scientific discoveries remain publicly available rather than monopolized under patent protection (World Bank 2008). Even private agricultural R&D has slowed over the past decade, as a lagged response to declines in public agricultural R&D that generates the basic discoveries on which industry research typically builds. The notable exception has been maize, for which intellectual property rights and rapidly diffusing transgenic varieties have fostered robust private investment. Not coincidentally, there were limited maize price spikes and virtually no urban maize price riots in the world in the face of a historic drought in the midwestern United States in 2012.

The competitive landscape for the development and diffusion of new agricultural technologies has shifted markedly over the past generation. Private companies and national agricultural research systems in developing countries such as Brazil, China, and India now play a prominent role, and each is far more focused on advancing narrower objectives than was the donor-funded international agricultural research center network (Byerlee and Fischer 2002; Paarlberg 2010). Meanwhile, the financing and coordination of global public goods research have become increasingly problematic (Lele et al. 2003).

This matters especially because of widespread emphasis on "closing the yield gap," defined as the difference in productivity between the maximal output attainable given current technologies and management practices and realized production (Godfray et al. 2010). But there is often a major difference between maximal attainable agronomic output and optimal output given input and output prices and exogenously varying environmental conditions such as rainfall and pest and weed pressure. Often what seems like rampant technical inefficiency among smallholder farmers merely reflects inter-farm variation in conditions beyond farmers' control, and what inefficiency remains is commonly difficult to remedy cost-effectively (Sherlund et al. 2002; Barrett et al. 2010). The challenge of increasing food production is extraordinarily site-specific, depending on shifting agroecological and socioeconomic constraints. Thus even substantially accelerated production of global public agricultural R&D may not suffice in the low-income regions of Africa and Asia where the capacity to invest in adaptation is limited, yield gaps are large, and future demand expansion will be especially robust. Revitalized investment in R&D for farming systems productivity growth is essential, not only because it has been the primary path to supply expansion in recent decades but equally because productivity growth alleviates rather than intensifies competition for rural resources by encouraging sustainable agricultural intensification (Godfray et al. 2010). But because the lags between R&D investments and productivity growth are long, and diffusion is uneven, especially among the small farmers who produce most of the food in the low-income countries of sub-Saharan Africa and South Asia, enhanced agricultural R&D will be of limited usefulness in the coming decade.

A further complication to sole reliance on technological improvements is the increasingly widespread view that current intellectual property regimes actually impede innovation, adaptation, and diffusion. These effects reinforce heightened public distrust of biotechnology companies as pursuing commercial interests even when they conflict with the public good. That distrust has erupted into sharp ideological debates about genetically modified (GM) foods, not only between Europe and the United States, but equally within developing countries. Concerned about prospective human health impacts of GM maize, the government of Zambia rejected food aid shipments sent in 2002 in response to a serious drought that left roughly three million people facing severe food shortages. Debates about GM crops have become highly polarized and politicized, often impeding the prudent use of modern methods of animal and plant genetic modification necessary to address building food security pressures (Paarlberg 2008, 2010). Yet in 2011 the developing countries caught up with the developed countries in total hectares planted in GM crops, although this increase has been overwhelmingly industrial crops, chiefly cotton, rather than food or feed commodities (James 2011). Ideological battles over GM foods notwithstanding, there seems little prospect either that the advance of GM food production will cease or that it will foment sociopolitical instability, rather than mere political tirades.

Perhaps the most obvious way to improve food systems efficiency would be to reduce rates of food losses. While estimates of the food lost between harvest and consumption are stunningly imprecise, most of the literature suggests the volumes are huge: 15 to 50 percent worldwide (Ventour 2008; Gustavsson et al. 2011). Strategies to reduce losses will necessarily vary markedly between high-income countries, where most losses are more downstream, at retail and consumer stages, and low- and middle-income countries, where losses occur primarily upstream, on-farm or in wholesale processing, storage, and transport. By some estimates, the prospective gains are considerable. Gustavsson et al. (2011) estimate that consumer food waste in developed countries equals the entire food production of sub-Saharan Africa. As Rosegrant et al. (Chapter 2) point out, however, the key issue is the likely returns to investments in reducing postharvest losses relative to alternative uses of scarce resources on agricultural R&D or infrastructure improvements that reduce, not just losses, but also the costs of delivering food from farms to cities.

The expanding use of foods—vegetable oils for biodiesel, and cassava, maize, and sugarcane for ethanol—as feedstocks for biofuels production represents an even bigger, largely policy-driven loss of food from the global system, one driven in significant measure by subsidies, tax credits, and other energy policies in the United States, European Union, and other major markets. Food and energy markets are increasingly closely coupled—and probably irreversibly so. But policy changes can add important slack back into global and regional maize and cassava markets, especially.

Technological fixes are no panacea and the history of limited diffusion and unintended environmental and social consequences of Green Revolution advances serve as a caution. But there seem few better options than improvements in agricultural productivity through the development and diffusion of new technologies, perhaps especially those targeted at relieving increasing scarcity of land and water and at tolerating various abiotic and biotic stresses associated with climate change.

Policy Interventions to Temporarily Augment Supply

When prices spike, governments often face tremendous pressure to increase food supplies temporarily through policy interventions that reallocate food across time (through releases from strategic food reserves), countries (through export bans or import subsidies), or people (through social protection measures, especially food assistance programs). Three problems can arise, however, although these are most serious in the case of trade interventions, less so with storage, and least of all with respect to social protection, for reasons developed below. First, none of these policies increases global supply: they merely redistribute it. Where a price spike originates in a temporary market distortion—for example, a speculative bubble in commodity prices—interventions need not be a problem. Indeed, in principle they could offer a helpful fix to a temporary market failure. But if the core issue is depleted stocks due to a demand–supply imbalance, there is necessarily a beggar-thy-neighbor effect, taking food either from other countries' consumers or from future populations. And so long as global stocks-to-use ratios remain low, markets remain especially sensitive to any of a host of shocks, so the underlying problem does not really go away.

Because governments find it well-nigh impossible to distinguish between those two scenarios in real time, a second problem arises: the exporting of stress to a different group. When exporters suddenly renege on delivery contracts, causing shortages and magnifying price spikes in their import partners' markets, international discord is the predictable result. Such tensions rarely devolve into conflict, but they sow distrust. And the affected country can be stuck with a more serious food price shock—and prospective subsequent urban unrest and rural resource competition—than it would otherwise have had to contain.

Third, such interventions can breed dangerous complacency, distracting attention from the structural sources of price spikes and diverting precious resources from more fundamental and durable initiatives—such as agricultural R&D, reformed water pricing, land tenure, or intellectual property policies—that only pay off with a longer lag. In an imaginary world with no opportunity cost to high-level policymaker time and attention, and much less investable fiscal or philanthropic resources, short-term solutions might be desirable. But as highlighted by the decades that food security concerns spent in the policy world's wilderness, it is foolhardy not to seize on more structural solutions in the rare moments when senior policymakers' attention focuses on these concerns.

Trade policy interventions such as export bans or import subsidies remove commodity supplies from the global market to the domestic market of the activist government. Such measures merely aggravate price increases in other, less insulated economies and thin global markets, fuelling increased price instability. Indeed, when many countries simultaneously use trade policy instruments to insulate domestic food markets, their interventions offset each other, rendering the exercise completely ineffective domestically but exacerbating global price volatility. Sharma (2011) reports that, of 105 countries surveyed, 31 percent enacted export restrictions between 2007 and 2010, and another half of the countries reduced food import taxes, which acts similarly to exporters enacting restrictions. Analysts' subsequent assessments uniformly suggest that export restrictions played a significant role in fuelling the global price spikes that came on the heels of these actions.

The use of trade policy instruments holds obvious appeal to states looking to placate restless urban constituents. But it becomes especially problematic for food-importing countries growing ever more dependent on reliable trade partners to ensure food access in the face of increasingly frequent and severe weather extremes, exhaustion of their domestic land and water resources, or both. That describes precisely events in 2010–11, when Russia and Ukraine suddenly enacted wheat export bans and duties, with devastating effects on their import clients in the Middle East and North Africa. International trade is supposed to dampen price fluctuations by spreading weather and other production risks across countries—a major reason why most economists advocate for agricultural trade liberalization. But when just a few major exporters dominate the global market—as is true in maize, rice, soy, and wheat markets—if exporters prove unreliable suppliers, importers have far less motivation to open their markets. The disruptive episode of sudden export restrictions and precipitous reductions of import duties by major importers in the 2007 to 2011 period sowed considerable distrust among government trade negotiators.

In order to prevent countries from inadvertently exporting prospective socio-political instability, multilateral arrangements to restrict actions that can disrupt trade's effective functioning as a shock absorber acquire added importance. But the agricultural trade liberalization disciplines under negotiation in the Doha Round of the World Trade Organization (WTO) began in 2001, when real food prices were near historic lows. Negotiators' focus fell squarely on import restrictions and export subsidies—and related food aid and domestic farm policies—perceived to push commodity surpluses onto global markets in order to keep domestic market prices artificially high for farmers. As food prices rose rapidly over the subsequent decade, hitting historic highs in 2011, the WTO and other international agreements on agricultural trade proved ill-suited to constrain governments suddenly aiming to use trade restrictions to do the opposite: to push domestic market prices down by evacuating food from the global market.

It is not just trade policies that matter. Farm policies aimed at domestic markets can have major international spillover effects. The impacts of US, EU, and Japanese policies are widely known and extensively discussed. Less well appreciated are the dramatic changes taking place in the developing world. For example, in China, real farm subsidies grew more than one thousandfold between 2002 and 2010, from 100 million to 123 billion yuan (Christiaensen, Chapter 17). If and when China relaxes its grains self-sufficiency policy to stem domestic food price inflation, Chinese farm policy could suddenly have significant effects, destabilizing global maize, rice, and soy markets. On top of international disputes over maritime resources and genetic material, policy responses to mounting food security pressures have real potential to fuel international tensions that could spark into more substantial unrest.

Because sudden export restrictions have made international trade appear an unreliable source of supply in times of high prices, many food-importing countries have begun building or rebuilding strategic grain reserves in the hopes of stabilizing prices domestically through releases from storage, especially in the event of an interruption of established import arrangements. There have also been widespread calls for multilateral emergency food security reserves to act on behalf of low-income or smaller countries that might not be able to establish and manage effectively domestic programs. Storage can indeed be valuable for managing intertemporal price fluctuations associated with commodity price bubbles. Buffer stock releases can lower and stabilize food prices, not just of the released commodity but also of substitutes (Barrett 1997). When price stabilization focuses on stabilization around longer-run trends in world market prices, well-run buffer stock schemes can be fiscally sustainable and perhaps foster greater macroeconomic stability (Timmer and Dawe 2007). But there is also quite a checkered history of mismanagement, and most designs are expensive, structurally unsustainable, and often more price destabilizing than a workably competitive storage market (Newbery and Stiglitz 1981; Knudsen and Nash 1990; Gilbert 1996; Wright 2012).

The economic implications of public storage schemes matter because government efforts to reduce food prices through buffer stock releases, if successful, necessarily stimulate increased food consumption, thereby aggravating the underlying problems of excess demand relative to supply and low stocks-to-use ratios. In a fundamental sense, buffer stock releases merely kick the food price spike can down the road a bit. Like sudden trade policy adjustments, strategic grain reserve interventions have long held appeal because they offer political leaders a tool to relieve food price pressures temporarily, so that they are seen as taking action in response to consumer stress. But at best, such policies fail to offer longer-term fixes while pushing the problem into others' laps, and government storage and trade interventions often aggravate the underlying structural problem that ignited the food price spike in the first place. Pursuit of such strategies therefore often signals impending problems elsewhere or later.

Probably the most compelling argument for strategic reserves is to ensure that poor families retain access to adequate nourishment when food prices rise, thereby linking storage to effective social protection schemes. Beyond the humanitarian purposes of social protection programs, these also play a crucial role in breaking the link between food price spikes and sociopolitical unrest. Very few of the countries that experienced food price riots between 2007 and 2011 had effective food assistance or other social protection programs in place. Conversely, countries that had broad-based social protection programs—cash transfer, employment guarantee, and other schemes—in place, largely remained placid, even where rice or wheat are staples.

There are multiple approaches to food assistance, however, and they are not all equally effective. Price controls and consumer subsidies are blunt instruments commonly implemented by states as emergency measures, but they are almost always distributionally regressive, meaning that most benefits flow to better-off individuals for the simple reason that benefits are proportional to volume consumed and the better-off consume more than the poor. Moreover, they are generally not fiscally or operationally sustainable and thus sow the seeds of their own demise. And when price controls or large-scale subsidies have to be scaled back when the state finds it can no longer afford them, the sudden withdrawal of what has become an entitlement can cause unrest akin to that of a market price spike. The clearest example would be Egypt's 1977 bread riots, in which dozens of people died after the government removed subsidies on rice, flour, and cooking oil, before the army was dispatched and the subsidies reinstated in order to quell the unrest.

A key ingredient to impactful social protection is effective targeting of the needy (Barrett 2002). When targeted at vulnerable households and individuals, food assistance and related social protection programming can be fiscally sustainable, with minimal impacts on local (much less international) markets. By safeguarding the poor's access to food, these policies buffer the effects of market shocks on especially vulnerable populations. This has not only an economic effect but also a psychosocial one, as it makes tangible the state's support for those enduring hardships, making mobilization against a government that is committed to social protection less likely. Perhaps that is one reason why none of the roughly 20 countries that have nationwide conditional or unconditional cash transfer schemes erupted into food riots between 2007 and 2011, even though several of them have significant poor populations that spend a large share of disposable income on maize, rice, or wheat.

Many of the world's poorest countries lack either the financial or bureaucratic means to implement large-scale social protection programs. This is where international food assistance can play an especially valuable role, both in humanitarian relief and in increasing the legitimacy of the state (OECD 2008; Brinkman and Hendrix 2011). Perhaps the greatest challenge, besides shrinking real resources for international food assistance (Barrett et al. 2011), is the increased securitization of aid that may undermine its efficacy in helping promote sociopolitical stability while advancing food security objectives.

LOOKING FORWARD

Demand-side pressures due to population and income growth and urbanization will continue to build in the global food economy over the coming decade, especially in Asia and sub-Saharan Africa. After years of neglect of agriculture, accelerating supply expansion to match demand growth will take time. As the past five years have demonstrated, this ongoing structural imbalance will likely lead to rising real food prices punctuated by occasional food price spikes.

These price spikes are strongly associated with urban unrest where already-aggrieved populations lack adequate social safety nets and a political opposition is ready to mobilize consumer dissatisfaction into riots, if not outright rebellion. Although urban food riots rarely directly lead to state overthrow, governments can nonetheless be expected to take actions to try to avert food price spikes and the urban riot risks they run. Meanwhile, private producers and investors will predictably respond to price signals by seeking to expand profitable food production, secure inexpensive resource inputs, or both. Private and state responses to the price signals produced by immutable demand-side pressures lie at the heart of the complex relationship between food security and sociopolitical instability.

Where the dominant response becomes intensified competition for increasingly scarce productive resources, the potential for instability and even conflict becomes considerable. This will be especially true in regions where extreme weather events associated with climate change compound the resource competition. Grievances associated with resource reallocations widely perceived as illegitimate abrogations of established tenurial institutions are especially likely to fuel insurgencies and civil war, which typically have rural rather than urban roots. Rural resource competition induced by high food prices likely poses the greatest risk of serious sociopolitical instability associated with food insecurity.

Responses based on beggar-thy-neighbor strategies such as export bans can have similarly adverse consequences on sociopolitical stability. Because such policies destabilize global markets, they sow tension between traditional trade partners, as importing countries are left in the lurch by states that suddenly restrict exports. At least as importantly, sudden export restrictions—or relaxation of import restrictions—effectively displace the higher food prices that concerned activist governments into countries with less activist states while exacerbating international market price instability, thereby exporting the potential for urban unrest. But such measures hold political appeal because they can be implemented quickly and offer highly visible acts by states in solidarity with food consumers, thereby helping to preempt unrest—if only temporarily.

By contrast, investment in agricultural R&D, technology transfer, and social protection programs that increase food availability and access per capita without increasing the draw on land, water, and labor resources represents the most promising—but the slowest—responses to price signals that reflect looming food scarcity. There is hope in the example of the Green Revolution, which overcame a seemingly insurmountable explosion in human population growth to usher in an extended period of rapid improvements in global food security, contemporaneous with the Long Peace of the latter twentieth century. The big differences between the Green Revolution period and today are the need to focus far more intently on (i) sub-Saharan Africa and South Asia because those are the places where food demand growth and rural resource stress are likely to be greatest, (ii) careful conservation of increasingly scarce land and water resources through complementary (e.g., water pricing) policies, (iii) the complexities of intellectual property in genetic resources, (iv) ensuring poorer smallholders have equal and adequate access to improved technologies, and (v) the need for complementary social protection programming to ensure the poor's access to adequate food in periods when food prices do spike or extreme weather events disrupt food supplies or livelihoods. Unlike the Green Revolution era, high-income countries today also need to consider the implications of biofuels policies that encroach on feed and food supplies, potentially undermining food security and catalyzing some of the prospective destabilizing responses explored in this volume. In short, a broad range of coordinated options needs to be considered and pursued, as the most cost-effective routes to food security and sociopolitical stability vary enormously across countries and over time.

The core conclusion of the chapters collected in this volume is that there exist both quite reasonable hypotheses and some suggestive empirical evidence that food insecurity associated with high food prices can spark sociopolitical unrest, especially when governments try to augment supply so as to lower food prices or firms try to capitalize on high prices. This is another important reason to guard against renewed complacency about the evolving global food security challenge.

Past successes prove the potential of improvements in food production and distribution systems to reduce human suffering and maintain sociopolitical stability. But structural food demand and supply patterns pose significant risks over the coming decade, and the task of advancing food security will be more difficult in the twenty-first century than it was in the preceding one, in the face of climate change and with extra billions of better-off bellies to fill.

The looming food security challenge of the coming decades can nonetheless be met. The key question is how. Will it be through intensified competition for productive resources or through beggar-thy-neighbor policies that sow tension and risk conflict within and between states? Or will it come more through efforts to increase productivity per capita—via reduced postharvest losses, agricultural R&D, and rationalized biofuels policies to reduce the diversion of food and feed into liquid fuels production—combined with expanded social protection programs to ensure that displaced populations have reliable, affordable access to an adequate diet? The means by which governments, firms, and private philanthropies tackle the food security challenge of the coming decade will fundamentally shape the relationship between food security and sociopolitical stability.

ACKNOWLEDGMENTS

This chapter has benefitted from comments from a seminar audience at Harvard's Weatherhead Center for International Affairs, as well as insights shared by the various contributors to this volume and helpful comments from Arun Agrawal, Marc Bellemare, Nathan Black, Mark Cane, Sheri Englund, Erin Lentz, Travis Lybbert, Will Masters, Cynthia Mathys, Dan Maxwell, Ellen McCullough, Rob McLeman, Michael Mulford, Emmy Simmons, and Peter Timmer. This broader project was supported by the United States intelligence community. The findings, opinions, and recommendations expressed in this chapter, and especially any remaining errors, are mine alone and do not reflect the views of any organization or agency that supported this work nor those of anyone else involved in the project.

NOTES

1. This section draws heavily on Barrett (2010) and Barrett and Lentz (2010).
2. It is perhaps worth pointing out that, rather like food security, we refer to the desired state of being—stability—rather than the more common, less desirable

condition—instability—that characterizes many (and at some point in time, all) nations. We unapologetically promote the aspirational vision of food security and sociopolitical stability.

3. A few prominent examples include *The Economist* (2012), Friedman (2012), National Intelligence Council (2012a, b), Tuttle and Wedding (2012).
4. Some of the seminal research in this vein includes Scott (1976, 1985), Tilly (1978), Bates (1987), and Kalyvas (2006, 2007).

REFERENCES

Alston, J. M., C. Chan-Kang, M. C. Marra, P. G. Pardey, and T. J. Wyatt. 2000. *A Meta-analysis of the rates of return to agricultural R&D: Ex pede Herculem.* IFPRI Research Report No. 113. Washington, DC: International Food Policy Research Institute.

Arezki, R., and M. Brueckner. 2012. Effects of international food price shocks on political institutions in low-income countries: Evidence from an International Food Net-Export Price Index. Unpublished working paper.

Arezki, R., K. Deininger, and H. Selod. 2011. What drives the global "land rush"? World Bank Policy Research Working Paper 5864.

Barrett, C. B. 1997. Liberalization and food price distributions: ARCH-M evidence from Madagascar. *Food Policy 22*(2): 155–73.

———. 2002. Food security and food assistance programs. In *Handbook of Agricultural Economics*, vol. 2b, ed. B. L. Gardner and G. Rausser, 2103–2190. Amsterdam: Elsevier.

———. 2008. Smallholder market participation: Concepts and evidence from eastern and southern Africa. *Food Policy 33*(4): 299–317.

———. 2010. Measuring food insecurity. *Science 327*: 825–8.

———, and M. F. Bellemare. 2011. Why food price volatility doesn't matter. *Foreign Affairs.* Available at <http://www.foreignaffairs.com/articles/67981/christopher-b-barrett-and-marc-f-bellemare/why-food-price-volatility-doesnt-matter> (accessed April 27, 2013).

——— and J. Y. Hou. 2010. Reconsidering conventional explanations of the inverse productivity-size relationship. *World Development 38*(1): 88–97.

Barrett, C. B., A. Binder, and J. Steets. 2011. *Uniting on food assistance: The case for transatlantic cooperation.* London: Routledge.

Barrett, C. B., and E. C. Lentz. 2010. Food insecurity. In *The International Studies Encyclopedia*, vol. IV, ed. R. A. Denemark, 2291–2311. Chichester, UK: Wiley-Blackwell.

Bates, R. H. 1981. *Markets and states in tropical Africa.* Berkeley: University of California Press.

——— 1987. The agrarian origins of Mau Mau: A structural account. *Agricultural History 61*(1): 1–28.

Bellemare, M. F. 2012. Rising food prices, food price volatility, and social unrest. Unpublished working paper.

Binswanger, H., K. Deininger, and G. Feder. 1995. Power, distortions, revolt and reform in agricultural land relations. In *Handbook of development economics*, vol. 3B, ed. J. Behrman and T. N. Srinivasan, 2659–2772. Amsterdam: Elsevier.

Black, N. 2010. Change we can fight over: The relationship between arable land supply and substate conflict. *Strategic Insights* 9(1): 30–64.

Blattman, C., and E. Miguel. 2010. Civil war. *Journal of Economic Literature* 48(1): 3–57.

Boulding, K. 1978. *Stable peace.* Austin, TX: University of Texas Press.

Brinkman, H.-J., and C. S. Hendrix. 2011. *Food insecurity and violent conflict: Causes, consequences and addressing the challenges.* Occasional Paper 24. Rome: World Food Programme.

Byerlee, D., and K. Fischer. 2002. Accessing modern science: Policy and institutional options for agricultural biotechnology in developing countries. *World Development* 30(6): 931–48.

Carter, B. L., and R. H. Bates. 2012. Public policy, price shocks, and civil war in developing countries. Unpublished working paper.

Chen, S., N. Loayza, and M. Reynal-Querol. 2008. The aftermath of civil war. *World Bank Economic Review* 22(1): 63–85.

Cohen, M. J., and P. Pinstrup-Andersen. 1999. Food security and conflict. *Social Research* 66(1): 375–416.

Collier, P. 1999. On the economic consequences of civil war. *Oxford Economic Papers* 51(1): 168–83.

———, and A. Hoeffler. 2004. Greed and grievance in civil war. *Oxford Economic Papers* 56(4): 563–95.

Conference Board. 2012. *Global economics outlook 2012.* New York: The Conference Board.

Council for Agricultural Science and Technology (CAST). 1999. Animal agriculture and global food supply. Task Force Report No. 135. Ames, IA: CAST.

Dasgupta, P. 1993. *An inquiry into well-being and destitution.* Oxford, UK: Clarendon Press.

Deaton, A., and G. Laroque. 1992. On the behaviour of commodity prices. *Review of Economic Studies* 59(1): 1–23.

Defense Intelligence Agency (DIA). 2012. *Global water security.* Intelligence Community Assessment ICA 2012–08.

Dell, M., B. F. Jones, and B. A. Olken 2012. Temperature shocks and economic growth: Evidence from the last half century. *American Economic Journal: Macroeconomics* 4(3): 66–95.

Devereux, S. 1993. *Theories of famine.* New York, NY: Harvester Wheatsheaf.

———, B. Vaitla, and S. H. Swan. 2008. *Seasons of hunger: Fighting cycles of quiet starvation among the world's rural poor.* London: Pluto Press.

Evenson, R. E., and D. Gollin, eds. 2003. *Crop variety improvement and its effect on productivity: The impact of international agricultural research.* Wallingford, UK: CABI.

Fairhead, J., M. Leach, and I. Scoones. 2012. Green grabbing: A new appropriation of nature? *Journal of Peasant Studies* 39(2): 237–61.

Fogel, R. 2004. *The escape from hunger and premature death, 1700–2100.* Cambridge: Cambridge University Press.

Friedman, T. L. 2012. The other Arab Spring. *New York Times.* April 7.

Fuglie, K. 2010. Total factor productivity in the global agricultural economy: Evidence from FAO Data. In *The Shifting Patterns of Agricultural Production and Productivity Worldwide*, ed. Julian M. Alston, Bruce A. Babcock, and Philip G. Pardey. Ames, Iowa: The Midwest Agribusiness Trade Research and Information Center.

Gaddis, J. L. 1989. *The long peace: Inquiries into the history of the Cold War.* Oxford: Oxford University Press.

Gates, S., H. Hegre, H. M. Nygård, and H. Strand. 2012. Development consequences of armed conflict. *World Development 40*(9): 1713–22.

Gilbert, C. L. 1996. International commodity agreements: An obituary notice. *World Development 24*(1): 1–19.

Godfray, H. C. J., J. R. Beddington, I. R. Crute, L. Haddad, D. Lawrence, J. F. Muir, J. Pretty, S. Robinson, S. M. Thomas, and C. Toulmin. 2010. Food security: The challenge of feeding 9 billion people. *Science 27* (5967): 812–18.

Gustavsson, J., C. Cederberg, A. Meybeck, U. Sonesson, and R. van Otterdijk. 2011. *Global food losses and food waste.* Rome: FAO.

Haro, G. O., G. J. Doyo, and J. G McPeak. 2005. Linkages between community, environmental, and conflict management: Experiences from Northern Kenya. *World Development 33*(2): 285–99.

Hsiang, S., and M. Burke. 2012. Climate, conflict, and social stability: What do the data say? Unpublished working paper.

Hsiang, S. M., K. C. Meng, and M. A. Cane. 2011. Civil conflicts are associated with the global climate. *Nature 476*: 438–41.

Ivanic, M., W. Martin, and H. Zaman. 2012. Estimating the short-run poverty impacts of the 2010–11 surge in food prices. *World Development 40*(11): 2302–17.

James, C. 2011. *Global status of commercialized biotech/GM Crops: 2011.* ISAAA Brief No. 43. Ithaca, NY: ISAAA.

Kalyvas, S. 2006. *The logic of violence in civil war.* Cambridge: Cambridge University Press.

Kalyvas, S. 2007. Civil wars. In *Handbook of Political Science*, ed. C. Boix and S. Stokes. New York: Oxford University Press.

Knudsen, O., and J. Nash. 1990. Domestic price stabilization schemes in developing countries. *Economic Development and Cultural Change 38*(3): 539–58.

Lagi, M., K. Z. Bertrand and Y. Bar-Yam. 2011. *The food crises and political instability in North Africa and the Middle East.* Cambridge, MA: New England Complex Systems Institute.

Lele, U., C. Barrett, C. K. Eicher, B. Gardner, C. Gerrard, L. Kelly, W. Lesser et al. 2003. *The CGIAR at 31: An independent meta-evaluation of the Consultative Group on International Agricultural Research.* Washington, DC: World Bank.

Lipton, M. 1977. *Why poor people stay poor: Urban bias in world development.* Cambridge, MA: Harvard University Press.

Miguel, E., S. Satyanath, and E. Sergenti, E. 2004. Economic shocks and civil conflict: An instrumental variables approach. *Journal of Political Economy 112*(4): 725–753.

National Intelligence Council. 2012a. *Global food security: Key drivers—A conference report.* NICR 2012-05.

—— 2012b. *Global food security: Market forces and selected case studies.* NICR 2012–23.

Newbery, D. M. G., and J. E. Stiglitz. 1981. *The theory of commodity price stabilization.* Oxford, UK: Oxford University Press.

Nordås, R. and N. P. Gleditsch. 2007. Climate change and conflict. *Political Geography 26*(6): 627–38.

Organisation for Economic Co-operation and Development. 2008. *Aid for Food and Nutrition Security.* Paris: OECD.

Parrlberg, R. 2008. *Starved for science: How biotechnology is being kept out of Africa.* Cambridge, MA: Harvard University Press.

———2010. *Food politics: What everyone needs to know.* Oxford: Oxford University Press.

Pardey, P. G., N. Beintema, S. Dehmer, and S. Wood. 2006. *Agricultural research: A growing global divide?* Washington: International Food Policy Research Institute.

Patel, R., and P. McMichael. 2009. A political economy of the food riot. *Review* *32*(1): 9–35.

Pingali, P. 2012. Green Revolution: Impacts, limits, and the path ahead. *Proceedings of the National Academy of Sciences 109*(31): 12302–8.

Pinker, S. 2011. *The better angels of our nature: Why violence has declined.* New York: Viking.

Pinstrup-Andersen, P., ed. 1993. *The political economy of food and nutrition policies.* Baltimore: Johns Hopkins University Press.

Pomeroy, R., J. Parks, R. Pollnac, T. Campson, E. Genio, C. Marlessy, E. Holle et al. 2007. Fish wars: Conflict and collaboration in fisheries management in Southeast Asia. *Marine Policy 31*(6): 645–56.

Raitzer, D. A., and T. G. Kelley. 2008. Benefit–cost meta-analysis of investment in the Inter-national Agricultural Research Centers of the CGIAR. *Agricultural Systems* *96*(1): 108–23.

Ravallion, M. 1997. Famines and economics. *Journal of Economic Literature* *35*: 1205–42.

Reijinders, L., and S. Soret. 2003. Quantification of the environmental impact of different dietary protein sources. *American Journal of Clinical Nutrition 78*(suppl): 664S–8S.

Rockmore, M. E. 2012. The cost of fear: The welfare effects of the risk of violence in northern Uganda. Unpublished working paper.

Scheffran, J., M. Brzoska, J. Kominek, P. M. Link, and J. Schilling. 2012. Climate change and violent conflict. *Science 336*: 869–71.

Scott, J. C. 1976. *Moral economy of the peasant: Rebellion and subsistence in Southeast Asia.* New Haven: Yale University Press.

———1985. *Weapons of the weak: Everyday forms of peasant resistance.* New Haven: Yale University Press.

Sen, A., 1981. *Poverty and famines.* Oxford, UK: Clarendon Press.

Sharma, R. 2011. *Food export restrictions: Review of the 2007–2010 experience and considerations for disciplining restrictive measures.* Rome: FAO.

Sherlund, S. M., C. B. Barrett, and A. A. Adesina 2002. Smallholder technical efficiency controlling for environmental production conditions. *Journal of Development Economics 69*(1): 85–101.

Smil, V. 2000. *Feeding the world: A challenge for the 21st century.* Cambridge, MA: MIT Press.

The Economist. 2012. Let them eat baklava. March 17 edition.

Tilly, C. 1978. *From mobilization to revolution.* New York: McGraw-Hill.

Timmer, C. P. 2012. Behavioral dimensions of food security. *Proceedings of the National Academy of Sciences 109*(31): 12315–20.

——, and D. Dawe. 2007. Managing food price instability in Asia: A macro food security perspective. *Asian Economic Journal 21*(1): 1–18.

Tuttle, J. N., and K. Wedding. 2012. Will food prices drive instability? In *2012 Global Forecast: Risk, Opportunity, and the Next Administration*, ed. C. Cohen and J. Gabel, *41–43*. Washington: Center for Strategic and International Studies. Available at <http://csis.org/files/publication/120417_gf_nesseth_wedding.pdf> (accessed April 27, 2013).

Ventour, L. 2008. *The food we waste.* Banbury, UK: Waste and Resources Action Programme.

Webb, P., J. Coates, E. A. Frongillo, B. L. Rogers, A. Swindale, and P. Bilinsky. 2006. Measuring household food insecurity: Why it's so important and yet so difficult to do. *The Journal of Nutrition 136*: 1404S–8S.

White, B., S. M. Borras Jr., R. Hall, I. Scoones, and W. Wolford. 2012. The new enclosures: Critical perspectives on corporate land deals. *Journal of Peasant Studies 39*(3–4): 619–47.

Wilkinson, S. I. 2009. Riots. *Annual review of political science 12*: 329–43.

Williams, J. C., and B. D. Wright. 1991. *Storage and commodity markets.* Cambridge: Cambridge University Press.

Wolford, W., and R. Nehring 2013. Moral economies of food security and protest in Latin America. Chapter 12, this volume.

World Bank. 2008. *World development report 2008: Agriculture for development.* Washington, DC: World Bank.

—— 2011. *World development report 2011: Conflict, security and development.* Washington, DC: World Bank.

Wright, B.D. 2012. International grain reserves and other instruments to address volatility in grain markets. *World Bank Research Observer 27*(2): 222–60.

2

The Future of the Global Food Economy: Scenarios for Supply, Demand, and Prices

MARK W. ROSEGRANT, SIMLA TOKGOZ, AND PRAPTI BHANDARY

The global food economy is in the midst of a major price reversal. Over the past several decades, real-world prices of cereals and meats have declined as technological advances and other factors expanded supply relative to demand. Now, with long-term changes in supply and demand and other factors, prices of cereals and meats are projected to increase through 2025. Several food price spikes have punctuated the change since 2007, motivating researchers and policymakers to weigh the factors behind high and volatile prices, their implications for long-term food security, and the connection between food prices and civil conflict. Ultimately, policymakers need an assessment of food security and how it might be improved through increased agricultural productivity or other measures.

Long-term structural changes in the global food economy, examined in this chapter, have been driving food price crises. Sharply rising food prices in 2007 and 2008 and again in 2010 and 2011 were caused by a number of factors, including rapid growth in demand for biofuels, bad weather, commodity speculation, and restrictive trade policies. Also in play were long-run trends in supply-and-demand fundamentals, including low investment in agricultural research and slow yield growth. These higher food prices, in addition to impeding progress on food security and causing direct economic effects, may lead to greater conflict or civil strife. Although causality between food prices and conflict is not conclusive in the literature, it is a significant concern for policymakers concerned with food security.

The remainder of the chapter builds scenarios using these structural changes and food security issues and assesses whether they will have significant effects on long-term food security. We examine global food system dynamics to 2025 through scenarios of alternative futures for agricultural supply and demand and food using the International Food Policy Research Institute's (IFPRI) International Model for Policy Analysis of Agricultural Commodities and Trade (IMPACT) global food model. A baseline scenario shows that increasing prices are projected to continue for more than a decade, with negative effects on food security. We then offer several policy scenarios—specifically increased investment in agricultural research and/or reduced postharvest losses—as alternatives to the baseline. The results illustrate how levels of food supply, demand, and prices will likely change assuming increased investment in research and agricultural productivity growth and a reduction of postharvest food losses. With these hopeful steps forward, indicators for food security in the developing world show improvement over the next decade and beyond.

RECENT FOOD PRICE SPIKES

A variety of factors—and not a single, dominant cause—converged to drive up food prices from 2007 through 2011. Long-run demand and supply trends helped push up prices, with their effects magnified by poor weather, a sharp increase in use of corn for biofuel production, and linkages between the corn and oil markets. Increased market volatility encouraged worried buyers to purchase commodities ahead, while offering potential profit opportunities to financial traders. To limit the adverse impact of rising prices on consumers, some producing countries banned exports, which put additional upward pressure on world commodity prices.

When analyzing commodity and food price changes between 2007 and 2011, many researchers have identified low stocks relative to use as a primary contributor (Piesse and Thirtle 2009; Trostle 2008). The stock-to-use (s/u) ratio captures supply and demand in a single indicator to reflect either surplus or tightness in a particular market. The economic concept of the s/u ratio is simple. As food commodity stocks (measured at the end of the marketing year) decline relative to use for that year, the commodity becomes more valuable, with buyers willing to pay more and sellers reluctant to sell. For example, between 2000 and 2008, global stocks of grains and oilseeds relative to use were cut nearly in half to the lowest level in more than 30 years, while prices more than doubled (Piesse and Thirtle 2009; Trostle 2008). This inverse relationship between s/u and price is particularly strong for corn, a leading agricultural commodity, and differs depending on the time period.

For the decade immediately prior to the 2008 food price spike, the consensus among researchers is that demand growth outstripped supply growth, resulting in a declining s/u ratio and stronger prices for most commodities. This contrasted with the period between 1985 and 1990, when supply mostly kept pace with demand. After the sharp run-up in 2008, prices retreated as demand softened, but the market raced ahead again in 2010 and 2011 when crops in the United States and elsewhere did not meet expectations for stock rebuilding (Trostle et al. 2011; USDA 2011).[1]

A number of demand factors contributed to the price spikes and helped maintain commodity prices above historical levels. These factors include income and population growth, biofuel expansion, and the declining value of the dollar. Supply factors include slow growth in global agriculture production, low investment in agricultural research,[2] and the rising cost of energy.

Once the market fundamentals were in place in 2008, several additional factors emerged as market uncertainty grew. Volatility in commodity markets attracted outside money from other investment sectors not faring as well, including equity markets and housing. Many researchers who studied the 2008 commodity price spike identified low commodity stocks as a contributing factor for a substantial increase in speculative activity in markets for major crops, including wheat, corn, and soybeans. Some have concluded that the "artificial demand created by investors' speculation in commodities futures put tremendous upward price pressure on food and energy commodities" (Mittal 2009, 17).

For rice, a commodity for which trading accounts for only 7 percent of global output, the market suffers from "thinness" that can result in highly volatile markets. Further exacerbating the explosive rice market in 2008 was the activation of export controls by a number of exporters, including India (November 2007), Vietnam and Egypt (January 2008), China (January 2008), and Cambodia (March 2008) (Headey et al. 2009).[3]

The initial price spike lasted approximately a year, starting in June 2007. By late 2008, food commodity prices had declined to mid-2007 levels as global crop production expanded and a world economic recession reduced demand. Commodity stocks increased relative to use, but by 2010, crop markets began strengthening again as poor weather in major producing counties (e.g., Russia and Australia) reduced output of wheat and other crops. In addition, improved economic growth in lesser-developed countries and rising oil prices helped drive up commodity prices further by mid-2011 (Trostle 2012).

Many of the same factors from the 2008 spike also contributed to the 2011 spike, although the timing and sequence varied (Trostle et al. 2011). These included global growth in population and income, rising energy prices and expanding global biofuel production, depreciation of the US dollar and slower growth in agricultural productivity, short-term weather-related production shortfalls, and subsequent decline in stocks. In 2010 and 2011, poor weather was especially important, as was a substantial increase in Chinese demand for soybeans. Market fundamentals proved to be critical to overall price levels in 2011 (Abbott et al. 2011).

The emerging consensus among researchers is that these food price spikes were caused by many interconnected factors. A summary analysis stated, "the more one assesses this crisis, the more one concludes that it is the result of a complex set of interacting factors rather than any single factor" (Headey and Fan 2010, xiii). This contrasts with the same authors' earlier analysis that identified the oil price–biofuels nexus as the driving force behind the surge in food prices (Headey et al. 2009).

Similarly, a 2008 analysis emphasized the importance of exchange rates and oil prices on the price of corn and overall commodity prices: "The reality is that most of the increase in corn demand has been driven by the higher oil price and the fall in the U.S.$" (Abbott et al. 2008, 48). In contrast, a subsequent analysis conducted in 2011 listed exchange rates as only the sixth factor among eight that led to high food prices in 2008 (Abbott et al. 2011). The first five were: (1) supply-utilization shocks (weather, production shortfalls, low stocks), (2) third-world income and population growth and resulting dietary transitions, (3) long-run production trends and declining investments in agricultural research, (4) biofuels and the link between corn and crude oil, and (5) export restrictions and trade policy responses.

While acknowledging the importance of supply and demand factors, others have concluded that at least one-half of the 2008 price spike was the result of a speculative bubble in the case of wheat, corn, and soybeans, and a market panic in the case of rice (Piesse and Thirtle 2009).[4] Speculation was also identified as a key factor in the price spike, along with hoarding and the lack of a commodity reserve that could have dampened volatile prices (von Braun and Torero 2009).

The factors behind food price spikes from 2007 through 2011 can be grouped by time period (short run vs. long run), as well as by origin (either market-driven or policy-related). The primary factors appear to be low stocks relative to use, with demand strengthening considerably from income and population growth, and biofuels expansion due to energy policies. Also key were rising energy prices and their subsequent impact on the cost of production and on demand for biofuel as a substitute for petroleum-based motor fuel. Low investment in research slowed growth in yields and dampened supply while bad weather created short-term supply shocks. A number of other factors also contributed in upward price pressure at various times, such as export restrictions and pre-emptive buying in the short term and a decline in the US dollar over the longer term. Finally, hoarding and financial speculation likely added to price volatility as market uncertainty reached its peak.

PRICE SPIKES, FOOD INSECURITY, AND SOCIAL CONFLICT

The food price spikes in the late 2000s caught the world's attention, particularly when sharp increases in food and fuel prices in 2008 coincided with street demonstrations and riots in many countries. For 2008 and the two preceding years, researchers identified a significant number of countries (totaling 54) with protests during what was called the global food crisis (Benson et al. 2008). Violent protests occurred in 21 countries, and nonviolent protests occurred in 44 countries. Both types of protest took place in 11 countries. In a separate analysis, developing countries with low government effectiveness experienced more food price protests between 2007 and 2008 than countries with high government effectiveness (World Bank 2011a). Although the incidence of violent protests was much higher in countries with less capable governance, many factors could be causing or contributing to these protests, such as government response tactics, rather than the initial food price spike.

Data on food riots and food prices have tracked together in recent years. Agricultural commodity prices started strengthening in international markets in 2006. In the latter half of 2007, as prices continued to rise, two or fewer food price riots per month were recorded (based on World Food Programme data, as reported in Brinkman and Hendrix 2011). As prices peaked and remained high during mid-2008, the number of riots increased dramatically, with a cumulative total of 84 by August 2008. Subsequently, both prices and the monthly number of protests declined.[5]

Several researchers have studied the connection between food price shocks and conflict, finding at least some relationship between food prices and conflict. According to Dell et al. (2008), higher food prices lead to income declines and an increase in political instability, but only for poor countries. Researchers also found a positive and significant relationship between weather shocks (affecting food availability, prices, and real income) and the probability of suffering government repression or a civil war (Besley and Persson 2009). Arezki and Brückner (2011) evaluated a constructed food price index and political variables, including data on riots and anti-government demonstrations and measures of civil unrest. Using data from 61 countries over the period 1970 to 2007, they found a direct connection between food price shocks and an increased likelihood of civil conflict, including riots and demonstrations.

Other researchers have broadened the analysis by considering government responses or underlying policies that affect local prices, and consequently influence outcomes and the linkage between food price shocks and conflict. Carter and Bates (2012) evaluated data from 30 developing countries for the time period 1961 to 2001, concluding that when governments mitigate the impact of food price shocks on urban consumers, the apparent relationship between food price shocks and civil war disappears. Moreover, when the urban consumers can expect a favorable response, the protests only serve as a motivation for a policy response rather than as a prelude to something more serious, such as violent demonstrations or even civil war.

Many in the international development community see war and conflict as a development issue, with a war or conflict severely damaging the local economy, which in turn leads to forced migration and dislocation, and ultimately acute food insecurity. Brinkman and Hendrix (2011) ask if it could be the other way around, with food insecurity causing conflict. Their answer, based on a review of the literature, is "a highly qualified yes," especially for intrastate conflict. The primary reason is that insecurity itself heightens the risk of democratic breakdown and civil conflict. The linkage connecting food insecurity to conflict is contingent on levels of economic development (a stronger linkage for poorer countries), existing political institutions, and other factors. The researchers say establishing causation directly is elusive, considering a lack of evidence for explaining individual behavior. The debate over cause and effect is ongoing.

Policies can nevertheless be implemented to reduce price variability. Less costly forms of stabilization, at least in terms of government outlays, include reducing import tariffs (and quotas) to lower prices and restricting exports to increase food availability. However, these types of policy responses, while perhaps helping an individual country's consumers in the short run, can lead to increased international price volatility, with potential for disproportionate adverse impacts on other countries that also may be experiencing food insecurity.

The World Bank (2011a) has called for international organizations and others to coordinate their activities to compile useful, real-time information for policymakers examining the current availability of key resources, including food, water, oil, and land. Such an approach could assess trends in vulnerability to scarcity among poor people and regions affected by violence and food insecurity. It could also provide effective early warning systems on food and agriculture, monitoring food insecurity and potential conflict—although convincing countries to share this information or allow others to monitor their resources might be a challenge.

ESTIMATING FOOD LOSSES AND IMPROVING RECOVERY

Concerns for improving food security—as well as economic efficiency—have brought increasing attention to food losses and their recovery. The motivation is to increase food availability, reduce food insecurity, improve resource utilization, and lower overall food costs. The literature identifies significant food losses in developing countries, stemming mostly from the farm, processing, and transport. Losses have been estimated to be fairly high for fruits and vegetables, reaching 50 percent or more, and somewhat lower for grains. In contrast, losses can be lower in developed countries and typically occur in the marketing (or value) chain or "post-plate" by consumers. Newly completed research concludes that percentage losses in certain cases are in the single digits and well below generally perceived levels.

Food losses refer to any decrease in food mass throughout the edible food supply chain. They can take place during production (for example, farm loss from damage or spillage during harvest), postharvest and processing stages, distribution, retail sale, and home consumption (including spoilage and table waste). Postharvest losses typically are due to spillage ahead of processing or degradation from insects or microorganisms during storage.

Estimating losses can be difficult because sources are hard to trace, the sources themselves may not be very reliable, or there may be problems with inferential or other types of methods (World Bank 2011b; Hodges et al. 2010). Moreover, loss can depend greatly on weather conditions. For example, during the production stage, losses can increase sharply if harvest is delayed due to poor weather (Smil 2004).

Abstracting from these difficulties, a recent review of global food losses by Gustavsson et al. (2011) attempts to summarize losses along the entire food chain, from production through consumption, and across major commodity groups and regions of the world. In developing countries, food is lost mostly in the early or middle stages of the food chain, either on the farm or during processing and distribution, primarily due to limitations in harvesting techniques, storage/cooling facilities, and the market/transport system overall. In developed countries, losses are greatest after the production stage, caused primarily by heavy culling to meet fruit and vegetable quality standards and consumers throwing out unwanted food because they lack economic incentives to minimize loss.

Worldwide losses are estimated at approximately 20 to 30 percent for cereal crops and generally much higher for the more perishable fruits and vegetables, at more than 40 percent (see Figure 2.1). A number of other studies have a more narrow geographic or commodity focus, but generally indicate a similar magnitude of loss. In eastern and southern Africa, the range of post-harvest loss for grain is between 10 percent and 20 percent depending on the crop and season, with a weighted average of 13.5 percent (World Bank 2011b). A compilation of estimates by Ruiz and Ringler (undated) shows a range of estimates for Asia and Africa (see Figure 2.2). Additional estimates for grain can be found in Smil (2004), Lundqvist et al. (2008), Parfitt et al. (2010).

For fruits and vegetables, Kader (2005) estimates that about one-third of global production is wasted or not consumed. Spoilage of fruits and vegetables is reported to be as high as 30 percent in India because the postharvest infrastructure is weak (Swaminathan 2006 as cited in Lundqvist et al. 2008).

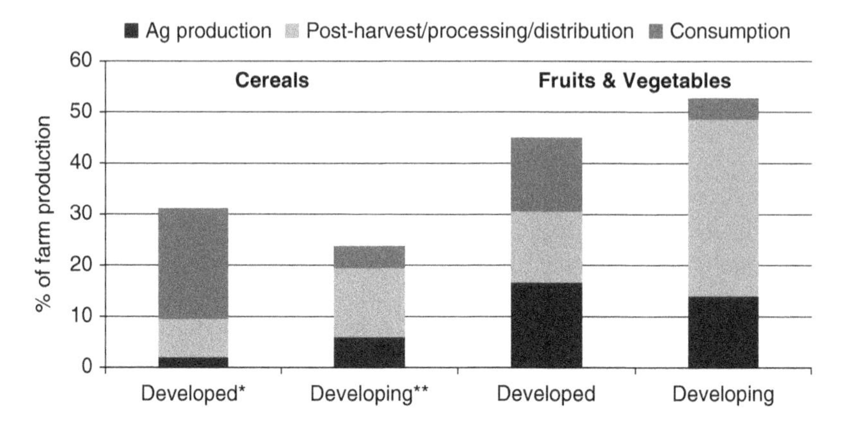

Fig. 2.1. Food losses as a share of farm production

*North America, Oceania, Europe, and Industrialized Asia. **Africa and South/Southeast Asia.

Source: Author's calculations using data in Gustavsson et al. (2011). Data have been converted from percentage loss of product entering the stage of value chain to percentage loss of initial farm production.

Fig. 2.2. Range of post-harvest loss estimates

Note: Excludes maize estimate in Ghana reported at 66%.

Source: Constructed by author using data from Ruiz and Ringler (undated compilation of multiple studies).

Similarly, losses in a variety of Asian countries range from 16 percent to 50 percent, as reported in Parfitt et al. (2010) and Choudhury (2006).[6]

In contrast to most published research, forthcoming work by Reardon et al. (2012) indicates that the levels of wastage in the value (marketing) chain are much lower than popular or scholarly perception. In Bangladesh, total quantities of potatoes wasted in the entire value chain (not used for consumption) are only 5.2 percent in the harvest period and 6.4 percent in the off-season. Even lower numbers are estimated in India (3.2 percent and 3.3 percent, respectively). The share of physical waste in China is estimated to be slightly higher at 5.35 percent, possibly because of the significantly longer transport distances. The results have implications for potential investments of postharvest technologies aimed at reducing losses in the value chains, as low levels of loss might indicate little or no return on technologies or policies for further loss reductions.

A substantial and contentious literature exists on food loss, but to address the vexing problem of food insecurity and need for making the best use of global resources, researchers and policymakers must ask of any proposal: does the economic benefit of the prescribed approach exceed its cost? As Kader (2005, 2170) points out, "It is not economical or practical to aim for 0% losses, but an acceptable loss level for each commodity-production area and season combination can be identified on the basis of cost-benefit analysis (return on investment evaluations)." It is clear that an "acceptable" level of loss has yet to be determined in the literature.

Several studies do, however, consider economic costs due to postharvest losses for certain crops and countries. Komen et al. (2010) evaluates economic costs of losses of maize in Kenya, and concludes that if losses are not managed, storing the crop for selling after harvest is risky due to price fluctuations and potential for losses due to rot, insects, and rodents. For maize in Uganda, the use of an improved maize sheller and crib reduced postharvest losses and provided other benefits, resulting in a favorable benefit–cost ratio of between 4.3 and 5.5 (Mwebaze and Mugisha 2011). Separately, a large compendium on postharvest operations includes a short section on economic considerations for farmers, outlining an analysis to determine whether an improved postharvest system would provide sufficient benefit for the additional costs (Mejia 2003).

The report by the World Bank (2011b) repeats the important point of not aiming for zero losses, and estimates dollar amounts of postharvest losses, which appear to be significant in absolute terms. However, the report does not compare the figures with other investments or pursue a marginal benefit/cost analysis that would help guide the identification of optimal strategies. Likewise, Gustavsson et al. (2011) estimate sizeable losses but do not consider whether they are economically significant based on a cost/benefit analysis. Smil (2004) concludes that the major challenge is *not* to use more investment and greater inputs to increase production, but instead to pick up the slack in the current system by reducing food losses. This approach, while popular, should be subjected to a marginal cost-benefit analysis.[7]

THE GLOBAL FOOD ECONOMY: SCENARIOS AND SOLUTIONS

World agricultural markets have gone through many transformations in demand dynamics and supply constraints since the mid-twentieth century. These changes are reflected in the prices of major agricultural commodities

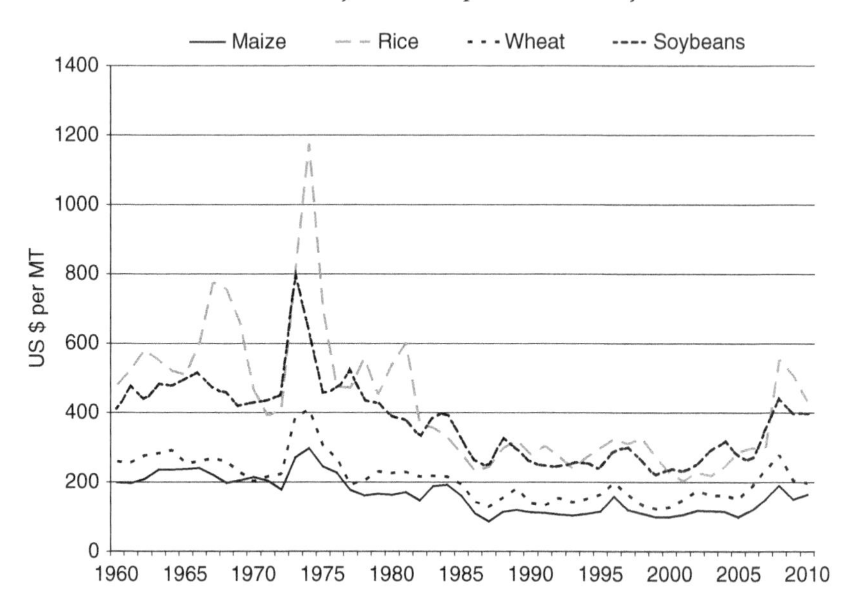

Fig. 2.3. Historical real price trend of major agricultural commodities in 2005 US$
Source: Authors' computations based on World Bank (2011c)

as seen in Figure 2.3. Crop prices (wheat, rice, maize, soybeans) in 2005 US$ showed a steep downward trend in real terms from the 1960s to the early 1990s (World Bank 2011c). Prices of these crops in real terms leveled off beginning in the early 1990s. Past price spikes have been short term. Starting in 2007, the prices of major grains increased dramatically in real terms and reached their peak in 2008. While real prices of these commodities declined in 2009 and 2010, they did not go back to their previous levels, but increased again in 2011.

The main reason behind the recent food price spikes is the rate of demand growth far outpacing the rate of supply growth. These price hikes have caused concerns about food security in developing countries, where the consumers spend a relatively larger share of their income on food purchases. In this environment, how can food production be increased in order to meet the new levels of food demand? There are three sources of growth in crop production: increased exploitation of worldwide arable land and water resources, yield increases, and irrigation investments.

Productivity growth is a critical component of agricultural supply increase. A number of key factors affect crop yields, including climatic, environmental, technological, economic, and policy conditions. Development of new varieties, technological diffusion, input use such as fertilizer and irrigation, land improvements, adoption of conservation tillage techniques, denser planting, earlier planting, irrigation, pest control, and weed control all increase crop yields. Land degradation, adverse climate conditions, and limited resource conditions reduce crop yields.

Crop yields have risen significantly in the past decades, yet the growth rates of yields have slowed down in some countries for some major crops (FAOSTAT 2011). For example, the world annual average yield growth rates between 1970 and 1990 for maize, rice, and wheat were 2.68 percent, 2.02 percent, and 2.90 percent respectively. This rate of growth dropped to 1.90 percent, 1.14 percent, and 1.19 percent for the 1990 to 2010 period. This led to a debate on whether a yield plateau has been reached in many countries, although the literature shows both sides of the argument. For many developing economies, there is still room for yield growth through increased input use, technological change, and better farm management.

There is an extensive literature focused on understanding the determinants of the gains in productivity in the agricultural sector (Alston et al. 2000, 2010; Evenson 2001). The studies suggest that public and private research and development (R&D) investments played a crucial role in realizing productivity growth. Biotechnology is the most recent result of these R&D investments and it has contributed to gains in productivity and reduced costs of production in the agricultural sector. Most of these R&D investments occurred in developed countries, such as the United States and Europe. At the same time, diffusion of new technology from countries where it originated to new countries played a role in increasing productivity in the world.

IMPACT Baseline

To provide a baseline for analyzing alternative scenarios in the subsequent section, we utilize the IMPACT model: a partial equilibrium, multi-commodity, multi-country model that generates projections of global food supply, demand, trade, and prices (Rosegrant et al. 2008). IMPACT covers over 46 crops and livestock commodities and includes 115 countries/regions, with each country linked to the rest of the world through international trade and 281 food-producing units (grouped according to political boundaries and major river basins). Demand is a function of prices, income, and population growth. Crop production is determined by crop and input prices, the rate of productivity growth, and water availability. The baseline scenario assumes a continuation of current trends and existing plans in agricultural policies and investments in agricultural productivity growth. Population projections are the medium variant population growth rate projections from the population statistics division of the UN, and income projections are estimated by the authors, drawing upon the Millennium Ecosystem Assessment (2005).

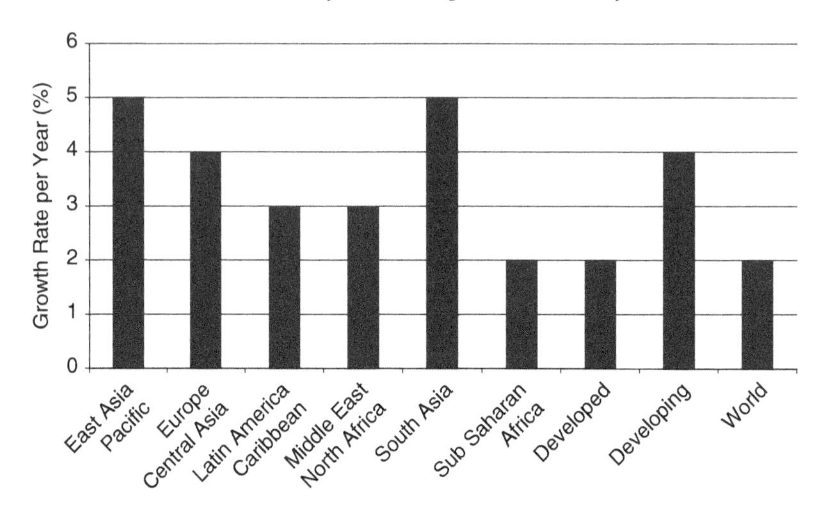

Fig. 2.4. Annual average growth rate in per capita GDP between 2010 and 2025—baseline projections

Source: IFPRI IMPACT Model Projections 2011

We have used IMPACT to generate a baseline for global agricultural markets through 2025. Major drivers of this baseline are income growth, population increase, productivity gains in many agricultural activities, and biofuel sector expansion.

As seen in Figure 2.4, annual average growth rate for GDP per capita is highest for the East Asia and Pacific region and South Asia region, followed by Europe and Central Asia. This income growth leads to changes in consumption patterns in these regions, with consumers eating more meat and dairy products. Population growth also increases food demand in the projection period. The highest population growth rate is observed in sub-Saharan Africa, followed by the Middle East and North Africa.

Pressure to meet the growing demand increases on producers. There are multiple venues for supply increase: intensification, extensification, and irrigation investments. As seen in Figure 2.5, although world crop yields continued to grow in the past decades, the growth rates have declined over time. This is particularly observed for wheat and maize. IMPACT projects continued to yield growth for major crops, but at a rate slower than observed in the past. Figure 2.6 presents annual average growth rates for cereals between 2010 and 2025. We observe that these growth rates are below 2 percent for this period.

The baseline IMPACT model assumes annual global public investment in agricultural R&D of $5 billion per year, computed based on recent trends in actual and projected research expenditure and the growth of crop and animal yield and numbers.

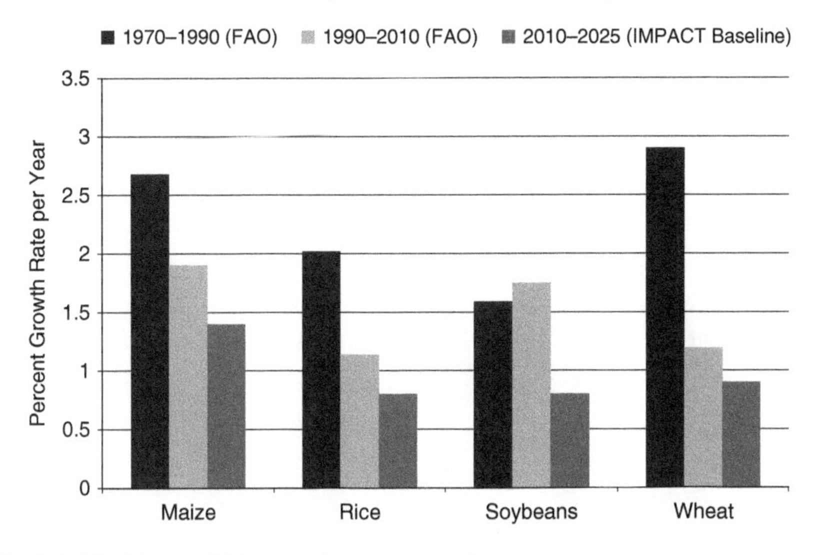

Fig. 2.5. World crop yields annual average growth rate
Source: IFPRI IMPACT Model Projections 2011

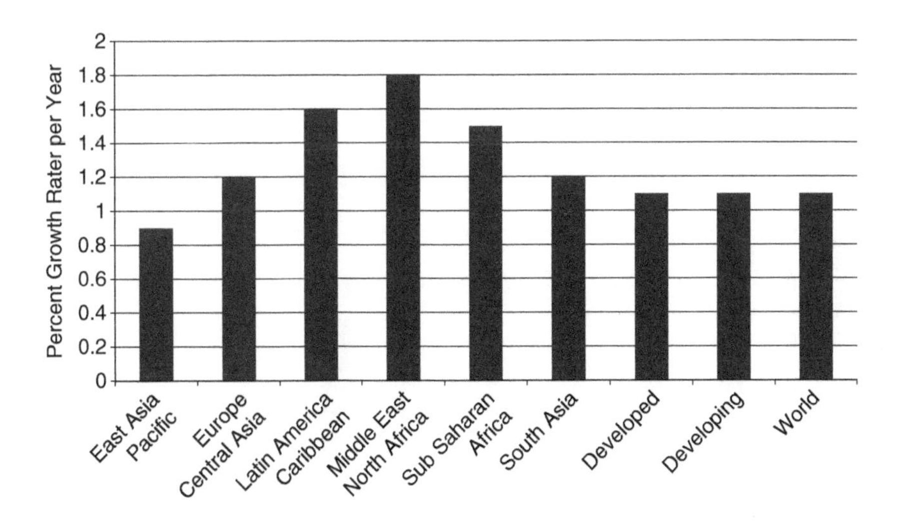

Fig. 2.6. Annual average growth rate of total cereal yields, 2010 to 2025—regions
Source: IFPRI IMPACT Model Projections 2011

As production responds to higher prices, harvested crop area increases for most of the crops. In terms of crop area, world maize, wheat, soybeans, and sorghum area expands between 2010 and 2025, whereas rice and other grains area declines slightly (Figure 2.7). Cereal area expands in the projection period for all regions, except South Asia. Overall, world harvested crop area increases, putting additional pressure on limited natural resources and water.

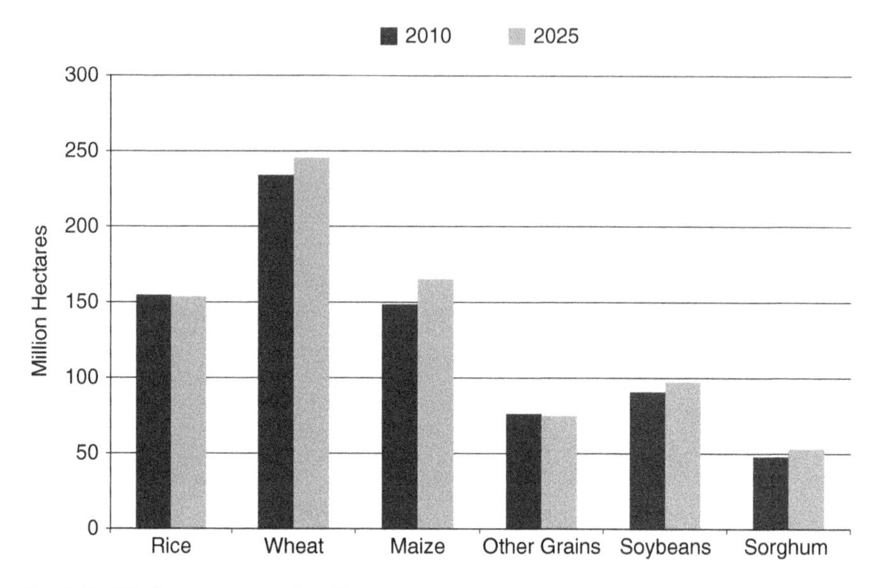

Fig. 2.7. World crop area—baseline projections
Source: IFPRI IMPACT Model Projections 2011

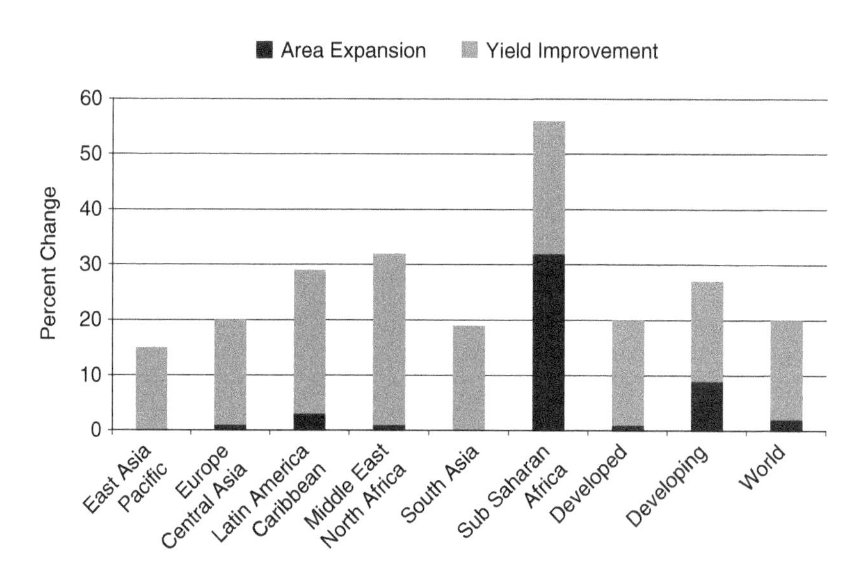

Fig. 2.8. Sources of cereal production growth, 2010 to 2025—regions
Source: IFPRI IMPACT Model Projections 2011

We see in Figure 2.8 that the main source of supply increase is yield improvement for all regions in the world, except in the sub-Saharan Africa region. This highlights the importance of productivity growth in the face of limitations on the potentially available arable land.

Significant demand pressure explains the crop price increases (Figure 2.9) despite higher supply. The highest increase is seen in maize, followed by soybeans and sorghum. Higher crop prices also increase feedstock costs for livestock producers, leading to higher prices for meat and milk (Figure 2.10). These price increases are because of biofuel sector expansion, income growth

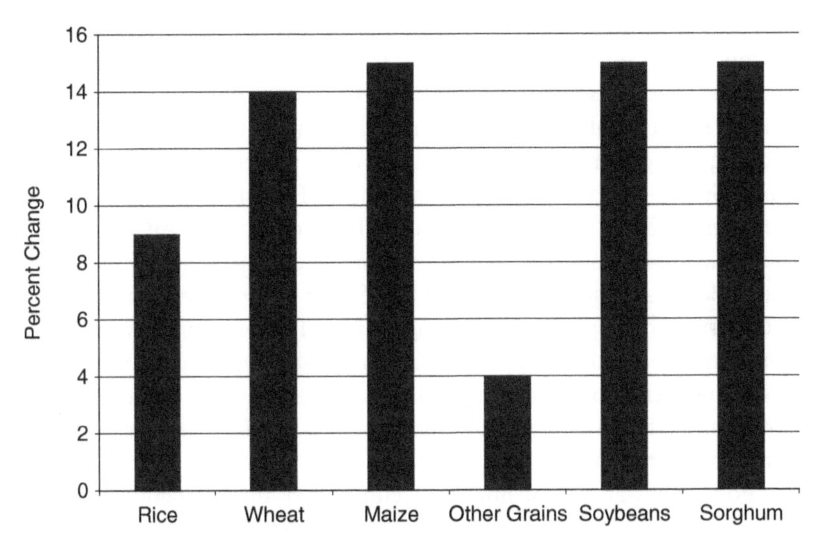

Fig. 2.9. Percent change in world prices of crops between 2010 and 2025
Source: IFPRI IMPACT Model Projections 2011

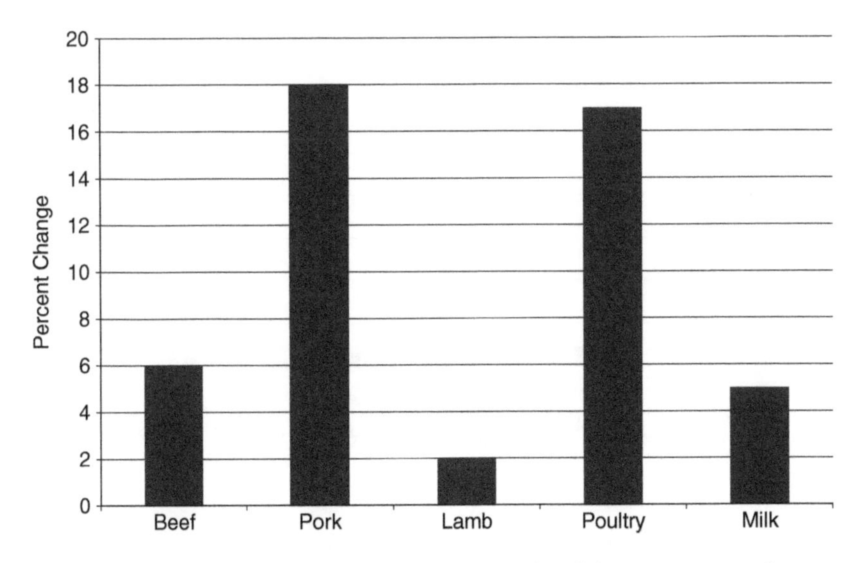

Fig. 2.10. Percent change in world prices of meat and milk between 2010 and 2025
Source: IFPRI IMPACT Model Projections 2011

in Asian countries, and worldwide population growth. Price increases show some rate differences due to different rates of growth in demand and supply. Pork and poultry prices increase most, since demand growth outstrips supply growth for these commodities.

Per capita food demand for cereals (direct human consumption) reveals a more static picture for the world, despite regional differences. Food consumption of coarse grains is much less than its use as livestock feed, although in Africa food consumption of coarse grains is higher than elsewhere in the world. East Asia and Pacific, South Asia, and Latin America and the Caribbean regions experience slight negative growth in cereal consumption due to income growth and consequent dietary pattern changes towards protein rich goods like meat and milk. Sub-Saharan Africa region and Middle East and North Africa region, on the other hand, are projected to increase their per capita cereal food demand.

The impact of income growth is most clearly seen in per capita food demand for meat. East Asia and Pacific region and South Asia region demand growth rates outstrip other regions, in keeping with their rapid growth in per capita income compared with other developing and developed regions. Sub-Saharan Africa is another region that shows large increases in per capita consumption of meat, with demand growing steadily from relatively low levels owing to steady income growth over the period.

The population at risk of hunger declines over the projection period for most of the regions (Figure 2.11), with the largest decline in the East Asia and Pacific region.[8] We also observe a decline in the number of malnourished children over the years due to an increase in food consumption[9]. The largest decline is again in the East Asia and Pacific region.

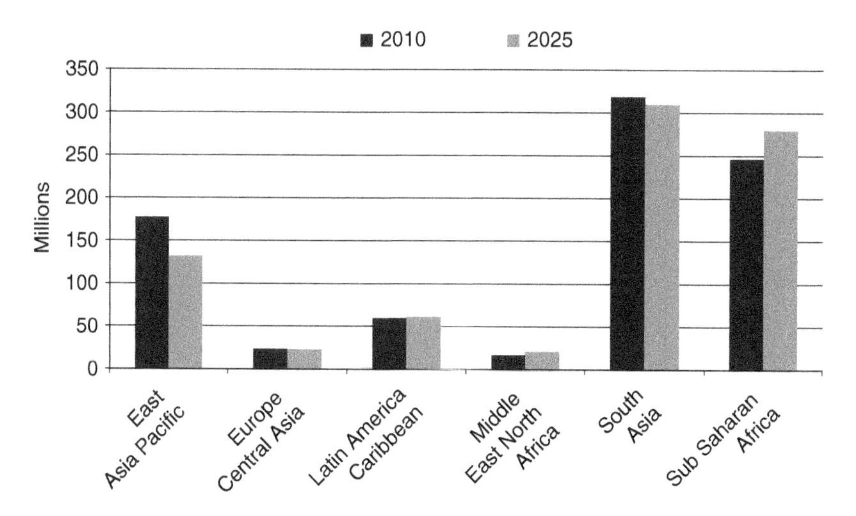

Fig. 2.11. Population at the risk of hunger—baseline projections

Source: IFPRI IMPACT Model Projections 2011

Alternative Policy/Investment Scenarios and Results

After estimating the baseline of the global food economy, constructed with specific assumptions, we can project alternative outcomes (scenarios) using the IMPACT model by changing certain baseline assumptions. We focus on three types of policies and investments that can be altered to enhance agricultural productivity, food security, and environmental sustainability: (1) agricultural research, (2) efficiency of agricultural research (higher yield gains per dollar of research investment or elasticity of research), and (3) irrigation infrastructure. Key elements of the scenarios are presented in Table 2.1. We also consider the impact of a reduction in post-harvest losses as an additional scenario. We offer these scenarios so policymakers can view the potential impacts of policy change without the complication of alternative economic growth scenarios. As such, the underlying expectations for global or country-specific economic growth and other baseline assumptions remain unchanged.

Scenario 1: Increased Investment in Agricultural Research

Using expert opinion and literature reviews, we examine the impact of increased investment in agricultural research on yields. To analyze the impact of increased investments in agricultural R&D, we increase the annual global

Table 2.1. Key elements of scenario definitions

	Change from baseline (all changes begin in 2010 unless otherwise noted)		
	Scenario 1: Increased Investment in Ag. Research *(INC AG RES)*	Scenario 2: Increased Investment in Ag. Research with Efficiency of Research *(INC AG RES w/EFF)*	Scenario 3: Increased Ag. Research with Efficiency of Research & Irrigation Expansion *(INC AG RES w/EFF & IRR EXP)*
Livestock numbers growth	+ 0.3	+ 0.3	+ 0.3
Livestock yield growth	No change	+ 0.3 beginning 2015, + 0.5 beginning 2030	+ 0.3 beginning 2015, + 0.5 beginning 2030
Food crops yield growth	+ 0.6	+ 0.6+0.78 = +1.38 beginning 2015, +0.6+0.9 = +1.5 beginning 2030	+ 0.6+0.78 = +1.38 beginning 2015, +0.6+0.9 = +1.5 beginning 2030
Irrigated area growth	No change	No change	+0.30
Rain-fed area growth	No change	No change	−0.05

Source: IFPRI IMPACT projections 2011

public investment in agricultural R&D from $5 billion per year to $16 billion. We compute the estimated increases in agricultural investment based on the alternative yield scenarios and the elasticities of yield with respect to agricultural research expenditures (Rosegrant et al. 2008).

This scenario assumes an increase in all crop yield growth rates over the baseline and an increase in livestock numbers growth rates. The livestock number growth rate is increased by 0.3 from 2010 to 2050 and the crop yield growth rate is increased by 0.6 over the same period for all countries in IMPACT. With this scenario, we examine increased investments in agricultural research and their impact on the production of agricultural commodities.

IMPACT model results indicate increased investment in agricultural research in the crop and livestock sectors leads to relatively higher crop yields and growth in livestock numbers. This, in turn, boosts agricultural production and results in lower agricultural commodity prices (Table 2.2). International commodity prices for maize, rice, and wheat in 2025 are 7 percent, 10 percent, and 8 percent lower compared to the 2025 baseline values. Livestock prices are reduced less: by 6 percent for beef, 6 percent for pork, 8 percent for poultry, and 7 percent for sheep and goat.

Developing-country net food imports are affected by higher investments in agricultural R&D: the developing-country net import volume for cereals would decline from 97.4 million metric tons under the baseline (2025 value) to 96.5 million metric tons under the increased agricultural investment scenario (Table 2.3) due to higher domestic supply. On the livestock product side, developing countries would slightly increase their net imports of meat—from 2.5 million metric tons in baseline to 3.3 million metric tons in scenario with lower world prices.

Lower world prices and higher domestic and international supply mean consumers can buy more food: cereals and meat. These changes are reflected in food security indicators. In this scenario, we observe a lower number of malnourished children in all regions. The highest decline is in the Middle East and North Africa, followed by East Asia and Pacific. Furthermore, the population at risk of hunger declines significantly in all regions. For South Asia, this decline is 16 percent, followed by Latin America and the Caribbean at 14 percent.

Scenario 2: Increased Investment in Agricultural Research with Efficiency of Research

In this scenario, we combine more agricultural research with improvements in efficiency of research, which further boosts productivity. To analyze the impact of increased investments in agricultural R&D with efficiency of research, we increase the annual global public investment in agricultural R&D from $5 billion per year to a total investment of $20 billion per year.

Table 2.2. Projected change in world commodity prices presented as the percent change between baseline and scenario for 2025

	World Commodity Prices			
Commodity	Scenario 1:	Scenario 2:	Scenario 3:	Scenario 4:
Beef	−6%	−9%	−10%	−3%
Pork	−6%	−9%	−9%	−3%
Lamb	−7%	−11%	−11%	−2%
Poultry	−8%	−10%	−11%	−4%
Milk	−6%	−9%	−9%	−2%
Rice	−10%	−12%	−15%	−8%
Wheat	−8%	−10%	−11%	−9%
Maize	−7%	−8%	−9%	−10%
Other Grains[a]	−7%	−8%	−8%	−11%
Millet	−9%	−10%	−11%	−10%
Sorghum	−8%	−9%	−9%	−8%
Rapeseed	−8%	−10%	−9%	−5%
Rapeseed Meal	−1%	1%	1%	−2%
Rapeseed Oil	−8%	−10%	−10%	−5%
Soybeans	−9%	−10%	−10%	−5%
Soybean Meal	−3%	−2%	−2%	−3%
Soybean Oil	−9%	−12%	−12%	−5%
Vegetables	−9%	−10%	−12%	−6%
Tropical & sub-Tropical Fruits	−9%	−11%	−11%	−7%
Temperate Fruits	−10%	−12%	−13%	−9%
Sugar	−10%	−11%	−13%	−6%

[a] Other grains include barley and rye
Source: IFPRI IMPACT projections 2011

The crop yield growth rate is increased by 0.6 from 2010 to 2015, by 0.78 from 2015 to 2030, and by 0.9 from 2030 to 2050. The livestock numbers growth rate increases by 0.3 from 2010 to 2050, whereas the livestock yield growth rate increases by 0.3 from 2015 to 2030 and further by 0.5 from 2030 to 2050. These changes are applied to all countries in IMPACT.

Results indicate that the effect on world prices is greater than for Scenario 1: maize declines 8 percent, rice declines 12 percent, and wheat declines

Table 2.3. World net exports for baseline and scenarios for 2025 (thousand metric tons)

		Scenario 1	Scenario 2	Scenario 3
Region	Baseline	INC AG RES	INC AG RES w/EFF	INC AG RES w/EFF & IRR EXP
Latin America and the Caribbean				
Cereals	−11,029	−11,974	−11,939	−13,886
Meat	13,091	12,860	12,656	12,646
East Asia and Pacific				
Cereals	−56,344	−55,254	−55,746	−49,031
Meat	−7,459	−7,543	−7,340	−7,371
Europe and Central Asia				
Cereals	88,011	87,761	87,278	84,574
Meat	−2,541	−2,483	−2,460	−2,452
Middle East and North Africa				
Cereals	−40,292	−39,289	−39,291	−37,474
Meat	−1,274	−1,332	−1,355	−1,353
South Asia				
Cereals	−23,072	−20,206	−19,281	−14,016
Meat	−683	−832	−903	−908
Sub-Saharan Africa				
Cereals	−51,716	−54,572	−54,722	−57,443
Meat	−2,828	−3,100	−3,205	−3,215
Developed				
Cereals	97,378	96,543	96,731	90,322
Meat	2,512	3,261	3,443	3,488
Developing				
Cereals	−97,378	−96,543	−96,731	−90,322
Meat	−2,512	−3,261	−3,443	−3,488

Source: IFPRI IMPACT projections 2011

10 percent. Lower feedstock costs and higher productivity increase the meat and dairy supply, leading to much lower prices for beef, pork, poultry, and milk in world markets.

Higher supply associated with increased agricultural investment, paired with efficiency of research, leads to some regions increasing their domestic supply and lowering their net imports, such as East Asia and Pacific for meat and cereals, Europe and Central Asia for meat, Middle East and North Africa for cereals, and South Asia for cereals. In some regions, lower world prices induce higher net imports despite a higher domestic supply. This is the case for Latin America and the Caribbean for cereals, Middle East and North Africa for meat, and sub-Saharan Africa for meat and cereals.

Higher available supply and lower prices induce consumers to increase food consumption, improving food security indicators. The number of malnourished children decreases by 10 percent in Middle East and North Africa, 9 percent in East Asia and Pacific, and 7 percent in Latin America and the Caribbean (Table 2.4). The population at risk of hunger falls significantly: as much as 19 percent in the South Asia region and 17 percent in Latin America and the Caribbean region (Table 2.5).

Table 2.4. Projected change in food security indicators presented as the percent change between the baseline and scenarios for 2025

	Number of Malnourished Children			
	Scenario 1	Scenario 2	Scenario 3	Scenario 4
Region	INC AG RES(%)	INC AG RES w/ EFF (%)	INC AG RES w/ EFF & IRR EXP (%)	POST-HARVEST LOSS (%)
East Asia and Pacific	−7%	−9%	−11%	−4%
Europe and Central Asia	−5%	−6%	−7%	−4%
Latin America and the Caribbean	−5%	−7%	−8%	−4%
Middle East and North Africa	−8%	−10%	−11%	−7%
South Asia	−2%	−2%	−3%	−1%
Sub-Saharan Africa	−3%	−4%	−4%	−3%
Developing	−3%	−4%	−5%	−2%
World	−3%	−4%	−5%	−2%

Source: IFPRI IMPACT projections 2011

Table 2.5. Projected change in food security indicators presented as the percent change between the baseline and scenarios for 2025

	Population at Risk of Hunger			
	Scenario 1	Scenario 2	Scenario 3	Scenario 4
Region	INC AG RES(%)	INC AG RES w/ EFF (%)	INC AG RES w/ EFF & IRR EXP (%)	POST-HARVEST LOSS (%)
East Asia and Pacific	−7%	−8%	−10%	−4%
Europe and Central Asia	−3%	−3%	−4%	−2%
Latin America and the Caribbean	−14%	−17%	−18%	−11%
Middle East and North Africa	−6%	−7%	−9%	−5%
South Asia	−16%	−19%	−23%	−11%
Sub-Saharan Africa	−13%	−16%	−17%	−11%
Developed	−6%	−8%	−8%	−4%
Developing	−13%	−15%	−18%	−10%
World	−13%	−15%	−17%	−9%

Source: IFPRI IMPACT projections 2011

Scenario 3: Increased Agricultural Research with Efficiency of Research and Irrigation Expansion

Irrigated agriculture has a major role in boosting agricultural yields and outputs and has made it possible to feed the world's growing population. It has helped maintain food production levels and contributed to price stability through greater control over production and scope for crop diversification. In developing countries, irrigation development has been particularly vital in achieving food security, especially as an important component of the Green Revolution technology package, both locally, through increased income and improved health and nutrition, and nationally, by bridging the gap between production and demand.

This scenario is designed to highlight the significance of expansion of irrigation for enhancing agricultural productivity. Here we combine specifications in Scenario 2 with accelerated investments in irrigation infrastructure in the form of increased expansion of harvested irrigated area growth rates by 0.3. A correlated assumption for this investment is that rain-fed area growth rates decrease by 0.05 as compared to the baseline. These changes are applied to all countries in IMPACT.

The IMPACT results show that prices decline for crops such as rice, vegetables, temperate fruits, wheat, maize, and millet relative to Scenario 2. However, the price impacts in general are relatively low because of the already high productivity outcomes achieved from direct investment in agricultural R&D and because the rain-fed area growth rate is lowered under this scenario, as some of the expansion in irrigated area displaces existing rain-fed areas.

East Asia and Pacific and South Asia lower their cereal net imports, since now they produce more cereals with higher average yields. This leaves more cereal in world markets to be imported by other regions, such as Latin America and the Caribbean and sub-Saharan Africa.

There is some improvement in food security indicators relative to Scenario 2. This is largest for regions such as East Asia and Pacific, South Asia, and the Middle East and North Africa.

Scenario 4: Post-harvest Loss

In this scenario, we analyze the implications of reductions in post-harvest losses in agriculture. To do that we simulate a reduction in post-harvest losses by 10 percentage points for all food crops, by 20 percentage points for fruits and vegetables, and 10 percentage points for livestock over a 25-year period. As described earlier, estimates of post-harvest losses vary widely, and this scenario uses conservative estimates.

This scenario highlights the impacts of reduction of post-harvest losses. Reducing post-harvest losses by 10 percentage points for all food crops and livestock and by 20 percentage points for fruits and vegetables leads to a higher food supply available in markets and lower prices. The results show that the decrease in prices for Scenario 4 is generally lower compared to our previous three scenarios. Price reductions of major commodities like rice and wheat are comparable to the reductions in Scenario 1 for increased investment in agricultural research. However, prices of maize and other grains are reduced by 10 to 11 percent, which is higher compared to the other scenarios. Decrease in meat prices is also the lowest for this scenario.

This scenario has the least improvement in the number of malnourished children relative to the other scenarios. The highest decrease in number of malnourished children can be seen in the Middle East and North Africa, while South Asia has the lowest.

Like our previous scenarios, we observe some reduction in the population at risk of hunger. However, similar to the malnutrition numbers, the decline in population at risk of hunger is the lowest for this scenario compared to the others. The highest reductions are seen in Latin America and the Caribbean, South Asia, and sub-Saharan Africa.

Welfare analysis

We conduct a welfare and benefit–cost analysis to assess the welfare impact of increased investments in agricultural R&D (Scenario 1 and Scenario 2). We calculate changes in consumer surplus, producer surplus, and net surplus arising from the investment-induced changes in crop yields, production, and food prices to compute the welfare gains. Next, we compute the benefit–cost ratio, which is the ratio of net present value of the net surplus to the net present value of the investment costs. To evaluate the benefits to society on the consumer- and producer-side, we then calculate the welfare components using a traditional economic welfare analysis approach.[10] To compare a technology's overall impact in the agricultural sector, we combine the total changes in consumer and producer surplus, yielding a benefit flow, which is used in a benefit–cost analysis.

As prices of crop and livestock decline by far more than the increase in productivity growth, there is a 4 percent decline in producer surplus in Scenario 1 and a 5 percent decline in Scenario 2. On the other hand, due to lower prices and higher consumption, consumers (including net-consuming farmers in developing countries) benefit substantially. As a result, consumer surplus rises significantly, by 15 percent for Scenario 1 and 20 percent for Scenario 2. Scenario 1 predicts that between 2010 and 2025 producer surplus decreases by 2.51 percent and consumer surplus increases by 8.89 percent. Scenario 2 shows that over the same time frame, producer surplus decreases by 2.82 percent and consumer surplus increases by 10.26 percent. Globally, the additional investment in agricultural R&D raises total welfare by 2.05 percent, yielding a net present value of benefits of $827 billion for Scenario 1. As for Scenario 2, there is a 2.41 percent increase in welfare as a result of the additional investment, giving a net present value of benefits of $974 billion.

The rates of return are positive for both scenarios. The internal rate of return to increased investments for Scenario 1 is 64 percent with a benefit–cost ratio of 6.61 indicating the high returns to expanded investment in agricultural R&D. In Scenario 2, the internal rate of return to increased investments is slightly lower at 51 percent; the benefit-cost ratio is comparable at 5.35.

KEEPING FUTURE FOOD PRICES IN CHECK

The tight food markets and price spikes of 2007 to 2011 were caused by a number of factors, including low food stocks, rapid growth in demand for biofuels, bad weather, commodity speculation, and trade restrictions. Fundamental market factors including increased demand for meat, livestock feed, and rice and wheat due to rapid economic growth and urbanization, particularly in Asia and Africa, have also contributed to a tighter world food supply and demand situation. In the longer term, climate change, growing water scarcity, and worsening water quality will be major challenges to food security. Examining these new global food system realities through the lens of scenarios of alternative futures for agricultural supply and demand and food indicates that—if current policies continue—real-world prices of most cereals and meats will rise. The substantial increase in food prices will cause relatively slow growth in calorie consumption, with both direct price impacts and reductions in real incomes for poor consumers who spend a large share of their income on food. This in turn will contribute to only slow improvement in food security for the poor in most regions of the world.

The evidence on the relationship between food insecurity and conflict is mixed. But, especially for intrastate conflict, it does appear that food insecurity increases conflict. The primary reason is that insecurity itself heightens the risk of democratic breakdown and civil conflict. The linkage connecting food insecurity to conflict is contingent on levels of economic development (a stronger linkage for poorer countries), existing political institutions, and other factors. Given the projected long-term higher prices and slow improvement in food security, the baseline scenario indicates a continued potential for protests or conflict, with an increasing probability to the extent that there is causality between food price insecurity and conflict.

On a hopeful note, increased investment in agricultural research can significantly reduce projected food prices relative to the baseline, with resulting improved food consumption and reductions in the number of malnourished children. The rates of return for agricultural research are high. A scenario simulating large reductions in postharvest food losses also contributes to lower food prices, higher food availability, and improved food security, although the impacts are not as positive as for increased agricultural research. In addition, the gains from post-harvest loss reduction are a one-off gain, which cannot expand beyond the economically recoverable losses, while productivity growth can be sustained over a longer period of time with appropriate investments.

The data on the magnitude of real-world post-harvest losses remains contested, with significant differences between more macro-level estimates (which tend to be high) and recent micro-level estimates (which are lower). To derive better estimates of the potential benefits and policy response, researchers need to create better evidence of loss along the value chain for key commodities, and properly analyze the cost of reducing the losses relative to research on yield improvements and other efforts to encourage more efficient use of resources.

For policymakers concerned about food security, improving life for the world's poor will likely involve greater investment in agricultural R&D and increased levels of irrigation. Reduced post-harvest losses could also play a role in increasing food supplies and reducing price pressure on the global food market.

NOTES

1. See Figure 5 in Trostle et al. (2011) for an annotated timeline of factors affecting the price spike.
2. Low investment in agricultural research is also often cited as a reason for slow yield growth and for creating a mismatch with long-term growth in demand (Piesse and Thirtle 2009). Large commodity supplies in the 1980s, due in part to agricultural policies in developed countries, had a two-fold effect. First, the resulting stable or low world prices reduced incentives for developing countries to produce crops and develop their agricultural sectors. Second, investments in agricultural research seemed unattractive if they were to make agricultural surpluses grow even larger. As a result, agriculture sectors in both developing and developed countries were not prepared for future expansion in demand and periodic shortages.
3. See Figure 2 in Headey et al. (2009) for an annotated timeline of export controls and the price of rice during 2007–09.
4. Speculation can be considered a characteristic of demand (rather than demand itself) that has potential for temporarily distorting demand signals and does not have long-term effects. This might preclude some researchers from mentioning it as a factor.
5. A monthly timeline on food riot protests through May 2008 is also available in von Braun (2008).
6. For developed countries, a detailed study on the United States estimates total losses from the edible food supply at 26 percent, on average (Hodges et al. 2010). Loss was greatest for vegetables at 34 percent and lowest for processed fruits and vegetables at 10 percent. The researchers point out that the food industry (and consumers, for that matter) will minimize post-harvest losses when they have financial incentives to do so, such as increased disposal fees for excess production or factory food waste. Also, most consumers in developed countries do not appear to be concerned with food waste because food is relatively abundant and inexpensive relative to income levels. Other studies on the United States, as summarized in Lundqvist et al. (2008), report similar figures.
7. The concern of identifying and measuring the costs of intervention to prevent post-harvest losses is not new. Thirty years ago, Greeley (1982) pointed out that the "single-minded concern with loss reduction" results in a failure to demonstrate rigorously the economic viability of any new innovation or proposed adoption of current technology (1). An analysis of possible alternatives, including research or policies outside the scope of food loss reductions would be prudent for policymakers and those investing in strategies to reduce global food insecurity. To derive a better policy response, it is critical that researchers create better evidence of loss along the value chain for key commodities and properly analyze the cost of reducing the losses relative to research on yield improvements, for example, along with other efforts to encourage more efficient use of resources.
8. Population at risk of hunger is computed using the share of population at risk of hunger, which is based on a strong empirical correlation between the share of malnourished within the total population and the relative availability of food, and

is adapted from the work done by Fischer et al. in the IIASA World Food System used by IIASA and FAO (Fischer et al. 2005).

9. The percentage of malnourished children under the age of five is estimated from the average per capita calorie consumption, female access to secondary education, the quality of maternal and child care, and health and sanitation (see Rosegrant et al. 2008 for details).
10. Since IMPACT has demand curves with demand elasticities, on the consumer side this calculation is straightforward as it allows us to calculate the consumer surplus. On the producer side, the calculation is not so direct, as the quantity supplied of each commodity is an area-yield equation, and does not represent the traditional supply curve that reflects the producer's marginal cost curve. Therefore, we have synthesized supply curves by land-type for each activity from the area and yield functions, calculated the producer surplus for each of these supply curves, and then aggregated to the national level.

REFERENCES

Abbott, P., C. Hurt, and W. Tyner. 2008. What's driving food prices? Farm Foundation Issue Report.

———. 2011. What's driving food prices in 2011? Farm Foundation Issue Report. Available at http://www.farmfoundation.org/news/articlefiles/105-FoodPrices_web.pdf (accessed March 20, 1213).

Alston, J. M., M. A. Andersen, J. S. James, P. G. Pardey. 2010. U.S. agricultural productivity growth and the benefits from public R&D spending. *Natural Resource Management and Policy 34*:504.

Alston, J. M., T. J. Wyatt, P. G. Pardey, M. C. Marra, C. Chan-Kang. 2000. A meta-analysis of rates of return to agricultural R&D: Ex pede Herculem? IFPRI Research Report 113. Washington, DC: International Food Policy Research Institute.

Arezki, R. and M. Brückner. 2011. Food prices and political instability. IMF Working Paper WP/11/62. Washington, DC: International Monetary Fund.

Benson, T., N. Minot, J. Pender, M. Robles, and J. von Braun. 2008. *Global food crises: Monitoring and assessing impact to inform policy responses.* IFPRI Food Policy Report. Washington, DC: International Food Policy Research Institute.

Besley, T. and T. Persson. 2009. Repression or civil war? *American Economic Review* 99: 292–97.

Brinkman, H. and C. Hendrix. 2011. *Food insecurity and violent conflict: Causes, consequences, and addressing the challenges.* WFP Occasional Paper No. 24. Rome: World Food Programme.

Carter, B. L. and R. Bates. 2012. Public policy, price shocks, and civil war in developing countries. Working Paper 2012–0001. Weatherhead Center for International Affairs. Boston, MA: Harvard University.

Choudhury, M. L. 2006. Recent developments in reducing postharvest losses in the Asia–Pacific region. In *Postharvest management of fruits and vegetables in the Asia-Pacific region*, ed. R. Rolle, 15–22. Rome: Food and Agriculture Organization and Asian Productivity Organization.

Dell, M., B. Jones, and B. Olken. 2008. Climate change and economic growth: Evidence from the last half century. NBER Working Paper Series, 14312.

Evenson R. 2001. Economic impacts of agricultural research and extension. In *Handbook of agricultural economics*, ed. Gardner B., and G. Rausser, 573–628, Amsterdam: Elsevier Science.

FAO. 2011. FAOSTAT database. Rome: Food and Agriculture Organization of the United Nations.

Fischer, G., M. Shah, F. N. Tubiello, and H. van Velhuizen. 2005. Socio-economic and climate change impacts on agriculture: An integrated assessment. *Philosophical Transactions of the Royal Society B 360*: 2067–83.

Greeley, M. 1982. Farm-level post-harvest food losses: The myth of the soft third option. *Feeding the hungry: A role for post-harvest technology?* IDS Sussex Bulletin *13*(3). Institute of Development Studies.

Gustavsson, J., C. Cederberg, and U. Sonesson. 2011. *Global food losses and food waste: Extent, causes and prevention.* Rome: Food and Agriculture Organization.

Headey, D. and S. Fan. 2010. *Reflections on the global food crisis: How did it happen? How has it hurt? And how can we prevent the next one?* IFPRI Research Monograph 165. Washington, DC: International Food Policy Research Institute.

Headey, D., S. Malaiyandi, and S. Fan. 2009. Navigating the perfect storm: Reflections on the food, energy, and financial crisis. IFPRI Discussion Paper 00889. Washington, DC: International Food Policy Research Institute. Available online: http://www.ifpri.org/publication/navigating-perfect-storm (accessed March 20, 2013).

Hodges, R., J. Buzby, and B. Bennett. 2010. Postharvest losses and waste in developed and less developed countries: Opportunities to improve resource use. *Journal of Agricultural Science 149*: 37–45

International Food Policy Research Institute. 2001. IMPACT projections. Washington, DC: International Food Policy Research Institute.

Kader, A. A. 2005. Increasing food availability by reducing postharvest losses of fresh produce. *Proceedings of the 5th International Postharvest Symposium.* ed. F. Mencarelli and P. Tonutti. Acta Horticulturae, 682, International Society for Horticultural Science.

Komen, J. J., C. M. Mutoko, J. M. Wanyama, S. C. Rono, and L. O. Mose. 2010. Economics of post-harvest maize grain losses in Trans Nzoia and Uasin Gishu districts of north-west Kenya. *Proceedings of the 12th Kari Biennial Scientific Conference: Transforming Agriculture for Improved Livelihoods through Agricultural Product Value Chains.* Kenya Agricultural Research Institute.

Lundqvist, J., C. de Fraiture, and D. Molden. 2008. *Saving water: From field to fork: Curbing losses and wastage in the food chain.* SIWI Policy Brief. Stockholm: Stockholm International Water Institute.

Mejia, D. 2003. *Maize: Post-harvest operation.* Information Network on Post-Harvest Operations. Rome: Food and Agriculture Organization.

Millennium Ecosystem Assessment. 2005. *Ecosystems and human well-being: General synthesis* (Millennium Ecosystem Assessment Series). Washington, DC: Island Press.

Mittal, A. 2009. The blame game. In *The global food crisis*, ed. J. Clapp and M. Cohen, 13–28. Canada: The Centre for International Governance Innovation and Wilfrid Laurier University Press.

Mwebaze, P. and J. Mugisha. 2011. Adoption, utilisation and economic impacts of improved post-harvest technologies in maize production in Kapchorwa District, Uganda. *International Journal of Postharvest Technology and Innovation 2*: 301–27.

Parfitt, J., M. Barthel, and S. Macnaughton. 2010. Food waste within food supply chains: Quantification and potential for change to 2050. *Philosophical Transactions of the Royal Society 365*: 3065–81.

Piesse, J. and C. Thirtle. 2009. Three bubbles and a panic: An explanatory review of recent food commodity price events. *Food Policy 342*: 119–29.

Reardon, T., K. Chen, and B. Minten. 2012. The quiet revolution in staple food value chains in Asia: Enter the dragon, the elephant, and the tiger. Mandaluyong City, Philippines: Asian Development Bank.

Rosegrant, M. W., S. Msangi, C. Ringler, T. B. Sulser, T. Zhu, and S. A. Cline. 2008. International Model for Policy Analysis of Agricultural Commodities and Trade (IMPACT): Model description. Washington, DC: International Food Policy Research Institute.

Ruiz and Ringler. Undated. Crop post-harvest losses: Myth or reality (paper draft). Washington, DC: International Food Policy Research Institute.

Smil, V. 2004. Improving efficiency and reducing waste in our food system. *Environmental Sciences 1*: 17–26.

Swaminathan, M. S. 2006. 2006–07: Year of agricultural renewal. 93 Indian Science Congress in Hyderabad, Public Lecture, January 4.

Trostle, R. 2008. Global agricultural supply and demand: Factors contributing to the recent increase in food commodity prices. Outlook Report No. WRS-0801, Washington, DC: Economic Research Service, USDA.

——. 2012. Food commodity prices: Past developments and future prospects. Presentation delivered at USDA Outlook Forum. February 23.

——, D. Marti, S. Rosen, and P. Westcott. 2011. Why have food commodity prices risen again? WRS-1103. Washington, DC: Economic Research Service, USDA.

US Department of Agriculture, World Agricultural Outlook Board. May 11, 2011. World agricultural supply and demand estimates. Washington, DC: USDA.

von Braun, J. 2008. The world food crisis: Political and economic consequences and needed actions. Presentation to the Ministry of Foreign Affairs, Stockholm, Sweden, September 22.

—— and M. Torero. 2009. Exploring the price spike. *Choices 24*: 16–21.

World Bank. 2011a. *World development report 2011: Conflict, security, and development.* Washington, DC: World Bank.

——. Food and Agriculture Organization. 2011b. *Missing food: The case of post-harvest grain losses in sub-Saharan Africa.* Report No. 60371–AFR. Washington, DC: Natural Resources Institute and the Food and Agriculture Organization.

——. 2011c. *GEM commodities.* Available online at http://data.worldbank.org/data-catalog/commodity-price-data (accessed October 11, 2011).

3

What Do We Know About the Climate of the Next Decade?

MARK A. CANE AND DONG EUN LEE

The years from 1876 to 1879 were a time of global catastrophe. Severe drought in China, India, Ethiopia, the Nordeste region of Brazil, and low Nile flow in Egypt led to widespread famine and disease, resulting in at least 17 million deaths worldwide—and perhaps twice that number (Davis 2001, 7). Based on excellent British records, it is fairly certain that rainfall in India in 1877 was less than at any time in the past 150 years. Although the droughts were especially severe, climate variations were not solely responsible for the appalling loss of life. Acknowledging the role of climate does not exonerate the Raj. During this period grain continued to be exported from India. British policy relied on railroads to distribute food in time of famine, and then only to those who worked. People were herded into camps in order to facilitate this policy, and malaria spread. The greatest loss of life was due to malaria and other diseases, rather than starvation (Whitcombe 1993).

In India, the drought precipitated a food shortage, leading to famine and epidemics among a weakened population. The variation in climate also had a direct influence on the malaria cycle by changing temperature and water-related factors, including rainfall, standing water, and stream flow. The malaria epidemic was worst after the drought was over because mosquitoes recover more rapidly from the drought than their predators, so the vector population was likely to have been larger than normal (Bouma and van der Kayy 1994, 1996). British interests favored a cash crop economy, reducing the flexibility of a society previously able to respond to drought by growing food for local use and by foraging in nearby forests. The immediate response of the Raj to the crisis—creating camps—made things worse. In complex ways, climate variability can substantially impact the resilience of a social system.

Many authors have argued for a connection between climate and the collapse of civilizations or less calamitous but still important historical events.

The books of Brian Fagan offer copious examples; in *Floods, Famines, and Emperors: El Niño and the Fate of Civilizations* he discusses the fall of the Old Kingdom in ancient Egypt, the Moche society of Peru, and the Maya of lowland Central America among others. While Fagan is perhaps overly generous in labeling all the climate changes in question as "El Niño," there is no doubt that these collapses coincided with strong climate changes. The most common explanation by Fagan and others for the link is a drought leading to food shortages and famine that place an unsupportable stress on an already weakened sociopolitical entity. Drought is the usual culprit, but catastrophic flooding destroyed Moche infrastructure and agriculture. Climate doubters emphasize the pre-existing weaknesses in these civilizations and tend to regard the concurrent climate changes as mere coincidence.

There is scholarly and other literature championing the impact of climate on food security and sociopolitical stability in contemporary times. Sahelian drought has been invoked as a cause of the conflict in Darfur. Hsiang et al. (2011) present quantitative evidence that over the past 60 years conflicts worldwide increase in an El Niño year. They do not establish a clear mechanistic linkage from El Niño to conflict, but offer a number of possible routes, food security prominent among them. They also point out that agricultural workers idled by droughts or other conditions are more susceptible to a call to arms as a means of redressing long-standing grievances, or simply as a chance to improve their lot with the spoils of war.

There is another lesson to be learned from the events of the 1870s and other instances of climate impacts on human affairs. Quite often, local climatic disasters are embedded in shifts of the global patterns of climate variation. The El Niño and its associated Southern Oscillation (ENSO) are the largest and best-known examples of such shifts. Though smaller in amplitude than the regular seasonal cycle, the unpredictable interannual variability is far more difficult for societies to cope with. ENSO influences the frequency of forest fires, hurricanes, severe winter storms, and other localized climatic events. The exceptionally strong El Niño in 1877 was undoubtedly a cause of the globally dispersed regional droughts mentioned above, all of which are typical occurrences in an El Niño year. Yancheva et al. (2007) find simultaneous changes in the Caribbean and in China of the position of the Intertropical Convergence Zone in the eighth through the tenth centuries AD. The droughts brought on by these changes coincide with the simultaneous rise and falls of the Classic Mayan civilizations and the Tang dynasty. At that time, these two sophisticated societies were unaware of each other's existence. In today's interconnected world, however, the impact of simultaneous global climate change could prove explosive.

The question at hand is the climate of the next decade, especially as it might impact food security and sociopolitical stability. The foregoing should make it plain that this will involve natural climate variations as much or more than anthropogenic climate change.

Even though natural variability will dominate the climate of the next decade, there are regions where the trends induced by anthropogenic climate change will have a detectable impact. The most prominent climate signals already evident in observational data and clearly attributable to anthropogenic climate change are the rise in mean global temperature and the amplification of the warming signal in Northern Hemisphere polar regions (IPCC 2007). These are clear warning signs of global problems to come, but neither is terribly relevant to food security issues in the next decade. The changes that are relevant are subtler, but no less real.

Extreme events—droughts, pluvials, heat waves, frosts, tropical cyclones, and severe storms in midlatitudes—are also unquestionably important for food security. It is virtually impossible to attribute a single event, such as Hurricane Katrina or the Russian summer heat wave of 2010, to anthropogenic climate change, but anthropogenic climate change does shift the odds of occurrence for extreme events.

There is no reason to doubt that anthropogenic climate change will entail changes in the mean climate state. IPCC models indicate that the Hadley Cell will expand (Lu et al. 2007), the amount of moisture in the atmosphere will increase, and the jet steams will be displaced (IPCC WG1 2007; Lu et al. 2008). Such changes in the mean may result in changes in the characteristic patterns of natural variability. Remote effects ("teleconnections") of natural modes like ENSO are transmitted via atmospheric wave-guides that will change when the mean state changes, so these remote influences might be altered. For example, the influence of the North Atlantic Oscillation (NAO) on the Mediterranean might be weakened, or the intensity of ENSO-related floods might change.

"Natural variability" can be broadly divided into variability internal to the climate system and variability generated by external forcing, which includes changes in solar radiance and in natural aerosol from volcanic eruptions.[1] There is some skill in predicting solar cycles a decade ahead, but we have no ability to predict volcanic eruptions, and at their historical worst they make the anthropogenic climate change expected in the next decade seem trivial. The Tambora eruption of 1815 caused the crop failures of the "year without a summer" (Stommel and Stommel 1983) and the eruption of 1258 was far stronger (Stothers 2000). Such powerful eruptions are highly unlikely in the next—or any—decade, but they cannot be ruled out, and far weaker but still potent events, such as the 1991 Pinatubo eruption, would be enough to alter aspects of the world's climate and put a temporary halt to global warming. We cannot foresee the volcanic eruptions of the next decade, but we could use the observational record of the past millennium to estimate the probability of eruptions and their consequences.

The question of whether there are distinct physics determining decadal variability has some bearing on how much accuracy we might expect from decadal predictions. Decadal climate prediction is a field in its infancy, forced into the world somewhat prematurely by the needs of policymakers. It aims to cover the gap between seasonal to interannual prediction with lead times of two years or less and projections of climate change a century ahead.

Because natural variability as well as greenhouse gases and aerosols will influence future decades, predictions must start from a specification of the current state of the climate system in addition to accounting for the effects of anthropogenic forcing. The hope is that the extra information provided by a good enough estimate of the initial state will allow the models to track natural variability, resulting in more accurate forecasts of the next decade or two. This hope is more likely to be realized if rather than being random, decadal variability arises from physical modes evolving deterministically over a decade.

The climate of the next decade will not fully determine crop yields, food prices, food security, or sociopolitical stability, but it will be influential. Over the next ten years, climate variations will be dominated by natural climate variability, with anthropogenic climate change having a secondary effect. For this next decade, the greatest climate concerns for agriculture are not secular trends but extreme events, including droughts, floods, and extended periods of extreme temperatures, as well as catastrophic events such as tropical cyclones. The climate system is chaotic, and to some degree extreme events are effectively random and hence unpredictable. However, in recent decades climate science has shown that large-scale climate patterns alter the probability of these events in many regions.

EL NIÑO AND THE SOUTHERN OSCILLATION (ENSO)

The term "El Niño" was used long ago by Peruvian fishermen for the annual warming of coastal waters that occurs around Christmas time. We now use the term for a broad warming of the eastern equatorial Pacific, though there is no accepted definition of how much of a warming over exactly what region it takes to qualify as an "El Niño." Figure 3.1 shows two widely used indices for ENSO over the past century and a half. NINO3, which refers to the sea surface temperature (SST) anomaly in the NINO3 region (90W–150W, 5S–5N) of the eastern equatorial Pacific, is a commonly used index of El Niño. The "Southern Oscillation" is a seesawing of atmospheric mass, and hence of sea-level pressure (SLP), between the eastern and western Pacific. It is most often indexed by the Southern Oscillation Index (SOI), a normalized SLP difference between Darwin, Australia and Tahiti. Figure 3.1 simply uses SLP at Darwin, which is almost the same because the anti-correlation between Tahiti and Darwin is so high. There is a striking similarity between the SLP and SST measures, one atmospheric and one oceanic, widely separated in space. Clearly, they

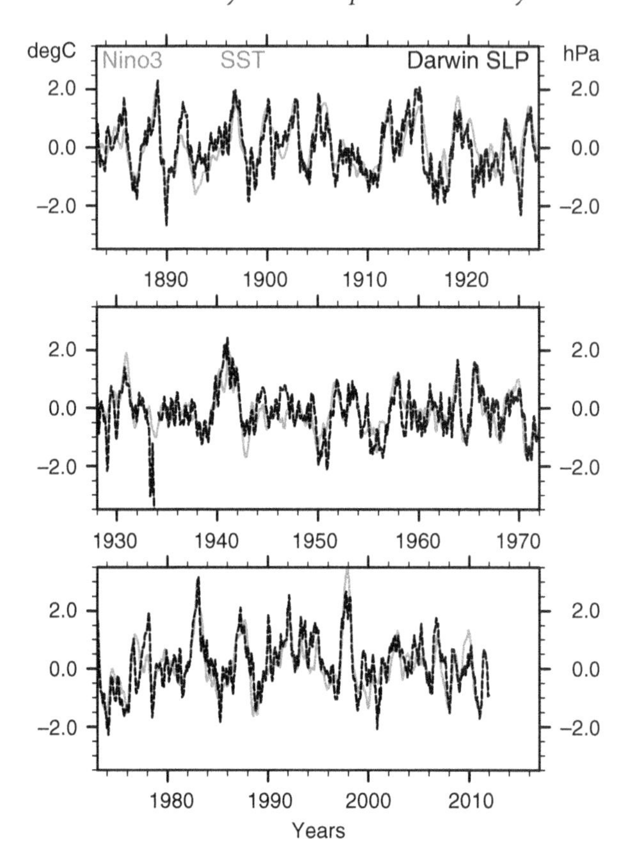

Fig. 3.1. Measures of El Niño and of the Southern Oscillation, 1866–2003

Note: The grey curve is a commonly used index of El Niño, the sea-surface temperature (SST) anomaly in the NINO3 region of the eastern equatorial Pacific (90°W–150°W, 5°S–5°N). The black curve is the sea-level pressure (SLP) at Darwin, Australia, an index of the atmospheric Southern Oscillation. Both are 3-month running mean anomalies from the long-term average. The close relationship between the two indices is evident. (Departures in the earliest part of the record are more likely due to data quality problems than to real structural changes in ENSO.)

Source: Drawn from Kaplan SST and Darwin SLP data at <http://iridl.ldeo.columbia.edu> (accessed April 21, 2013).

index the same phenomenon. Some periods such as the last decades of the nineteenth or twentieth centuries are marked by numerous high amplitude oscillations, while others, such as the 1930s, are rather quiet. The strong positive swings (El Niño events) in 1877, and in the years around 1890, and again around 1900 were times of widespread drought and famine.

Our understanding of the ENSO cycle is built upon Jacob Bjerknes' (1969, 1972) brilliant insights from his studies of the scant observational data then available.[2] Bjerknes did more than point out the empirical relation between the oceanic El Niño and the atmospheric Southern Oscillation. In Bjerknes' account of the connection between the ocean and atmosphere, the fishermen's coastal El Niño is incidental to the important oceanic change, the warming of the tropical Pacific over a quarter of the circumference of the Earth. ENSO is generated and maintained by two-way interactions between the ocean and atmosphere in the equatorial Pacific, according to Bjerknes. Work in the 1970s and 1980s, especially under the auspices of the international Tropical Ocean Global Atmosphere (TOGA) Programme, provided theoretical and observational support for Bjerknes' concept. Bjerknes' theory accounted for the existence of extreme warm (El Niño) states and cold (normal, or in the extreme, La Niña) states, but stopped short of explaining the oscillations between states. The required addition, equatorial ocean dynamics, was made by Klaus Wyrtki (1975, 1979) based on data from a network of Pacific island tide gauges. The first model to successfully simulate ENSO, and, soon afterwards, to predict ENSO, that of Zebiak and Cane (1987), was based explicitly on the Bjerknes-Wyrtki hypothesis.

The Zebiak-Cane numerical ENSO model depicts in a simplified manner the evolution of the tropical Pacific Ocean and overlying atmosphere. It is a dynamic model derived from the governing physical equations, in contrast to a statistical model built from a sequence of observations. Analysis of the model helped in developing a now widely accepted theory that treats ENSO as an internal mode of oscillation of the coupled atmosphere–ocean system, perpetuated by a continuous imbalance between the tightly coupled surface winds and temperatures on the one hand, and the more sluggish subsurface heat reservoir on the other.[3] One of the most significant results of the model simulations was the recurrence of ENSO at irregular intervals due solely to internal processes. This shows that solar variations or volcanic eruptions or other forcings external to the climate system are not required to drive the ENSO cycle, but it does not rule out the possibility that such external factors might influence it. Indeed, it turns out that they do (see Mann et al. 2005; Emile-Geay et al. 2007, 2008).

Bjerknes located the source of ENSO in a tropical Pacific coupling between El Niño and the Southern Oscillation, but he also proposed that the changes in atmospheric heating associated with tropical Pacific SST anomalies cause changes in midlatitude circulation patterns. Figure 3.2 is a version of a well-known diagram of the global influence of an ENSO warm event, an El Niño (after Ropelewski and Halpert 1987; all of the relationships discussed below may be found in that paper or Ropelewski and Halpert 1996). As a first approximation, one may say that ENSO cold events, often called "La Niña," have the opposite effects, but there are significant exceptions. Typically, the effects of ENSO events are strongest and most reliable in the tropical Pacific genesis region and contiguous continents. When there is a warm event one can be fairly certain of heavy rains in Peru, drought in Indonesia and New Guinea, and drought and fewer typhoons in Australia. Typical consequences

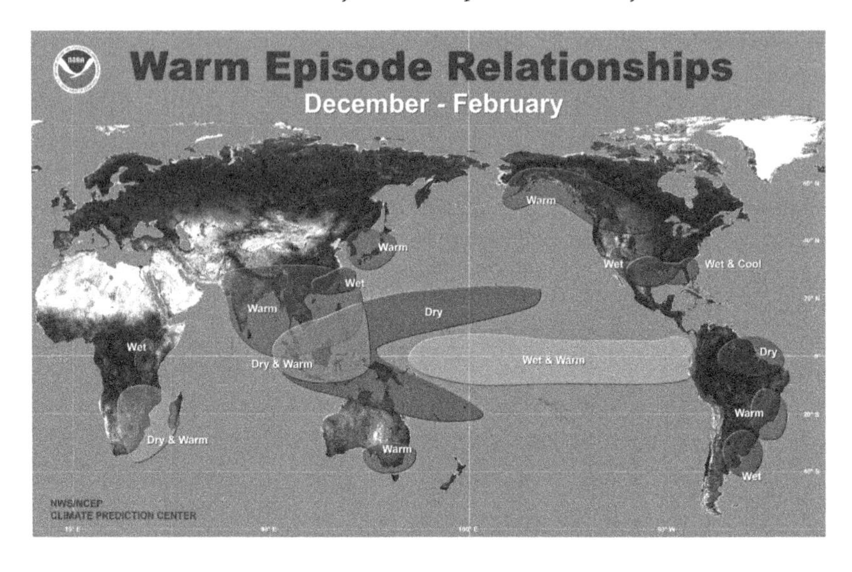

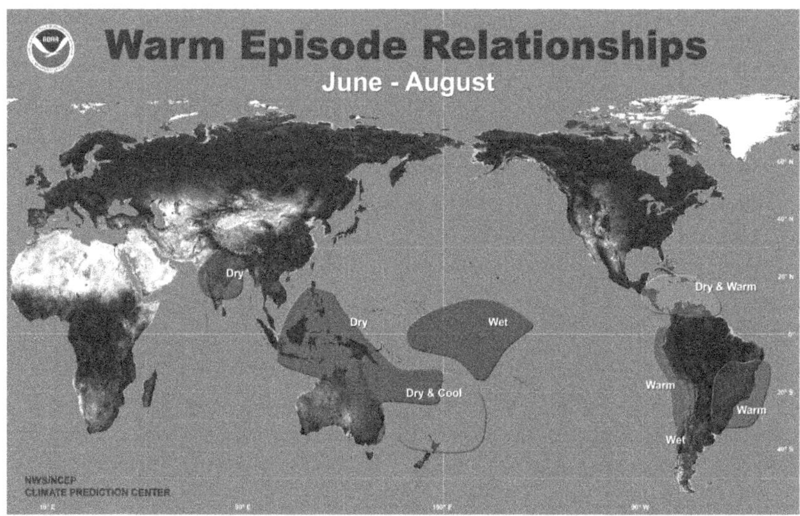

Fig. 3.2. Warm event (El Niño) relationships

Source: Ropelewski and Halpert (1987). Available at <http://www.cpc.ncep.noaa.gov/products/analysis_monitoring/impacts/warm.gif> (accessed April 21, 2013).

are somewhat less reliable in the global tropics, but still highly likely. Thus, there is frequent concurrence of ENSO warm events and a poor monsoon in India, below normal rains in southern Africa, resulting in poor maize yield in Zimbabwe (Cane et al. 1994), flooding in East Africa, drought in the Nordeste of Brazil, and reduced numbers of hurricanes in the Atlantic.

ENSO influence beyond the tropics is less certain. Outside of the tropics an ENSO event should be thought of as biasing the system toward certain preferred outcomes rather than as a certain cause. Many of the more reliable midlatitude effects occur in the Americas. With warm (El Niño) events heavy rains in the Great Basin region of the United States are more likely, and with cold (La Niña) events, midwestern drought (e.g., 1988) and lower corn yields (Phillips et al. 1999) are more likely. It is more likely still that the southwestern United States and northern Mexico will experience drought with a La Niña, and above average rainfall with an El Niño. Similarly, an El Niño is likely to bring heavy rains to Uruguay, southern Brazil, and northern Argentina, while a La Niña brings below average rainfall. Certain patterns become more likely to persist, altering the paths of hurricanes, typhoons, and winter storms. For example, typhoons are more likely to make landfall near Shanghai in an El Niño year.

Another way to say that not all ENSO connections are equally strong and reliable is the more general statement that the global impacts of each ENSO event are different. The differences are not related in any obvious way to the magnitude of the events. For example, the Indian monsoon was normal in 1997 despite the very strong El Niño, while in 2002 the monsoon was very poor although the 2002 El Niño was quite moderate. Understanding of these differences is limited: they have hardly been classified satisfactorily, let alone explained in physical terms. However, it is known that the global response is sensitive to the location and strength of the atmospheric heating in the tropics (e.g., Hoerling et al. 1997).

A second reason that the global impacts of each ENSO event are unique is that the atmosphere is a chaotic, dynamic system, which means that small changes in boundary and initial conditions can be amplified to give very large differences in the future state of the atmosphere. One result of this chaos is "weather"—that is, variability on a timescale much shorter than the timescale of what we call "climate variations." We may think of weather (and other short-time variability) as random "noise" in the climate system that makes the system unpredictable: even with SST and other boundary conditions fixed, the atmosphere may evolve into very different states if started with only slight differences in its initial state. The higher level of weather "noise" in the extratropics is a reason why ENSO impacts are more reliable in the tropics.

PREDICTING ENSO IMPACTS

ENSO theory has a number of implications for prediction. First, since the essential interactions take place in the tropical Pacific, data from that region alone may be sufficient for forecasting. Second, the memory of the coupled system resides in the ocean. Anomalies in the atmosphere are dissipated far too quickly to persist from one El Niño event to the next. The surface layers of the ocean are also too transitory. Hence the memory must be in the subsurface ocean thermal structure. The crucial set of information for El Niño forecasts is the spatial variation of the depth of the thermocline in the tropical Pacific Ocean. The thermocline is the thin region of rapid temperature change separating the warm waters of the upper ocean from the cold waters of the abyssal ocean.

Since 1985, our group at Lamont has used the Zebiak–Cane model (known as the Lamont model in the forecasting realm) to predict El Niño (Cane et al. 1986). The only data going into our first forecasts were observations of surface winds over the ocean. Using the wind data as a forcing field, we ran the ocean component of the model to generate currents, thermocline depths, and sea surface temperatures that served as initial conditions for forecasts, a step made necessary by the lack of direct observations of oceanic variables. Each forecast then consisted of choosing the conditions corresponding to a particular time, and running the coupled model ahead to predict the evolution of the combined ocean–atmosphere system. By making predictions for past times—"hindcasts" or "retrospective forecasts"—we could compare forecasts directly with reality. The results clearly demonstrated predictive skill, setting the stage for the first predictions of the future, made in early 1986, which called unambiguously for an El Niño occurrence later that year (Cane et al. 1986). The moderate El Niño that developed later in 1986 matched the forecast well enough to score it a success, although differences in timing and other details show that the prediction scheme was far from perfect.

In the 1980s the Lamont model was the only physically based forecasting system with this level of accuracy, but now several dozen models with varying degrees of complexity make routine ENSO forecasts.[4] These models can be divided into three categories: purely statistical models, physical ocean–statistical atmosphere hybrid models, and fully physical ocean–atmosphere coupled models. There have been a number of reviews of ENSO forecast skill over the years; Barnston et al. (2012) is the most recent. The general conclusions are that the predictions have useful skill at least two seasons ahead, and that the skills of the physical and statistical models are comparable, though there is a suggestion that the physical models have greater skill at longer leads. The ensemble mean forecast across all prediction systems has markedly greater skill than any single forecast (Tippett and Barnston 2008).

Most modelers believe that only for the past few decades is the observational data adequate for model initialization. Thus, the periods of retrospective forecasting are too short to distinguish among the skill scores of different prediction systems or to allow a confident estimate of our overall ability to predict ENSO. However, Chen et al. (2004) performed a retrospective forecast experiment spanning the past one and a half centuries that uses only reconstructed SST data for model initialization. Each month all other initial fields (winds, thermocline depth, etc.) are created by the model from these SSTs and the model's state at the previous month. At a six-month lead, the model is able to predict most of the warm and cold events that occurred during this long period, especially the larger El Niño and La Niña events, though the model had difficulty with small events and no-shows (see Chen et al. 2004, Fig. 1).

If the reaction to the 1986 forecasts was surprise that it could be done at all, the question now is why we aren't doing better. The factors now limiting the accuracy of forecasts are inherent limits to predictability, flaws in the models, gaps in the observing system, and flaws in the data assimilation systems used to introduce the data into the models.

ENSO is surely predictable, but how predictable is it? Is there much more room for improvement of our predictive skill? One cause of uncertainty is in the ways we estimate ENSO's predictability. In weather forecasting, predictability is estimated using twin-model experiments in which initial conditions are perturbed slightly from one run to the next, and the separation of the evolved atmospheric states tells us how fast small initial errors grow. This is error growth in the *model*, but weather-forecasting models are realistic enough to make it a reliable measure for error growth in nature. ENSO models however, have not been shown to be realistic enough for this purpose and the answer is model dependent. Estimates of El Niño's predictability based on the retrospective predictions over one to three decades available for most models encompass a relatively small number of events and so are quite uncertain. The uncertainty is increased by the fact that ENSO predictability varies from decade to decade (Chen et al. 1995; Balmaseda et al. 1995; Kirtman and Schopf 1998).

As evident in Figure 3.3, the predictive skill of the Lamont model varies substantially, especially at longer lead times. The periods with the highest overall scores, 1876 to 1895 and 1976 to 1995, are dominated by strong and regular ENSO events. The periods of lower skill have fewer and smaller events to predict. For example, during the 1936 to 1955 period, when the predictability was the lowest by all measures, the only strong El Niño is the prolonged warm event in 1940–42 (see Figure 3.1). Though data coverage was reduced in the years of the Great Depression and Second World War, it was not worse than in the nineteenth century, so for the results shown here we may conclude that temporal variations in predictability outweigh variations in data availability. This is only one model and one data assimilation method, and it does not rule out the likely possibility that observations of the subsurface ocean would improve forecasts, especially in the less active periods. Regardless, the predictability of ENSO is limited in principle by the chaotic and noisy nature of the climate system, so forecasts are most correctly presented as a probability distribution of possible future states.

Uncertainties as to the inherent limits to predictability notwithstanding, it appears that our current level of predictive skill is far from those limits. Our

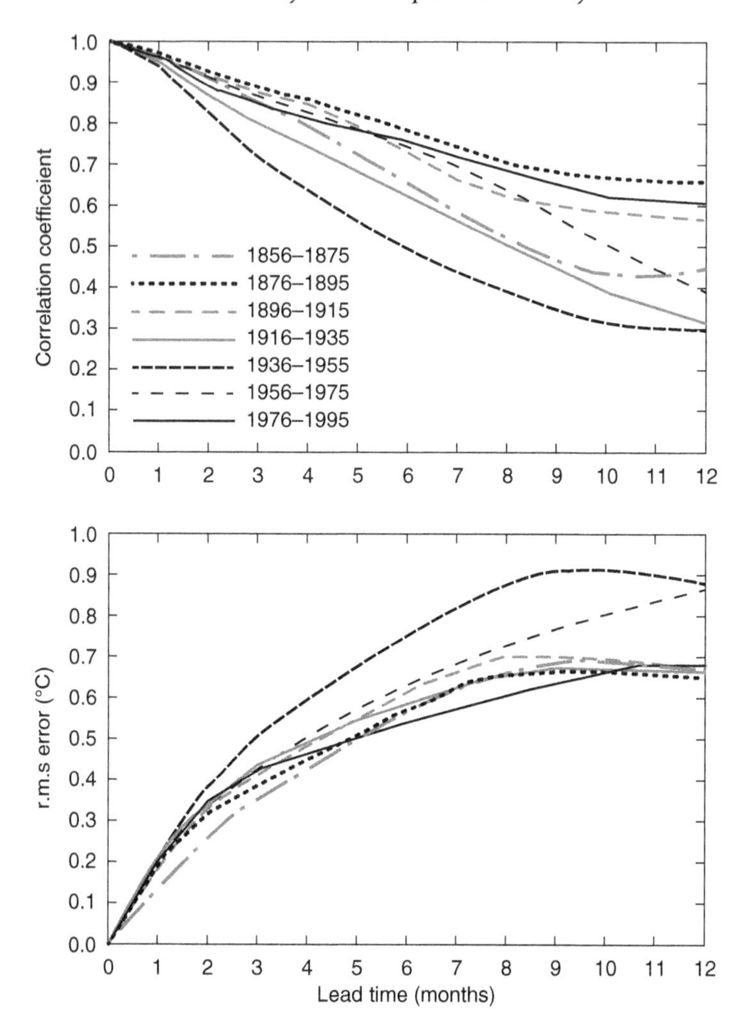

Fig. 3.3. Anomaly correlations (top) and rms (root-mean-square) errors (bottom) between observed and predicted NINO3.4 index

Note: Recorded as a function of lead-time, for seven consecutive 20-year periods since 1856.

Source: Chen et al. 2005.

task then is to improve our observing systems, models, and data assimilation methods. Tremendous efforts have been made in all these areas in the last two decades. Observation networks such as Tropical Atmosphere Ocean (TAO) array and satellite altimetry and scatterometry missions have proven invaluable for ENSO monitoring and forecasting. The current observing system does not seem to be the principal limitation on our skill (Stockdale et al. 2011), though constant vigilance is needed to ensure that it does not deteriorate.

Regional and global models with different degrees of complexity have been significantly improved in terms of both physics and computational capability, but these models are still notably imperfect, and their ability to simulate ENSO is not among their stronger points.[5]

The area with the greatest near-term potential for improving skill is data assimilation, the process of creating an initial state for the forecast by combining observational data with a model-generated first guess. While there is 60 years of data assimilation experience for weather forecasting, application from seasonal to interannual forecasting presents new challenges because of the necessity of initializing the ocean state as well as the atmosphere (Chen and Cane 2008). In addition to the new technical questions the ocean adds, there is the fact that observations of the subsurface ocean are few and far between, especially if one would like to "practice" forecasting by going back earlier than the last two decades. Even in the hypothetical case that data coverage is so good that the state of the atmosphere and ocean is very well known, it would still be a challenge to determine the initial state that yields the best forecast. The problem is that the model is not perfect, so that even if the data were so good that we could start it in the same place as nature, it would insist on evolving differently. One telltale of model imperfection is that the average state of a model over a long time—its climatology—differs from nature's climatology. In other words, the model has systematic biases. Unfortunately, it is not easy to characterize model bias in ways that lend themselves to improved data assimilation, and only a few studies have addressed this issue for ENSO forecasting (Chen and Cane 2008). More research is needed in the analysis of the pattern, nature, and statistics of the biases, and in the implementation of proper bias correction schemes for coupled General Circulation Models (GCMs).

Predicting the state of the tropical Pacific is only the beginning for predicting socially relevant climate variations, such as those affecting agriculture. Given the correct SSTs in the tropical Pacific, atmospheric models are quite good at capturing the "teleconnections" to land areas depicted in Figure 3.2. Figure 3.2 derives from a statistical analysis of data, so statistical models also capture these relations. Yet these teleconnections are not certain consequences of an ENSO event; rather the ENSO cycle shifts their probability of occurrence. In an El Niño year there is a very high probability of drought in Indonesia, while in India, the mean prediction would also be below average monsoon rainfall, but the distribution would be wider, reflecting the reality that in almost 40 percent of El Niño events rainfall is close to the average. Finally, not all seasonal to interannual predictability stems from ENSO. SST patterns in the other tropical oceans contribute, especially to regions adjacent to these oceans (Goddard et al. 2001). Insofar as these SSTs are predictable, however, it is primarily because of the influence of ENSO.

DECADAL VARIABILITY

There are many contrasts between decadal variability and ENSO in addition to the obvious one of timescale. Most telling, we have no accepted theories for decadal variability and no evidence that socially useful predictions are possible. In fact, it is not firmly established that there is such a thing as variations in climate with particular physical mechanisms that favor a decadal timescale. Perhaps we have identified certain variations as decadal because the short instrumental record does not allow us to see that the same patterns occur at longer intervals. On the other hand, there is value in describing these modes and their associated impacts, and we have not ruled out the possibility that they are predictable.

Surface temperature anomalies on decadal timescales are no larger than a few tenths of a degree Celsius. It might seem that such puny anomalies could not matter, but there is ample evidence that they do. Figure 3.4, adapted from Seager et al. (2005) shows that the Dust Bowl drought can be simulated by an atmospheric model forced by SSTs of this amplitude—though they must have the right pattern.[6] The anomalies that matter most are the La Niña-like pattern in the tropical Pacific; the lower right panel shows how much of the drought over North America is captured when the model is forced by anomalies in that region only. Next in importance is the warming in the North Atlantic, which resembles a positive Atlantic Multidecadal Oscillation, which is discussed below.

The literature on low-frequency climate variability is typically organized around a small set of variously named patterns with acronyms: El Niño–Southern Oscillation (ENSO); North Atlantic Oscillation (NAO), which for some is equivalent to the Arctic Oscillation (AO) or the Northern Annular Mode (NAM); Pacific Decadal Oscillation (PDO), Pacific Decadal Variability (PDV), or Interdecadal Pacific Oscillation (IPO); Atlantic Multidecadal Oscillation (AMO) or Atlantic Multidecadal Variability (AMV); Southern Annular Mode (SAM); and more. A few low-frequency climate variations, such as monsoon variability, have been studied extensively despite lacking the grace note of an acronym.

The PDO patterns of SST, SLP, and winds are compared with ENSO in Figure 3.5 (Mantua et al. 1997; Mantua and Hare 2002). The PDO amplitude is relatively greater in the North Pacific and smaller in the tropics, but the overall similarity in pattern suggests dynamic similarities (Alexander et al. 2002). When temperatures are colder in the central North Pacific, they are warmer in the eastern tropical Pacific and along the coast of North America. The time series are quite distinct: the PDO is dominated by variability at timescales longer than a decade and ENSO is interannual. In the twentieth century, PDO variability is greatest at periodicities of 15 to 25 years and 50 to 70 years, so we cannot say that the PDO has a single characteristic frequency. One view

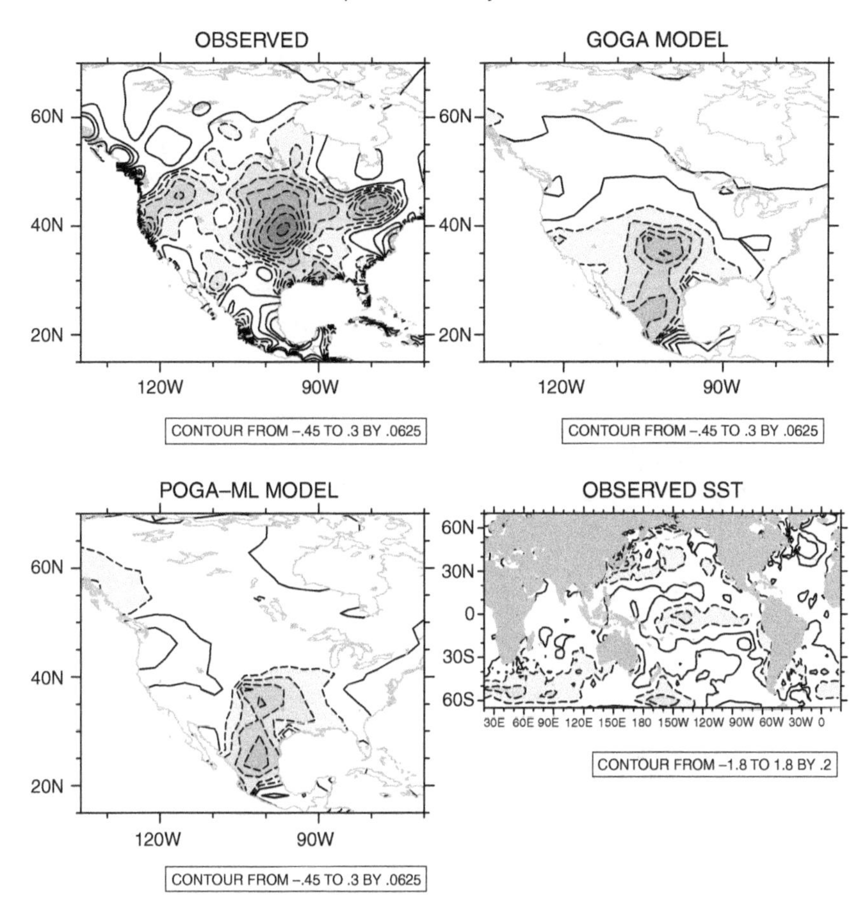

Fig. 3.4. Precipitation during the 1932–39 Dust Bowl drought

Note: (Upper left) The observed precipitation anomaly for the period 1932–1939. (Upper right) The precipitation anomaly from the GOGA ensemble of Atmospheric GCM simulations forced by the global SST anomaly pattern shown in the lower right hand panel. (Lower left) the precipitation anomaly from the POGA-ML ensemble of atmospheric simulations forced by the SST pattern in the tropical Pacific between 20N and 20S. Ocean temperatures elsewhere are calculated from a simple mixed layer model. (Lower right) The observed SST anomaly. Precipitation in mm/month; SST in °C.

Source: After Seager et al. 2005; data courtesy of Naomi Henderson and Richard Seager; the observational precipitation data used is CRU-TS3.1.

of Figure 3.5 sees only two full PDO cycles: cool tropical phases from 1890 to 1924 and again from 1947 to 1976, with switches to warm phases in about 1925 and 1977. A "climate regime shift," particularly in the Pacific, is often said to have occurred from 1976 to 1977 (e.g., Trenberth 1990). This shift has been observed in Pacific marine ecosystems; Alaskan salmon production is something of a poster child for the PDO (see Mantua and Hare 2002 and references

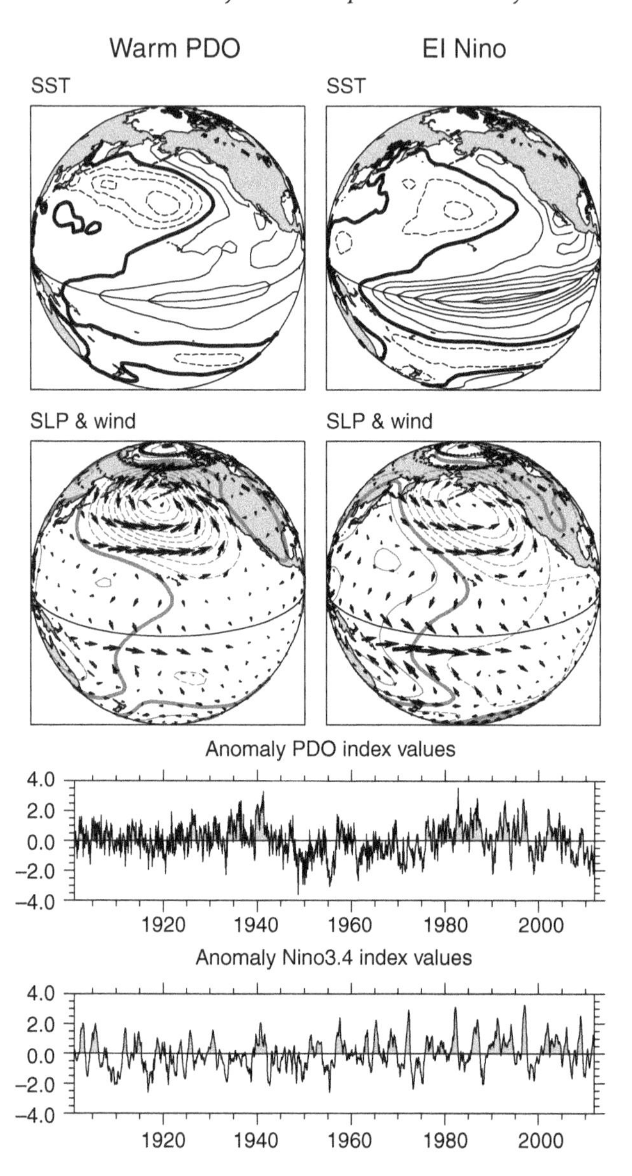

Fig. 3.5. The sea surface temperature (SST), sea level pressure (SLP), and wind patterns associated with the PDO (Pacific Decadal Oscillation) positive phase (left) and the ENSO warm phase (right).

Note: The contour interval (CI) is 0.2°C for SST and 0.4 hPa for SLP. The time series is based on monthly means.

Source: Top figure is ERSSTv3 SST; SLP and winds from NCEP Reanalysis; all available at http://iridl.ldeo.columbia.edu (accessed March 21, 2013). The time series data is from http://jisao.washington.edu/pdo/PDO.latest (accessed March 21, 2013).

therein). The warm phase of the PDO depicted in Figure 3.5 is usually accompanied by winter season (October to March) anomalies reminiscent of El Niño anomalies. These include warmer air temperatures and below average snow pack and stream flow in northwestern North America, cooler temperatures in the southeastern United States, above average precipitation in the southern United States and northern Mexico, and below average rainfall in eastern Australia (Power et al. 1999).

The NAO has been known far longer than the PDO and has a far larger literature. The climatic patterns over the North Atlantic are shown in Figure 3.6 along with a time series of a standard NAO index (Figure 3.6b), the SLP difference between the Azores and Iceland. Pressure at these two places tends to fluctuate

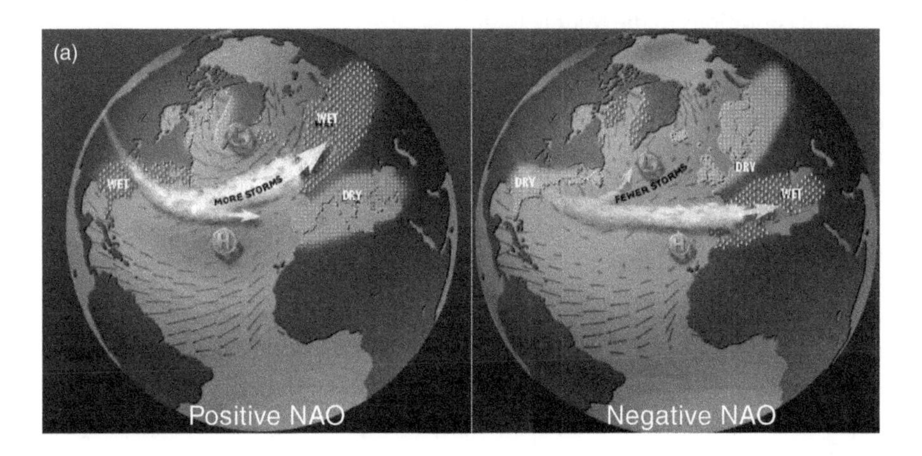

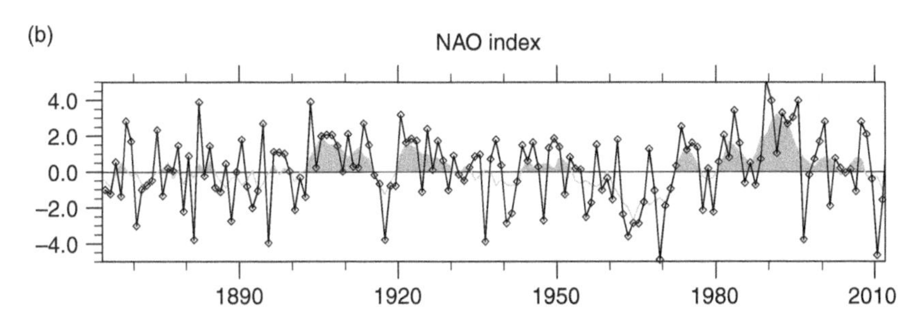

Fig. 3.6. The sea surface temperature (SST), sea level pressure (SLP), and storm track locations associated with positive (top left) and negative (top right) phases of the North Atlantic Oscillation (NAO). The time series (bot) is an NAO index.

Note: Wet and dry anomalies are labeled. The NAO index is the winter (December thru March) difference of normalized sea level pressure (SLP) between Lisbon, Portugal, and Stykkisholmur/Reykjavik, Iceland. The black curve is the DJFM average for every year; the gray curve is a 5-year running average.

Source: (top) <http://www.ldeo.columbia.edu/res/pi/NAO/> (accessed March 21, 2013). (bottom) <http://climatedataguide.ucar.edu/guidance/hurrell-north-atlantic-oscillation-nao-index-station-based> (accessed March 21, 2013).

out of phase: when the Icelandic low is anomalously low and the Azores high is anomalously high, the winds over the Atlantic tend to be stronger than normal and the enhanced southwest to northeast flow brings warm wet weather to northern Europe. In this positive phase of the NAO the Mediterranean region tends to be dry because the storm tracks are diverted away from it. Recent droughts and forest fires on the Iberian Peninsula are consistent with the positive NAO pattern. Positive NAO winters on the east coast of the United States are relatively mild and wet, while in the negative phase the east coast of the United States has more cold air outbreaks. This is because the storm tracks take a more southerly route, which also brings more precipitation to the Mediterranean. Mediterranean harvest yields of grapes and olives have been shown to depend significantly on the NAO. More generally, an upward trend in the NAO brings drought conditions to the Middle East.

While its intraseasonal variability has been attributed to weather "noise," the origin of the NAO's interannual persistence, which leads to an apparent decadal fluctuation, remains unclear (Kushnir et al. 2006). Over a considerable time there were many attempts to show that North Atlantic SST anomalies are essential in creating this enhanced persistence or even an oscillation in atmospheric pressure; they have largely failed. Barsugli and Battisti (1998) argue that the North Atlantic SSTs do matter in that the NAO SST pattern is the one that least damps what is in essence an *atmospheric* mode (also see the discussion in Kushnir et al. 2006). The variability in an index of the NAO is not different from a power law distribution; its spectrum has no significant peaks suggestive of an oscillatory process (Wunsch 1999). This fact supports the notion that the NAO is a consequence of a favored atmospheric pattern generated by synoptic weather "noise."

The AMO is a mode in which North Atlantic SSTs are warmer than normal almost everywhere or colder than normal almost everywhere. When the North Atlantic is warm, northern Europe tends to be warmer and wetter, and there is typically less rain over the United States, especially the eastern half of the country (Enfield et al. 2001; Sutton and Hodson 2005). Note the similarities to the positive phase of the NAO. A warmer tropical North Atlantic (as occurs with a positive AMO) results in more rainfall in the Sahel (Giannini et al. 2003; Biasutti and Giannini 2006). A weaker Indian monsoon has been linked to colder temperatures in the North Atlantic related to both NAO and AMO negative phases; Goswami et al. (2006) propose that the link is via changes in tropospheric temperatures over Eurasia.

A positive AMO has been shown to be strongly associated with an increase in Atlantic hurricane activity (Kerr 2005), which has created an intense debate over whether or not the current warm AMO is part of the global man-made warming signal (Mann and Emanuel 2006) or whether it is a internal mode of climate variability (Goldenberg et al. 2001). This attribution question requires separating the internal AMO mode from the climate response forced by solar variations, aerosols, or greenhouse gases (Trenberth and Shea 2006; Zhang et al. 2007; Knight 2009). Approaches range from simply removing the linear trend from SSTs averaged over the North Atlantic Basin (Enfield et al. 2001) to sophisticated statistical methods (Schneider and Held 2001; Ting et al. 2009; Del Sole et al. 2011). The issue has arisen again very recently in the assertion of Booth et al. (2012) that the low frequency variability in the Atlantic sector during the twentieth century was a consequence of variations in radiative properties of clouds induced by aerosols. This conclusion, which is drawn from simulations with a model using a new and very strong representation of this "indirect effect" of aerosols on the Earth's heat budget, will surely be challenged.

ANTHROPOGENIC CLIMATE CHANGE

Disputes about whether recent warming trends in the North Atlantic are attributable to natural or human causes notwithstanding, few would disagree with the anemic but correct claim that both are active. There is no doubt that the world has warmed over the past century, especially during the past few decades. No one should doubt that human activity is largely responsible: the evidence is too abundant and too pertinent (e.g., in IPCC 2007). Though models do not agree on the precise regional distribution of temperature changes, certain robust projections appear in observational data by the end of the twentieth century (Table SPM.2 in the Working Group I Report of IPCC 2007). Over most land areas these include warmer and fewer cold days and nights; more frequent hot days and nights; more frequent heat waves.

Unlike temperature, anthropogenic influences on precipitation are not obvious. As the air warms, the amount of water vapor the air holds goes up 7 percent per °C, but precipitation does not increase nearly as fast. Since the atmosphere stores very little water, the amount of global precipitation must equal the amount of evaporation. Global evaporation at the surface is constrained by the Earth's energy budget and is estimated to be about 2 percent per °C (Allen and Ingram 2002; Held and Soden 2006). One consequence of the far more rapid increase of water held by the atmosphere is that individual rain events will become more intense. While this trend is already seen in twentieth-century observational records (Table SPM.2 in the Working Group I Report of IPCC 2007), there is little else conclusive to be said about precipitation changes. The IPCC WG 1 report (2007, 23) says: "Significantly increased precipitation has been observed in the eastern parts of North and South America, northern Europe and northern and central Asia. Drying has been observed in the Sahel, the Mediterranean, southern Africa and parts of southern Asia. Precipitation is highly variable spatially and temporally, and robust long-term trends have not been established for other large regions." Even the cited trends cannot be firmly attributed to anthropogenic influences: natural causes cannot be ruled out. For example, Sahel rainfall is closely related to the position of the Intertropical Convergence Zone (ITCZ) in the Atlantic, and this in turn is strongly affected by the SST distribution in the Atlantic, including the AMO (Giannini et al. 2003; Zhang and Delworth 2006).

We can make a few general statements about the likely changes in precipitation. It is likely that the "wet places get wetter and the dry places get dryer" (Held and Soden 2006). This follows if, as expected with anthropogenic climate change, the amount of moisture in the atmosphere increases markedly while the circulation changes only slightly (Allen and Ingram 2002; Held and Soden 2006). The places where it rains are places where moisture converges in the atmosphere, and more moisture and about the same convergence means more convergence of moisture and hence more rain. Similarly, the dry places, which are the places where moisture diverges away, stand to lose more heavily.

A relatively robust result in IPCC AR4 models is the expansion of the circulation poleward, so that the midlatitude storm tracks move poleward, and the subtropical belt where air sinks expands as well as moving poleward. These subtropical regions contain most of the world's deserts and semi-arid regions. This, together with the "dry places get dryer" argument, leads one to expect increased droughts in, among other places, the Mediterranean region, the southwestern United States and northern Mexico, and much of Australia. Allen et al. (2012) summarize the observational evidence that the expansion is already occurring, but also show that it may be more a consequence of changes in tropospheric ozone and black carbon than greenhouse gases. Biasutti and Giannini (2006) present evidence that the late twentieth-century drying is more a response to aerosols than to greenhouse gases. These culprits are all anthropogenic in origin, though not the usual suspects.

PREDICTING DECADAL CLIMATE

Forecasting the climate of the next decade will demand some skill at forecasting interannual and decadal natural variability, as well as changes due to anthropogenic influences. Natural variability includes changes in external radiative forcing due to solar variations and volcanic eruptions as well as changes internal to the climate system. Anthropogenic influences include activities that alter atmospheric ozone and aerosols, as well as greenhouse gases. Other human influences include land-use changes, though additional changes in the next decade seem unlikely to have an appreciable effect on large regions.

Most skill in predicting interannual climate variability stems from our ability to predict ENSO. Prediction of ENSO a year or two ahead rests on a good theoretical understanding of ENSO dynamics. In particular, the upper levels of the tropical Pacific Ocean provide the "inertia" so that the ENSO cycle continues into the future. However, there is no demonstrated skill in predicting ENSO up to a decade ahead, and our limited understanding of the intrinsic predictability of ENSO does not rule out the possibility that this is impossible in theory as well as in our now primitive practice. We lack a decent theoretical understanding of decadal and multidecadal variations such as the PDO and AMO, so we do not know if their past behavior determines their future evolution. The leading view at present is that they are not self-sustained oscillations, but perhaps damped ocean modes driven by atmospheric noise, a view that would seem to limit hopes for predictability. That they do not exhibit a preferred timescale is another blow to optimism. Still, it may be that once one of these noise-driven modes is set decisively in motion, it will follow through in a predictable pattern. The greatest hope lies in the North Atlantic, where the meridional overturning circulation (AMOC), which models simulate with apparent success, might be the internal ocean mode that provides the inertia to impart predictability to the AMO.

As with ENSO, decadal predictions can be made with statistical models, dynamical models, or hybrids. Lean and Rind (2009) use linear regression models trained on historical data to predict surface temperature at each 5°x5° square on the Earth's surface, as well as the mean global temperature. The predictors are ENSO, solar and volcanic activity, and a measure of anthropogenic influence that attempts to account for greenhouse gases, land use, snow albedo changes, and tropospheric aerosols (Hansen et al. 2007).

The Lean and Rind statistical model pays no attention to decadal modes of variability, whereas forecasts with two dynamical models that greatly stimulated enthusiasm for decadal forecasting were looking to the AMOC/AMO mode as a source of skill (Smith et al. 2007; Keenlyside et al. 2008). The retrospective forecasts they report do show skill relative to the most basic forecasts, ones that either assume the persistence of existing conditions, or forecast climatological conditions. It is noteworthy that Smith et al. (2007) predict that the next five years will be warmer than the past decade, while Keenlyside et al. (2008) predict the opposite. Moreover, neither study is persuasive in showing that using information about the present state of the climate provides skill beyond that attributable to anthropogenic forcing.

The forecasts of Smith et al. (2007), for example, use the data assimilation system DePreSys to incorporate the observed state of the atmosphere and ocean in order to predict internal variability. They also estimate changes in anthropogenic sources of greenhouse gases and aerosol concentrations, as well as forecasts of changes in solar irradiance and volcanic aerosol. Their NoAssim forecasts differ from the DePreSys forecasts in that they do not assimilate the observed state of the atmosphere or ocean. Results from the two sets of forecasts offer just the slightest hint that assimilation might be helpful (see Smith et al. 2007, Fig. 3).[7]

Why isn't assimilation more beneficial? The reasons are similar to the ones that limit ENSO prediction. Starting from an observed initial state is essential to predicting the evolution of natural variability internal to the climate system such as the AMO or PDO, but it is not at all established that much predictability is even theoretically possible. Capturing the current state of these decadal modes requires knowledge of the ocean's state, and observations of the subsurface ocean are very sparse. The deployment of a global network of Argos floats within the past decade has vastly improved matters, but that still leaves us without much knowledge of earlier times to practice our hindcasts. Initializing the ocean state for prediction purposes is a very new endeavor, and current efforts are surely far from optimal. Finally, and perhaps most important, climate models are imperfect and have their own climate, biased away from nature's climate. When started from an initial state that is realistic they may evolve natural anomalies somewhat correctly, but they will also be moving away from nature's climate toward their own. That error may overwhelm any gains from starting with correct climate anomalies.

As part of the IPCC AR5 there will soon be a large number of studies in the same vein as Smith et al. (2007). For now, there are two actual forecasts in the literature in very different styles. Seager et al. (2004) used the Zebiak–Cane intermediate complexity ENSO model to make the prediction that an ENSO index averaged for the period 1998 to 2013 would be colder than in the previous 15 years.[8] The prediction is based solely on projecting internal variability: it does not explicitly account for anthropogenic or other external forcing. This work followed up on a theoretical study of decadal predictability in the Zebiak–Cane model (Karspeck et al. 2004). That study used an "identical twin" methodology in which a model simulation substitutes for reality. Karspeck et al. (2004) found some predictability, but even in this idealized setting it was only slightly better than chance—too small an advantage to be of much practical use.

Hoerling et al. (2011) predicted the mean decadal climate over North America for 2011 through 2020 in response to anthropogenic greenhouse gases. The methodology is complex, involving an ensemble of runs with a number of different models and corrections of model biases using several observational data sets for the twentieth century. SSTs from existing IPCC coupled model runs (CMIP3) together with two bias corrected versions of SST for 2011 to 2020 are used as boundary forcing for a multimodel set of Atmospheric General Circulation Model runs. Temperatures are warmer almost everywhere, markedly so in Alaska and northern Canada. The latter is largely a response to reduced summer sea ice in the Arctic. Precipitation generally increases over Canada and Alaska, and is reduced over much of the United States. In this study, the internal variability of the climate system is regarded not as something to be predicted, but as noise that moves the actual

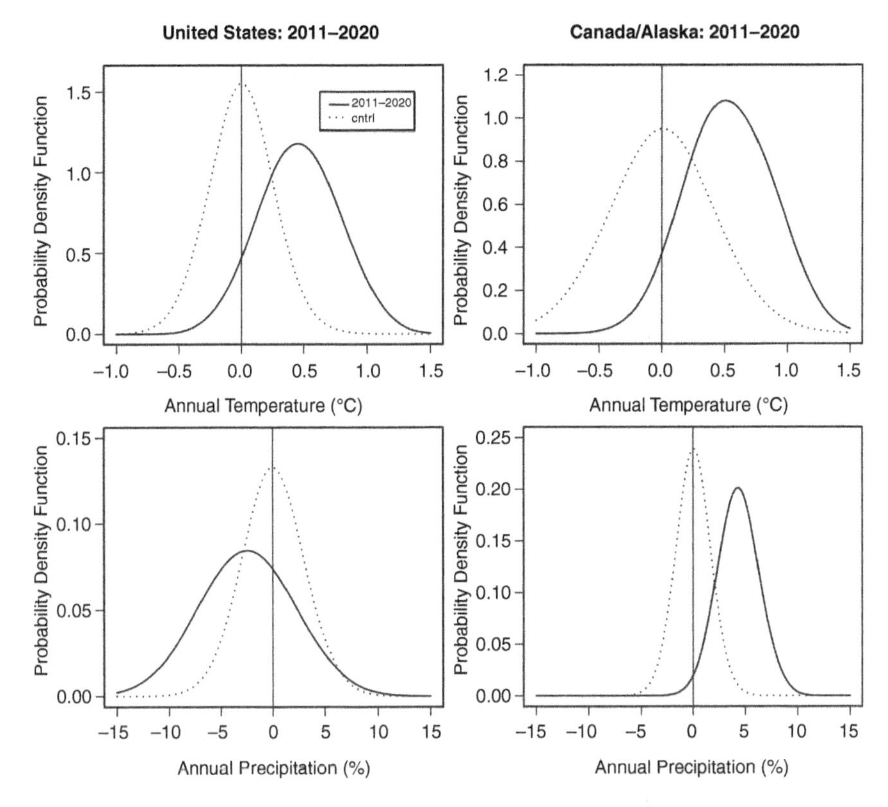

Fig. 3.7. Probabilistic forecasts (solid lines) for the 2011–20 decadal anomalies of (left) contiguous U.S. (top) surface temperature and (bottom) precipitation and (right) Canada/Alaska (top) surface temperature and (bottom) precipitation. Preindustrial climatologies are shown as dotted lines.

Note: The curves are based on commingling the PDFs of the forced responses to our scenarios of anthropogenic changes in ocean boundary conditions and the decadal climate conditions resulting from natural, internal decadal SST variability during the twentieth century. It is assumed that the North American anomalies resulting from natural decadal SST conditions and those from the anthropogenic change component are linearly additive. All departures are relative to the 1971–2000 reference. Dotted-line PDFs illustrate the statistics of decadal climate anomalies derived from the roughly 8,000 independent samples of 10-yr averages calculated from preindustrial CMIP3 simulations.

Source: Hoerling et al. 2011, Figure 4, © American Meteorological Society. Used with permission.

climate away from the forced response. Figure 3.7 shows distributions of possible states obtained by adding this internal noise to the forced response together with the distributions from simulations of a preindustrial world lacking anthropogenic forcing. There is a high probability of warmer temperatures in the United States (94 percent probability) and Canada (98 percent) and of above average precipitation in Canada (99 percent), while dryer conditions over the United States are likely (75 percent). In other words, the forced signal is sufficiently greater than the decadal climate variability to give a highly significant probability of climate change for all but United States precipitation. Note that these are forecasts for decadal averages, not for the individual years of a decade. Any given year would have a large component of internal variability. While there is some predictability for a year or two ahead due largely to ENSO, at longer leads there is no demonstrated predictive skill.

In all regions and globally, almost all predictive skill for the next decade derives from the anthropogenically forced trend. There could be some additional skill from extrapolating solar activity forward in time, but there is nothing to be done for volcanic aerosol—except to let it add to the uncertainty of the forecast. The internal variability of the climate system in modes such as ENSO, AMO, and PDO is also primarily a source of climate noise, adding uncertainty to forecasts, though there is some reason for optimism, especially for the North Atlantic sector.

THREE RECENT DROUGHTS: THE SOUTHWESTERN UNITED STATES, THE MEDITERRANEAN, AND EAST AFRICA

None of these extended droughts was predicted, and we cannot say if any of them might have been predictable. Since 1998 there has been a serious drought in the southwestern United States and northern Mexico. On the one hand, this might be taken as confirmation of the global warming projections that the subtropical zones will expand and become more arid. Seager et al. (2007) show precipitation minus evaporation in the southwestern United States as projected by four different IPCC models. Remarkably, all indicate a shift to a more arid regime circa 1998. On the other hand, there has been a prevalent La Niña pattern in the equatorial Pacific since the end of the 1997–98 El Niño, and southwestern drought is an expected consequence of this SST configuration. We do not know if the drought should be attributed solely to natural decadal variations resulting in the La Niña-like pattern of tropical Pacific SSTs—a pattern unlike the one predicted by IPCC climate projections—or whether it is at least in part due to anthropogenic climate change.

The Mediterranean has also been gripped by drought in recent years, raising the question of whether this drought might have contributed to the Arab Spring uprisings, including the ongoing conflict in Syria. The Mediterranean is another subtropical region expected to become dryer under the influence of anthropogenic climate change. However, it is also a region prone to drought when a positive phase of the NAO prevails. If anthropogenic climate change is the cause, then the drought may be expected to persist, but if the NAO is controlling, the Mediterranean would become wetter again as the NAO cycle shifts toward a negative state. Two recent studies (Hoerling et al. 2011; Kelley et al. 2011) used models to address this attribution question. These studies show that the larger share of Mediterranean drying in recent decades is associated with natural variability, especially the NAO, but that in a few regions—southern Spain, the Atlas Mountains, and the Middle East—human influence is at least comparable. This suggests that current drought conditions in Syria are attributable to anthropogenic climate change, and that future droughts will be a common occurrence in the Middle East.

East Africa has also suffered a decade-long drought, with disastrous consequences for food security in the region. Widespread famine in Somalia has led to a migration populating what is now the world's largest refugee camp in Kenya. According to the IPCC (2007) assessment, the expected consequence of anthropogenic climate change is above average rainfall in this region. Since 1998 the equatorial Pacific has been dominated by a pattern that resembles La Niña in having a strengthened zonal SST gradient, and one expected consequence of a La Niña event would be reduced rainfall in East Africa. Complicating the picture is the fact that the established link to La Niña and ENSO is with the short (October–December) rains, but in the current drought the long (March–May) rains have also been poor, failing disastrously in 2010 and 2011 (Lyon and DeWitt 2012 and references therein). Moreover, the ENSO link to the short rains is not directly with the Pacific but via ENSO-induced changes in Indian Ocean SSTs (Goddard and Graham 1999), whereas Lyon and DeWitt (2012) show that the long rains are linked to changes in western Pacific SSTs. At present we cannot say whether the East African drought is a consequence of natural variability overwhelming the response to greenhouse gas forcing, or whether the IPCC models, which tend to weaken the zonal SST gradient in the Pacific, are failing to capture the correct response to human influences on the climate system. If we attain some skill in decadal prediction of SSTs in the tropical Pacific and the North Atlantic, then perhaps droughts like these might prove to be predictable.

CLIMATE AND FOOD SECURITY: LOOKING FORWARD

A number of historical cases and some quantitative studies strongly suggest that sociopolitical stability can be influenced by climate changes, but the mechanistic links between them are far from clear. The most frequently cited is food security, as there is no doubt that food production is affected by variations in climate. Of course, food security is important in itself, regardless of whether or not its loss leads to sociopolitical instability. The same may be said for a number of other possible links that can connect climate to sociopolitical instability, including increased disease burdens, increased unemployment (of agricultural workers when a drought ruins crops, for example), destruction of transportation and other infrastructure by flooding, and elevated temperatures causing more crime and conflict and decreasing industrial productivity (Hsiang 2010).

Climate alone is not sufficient to generate food insecurity or sociopolitical instability. Other conditions must be met. Amartya Sen has taught us that famines are not so much a consequence of climatological drought as they are a consequence of entitlement, of access to resources. The most disastrous climate event of the past 150 years was probably the 1877 El Niño, which triggered widespread famines. The meteorological drought in India caused food shortages, but the policies of the British Raj transformed a climate event into the loss of millions of lives. The current famine in Somalia is a stark example. The 2012 drought has caused crop failure and food shortages, but it is the anarchic warlordism that is blocking relief efforts that could prevent starvation. The summer of 2012 brought extreme drought to the midwestern United States, but no one expects the crop failure to lead to food shortages there: prices will rise, but Americans will not starve.

Global impacts are more difficult to foresee and will depend on how crops fare elsewhere. The Middle East has come to rely on the Russian and Ukrainian wheat crops, and it is plausible though unproven that the failure of these crops in the Russian heat wave of 2010 had some influence on the Arab Spring in 2011. One of the lessons of climate science is that many climate anomalies have widespread impacts, so strong effects will rarely be confined to a single region. While climate events will lead to the worst outcomes in places where the society is internally vulnerable, external climate variations can also push such societies over the edge. For example, although conflicts occur in places that are conflict-prone for societal reasons, Hsiang et al. (2011) have shown that the number of civil conflicts in the world increase in an El Niño year.

What can be said about the climate of the next decade, about its impacts on agriculture and other human activities, and about the implications for food security and sociopolitical stability? Far less than we would like.

We wish we could predict the climate variations of the next decade with some certainty. Unfortunately, our ability to do this is quite limited, because our knowledge and prediction tools are limited, but more importantly, because the climate system is chaotic. The climate of the next decade will be a combination of natural interannual variability, natural variability on decadal and other long time scales, and the response to anthropogenic forcing. The last includes industrial aerosols and land use changes in addition to the main actor, greenhouse gases. Natural changes may be internal to the climate system or forced by external factors, specifically solar radiation and volcanic aerosols. These external factors do matter for the climate of the next decade, and while we can probably extrapolate solar activity a decade ahead, we have no ability to predict volcanic eruptions.

The main factor in interannual variability is ENSO, which is reasonably well understood and can be predicted. Current predictive skill extends for a year or two ahead at best, and although ENSO forecasting may improve, it is unlikely that there is even the theoretical possibility of predictability a decade ahead. The best hope seems to lie in the North Atlantic, where models simulate variability with some realism, and where there is a plausible role for the ocean, with its long memory. Anthropogenic influences do appear to be predictable a decade ahead, and the climate response over this time is nearly invariant under different scenarios of greenhouse gas emissions. Over much of the globe this predictable effect dominates over internal variability in decadal averages, though not for year-to-year variations. It does introduce a bias that alters the probability of occurrence of extreme events like warm spells and droughts. One of the more likely influences is the expansion and intensification of drying in subtropical regions, including the Mediterranean and the southwestern United States/northern Mexico.

The expansion of the subtropical dry belt is one of the more certain changes to be expected in the next decade, but in order to assess potential climate impacts on agriculture and other sectors we would like to have a globally complete, quantitative set of regional climate changes. Most important would be extreme events: extended heat waves, droughts, floods, severe storms, and tropical cyclones. We know it is impossible to predict such events very far ahead, but by knowing the influences of anthropogenic climate change, of interannual variations (ENSO, primarily), and longer period climate variations such as the AMO or PDO, we can estimate the expected changes in the probability of occurrence of extreme events.

The next years will bring a substantial increase in the number of decadal forecasts. A deeper understanding and some improvement in forecast skill is a likely outcome, although we will probably not gain a useful ability to predict decadal variations internal to the climate system, such as the AMO. At a minimum, we can expect a better depiction of the regional characteristics of climate noise and a better accounting of the probabilities of extreme events. This climate information may improve assessment of the climate risk for food security and could be used as input to crop models, hydrologic models, and other sectoral models to provide a best assessment of the climate risk for food security and other factors that influence sociopolitical stability.

ACKNOWLEDGMENTS

This work was supported by grant DE-SC0005108 from the Department of Energy and NOAA grant NA08OAR4320912. Thanks to Yochanan Kushnir for teaching us so much about decadal variability and for his comments on a draft version, and to Arthur Green, Lisa Goddard, Colin Kelley, Naomi Henderson, Richard Seager, and Mingfang Ting for valuable discussions.

NOTES

1. It also includes the changes in Earth's orbit that were responsible for the cycles of ice ages until human interference in the climate system rendered them obsolete. Orbital changes will continue to influence climate, but their timescales of tens of thousands of years are not relevant here.
2. Cane (1986) gives a historical account of ENSO theory.
3. Cane (2005) offers a relatively accessible description of the mechanism that maintains the ENSO cycle; Sarachik and Cane (2010) is a comprehensive account.
4. Chen and Cane (2008) has a brief account; current forecast information is available at http://iri.columbia.edu (accessed March 21, 2013). See "Forecast Plume" for the individual predictions of the entire suite of models. Operational forecasts by many groups throughout the world can also be found in the quarterly Experimental Long Lead Forecast Bulletin at http://www.iges.org/ellfb (accessed March 21, 2013).
5. See section 7.8 of Sarachik and Cane (2010) and references therein for a review covering the CMIP3 models used in the IPCC Fourth Assessment; similar evaluations of the CMIP5 models used in the Fifth Assessment will be available by the time this goes to press.
6. Cook et al. (2008) show that including the dust of the Dust Bowl improves the simulation, especially over the northwestern US.
7. Oldenborgh et al. (2012) carried out a similar study but with an ensemble of four forecast models. They too find that beyond the first year there is no skill in predictions of the variations of global mean temperature about the trend due to the increase in greenhouse gases and other anthropogenic influences.
8. As of January 2013 we can say that this forecast has proven to be correct.

REFERENCES

Alexander, M. A., I. Bladé, M. Newman, J. R. Lanzante, N.-C. Lau, and J. D. Scott. 2002. The atmospheric bridge: The influence of ENSO teleconnections on air–sea interaction over the global oceans. *Journal of Climate 15*: 2205–31.

Allen, M. R., and W. J. Ingram. 2002. Constraints on future changes in climate and the hydrologic cycle. *Nature 419*: 224–32.

Allen, R. J., S. C. Sherwood, J. R. Norris, and C. S. Zender. 2012. Recent Northern Hemisphere tropical expansion primarily driven by black carbon and tropospheric ozone. *Nature 485*: 350–4. doi:10.1038/nature11097.

Balmaseda, M. A., M. K. Davey, and D. L. T. Anderson. 1995. Decadal and seasonal dependence of ENSO prediction skill. *Journal of Climate 8*: 2705–15.

Barnston, A. G., M. K. Tippett, M. L. L'Heureux, S. Li, and D. G. DeWitt. 2012. Skill of real-time seasonal ENSO Model predictions during 2002–11: Is our capability increasing? *Bulletin of the American Meteorological. Society 93*: 631–51.

Barsugli, J. J., and D. S. Battisti. 1998. The basic effects of atmosphere–ocean thermal coupling on midlatitude variability. *Journal of the Atmospheric Sciences 55*: 477–93.

Biasutti, M., and A. Giannini. 2006. Robust Sahel drying in response to late 20th century forcings. *Geophysical Research Letters 33*:L11706. doi:10.1029/2006GL026067.
Bjerknes, J. 1969. Atmospheric teleconnections from the equatorial Pacific. *Monthly Weather Review 97*: 163–72.
———. 1972. Large-scale atmospheric response to the 1964–65 Pacific equatorial warming. *Journal of Physical Oceanography 15*: 1255–73.
Booth, B. B. B., N. J. Dunstone, P. R. Halloran, T. Andrews, and N. Bellouin. 2012. Aerosols implicated as a prime driver of twentieth-century North Atlantic climate variability. *Nature 484*: 228–32. doi:10.1038/nature10946.
Bouma, M. J., and J. J. van der Kayy. 1994. Epidemic malaria in India and the El Niño Southern Oscillation: Health and climate change. *Lancet 344*: 1389.
———. 1996. The El Niño Southern Oscillation and the historic malaria epidemics on the Indian subcontinent and Sri Lanka: An early warning system for future epidemics? *Tropical Medicine and International Health 1*: 86–96.
Cane, M. A. 1986. El Niño. *Annual Review of Earth and Planetary Sciences 14*: 43–70.
———. 2005. The evolution of El Niño, past and future. *Earth and Planetary Science Letters 104*: 1–10.
———, G. Eshel, and R. W. Buckland. 1994. Forecasting maize yield in Zimbabwe with Eastern equatorial Pacific sea surface temperature. *Nature 370*: 204–205.
———, S. E. Zebiak, and S. C. Dolan. 1986. Experimental forecasts of El-Niño. *Nature 321*(6073): 827–32.
Chen, D., and M. A. Cane. 2008. El Niño prediction and predictability. *Journal of Computational Physics 227*(7): 3625–40. doi:10.1016/j.jcp.2007.05.014.
———, A. Kaplan, S. E. Zebiak, and D. Huang. 2004. Predictability of El Niño over the past 148 years. *Nature 42*: 733–36.
———, S. E. Zebiak, A. J. Busalacchi, and M. A. Cane. 1995. An improved procedure for El Niño forecasting: Implications for predictability. *Science 269*: 1699–702.
Cook, B. I., R. L. Miller, and R. Seager. 2008. Dust and sea surface temperature forcing of the 1930s "Dust Bowl" drought. *Geophysical Research Letters 35*: L08710. doi:10.1029/2008GL033486.
Davis, M. 2001. *Late Victorian holocausts: El Niño famines and the making of the Third World.* London: Verso.
Del Sole, T., M. K. Tippett, and J. Shukla. 2011. A significant component of unforced multidecadal variability in twentieth century global warming. *Journal of Climate 24*: 909–25
Emile-Geay, J., M. A. Cane, R. Seager, A. Kaplan, and P. Almasi. 2007 El Niño as a mediator of the solar influence on climate. *Paleoceanography 22*: PA3210. doi:10.1029/2006PA001304.
Emile-Geay, J., R. Seager, M. A. Cane, E. R. Cook, and G. H. Haug. 2008. Volcanoes and ENSO over the last millennium. *Journal of Climate 21*: 3134–48.
Enfield, D. B., A. M. Mestas-Nuñez, and P. J. Trimble. 2001 The Atlantic multidecadal oscillation and its relation to rainfall and river flows in the continental U.S. *Journal of Geophysical Research 28*: 2077–80.
Giannini, A., R. Saravanan, and P. Chang. 2003. Oceanic forcing of Sahel rainfall on interannual to interdecadal time scales. *Science 302*: 1027–30.

Goddard, L., and N. E. Graham. 1999. The importance of the Indian Ocean for simulating precipitation anomalies over eastern and southern Africa. *Journal of Geophysical Research 104*: 19099–116.

Goddard, L., S. J. Mason, S. E. Zebiak, C. F. Ropelewski, R. Basher, and M. A. Cane. 2001. Current approaches to seasonal to interannual climate predictions. *International Journal of Climatology 21*: 1111–52.

Goldenberg, S. B., C. W. Landsea, A. M. Mestas-Nuñez, and W. M. Gray. 2001. The recent increase in Atlantic hurricane activity: Causes and implications. *Science 293*: 474–49.

Goswami, B. N., M. S. Madhusoodanan, C. P. Neema, and D. Sengupta. 2006. A physical mechanism for North Atlantic SST influence on the Indian summer monsoon. *Geophysical Research Letters 33*: L02706. doi:10.1029/2005GL024803.

Hansen, J., M. Sato, R. Ruedy, P. Kharecha, A. Lacis, R. L. Miller, L. Nazarenko et al. 2007. Climate simulations for 1880–2003 with GISS model E. *Climate Dynamics 29*: 661–96. doi:10.1007/s00382-007-0255-8.

Held, I. M. and B. J. Soden. 2006. Robust responses of the hydrological cycle to global warming. *Journal of Climate 19*: 5686–99.

Hoerling, M., J. Hurrell, A. Kumar, L. Terray, J. Eischeid, P. Pegion, T. Zhang, X. Quan, T. Xu. 2011. On North American decadal climate for 2011–20. *Journal of Climate 24*: 4519–28. doi:10.1175/2011JCLI4137.1.

Hoerling, M. P., A. Kumar, and M. Zhong. 1997. El Niño, La Niña, and the nonlinearity of their teleconnections. *Journal of Climate 10*: 1769–86.

Hsiang, S. 2010. Temperatures and cyclones strongly associated with economic production in the Caribbean and Central America. *Proceedings of the National Academy of Sciences 107*: 15367–72.

———, K. Meng, and M. A. Cane. 2011. Civil conflicts are associated with the global climate. *Nature 476*: 438–41. doi:10.1038/nature10311.

IPCC. 2007. *Climate change 2007: The physical science basis.* Eds. S. Solomon, D. Qin, M. Manning, Z. Chen, M. Marquis, K. B. Averyt, M. Tignor, and H. L. Miller. Contribution of Working Group I to the Fourth Assessment Report of the Intergovernmental Panel on Climate Change. Cambridge and New York: Cambridge University Press.

Karspeck, A., R. Seager, and M. A. Cane. 2004. Predictability of tropical Pacific decadal variability in an intermediate model. *Journal of Climate 17*: 2842–50.

Keenlyside N. S., M. Latif, J. Jungclaus, L. Kornblueh, and E. Roeckner. 2008. Advancing decadal-scale climate prediction in the North Atlantic sector. *Nature 453*: 84–88. doi: 10.1038/nature06921.

Kelley, C., M. Ting, R. Seager, and Y. Kushnir. 2011. The relative contributions of radiative forcing and internal climate variability to the late 20th century drying of the Mediterranean region. *Climate Dynamics.* doi: 10.1007/s00382-011-1221-z.

Kerr, D. 2005. Atlantic climate pacemaker for millennia past, decades hence? *Science 309*: 41–42.

Kirtman, B. P., and P. S. Schopf. 1998. Decadal variability in ENSO predictability and prediction. *Journal of Climate 11*: 2804–22.

Knight, J. R. 2009. The Atlantic Multidecadal Oscillation inferred from the forced climate response in coupled general circulation models. *Journal of Climate 22*: 1610–25.

Kushnir, Y., W. A. Robinson, P. Chang, and A. W. Robertson. 2006. The physical basis for predicting Atlantic sector seasonal-to-interannual climate variability. *Journal of Climate 19*: 5949–70.

Lean, J. L., and D. H. Rind. 2009. How will Earth's surface temperature change in future decades? *Geophysical Research Letters 36*: L15708. doi:10.1029/2009GL038932.

Lu, J., G. Chen, and D. M. W. Frierson. 2008. Response of the Zonal Mean Atmospheric Circulation to El Niño versus global warming. *Journal of Climate 21*:5835–51.

Lu, J., G. A. Vecchi, and T. Reichler. 2007. Expansion of the Hadley cell under global warming. *Geophysical Research Letters 34*: L06805. doi:10.1029/2006GL028443.

Lyon, B., and D. G. DeWitt. 2012. A recent and abrupt decline in the East African long rains. *Geophysical Research Letters 39*: L02702. doi:10.1029/2011GL050337.

Mann, M. E., M. A. Cane, S. E. Zebiak, and A. Clement. 2005. Volcanic and solar forcing of El Niño over the past 1000 years. *Journal of Climate. 18*: 447–56.

——, and K. A. Emanuel. 2006. Atlantic hurricane trends linked to climate change. *EOS, Transactions American Geophysical Union 87*: 233.

Mantua, N. J., S. R. Hare, Y. Zhang, J. M. Wallace, and R. C. Francis. 1997. A Pacific interdecadal climate oscillation with impacts on salmon production. *Bulletin of the American Meteorological Society 78*:1069–79.

Mantua, N. J., and S. R. Hare. 2002. The Pacific Decadal Oscillation. *Journal of Oceanography 58*: 35–44.

Oldenborgh, G. J. van, F. J. Doblas-Reyes, B. Wouters, and W. Hazeleger. 2012. Skill in the trend and internal variability in a multi-model decadal prediction ensemble. *Climate Dynamics 38*(7): 1263–80. doi:10.1007/s00382-012-1313-4.

Phillips, J., B. Rajagopalan, M. A. Cane, and C. Rosenzweig. 1999. The role of ENSO in determining climate and maize yield variability in the U.S. cornbelt. *International Journal of Climatology 19*: 877–88.

Power, S., T. Casey, C. Folland, A. Colman, and V. Mehta. 1999. Inter-decadal modulation of the impact of ENSO on Australia. *Climate Dynamics 15*:319–24.

Ropelewski, C. F., and M. S. Halpert. 1987. Global and regional scale precipitation patterns associated with the El Niño/Southern Oscillation. *Monthly Weather Review 115*: 1606–26.

——. 1996. Quantifying Southern Oscillation–precipitation relationships. *Journal of Climate. 9*: 1043–59.

Sarachik, E. S., and M. A. Cane. 2010. *The El Niño-Southern Oscillation phenomenon.* London: Cambridge University Press.

Schneider, T., and I. M. Held. 2001. Discriminants of twentieth-century changes in earth surface temperatures. *Journal of Climate 14*: 249–54.

Seager, R., A. Karspeck, M.A. Cane, Y. Kushnir, A. Giannini, A. Kaplan, B. Kerman, and J. Velez. 2004. Predicting Pacific decadal variability. In *Earth's Climate: The Ocean–Atmosphere Interaction*, ed. C. Wang, S.-P. Xie, and J. A. Carton, 115–30. Washington DC: American Geophysical Union.

Seager, R., Y. Kushnir, C. Herweijer, N. Naik, and J. Miller. 2005. Modeling of tropical forcing of persistent droughts and pluvials over western North America: 1856–2000. *Journal of Climate. 18*: 4065–88.

Seager, R., M. Ting, I. M. Held, Y. Kushnir, J. Lu, G. Vecchi, H.-P. Huang et al. 2007. Model projections of an imminent transition to a more arid climate in southwestern North America. *Science 316*: 1181–84.

Smith, D. M., S. Cusack, A. W. Colman, C. K. Folland, G. R. Harris, J. M. Murphy. 2007. Improved surface temperature prediction for the coming decade from a global climate model. *Science 317*: 796–99. doi: 10.1126/science.1139540.

Stockdale, T. N., D. L. T. Anderson, M. A. Balmaseda, F. Doblas-Reyes, L. Ferranti, K. Mogensen, T. N. Palmer, F. Molteni and F. Vitart. 2011. ECMWF Seasonal Forecast System 3 and its prediction of sea surface temperature. *Climate Dynamics 37*: 455–71. doi:10.1007/s00382-010-0947-3.

Stommel, H., and E. Stommel. 1983. *Volcano weather: The story of 1816, the year without a summer.* Newport: Seven Seas Press.

Stothers, R. B. 2000. Climatic and demographic consequences of the massive volcanic eruption of 1258. *Climatic Change. 45*: 36D374.

Sutton, R. T., and D. L. R. Hodson. 2005. Atlantic Ocean forcing of North American and European summer climate. *Science 309*: 115–18.

Ting, M., Y. Kushnir, R. Seager, and C. Li. 2009. Forced and internal 20th century SST trends in the North Atlantic. *Journal of Climate 22*: 1469–81.

Tippett, M. K., and A. G. Barnston. 2008. Skill of multimodel ENSO probability forecasts. *Monthly Weather Review 136*: 3933–46.

Trenberth, K. E. 1990. Recent observed interdecadal climate changes in the Northern Hemisphere. *Bulletin of the American Meteorological Society 71*: 988–93.

———, and D. J. Shea. 2006. Atlantic hurricanes and natural variability in 2005. *Geophysical Research Letters 33*: L12704. doi:10.1029/2006GL026894.

Whitcombe, E.1993. Famine mortality. *Economic and Political Weekly. 28*(23): 1169–84.

Wunsch, C. 1999. The interpretation of short climate records, with comments on the North Atlantic Oscillation and Southern Oscillation. *Bulletin of the American Meteorological Society 80*: 245–55.

Wyrtki, K. 1975. El Niño—the dynamic response of the equatorial Pacific Ocean to atmospheric forcing. *Journal of Physical Oceanography 5*: 572–74.

———. 1979. The response of sea surface topography to the 1976 El Niño. *Journal of Physical Oceanography 9*: 1223–31.

Yancheva, G., N. R. Nowaczyk, J. Mingram, P. Dulski, G. Schettler, J. F. W. Negendank, J. Liu, D. M. Sigman, L. C. Peterson, and G. H. Haug. 2007. Influence of the intertropical convergence zone on the East Asian monsoon. *Nature 445*: 74–77.

Zebiak, S. E., and M. A. Cane. 1987. A model El Niño–Southern Oscillation. *Monthly Weather Review 115*: 2262–78.

Zhang, R., and T. L. Delworth. 2006. Impact of the Atlantic Multidecadal Oscillation on North Pacific climate variability. *Geophysical Research Letters 34*. doi: 10.1029/2006GL028683.

4

The Global Land Rush

KLAUS DEININGER

The world first took notice of a renewed trend towards large-scale land acquisition and the challenges it poses when, in 2008, evidence of a Korean firm obtaining more than one million hectares in Madagascar, with virtually no compensation to local people, was widely circulated in the global press. The resulting controversy contributed to the collapse of the country's government and the withdrawal of the investment. Three factors have contributed to a surge in demand for land, especially in Africa: expectation of continued strong demand growth and increased price volatility in agricultural commodities, increased use of what might traditionally have been considered marginal lands for production of environmental services, and the fact that many actors in the financial sector consider land as an asset with very desirable properties in the current macro-economic environment.

Attitudes toward this trend vary widely. Denouncing the phenomenon as a modern "land grab," some point to the enormous risks and the growing body of case studies suggesting that many investments displace local populations, lack economic viability, and fail to comply with basic social and environmental safeguards (Pearce 2012). They point to historical precedents, the irony of envisaging large exports of food from countries that in some cases depend on regular food aid, and the risks inherent in trying to regulate such activities. Others, including host-country governments especially in Africa, point towards vast potential for import substitution, and note that, in an environment where yields are low and land still relatively abundant, investment in agriculture could provide an enormous boost to economic growth, poverty reduction, and food security. They note that in light of Africa's long history of underinvestment in agriculture, private investment in on-farm capital, infrastructure, and applied research will be indispensable, even if recent pledges for increased public investment in agriculture are realized. Indeed, some view the surge of agricultural investment as a structural break equivalent to industrial outsourcing in the 1980s, with pioneer investors heralding a much-needed corporatization of African agriculture well worth subsidizing through a range of instruments that may include cheap land (Collier and Venables 2011).

Against this background, this chapter contextualizes recent investments, and highlights important determinants of their impact by drawing on historical evidence of land acquisition and different regional patterns of agricultural expansion and large-scale farming. We assess the scope for area expansion, as well as intensification and yield growth on currently cultivated land in different countries, yielding a country typology that can be used to identify the countries where interest in large-scale land acquisition is most likely to materialize. Evidence on actual land transfers for specific countries where such data could be obtained and on land demand is broadly in line with this but also suggests that countries with weak land governance are disproportionately affected by the recent surge in investor demand.

Effective policies to improve land governance—those that enhance investment incentives, facilitate low-cost transfers of land to its most productive use, help delimit state land and protect environmentally sensitive areas, and allow land taxation that can help establish local public goods—are likely to help close yield gaps and improve the outcomes from land-related investment. In addition, countries where investor interest is likely to materialize will benefit from measures to allow clustering of investments in certain areas so as to realize synergies from provision of complementary infrastructure by the public and private sectors, public screening of investment proposals and monitoring of implementation, and support to agreed mechanisms for dispute resolution and arbitration. While many of these services are country-level responsibilities, the global community can lend support by disseminating information on desirable practices to increase transparency, provide good practice examples, and make it easier for responsible investors to obtain capital and other forms of support.

WHAT CAN WE LEARN FROM HISTORY?

Processes of large-scale land acquisition have historical precedents. Past spikes in land demand often coincided with technological improvements or reduced transport costs that significantly altered economic relationships. Review of historical experience points to three key lessons. First, failure to respect local rights to resources such as land, water, and environmental amenities often led to conflict and social tension that jeopardized investments' contribution to sustained improvement of welfare and their own economic viability. Second, absence of a proper regulatory framework, including clear property rights, price incentives, and complementary public goods has often encouraged acquisition of land for short-term extraction of resources instead of responsible investment. Finally, without mechanisms to charge investors the full opportunity cost of land or agile ways to transfer land to the highest bidder or the most productive use, land has been held for speculative purposes or owners have resorted to lobbying for subsidies to ensure the viability of their enterprises with undesirable consequences for overall economic growth.

During colonial expansion, demand for land was driven not so much by the desire to produce agricultural commodities as by the need to obtain labor and food for mines. Unequal patterns of land acquisition often gave rise to distortions that affected social and economic development in the long term (Binswanger et al. 1995). Even in subsequent phases when economic motives—such as elites' desire to "modernize" the economy by bringing "unused" land under production—dominated, outcomes were often unsatisfactory from a social or environmental perspective, thus failing to contribute to long-term development.

In many cases, the legal basis for large-scale acquisition of land by investors was the neglect of existing communal land rights based on the notion that land not obviously used for economic purposes was owned by the state and thus available for disposal. State powers often used the mechanism of requiring land registration within a short time to transfer nonregistered land to investors. Use of this measure was common in Central America, including in El Salvador and Guatemala, where abolition of communal land tenure deprived local people of a critical safety net (Lindo-Fuentes 1990) and led to widespread unrest.[1] The way in which land was expropriated led to sporadic eruptions of violence, and areas where land pressure had been particularly strong emerged as centers of a 1932 revolt during which some 10,000 to 20,000 peasants were killed (Mason 1986), with civil war and violence continuing into the 1990s.

Beyond fostering conflict and violence, the way in which land was transferred undermined the security of property rights and investment incentives. Landowners then resorted to political means to make ventures viable economically, using monopolies to keep unskilled labor plentiful and cheap, and political lobbying to establish policies favoring large farms, including subsidies to capital and tax relief. Where such measures were possible, development of human capital and development of democratic institutions were much delayed, compared to countries with smallholder production structures that resulted in better functioning factor markets (Nugent and Robinson 2010).[2] Microstudies illustrate that landowners' noncompetitive practices have been widespread and that these had implications for human capital acquisition and the functioning of capital markets (Rajan and Ramcharan 2011). This mirrors global evidence on the long-term impacts of an unequal asset distribution on agricultural production structure, economic outcomes, the political economy, and supply of public goods in other parts of the world such as Latin America (de Janvry 1981), especially Brazil and India (Banerjee and Iyer 2005; Iyer 2010).

RECENT REGIONAL EVIDENCE

Recent examples illustrate the importance of policies and institutional factors as determinants of the local impact of large-scale agricultural investment. While these examples point towards potentially large benefits from investment to make technology and markets available and increase productivity, they also highlight the importance of a regulatory environment to ensure this does not result in irreversible loss of biodiversity or inhumane treatment of workers.

In Africa, policies such as overvalued exchange rates, export tariffs, and low levels of public investment in rural areas have for a long time compromised the region's natural potential for progress through agro-investment. Elimination of many distortions in past decades has allowed agricultural growth to accelerate and paved the way for renewed investor interest in the continent. Africa's endowment of water and land implies that a better policy environment and business climate would allow profitable production of bulk commodities in addition to high-value export crops. Infrastructure constraints imply that initially supply would be limited to domestic and regional markets worth some $50 billion a year, which may in time provide a springboard for exports (World Bank 2009).

In past decades, externally driven agricultural ventures in Africa often ran into technical or managerial problems and were abandoned. In most cases, success with export agriculture was limited to higher-value crops (above a price of about $500 per ton) such as cotton, cocoa, or coffee, where good agro-ecological conditions, low (if any) compensation for land, and cheap labor helped offset the disadvantages arising from weak institutions, high transport costs, and ill-functioning output, input, and capital markets (World Bank 2009). Often investment focused on processing, marketing, and distribution, as in the case of Cote d'Ivoire which, in the 1980s, became the world's largest cocoa exporter based on small and medium-sized producers. Over two decades, some ten million hectares of forestland was converted to cocoa, mostly by attracting immigrants with the promise of ownership of part of the plantation once established. The closing of the land frontier and the ensuing competition for land led locals to renege on these promises, prompting conflict and a decade-long civil war that annihilated much of the progress made.

In contrast, little long-term investment is needed to grow bulk commodities, so insecure property rights and interference by rent-seeking elites or the state often led to eventually unsustainable ventures dominated by extractive motives. Drawing on colonial roots, a scheme for semi-mechanized sorghum and sesame production in Sudan expanded rapidly in the 1970s, when financing from the Gulf States aimed to transform Sudan into a regional breadbasket. Easy access to land and subsidized credit for machinery attracted a large number of civil servants and businessmen into the sector. Some 5.5 million hectares were converted to arable land according to official statistics, with up to 11 million hectares of additional land occupied informally (Government of Sudan 2009). Neglect of traditional land rights by small farmers and pastoralists led to conflict (Johnson 2003), undermining security of rights and investment incentives. The ensuing highly extractive mode of cultivation undermined ecological sustainability and economic competitiveness as investors obtained only a fraction of potential yields. In an agro-ecological setting comparable to Australia with yields of four tons per hectare, sorghum yields are only 0.5 tons per hectare. Conflict-induced insecurity of property rights undermined incentives to apply fertilizer, rotations, or livestock to maintain soil fertility, thus leading to declining yields in the long term.

Lack of managerial incentives was a key reason for the failure of one of the most spectacular—though ultimately unsuccessful—ventures to expand cultivated area in Eastern Europe. The Virgin Lands campaign aimed to draw large parts of Kazakhstan into large-scale production. The goal of bringing 13.5 million hectares into cultivation was reached by 1954 and a total of 55 million hectares was brought into state farms with sizes of up to 70,000 hectares. Despite large investments in roads, housing, schools, and other infrastructure, managerial and incentive problems, together with marginal agro-ecological suitability, led to low productivity. Dry spells and low yields due to exhaustion of soils caused growers to abandon much of the area, with lasting environmental damage (McCauley 1976). Since then, Eastern European countries have undergone major transitions to new agrarian structures and reemerged as grain exporters in global markets. The fact that they are likely to benefit from climate change may enhance their importance for global food security.

In Southeast Asia, expansion of oil palm cultivation, one of the most profitable land uses in the humid tropics—but also highly labor intensive (Butler and Laurence 2009)—provided ample scope for positive social impacts and economic diversification. The commercial oil palm industry was pioneered by Malaysia, but expanded to Indonesia, which in 2008 became now the world's largest producer, slightly ahead of Malaysia. The planted area in Indonesia more than doubled between 1997 and 2007, from about 2.9 million hectares to 6.3 million hectares, giving rise to significant gains in employment with an estimated 1.7 to 3 million jobs generated in the industry (Rist et al. 2010).

The full potential was, however, not always achieved due to lack of a regulatory framework so that firms could exercise market power and policies favoring extraction over investment. Many concessions, up to 12 million hectares by some estimates (Farigone et al. 2008), were deforested without ever having been planted to oil palm.[3] At the same time, smallholder cultivation, which currently accounts for about a third of production, has been constrained by neglect of local land rights, low-quality planting material, limited access to finance, and mills' market power in a noncompetitive market for fresh fruit. With continued increases in demand, oil palm area is expected to increase by 10 to 20 million hectares in the next decade. Whether this area comes from degraded land, usually portrayed as unproductive wasteland, or from expansion into forest, will have far-reaching implications (Fairhurst and McLaughlin 2009).

In Brazil, three processes of land expansion can be broadly distinguished. The best known—forest clearing in the Amazon basin for extensive livestock ranching and establishing land rights—is widely seen as having been accelerated by the requirement to clear forest in order to demonstrate "productive use." Expansion in the Amazon has been rapid: the cattle population more than doubled from 1990 to 2006 and pasture area expanded by 24 million hectares (Pacheco and Poccard Chapuis 2009). Net impacts were often negative, though, as most of the land deforested was of low quality and not put to productive use.

Building on decades of public research and a proactive policy to develop value chains, Brazil developed cutting-edge sugar and ethanol industries that now account for 20 and 34 percent of production and 38 and 74 percent of world trade in sugar and ethanol respectively. High sucrose varieties and advances in processing, including cogeneration, may further strengthen this global competitiveness. Some two-thirds of the area for sugarcane expansion was from former pastureland, 32 percent was converted from other crops, and only 2 percent from natural vegetation, suggesting that higher productivity can mitigate negative environmental effects of agricultural expansion, at least to some degree.

An expansion of soybeans and other crops in the *cerrado* (savannah) region was based on public investment in research and development that allowed cultivation of acid soils previously unsuitable for agriculture, use of appropriate varieties, and adoption of conservation tillage. Although a technological success, direct impacts on employment generation and rural poverty reduction remained limited as large-scale subsidized credit for large farmers led to mechanized rather than labor-intensive production (Rezende 2005). Public- and private-sector actors now recognize that agri-investment and expansion pose social and environmental challenges that require action to reduce detrimental impacts. These include rehabilitation of degraded lands, stricter enforcement of legal reserves (mandates to keep forested areas on agricultural properties), better delineation of protected areas, and environmental zoning.

IS CORPORATE FARMING VIABLE OR DESIRABLE?

Initial outlays for capital and possibly technology and marketing will favor large farms, but in the long run, farm size should be driven by the productivity of differently sized units and factor price ratios. Owner-operated smallholder farms have long displayed higher levels of productivity and have made an enormous contribution to poverty reduction. Except in highly distorted environments, farm sizes grow as a result of increases in nonagricultural wages and farmers' aspiration to a comparable standard of living. While recent technical developments can reduce the transaction cost of supervising hired labor, which was traditionally a key factor underlying the inverse relationship between farm size and productivity, these technologies may not be economical in low-wage settings. Thus, once initial setup costs have been amortized, owners may break up large farms into smaller units for production, possibly under outgrower arrangements to ensure integration with marketing and processing.

In most countries, agricultural production—distinct from marketing, processing, or distribution—is dominated by owner-operated units that combine ownership of the main means of production with management. Many studies have demonstrated that agricultural production has few technical economies of scale, so that a range of production forms can coexist. Owner-operated farms have three main advantages over those run on wage labor (Allen and Lueck 1998; Binswanger and Deininger 1997; Deininger 2003). First, as residual claimants to profit, owner-operators and family workers will be more likely to work harder than wage workers who require costly supervision. Supervision is particularly costly in agriculture due to the spatial dispersion of production and the need to adjust to micro-variations in soil and climate. Second, owner-operators have intimate knowledge of local soil and climate, often accumulated over generations, that allows them to tailor management to local conditions, and the flexibility to quickly adjust management decisions to site, seasonal, and market conditions. Finally, as family labor can more easily be reallocated to other tasks on and off the farm, family farms have greater flexibility to adjust the labor supply to seasonal variation in labor demands. The favorable conditions provide the basis for owner-operated smallholder farming as the predominant form of production, with an exceptionally large potential for employment generation and poverty reduction.[4]

Difficulties in accessing financial markets, reliance on capital goods such as machinery, and the ability to benefit from vertical integration may partly offset the advantages of owner-operated agricultural operations. High transaction costs of providing formal credit in rural markets imply that the unit costs of borrowing and lending decline with loan size, biasing lending against small farmers. Unless ways are found to provide small farmers with access to finance (such as through credit cooperatives), difficulties in obtaining financing may thus counteract family farmers' supervision cost advantages. Also, expensive machinery may reach its lowest cost of operation per unit area at a scale larger than the average size of operational holdings. While this could result in economies of scale and increase the optimum operational farm size, machine rental can help small farms use large machinery, circumventing this constraint for all but the most time-bound operations.

Due to the perishable and bulky nature of the raw produce, plantation crops such as sugar cane, bananas, and tea have long been recognized cases where the benefits from coordinating harvesting with processing provide advantages to large units of production. The need for close coordination of production and processing, often within 24 to 48 hours, to avoid deterioration of some harvest products requires tight adherence to delivery and harvesting schedules and transmits economies of scale in processing to the production stage (Binswanger and Rosenzweig 1986). For this reason, sugar factories and palm oil mills usually run their own plantations to ensure a base load for processing, with the optimum size of the catchment area depending on processing technology.[5] Concentrating production close to the point of processing is also important for less perishable crops where processing results in significant bulk reduction, although the scope for storing produce on the farm reduces the need to integrate production and processing. Compared to a multi-product agricultural farm, perennial crops that require year-round labor have developed highly structured, industrial-type production processes that facilitate labor supervision and management. Focus on a single crop with low seasonality that can provide year-round employment facilitates the hiring of permanent workers and managers with specialized skills and allows the development of specific systems for monitoring. This is similar to highly specialized stall-fed livestock operations in industrial countries which, for similar reasons, moved away from family farm to non-family corporate farming (MacDonald 2011).

The scope for mechanization and recent innovations in breeding can reduce or eliminate the supervision advantage of owner-operators, although doing so is not costless. For example, all elements of sugarcane production (including harvesting) can now be fully mechanized, a step that may reduce negative environmental effects from burning. Recent innovations in breeding, zero tillage, and information technology also make supervision easier. By facilitating standardization, these innovations allow supervision of operations over large spaces, possibly reducing owner-operator advantages. Pest-resistant and herbicide-tolerant varieties reduce the number of steps in the production process and the labor intensity of cultivation. The scope for substituting information technology and remotely sensed data on field conditions for personal observation increases managers' ability to supervise and make decisions remotely, although modern technology can be equally useful to organize smallholders and provide them access to market information.

Historically, the main reason for operational farm sizes to grow was operators' desire to obtain comparable levels of income with economic development and the movement of population out of agriculture, which implies that capital will increasingly substitute for labor and that operational farm size (not necessarily ownership) will increase over time in line with wage rates. As Figure 4.1 illustrates, both variables moved together closely in the United States for most of the twentieth century (Gardner 2002). Still, at least

Fig. 4.1. Relationship between farm size and wage for the US, 1900–2005

Source: Based on Gardner, B. L. 2002. U.S. *Agriculture in the 20th century: How it flourished and what it cost.* Cambridge, MA: Harvard University Press.

in annual crop production, even large farms in the United States are mostly owner-operated rather than company-owned. To assess whether the desire to obtain comparable agricultural wage levels with economic development will remain the most important driver of growth in farm size or whether this has drastically been changed by new technology, we briefly consider South America and Eastern Europe where large farms emerged for bulk commodities beyond the level that would be required to keep up with wage growth.

In South America, particularly Argentina, large farm management companies, often headed by extended family networks, rent large swaths of land and contract with machine operators. The model started during Argentina's financial crisis in 2001 when access to outside capital provided significant advantages. With clear property rights allowing easy contracting, several companies farm more than 100,000 hectares. Many of them are traded publicly, have vertically integrated into input supply and output markets, and operate across several countries. Access to a pool of qualified agronomists who undergo continued training and are organized hierarchically allows adoption of industrial methods of quality control and production at low cost, providing an additional advantage. Competitive land lease markets, with contracts renewed annually, imply that at least part of the savings is passed on to landowners, who generally receive lease payments above the return to land they would have been able to obtain by self-cultivation (Manciana 2009). The most interesting aspect is the emergence of farm management companies that own neither land nor machinery but have significant market power. Still, the operational units they manage are never above the 10,000 to 15,000 hectare range.

Even larger farming structures exist in Eastern Europe. In Russia, the 30 largest holdings farm 6.7 million hectares or 5.5 percent of the country's cultivated area, and in Ukraine, the largest 40 firms control 4.5 million hectares or 13.6 percent of the cultivated area (Lissitsa 2010). In an environment with deficient infrastructure and governance, highly concentrated and monopolized markets for inputs and output, and shallow domestic capital markets, large operators' market power, ability to control storage and access to export terminals, and, importantly, the ability to directly access foreign capital through stock market flotation, provide large operators with clear competitive advantages. While more detailed analysis of productive efficiency will be useful, descriptive evidence suggests that the largest farms are not the most efficient, supporting the notion that their emergence owes more to pervasive market imperfections than superior productivity and that improved factor markets could help to develop the "missing middle" of farms with a size of between 500 and 10,000 hectares, depending on the crop mix.

The establishment of "greenfield" investments is risky even under the best of circumstances. In fact, many well-intended or well-resourced schemes faltered, including the "bonanza farms" established in the Dakotas between 1860 and 1900 (Drache 1964), Brazilian rubber plantations established by Henry Ford in the 1920s (Grandin 2009), and large-scale agriculture in the Lakeland Downs of Australia's far northeast in the 1960s (Rogers 2008). In Africa, even not-for-profit institutions that could deploy capital to loss-making ventures for a long time only managed success rates below one-third (Johansen 1988). Such failure is not a problem, and may in fact generate important knowledge, if markets exist to transfer the land involved to better uses. However, well-functioning factor markets will be critical to facilitate this, because if land markets do not exist, as in many contemporary African countries, speculation and rent-seeking pose considerable dangers.

While mega-farms and large land deals have captured the headlines, production of many plantation crops has been shifting back to smallholders, sometimes under contract farming arrangements to provide access to capital and other markets. While political issues may have played a role, even banana production is now in many places contracted out to smallholders. With establishment costs already amortized, global tea production, especially of high quality teas commanding premium prices, has reverted to smallholders with factories, and marketing controlled by producer-owned cooperatives (Herath et al. 2009). Similarly, rubber is a labor-intensive crop that can be processed on farms by simple methods. In all major rubber-producing countries, once technology and basic infrastructure were in place, large plantations quickly reverted to small-scale production, and smallholders now produce 85 percent of global output (Hayami 2010). This is consistent with the notion that there are considerable economies of scale in initial development of technology and establishment of infrastructure in thinly populated, land-abundant areas.

Once establishment costs have been amortized, if factor markets are reasonably competitive and labor moves in, the advantages of large-scale production may be rather small.

WHAT IS THE POTENTIAL FOR OUTPUT GROWTH AND AREA EXPANSION?

Increasing demand for agricultural commodities can be met through more intensive use of land already under cultivation or by expansion into previously uncultivated areas. Experience suggests that the latter poses much greater economic and social risks and may not yield significant benefits unless overall yields are increased.

The starting point for gauging the potential supply of land for rainfed cultivation is an assessment of potential yields that can be achieved on a given plot (approximated by a pixel in the analysis). To do so, we use the global agro-ecological zoning methodology developed by the International Institute for Applied Systems Analysis (Fischer et al. 2002), which predicts potential yield for rainfed cultivation of seven key crops based on simulated plant growth at each stage of the vegetative cycle, taking into account factors including soil, temperature, precipitation, elevation, and slope, and allowing simulations for different climate change scenarios. Applying a price vector allows us to determine the crop that produces the highest revenue for any cell. For areas that are currently cultivated, the difference between possible output and what is attained provides an estimate of the yield gap. Although there are many reasons for the existence of such a gap, this hypothetical exercise can break this gap into various factors. If they are not currently cultivated, not designated as a protected area, non-forested, and have low population density so that whatever existing interests are displaced can be compensated, areas with high potential could be possible candidates for area expansion. To identify the relevant subsets, we use a number of data sets.[6] Note that all of these simulations focus on potential for rainfed cultivation. Inclusion of a hydrological model could yield more realistic outcomes at the cost of much greater complexity.

We obtain two main results: First, yield gaps vary widely across regions and are especially large for Africa. In fact, with the exception of South Africa, no country in sub-Saharan Africa realizes more than 25 percent of potential production. This implies vast potential to increase yields from smallholders by providing access to technology, infrastructure, and markets.[7] Policy measures to increase smallholder productivity could provide significant benefits to local populations and will in most cases involve much lower risks or costs than area expansion.

The seven countries with the largest amount of suitable but uncultivated land according to this measure (Sudan, Brazil, Australia, Russia, Argentina, Mozambique, and Democratic Republic of the Congo, in that order) account for 224 million hectares, or more than half of global availability. A total of 32 countries with more than three million hectares of land each make up more than 90 percent of available land. Of these, 16 are in Africa, eight in Latin America, three in Eastern Europe and Central Asia, and five in the rest of the world. More strikingly, many of the counties with ample amounts of suitable but uncultivated land have limited amounts of land under cultivation. For example, the area of currently uncultivated land suitable for cultivation is more than double what is currently cultivated in 11 countries and more than triple the currently cultivated area in six countries.

Using the above information to plot land availability against the yield gap in Figure 4.2 allows us to distinguish four categories of countries with different endowments and development potential:

Little land for expansion, low yield gap. This type includes Asian countries with high population density, such as China, Vietnam, Malaysia, Korea, and Japan, Western European countries, and some countries in the Middle East with limited land suitable for rainfed production. Agricultural growth was and will be led by highly productive smallholders. To meet expanding demand for horticultural and livestock products, private investors are likely to provide capital, technology, and market access through contractual arrangements even

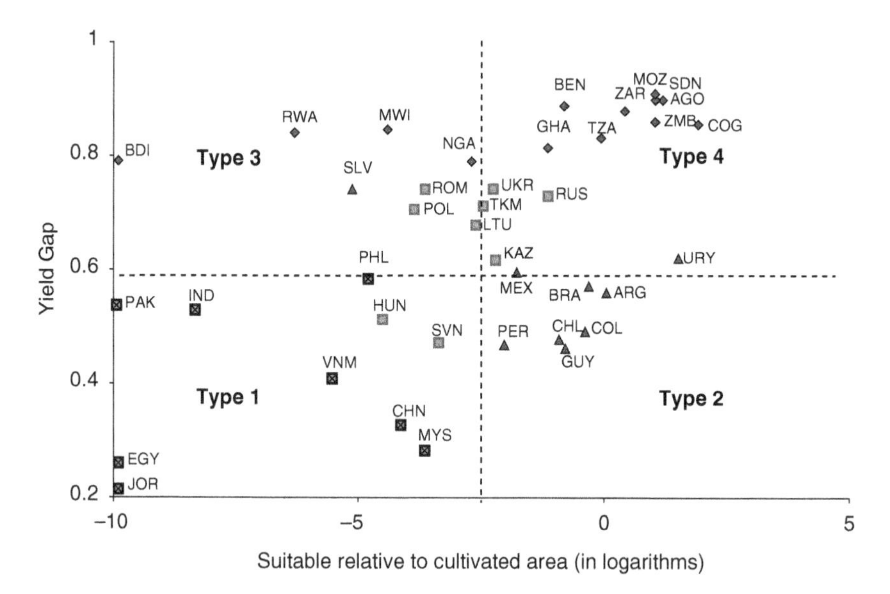

Fig. 4.2. Potential land availability vs. potential for increasing yields

Source: Deininger et al. (2011)

more than in the past. With economic growth, land consolidation—largely by entrepreneurial farmers leasing or buying plots from neighbors—will gradually increase farm sizes. This, as well as increased land use for nonfarm industries, urban expansion, and infrastructure, requires well-functioning land markets and institutions.

Suitable land available, low yield gap. Countries in this group have reasonably well-defined property rights, infrastructure access, and access to technology, often due to past investment in infrastructure, institutions, and human capital. It is in these cases where institutional investors have exploited opportunities for cropland expansion, in some cases leading to tremendous increases in land prices and where new forms of institutional arrangements (such as Argentina's farm management companies) emerged. If property rights are secure, markets function well, and areas with high social or environmental value are protected effectively (possibly using payments for environmental services), the public sector's role is mainly regulatory. But if land rights are insecure or ill-defined, large-scale land acquisition may threaten forests or lead to conflict with existing land users. A desire to avoid these negative outcomes has prompted some countries, such as Brazil, to launch large-scale programs to regularize land and tighten environmental regulation.

Little land available, high yield gap. This category includes the majority of developing countries with limited land and water availability. Large numbers of smallholders may be locked into poverty because the area currently cultivated remains far below the yield potential. Strategic options depend on the size and evolution of the nonagricultural sector. If the sector is small, higher agricultural productivity will be the only viable mechanism for rapid poverty reduction. This process will require public investment in technology, infrastructure, and market development to raise smallholder productivity, following the example of the Green Revolution in Asia. If incomes and employment in the nonagricultural sector grow rapidly, land markets work reasonably well, and population growth is low, there may be scope for land consolidation and an associated move to larger operational units with appropriate technology.

Suitable land available, high yield gap. This category includes sparsely populated countries in Africa (DR Congo, Sudan, Tanzania, Mozambique, and Zambia) with large tracts of land suitable for rainfed cultivation, but also large smallholder populations that only achieve a fraction of their potential productivity. In virtually all African countries where demand for land acquisition has recently increased dramatically, the level of productivity achieved by existing smallholder cultivators is less than 25 percent of potential. Capital-intensive activities with low labor absorption, such as annual crops using fully mechanized production, will be appropriate only if population density is low, the likelihood of in-migration is limited, and the nonagricultural sector can absorb expected growth of the labor force in the medium term. If institutions are available to deal with the associated risks, outside investment can help foster local development. The challenge for such countries is to establish mechanisms to attract capable investors, rather than speculators, to increase productivity and have benefits accrue to local populations. Rather than relying on ad hoc investor initiatives, provision of basic public services—such as infrastructure and technology—for areas and crops that fit well with the country's comparative advantage may be an option.

GLOBAL DEMAND FOR LAND ACQUISITION

The global demand for land reached historic heights in 2009 following the 2007 and 2008 commodity price spike. More importantly, the demand for land has to a large extent focused on countries where weak land governance creates scope for neglect of local rights and social and environmental safeguards. To mitigate these risks to human rights, country-led efforts to improve land governance will need to be complemented with private- and public-sector initiatives to monitor large-scale land acquisitions.

Obtaining consistent data on land transfers from official sources is impossible at present due to either confidentiality or because varying types of often inconsistent records are kept at different levels of government. Most research thus relies on reporting from secondary sources, with a strong emphasis on press reports (Alomar and Cousquer 2012). Following this approach, we use information from press reports to get a global picture of recent demand for large-scale land acquisition. We focus on reports published between October 1, 2008, and August 31, 2009, the height of the boom, as reported by the NGO Grain. For each of a total of 464 large-scale investments that remained once obvious duplicates had been eliminated, we coded origin and destination country, size, commodities involved, investor type, and whether any production activity started on the ground. The land matrix project, a more recent attempt to collect data on actual land deals, supposedly based on ground verification (Anseeuw et al. 2012), has attracted considerable criticism for erroneous reporting and double-counting.[8]

Table 4.1 compares regional rates of land expansion between 1961 and 2007 to the identified demand for land as measured in our data. Demand disproportionately focuses on Africa, where almost 70 percent of the area of interest to investors is located and where 2009 demand for land alone was larger than the agriculturally cultivated area of Belgium, Denmark, France, Germany, the Netherlands, and Switzerland combined.[9] These projects are large, with a median size of projects of about 40,000 hectares.

As Table 4.2 indicates, not all intended land investment projects result in actual transfers or actual land use, yet the size of recent land transfers exceeds the size of land expansion in the past by several times (Deininger et al. 2011).

Table 4.1. Historical land expansion and recent land demand

Region	*Cultivated land area*[a] *(millions of ha)*			*Annual change*		*Land demand 2009*[b]	
	1961	1997	2007	1961–1997	1997-2007	Mn ha	year eq.[c]
Sub-Saharan Africa	134.6	192.2	218.5	1.60	2.63	39.7	21.8
Latin America	102.6	160.9	168.0	1.62	0.71	3.2	2.2
East Asia & Pacific	183.9	235.7	262.8	1.44	2.72	8.0	4.6
South Asia	197.9	212.9	213.5	0.41	0.06	0.7	2.1
Oceania	34.0	42.8	46.7	0.25	0.38	0.0	0.2
Middle-East & North Africa	77.9	91.3	89.0	0.37	−0.23	1.4	5.9
Eastern Europe & C Asia	291.5	263.6	241.7	−0.77	−2.19	4.6	--
Western Europe	99.4	86.8	83.5	−0.35	−0.32	--	--
North America	235.3	232.5	225.3	−0.08	−0.72	0.2	--
World total	1357.1	1518.6	1549.0	4.49	3.04	57.8	13.9

Source: Own computation based on FAOSTAT and GRAIN.
[a] Cultivated area is land under arable or permanent crops. [b] Land demand 2009 refers to intended or actual land acquisitions based on media reports. [c] The last column ("year eq.") identifies this demand in terms of the number of years using average annual expansion in the 1961–2007 period.

Table 4.2. Extent of large land acquisitions in selected countries, 2004–09

Country	*Projects*	*Area (1,000 ha)*	*Median size (ha)*	*Area share domestic (%)*
Cambodia	61	958	8,985	70
Ethiopia	406	1,190	700	49
Liberia	17	1,602	59,374	7
Mozambique	405	2,670	2,225	53
Nigeria	115	793	1,500	97
Sudan	132	3,965	7,980	78

Source: Country inventories from Deininger et al. (2011)

Total transfers in 2004 to 2009 amounted to 4 million hectares in Sudan, 2.7 million in Mozambique, 1.2 million in Ethiopia, and 1.6 million in Liberia (mainly renegotiation of existing agreements). Virtually everywhere, local investors, rather than foreign ones, were dominant players, although some of them may of course be supported by, or act as a front for, outsiders. In most cases, expected job creation and net investment were either not recorded consistently or were low. Often, land was not fully used, as in Mozambique where an audit found that about half of transferred land was either entirely unused (34 percent) or noncompliant with the anticipated time schedule (15 percent). The amount of land transferred is also affected by policy: in Tanzania, where land rights are firmly vested with villages, less than 50,000 hectares were transferred to investors in the same time period.

Case studies of specific investments in DRC, Liberia, Mexico, Mozambique, Tanzania, Ukraine, and Zambia highlight that economically viable investments could provide benefits through four channels: social infrastructure, often supported by community development funds using land compensation; generating employment and jobs; providing access to markets and technology for local producers; and higher local or national tax revenue (Deininger and Byerlee 2011). The structure of benefits will have profound distributional consequences and should thus be considered in advance. In practice, benefits were limited and often land acquisition deprived local people, in particular the vulnerable, of their rights without clear compensation. Consultations, if conducted, were superficial and did not result in written agreements so that environmental and social safeguards were neglected, even if they had been considered initially.

To investigate what determines land demand and the direction of investment, we draw on the literature to assess factors that increase a country's attractiveness to investors. Traditional variables in "gravity models" include physical, cultural, and geopolitical proximity; we complement these factors with evidence on investor countries' dependence on food imports and target countries' potential for increased agricultural production. We also include standard investment climate measures for the quality of the legal and regulatory environment and an index of land governance which we construct from a cross-country database by the French Development Agency (de Crombrugghe et al. 2009). This index is interpreted as an indicator of overall tenure security of local users, with low values of the index describing countries with high levels of tenure insecurity.[10]

How good land governance will affect a country's attractiveness for agricultural investment is an empirical issue that is largely related to the balance between quick extraction and long-term investment. Security of property rights will be a key determinant of long-term investment decisions as investors will not tie up major resources in countries where weak or unclear rights may lead to conflict or expropriation (Schnitzer 1999) once investments are sunk. On the other hand, approval may be quicker and extractive exploitation easier where property rights and the state's capacity to enforce them are weak. Some investors are quite clear about their ability to enforce property rights through private militias, despite the problematic historical precedents.

For the unilateral case, results from regressions of the number of large-scale land acquisitions in a country of destination suggest a number of regularities.[11]

First, the potentially cultivable area outside of forests or the value of potential output of such areas (but not in forests) is highly significant throughout, pointing towards land availability as a key motivation for such deals. A 10 percent increase of potential area or output value would increase the number of projects by between 5.1 percent and 7.1 percent. Second, in contrast to the literature on foreign direct investment, indicators of good governance, such as investor protection and the rule of law, are insignificant. Third, an economically meaningful and robust negative association emerges between investor interest and land tenure security.[12] Thus, instead of being contingent on strong protection of rights, demand for land acquisition is more pronounced in countries with weak protection of land rights. Weak land governance may attract investors because it may allow them to fast-track proposals or to defend properties without relying on the state—although historically, such arrangements have often had unfavorable consequences.

IMPROVING LAND AND RESOURCE GOVERNANCE

With shifting demand for various types of agricultural produce making land increasingly valuable, institutional innovation in land rights and governance will be needed to ensure that this resource is put to its best use, even if a country does not have much uncultivated land available for expansion. Critical improvements include recognizing and mapping existing rights, eliminating the potential for using the powers of eminent domain to acquire land for subsequent transfer to private parties, establishing or operating property registries to document ownership and facilitate voluntary transfers of land at low cost, and taxing either land or profits from firm operation to fund public goods that will enhance local economic activity and thus create incentives for further investment.[13]

Recognition of existing rights. Even if it has been occupied by local communities for a long time, much land in Africa is legally considered state land that can be transferred to investors without first going through a time-consuming process of ascertaining or compensating existing use rights. Supporters of awarding state land to investors have argued that the urgency of attracting investment makes it impossible to wait until land rights are determined and mapped. At the same time, the need to deal with informal occupants on land that was granted as supposedly unoccupied state land has emerged as an issue leading to abandonment of high-profile investments. Failure to resolve this issue early may thus merely postpone problems and, in addition, reduce transparency, competition, and the perception of fairness.

Recent examples show that, if legal provisions are in place, rights to large areas of land can be adjudicated quickly, cost effectively, and in a way that includes land-use planning, thus identifying areas that could be made available to outsiders. Starting in the 1990s Mexico, with technology that was much less advanced than what is available today, mapped and adjudicated rights to close to 100 million hectares (an area larger than Spain and France together) in less than a decade. The program significantly improved governance by having more than 18,000 communities (*ejidos*) establish internal by-laws, providing a basis for representation and interaction with the outside world, separation of powers, and establishment of internal controls, with support from central institutions. Contrary to initial fears, this process did not result in a wave of land sales but allowed locals to enter into contracts and joint ventures with outsiders and increased productivity. The key mechanism to raise productivity was to facilitate increased migration, which allowed more able farmers to expand their area (de Janvry et al. 2012) and augmented local welfare through remittances (Valsecchi 2010). More recently, Rwanda demarcated the entire country (more than ten million parcels) in a participatory process.

Identifying state land. In many countries legal provisions require land intended to be transferred to investors to be first expropriated or converted into state land. As acquisition and divestiture of state land are key areas of concern in many countries, this opens the door to bad governance, but also means that even communities that are interested in transferring land to an investor or in establishing joint ventures are not able to do so or to benefit from the transfer directly. Good practice suggests that expropriation should be a last resort to prevent moral hazard and holdout problems by private owners, whereas transfers among private parties should be based on agreements between the parties. Legislation that requires expropriation as a precondition for transfers to investors or gives wide latitude to expropriate for transfer to private interests should be amended to give a clear rationale (e.g., in terms of environmental externalities) for declaring areas as state land, and follow this up with an inventory that unambiguously demarcates such lands on the ground, and with transparent mechanisms for divestiture of land that does not meet these criteria, with preference given to actual users.

Ensuring tenure security and allowing voluntary land transfers at low cost. To provide investment incentives and facilitate movement of land to its best use, it is critical that current and comprehensive information on assignment of property rights is broadly accessible, cost-effective, and verified on the ground. In addition to private land, coverage should include state and community land. Such documentation should allow low-cost registration of any transfers among private parties and include relevant contractual details. High levels of stamp duties, which in many instances act as a strong disincentive to the transfer of land to better uses or users should be lowered and be replaced with a regime of land taxation (which would provide incentives to bring land into use) or profit taxation with no loopholes.

Pricing and taxation to capture benefits from social infrastructure. The fact that, in many cases, investors pay lease fees below the opportunity cost of land discourages efficient land use and generation of local benefits. Taxation regimes to allow governments at different levels to benefit from increased land values due to public investment are crucial. While the potential for elite capture will have to be taken into account, the ability to raise revenue and decide on desired levels of service provision at the local level is a key feature of effective decentralization. Taxes on land or property are among the best sources of local revenues if local governments are allowed to retain a proper share of the tax revenues they collect, the technical means (for example, cadastres) are available, and clear principles are established for valuation to avoid arbitrariness.[14]

For countries that have significant amounts of unutilized land, improvements in land governance will be most effective if they are combined with ways to promote investment, and are in line with the country's comparative advantage and envisaged long-term evolution of operational structure. Improvements to achieve this include targeting investments to specific clusters, ex-ante screening of proposals, establishment of agile mechanisms for dispute resolution and arbitration, and options for investors to signal compliance with standards so as to allow them to access capital at lower cost.

Clustering and provision of complementary infrastructure. As the value of land that is not currently in use is largely determined by transport cost, targeting investment to areas with potential for complementary efforts (for example, in mining) to pay for the needed infrastructure can be a promising strategy. Land-related agricultural investment is sensitive to infrastructure access—including roads, markets, and technology—so the ad hoc and spatially dispersed approach to large-scale land-intensive investment that many countries have followed thus far is likely to bring limited benefits. Instead, areas with agricultural potential along existing or planned transport corridors can be a natural focal point toward which investment may be channeled. Efforts to fill in infrastructure, plan land use, and systematically document existing rights should initially be focused on such areas to reduce information cost and contribute to emergence of more transparent and competitive markets. Such clustering could also generate technology and marketing infrastructure that could help integrate smallholders into value chains.

Transparent screening of investment proposals. Local communities in many of the areas of interest to investors may lack the ability to assess the technical and economic viability of investments, identify key challenges associated with them, effectively negotiate complicated contracts, or enforce compliance with such contracts even if judicial infrastructure were accessible. Without ways to rigorously screen proposals for technical viability, this may result in ill-considered investments causing irreversible damage before closing down, bringing disappointment all around. A number of countries found such assistance to bring high returns. In Mexico, local communities' ability to draw on technical assistance and independent vetting of all contracts by the *Procuraduria Agraria* seems to significantly improve outcomes. Similarly, in Peru, a public auction process, together with independent professional vetting of proposals, helped attract investors by reducing red tape, and improved outcomes and local benefits. The institutions involved seem to function best if there is genuine and significant private-sector input, while avoiding potential conflicts of interest, with standardized contractual models to reduce transaction costs for all parties.[15]

Dispute resolution and arbitration. Many of the investments under discussion venture into novel territory in terms of technology and organization of production, and are conducted in an environment where rights are often vague and fluid. It would thus be surprising if their implementation did not involve some conflicts. The narrow margins from agricultural cultivation imply that any delays from such disputes can be very costly and potentially compromise the economic viability of an entire project. Avenues for quick and authoritative dispute resolution or bodies for arbitration that are recognized by both parties can thus be a key point affecting location decisions by investors. Having an agreed authority to arbitrate in case of conflict is one option that has been used effectively in other contexts. Similarly, criteria for failure of an investment should be clearly defined, along with ways to deal with existing assets without imposing undue burdens.

As large-scale farming is highly capital-intensive, the cost at which potential investors can obtain capital will affect their decisions. A rating or other system that factors risks—which may be specific to a country, a commodity, an investor, or a specific form of production—more systematically into the cost of capital could thus shape the nature and size of investment flowing into the sector. To allow this system to operate, responsible investors or countries must be able to signal their behavior to others. Voluntary commodity standards have been a widely used mechanism. Some of the lessons learned, such as the importance of third-party verification and participation by all relevant stakeholders in setting standards, could be drawn upon to monitor adherence to performance standards by agribusiness and countries, thereby reinforcing internationally recognized principles such as the voluntary guidelines of the Food and Agricultural Organization of the UN 2012.

The global community can play an important role in preventing a race to the bottom in extractive industries through independent review and monitoring that includes all stakeholders. The institutional environment and land governance at the country level are key factors. Relevant indicators could include the share of state land and the amount of state and individually owned land mapped and registered (possibly by gender), the size of areas expropriated and transferred to private parties, the number of land-related conflicts, and the number of registered land transactions (by size and ideally by gender). At the project level, these indicators could be matched with the amount invested, the number of jobs in different categories generated, the amount of resources transferred to local people, and the added value to the local or the global economy. Country investment authorities can track these indicators at low cost, which will not only allow early warning about dangers but will also help document investors' positive contributions to local economic development and welfare. Moving on these recommendations could greatly increase the likelihood of such investment helping countries and poor rural dwellers to realize—rather than undermine—their long-term economic potential.

ACKNOWLEDGMENTS

The paper benefited significantly from insightful comments by K. Anderson, C. Barrett, H. Binswanger, D. Byerlee, L. Christiaensen, C. Jackson, P. Jourdan, J. Lindsay, P. McMichael, W. Martin, B. Minten, C. Monga, A. Norton, A. Noman, H. Selod, J. Swinnen, and W. Wolford, as well as fruitful discussions during seminars at Cornell University, the World Bank, IAMO, the South African Ministry of Finance, and the African Union. The views expressed in this paper are those of the author and do not necessarily reflect those of the World Bank, its Executive Directors, or the countries they represent.

NOTES

1. In El Salvador, an 1856 decree states that communal land not at least two-thirds planted with coffee be considered underutilized to revert to the state for (free) allocation to investors willing to plant coffee. An 1882 land titling program then required occupants of communal (*ejido*) land to register their claims within 6 months by proving cultivation and paying a titling fee. Lands not claimed were to be sold at public auctions. As illiterate Indians were often unaware of these policies, extraordinary landownership concentration resulted, allegedly affecting some 40 percent of the country's territory (Lindo-Fuentes 1990). Communal land tenure was abolished in 1888. Similarly in Guatemala, an 1879 law gave users three months to register titles to their land, after which it would be declared abandoned, with much of the "abandoned" land then allocated to large coffee growers.
2. Even in cases such as the United States during the late 19th century, high land inequality at county level reduced taxes and thus investments in public education (Vollrath 2009).
3. Some 70 percent of Indonesia's oil palm plantations are on land previously part of the forest estate (Koh and Wilcove 2008). The government provided forested land almost for free. While timber sales were expected to finance planting and oil palm establishment, many companies allegedly used fictitious palm oil schemes to obtain logging licenses without ever establishing oil palm estates.
4. Smallholder-driven expansion of rice of about 10 million hectares since 1990 illustrates the scope for appropriate policies to set in motion a virtuous cycle of agro-export-led diversification. Policy reforms, clarification of property rights, and public investment were at the root of the shift from rice imports, for example the case of Vietnam becoming one of the largest exporters (Do and Iyer 2008).

5. Recent developments increased the scale of sugar mills to about 70,000 hectares from 20,000 hectares a decade ago, with a total investment of $1 billion.
6. Data used include the GLC2000 land cover, the IFPRI Agricultural Extent database, the FAO 2000 Global Forest Resources Assessment to identify land use, the 2009 World Database of Protected Areas to identify protected areas, and LANDSCAN 2003 data on population density to identify areas with less than 5, 10, or 25 persons per square kilometer, for example some 100, 50, or 20 hectares per household. Also, as market access will affect transport cost, we classify areas based on whether they are within six hours of an urban center with a population of at least 50,000 based on the World Bank's Global Mobility Database.
7. If Africa were to attain 80 percent of potential yield, a level usually considered economical, it could quadruple its maize output, equivalent to an area expansion of 90 million hectares at current yields.
8. See for example http://farmlandgrab.org/post/view/20405 (accessed March 23, 2013).
9. Note that the figure refers to overall demand for land deals rather than only actual transactions or area brought under production.
10. The main variables contributing to the first axis of the principal component analysis are listed below (contributions in brackets): "land tenure security" (16 percent), "public policies addressing land rights" (15 percent), "land ownership rights security" (14 percent), "diversity of tenure situations" (11 percent), "recognition by the State of the diversity of tenure situations" (10 percent), "scarcity of land-related conflicts" (10 percent), "traditional collective use and ownership" (9 percent), "significance of land use policies" (6 percent). This first axis captures 40 percent of variance.
11. See Arezki et al. 2011 for details.
12. Everything else constant, a one standard deviation drop in the land governance index—equivalent to the difference between Brazil and Angola—would be predicted to increase the number of investment projects by 33 percent.
13. See Deininger et al. 2011 for more details on a methodology to accomplish this.
14. Historically, use of land taxes has been effective in discouraging speculation and rent seeking in New Zealand, helping to break up a structure that had been dominated by large farms (Deininger and Byerlee 2012).
15. Peru established transparent auctions to transfer public land with strong and public technical vetting from the private sector. In the country's Pacific region, the transfer of 235,500 hectares of public land over the past 15 years in this way brought in $50 million in investment, creating the basis for the country's emergence as a major exporter of high-value horticultural products.

REFERENCES

Allen, D., and D. Lueck. 1998. The nature of the farm. *Journal of Law and Economics 41*(2): 343–86.

Alomar, R., and D. Cousquer. 2012. A proposal for setting up a global land purchase monitor. Washington, DC: Paper presented at the Annual Bank Conference on Land and Poverty 2012.

Anseeuw, W., W. Bache, T. Bru, M. Giger, J. Lay, P. Messerli, and K. Nolte. 2012. *Transnational land deals for agriculture in the global South: Analytical report based on the land matrix database.* Bern, Montpellier, Hamburg: CDE, CIRAD, CIGA, Bern University, ILC.

Arezki, R., K. Deininger, and H. Selod. 2011. What drives the global land rush. IMF Working Paper WP/11/251. Washington, DC: International Monetary Fund.

Banerjee, A., and L. Iyer. 2005. History, institutions, and economic performance: The legacy of colonial land tenure systems in India. *American Economic Review 95*(4): 1190–213.

Binswanger, H. P., and K. Deininger. 1997. Explaining agricultural and agrarian policies in developing countries. *Journal of Economic Literature 35*(4): 1958–2005.

——, and G. Feder. 1995. Power, distortions, revolt and reform in agricultural land relations. *Handbook of Development Economics 3B*: 2659–772.

Binswanger, H. P., and M. R. Rosenzweig. 1986. Behavioural and material determinants of production relations in agriculture. *Journal of Development Studies 22*(3): 503–39.

Butler, R. A., and W. F. Laurence. 2009. Is oil palm the next emerging threat to the Amazon? *Tropical Conservation Science 2*(1): 1–10

Collier, P., and A. J. Venables. 2011. Land deals in Africa: Pioneers and speculators. Discussion Paper 8644 Centre for Economic Policy Research, London.

de Crombrugghe, D., K. Farla, N. Meisel, C. de Neubourg, J. Ould Aoudia, and A. Szirmai. 2009. Institutional profiles database III. Technical report, Agence Francaise de Developpement (AFD), Maastricht Univeristy (UM), and CEPII, 2010. Presentation of the Institutional Profiles Database 2009.

de Janvry, A. 1981. *The agrarian question and reformism in Latin America.* Baltimore: Johns Hopkins University Press.

——, K. Emerick, M. Gonzalez-Navarro, and E. Sadoulet. 2012. Certified to migrate: Property rights and migration in rural Mexico. Working paper, University of California Berkley.

Deininger, K. 2003. *Land policies for growth and poverty reduction: A World Bank policy research report.* Oxford and New York: World Bank and Oxford University Press.

—— and D. Byerlee. 2011. *Rising global interest in farmland: Can it yield sustainable and equitable benefits?* Washington, DC: World Bank.

—— 2012. The rise of large farms in land abundant countries: Do they have a future? *World Development 40*(4): 701–14.

—— K., H. Selod, and A. Burns. 2011. *Improving governance of land and associated natural resources: The Land Governance Assessment Framework* Washington, DC: World Bank.

Do, Q. T., and L. Iyer. 2008. Land titling and rural transition in Vietnam. *Economic Development and Cultural Change 56*(3): 531–79.

Drache, H. M. 1964. *The day of the bonanza: A history of bonanza farming in the Red River Valley of the North.* Fargo, ND: Institute for Regional Studies.

Fairhurst, T., and D. McLaughlin. 2009. *Sustainable oil palm development on degraded land in Kalimantan.* Washington, DC: World Wildlife Fund.

Farigone, J., J. Hill, D. Tilman, S. Polasky, and P. Hawthorne. 2008. Land clearing and the biofuel carbon debt. *Science 319*(5867): 1235–8.

Fischer, G., H. V. Velthuizen, M. Shah, and F. Nachtergaele. 2002. *Global agro-ecological assessment for agriculture in the 21st century: Methodology and results.* Laxenburg and Rome: IIASA and FAO.

Food and Agricultural Organization of the UN. 2012. *Voluntary guidelines on the responsible governance of tenure of land, fisheries and forests in the context of national food security.* Rome: FAO.

Gardner, B. L. 2002. *US Agriculture in the 20th century: How it flourished and what it cost.* Cambridge, MA: Harvard University Press.

Government of Sudan. 2009. *Study on the sustainable development of semi-mechanized rainfed farming.* Khartoum: Ministry of Agriculture and Forestry.

Grandin, G. 2009. *Fordlandia: The rise and fall of Henry Ford's forgotten jungle city.* New York: Metropolitan Books.

Hayami, Y. 2010. Plantation agriculture. In *Handbook of agricultural economics*, ed. P. L. Pingali and R. E. Evenson, 3305–3322. North Holland: Elsevier.

Herath, D., and A. Weersink. 2009. From plantations to smallholder production: The role of policy in the reorganization of the Sri Lankan tea sector. *World Development 37*(11): 1759–72.

Iyer, L. 2010. Direct versus indirect colonial rule in India: Long-term consequences. *Review of Economics and Statistics 92*(4): 693–713.

Johansen, B. 1988. *The Islamic law on land tax and rent: The peasants' loss of property rights as interpreted in the Hanafite legal literature of the Mamluk and Ottoman periods.* Exeter Arabic and Islamic Series. London, New York and Sydney: Croom Helm. Distributed by Routledge, Chapman and Hall.

Johnson, D. H. 2003. *The root causes of Sudan's civil war.* Bloomington, IN: Indiana University Press.

Koh, L. P., and D. S. Wilcove. 2008. Is oil palm agriculture really destroying tropical biodiversity? *Conservation Letters 1*(2): 60–4.

Lindo-Fuentes, H. 1990. *Weak foundations: The economy of El Salvador in the nineteenth century.* Berkley: University of California Press.

Lissitsa, A. 2010. The emergence of large scale agricultural production in Ukraine: Lessons and perspectives. Paper presented at the Annual Bank Conference on Land Policy and Administration, Washington, DC, April 26 and 27, 2010.

McCauley, M. 1976. *Kruschev and the Development of Soviet Agriculture: The Virgin Land Programme 1953–1964.* London: Macmillan.

MacDonald, J. M. 2011. Why are farms getting larger? The case of the USA. Paper presented at the 51st Annual Conference, Halle/Saale, Sept. 28–30, 2011.

Manciana, E. 2009. Large scale acquisition of land rights for agricultural or natural resource-based use: Argentina. Draft paper, World Bank, Buenos Aires.

Mason, T.D. 1986. Land reform and the breakdown of clientelist politics in El Salvador. *Comparative Political Studies 18*(487): 516.

Nugent, J. B., and J. A. Robinson. 2010. Are factor endowments fate? *Revista de Historia Economica 28*(1): 45–82.

Pacheco, P., and R. Poccard Chapuis. 2009. *Cattle ranching development in the Brazilian Amazon: Emerging trends from increasing integration with markets.* Bogor, Indonesia: Center for International Forestry Research.

Pearce, F. 2012. *The Landgrabbers: The New Fight over who Owns the Earth.* London: Transworld Publishers.

Rajan, R., and R. Ramcharan. 2011. Land and credit: A study of the political economy of banking in the United States in the early 20th century. *The Journal of Finance* 66(6): 1895–931.

Rezende, G. C. D. 2005. Politicas trabalhista e fundiaria e seus efeitos adversos sobre o emprego agricoloa, a etrutura agraria e o desenvolvimento territorial rural no Brasil. Texto para discussao No 1108. Rio de Janeiro: Instituto de Pesquisa Economica Aplicada (IPEA).

Rist, L., L. Feintrenie, and P. Levang. 2010. The livelihood impacts of oil palm: Smallholders in Indonesia. *Biodiversity and Conservation 19* (2010): 1009–24.

Rogers, P. 2008. The "failure" of the Peak Downs scheme. *Australian Journal of Politics and History 10*(1): 81–115.

Schnitzer, M. 1999. Expropriation and control rights: A dynamic model of foreign direct investment. *International Journal of Industrial Organization 17*(8): 1113–37.

Valsecchi, M. 2010. Land certification and international migration: Evidence from Mexico. Working Papers in Economics No 440, University of Gothenburg.

Vollrath, D. 2009. The dual economy in long-run development. *Journal of Economic Growth 14*(4): 287–312.

World Bank. 2009. *Awakening Africa's sleeping giant: Prospects for competitive commercial agriculture in the Guinea Savannah zone and beyond.* Washington, DC: The World Bank.

5

Global Freshwater and Food Security in the Face of Potential Adversity

UPMANU LALL

In the next two decades, as the world's population is projected to pass nine billion, reliable and timely access to water of appropriate quality will prove essential for delivering high crop yields and meeting food production targets. Agriculture is the largest global user of water, accounting for approximately 85 percent of all water consumed. Consequently, the depletion and degradation of water resources in many countries threatens food production and security. Groundwater levels in northern China, South Asia, Mexico, southern Europe, western, southeastern, and midwestern United States, and the Middle East and North Africa have been declining for the past half-century, with the trend accelerating in the last two decades (Konikow and Kendy 2005; Wada et al. 2010). At the same time, serious cumulative effects of pollution and water withdrawal on the world's major rivers and freshwater lakes have emerged in areas including the Nile Basin, Lake Chad, the Amu and Syr Darya and the Aral Sea, the Murray-Darling basin in Australia, all major rivers in India and China, the Rio Grande, and the Mississippi and Colorado river basins. Local and national food security concerns have contributed to increases in regional agricultural intensity and to the increasing use of irrigation, fertilizers, and pesticides to achieve higher crop yields. Policies that subsidize physical and chemical agricultural inputs have led to inefficient resource use and widespread impacts on water systems. Correcting this situation is essential for regional and global water and food security in the twenty-first century.

Climate variability strongly impacts the supply of and demand for regional water resources. It directly influences water availability, and is a primary factor in low agricultural productivity or resource-induced conflict in areas that are water stressed or that experience high rainfall variability at seasonal and longer time scales. Changes in temperature, cloudiness, winds, and humidity also influence agricultural water demand. Many countries have invested in centrally financed and managed water storage and irrigation infrastructure and provide subsidies for irrigation water access in recognition of the importance of water for food production and for rural livelihoods,. Countries such as China and India made significant progress toward food security on the basis of such investments and policies, but persistent and deepening water scarcity now appears to threaten these gains. Droughts and floods in key growing regions have contributed to global food price shocks in recent years, leading to speculation that freshwater and food security may be compromised under future climate change scenarios, which project increased risk of water demand exceeding supply, flooding, and increasing aridity in the tropics and subtropics, where high population densities, poverty, and water and food security are already issues. Regional and transboundary conflict over access to declining and less reliable water resources also seems likely to increase as the climate changes.

The trends towards increasing groundwater use reflect efforts to buffer climate variability through irrigation during rainfall-deficient periods, as well as the lack of reliable access to surface freshwater resources such as canals, lakes, and rivers. Groundwater extraction for irrigation uses tremendous energy, especially in areas where water levels are dropping rapidly. Emerging regional water shortages constrain energy production, whether the objective is biofuel production, cooling for thermal power plants, or mineral processing and extraction. Consequently, many regions will compete progressively for water for agriculture and energy, leading to a global dialogue on the nexus of water, food, and energy.

WATER AS A RESOURCE

Water has some special attributes as a resource, especially in the context of agriculture and food security. The global hydrological cycle determines the availability of water, especially the annually or seasonally renewable supply at specific places in the world. This is an expression of climate and is quite variable in space and time. Average annual rainfall ranges from near zero in the desert regions to over 400 inches in the humid tropics. The probability of precipitation ranges from roughly equal for any calendar day in coastal, mid-latitude settings such as New York, to 10 to 15 rainy days during a 90-day monsoon season in the tropics. Precipitation that falls in a river basin constitutes the upper bound of its renewable water supply. At the land surface, precipitation is partitioned into infiltration or water that goes into the soil, and overland flow that is subsequently manifest as river flow. The near-surface soil moisture, which is available in the top one or two meters of soil, is the store that is directly available to vegetation and crops. If this is the only or primary source of water available for crops in a region, then the associated agriculture is considered rainfed. Plant growth rates, and hence crop yields, are typically proportional to the ability of the plant to maintain its potential rate of evapotranspiration given the ambient temperature, solar radiation, and wind. If the available soil moisture cannot support this rate, then the plant experiences water stress, and growth and crop yields are diminished.

Some investigators (e.g., Rockström et al. 2009) have termed the rainfed water available to a crop as "green water." There are virtually no regulations or pricing structures in place to limit access to green water collected on one's own property—it is typically free of cost to the user. However, as the land area under agriculture increases, and farmers employ local storage and tillage methods to increase the retention time for water on their land, the water balance for the river basin is affected, river flows may be reduced, and downstream users who rely on those flows will likely be impacted.

More than 98 percent of the world's liquid freshwater is stored as groundwater. Shallow groundwater sources lie at depths of less than 30 meters from the surface. They represent the limit to which manual or centrifugal pumps can extract water. Well-drilling costs also increase substantially beyond such depths. The age of shallow groundwater is typically one season to ten years, reflecting the average time it takes for the recharge to maintain the aquifer. Deeper groundwater bodies receive considerably less annual recharge, and require substantial energy to pump to the surface. Water at a depth of 200 meters may be more than a thousand years old. In short, soil moisture is replenished at daily or weekly time scales during the rainy season, shallow groundwater at decadal time scales, and deep groundwater at millennial time scales.

On the other hand, the largest volumes of freshwater on earth are held in groundwater—and much of that is deep. The naturally renewable supply is "free" in terms of soil moisture, but is progressively more expensive to develop and also smaller in volume as one goes to depth. It is useful to think of large-scale extraction from deep aquifers as similar to mining, since the rate of replenishment is typically smaller than the rate of extraction.

A useful metric of sustainability (Vörösmarty et al. 2000) for water is whether the average regional water use is less than two-fifths of the average annual renewable water supply. The remaining three-fifths is needed for ecosystem functions and as a buffer for dry years. In a variable climate, as resource use approaches the mean annual supply, the frequency and severity of disruptions will increase, leading to lower food productivity. The variability in the regional renewable water supply, in addition to the average supply, becomes important as a determinant of the storage needed to buffer against climate shocks. This variability is typically higher for within year and across year supply variations in the monsoonal climates of the tropics and subtropics, leading to a greater need for investment in buffering shortfalls in supply in these regions. However, to date, the per capita availability of constructed water storage—one of the buffering mechanisms—is dramatically higher in the more economically developed, high-latitude regions that typically have lower variability in the renewable water supply.

Irrigation using pumped groundwater or a system of surface water reservoirs and canals is called "blue water" (Rockström et al. 2009). This refers to a development of the natural water sources, and requires both investment and operational expenses. Surface reservoir and canal systems were developed in many countries post-World War II and contributed to agricultural gains. Many of these reservoir systems are multi-purpose: they provide irrigation, flood control, hydropower, navigation, fisheries, and recreational services. The irrigation releases are often routed through turbines leading to the ancillary production of energy.

Due to social and environmental issues associated with large dams and the diversion of flows, project cost over-runs, and poor management and control of delivered waters, with the exception of a few countries like China, organized opposition has led to a significant decline in such investments after 1980. Vörösmarty et al. (2010) argue that nearly 80 percent of the world's population is now exposed to threats to water security, and global biodiversity is threatened by the modification of over 65 percent of the world's river flows. In many cases, these negative impacts are synergistic. For instance, in the Tana River basin in Kenya, the modification of downstream flows by dams led to a significant decline in fisheries, which in turn led to a loss of the primary protein source for the regional population. The effectiveness of monitoring and enforcement of canal water rights has also emerged as a significant problem. Theft or use beyond allocated water from canals is common in many settings, and tail-enders on the canals rarely receive reliable supply outside the high-income countries. When the canal supply is unreliable, users are unwilling to pay, and effective cost recovery for the operation and maintenance of canal irrigation systems become a challenge, leading to their eventual degradation in much of the developing world.

A parallel trend since the 1980s is a dramatic increase in shallow and deep groundwater pumping in the emerging economies in the monsoonal regions with high climate variability—particularly South Asia and China. The initial investment is often private, and applies to well drilling and pump purchase. Improved access to energy sources through rural electrification and petroleum product supply chains has facilitated these developments.

Allocating and monitoring groundwater rights is not easy. Detailed hydrogeologic investigations to characterize the resource and to project possible adverse impacts do not come cheaply. Specialized institutions that can conduct such analyses and make them accessible to decision makers are also needed. Transaction costs associated with monitoring, metering, and billing millions of individual farmers in developing countries tend to be very high, and this has contributed to a trend toward fixed-charge pricing for electrical connections to power pumps in many locations. Given concerns for food security and the livelihood of smallholder farmers, the fixed charges are typically highly subsidized relative to the value of the electricity consumed. A natural consequence of the lack of a unit-cost pricing structure has been profligate use of groundwater, which in turn has affected the reliability of the regional electricity grid and encouraged inefficient water use in agriculture. Political factors help sustain the established system, since disrupting entitlements granted to a large segment of the population is politically unpopular. Governance of groundwater systems remains a global challenge.

Water development also leads to spatial and temporal externalities that have to be managed. Politically, water prices have been held low, so the economics of large-scale, long-distance water transfers—especially in an upstream or uphill direction—usually do not work. Exceptions exist: for instance in China, where the state chooses to invest in such measures and subsidize the result. The potential spatial impacts of development coupled with the cost of long-distance transfers limit regional water opportunities and translate into the primacy of local rather than global issues in many water security discussions. Interestingly, transboundary water conflicts are usually resolved with treaties and agreements, whereas local water issues can often emerge as strong political drivers and as agents of sectoral or group conflict. Droughts, energy prices, and disruptions of flow in combination with existing political instability are potential triggers of local and regional conflict.

"Virtual water trade" refers to the effective trade of water embedded in agricultural, energy, or other processed products that use water. Given that the water use intensity of different crops, textiles, manufactured goods, and energy products varies significantly, as does water availability, an effective spatial allocation of crops in line with water endowment and free trade of the resulting products could in theory reduce water and food stress. The current evidence, however, is that theory water exports may take place from water-stressed to relatively water-rich areas and are determined by productivity rather than stress.

As the Green Revolution advanced in the 1960s and beyond, agricultural research and production systems related to cereals—especially, rice, wheat, and maize—were significantly strengthened. The yields of these cereals are now nearly ten times higher in research settings than in many developing country field settings where large land areas are cultivated. Unfortunately, dryland cereals, such as sorghum and millet, and native cultivars in arid regions did not get a corresponding boost, and rice, wheat, and maize were propagated into many of the areas that were biophysically not well suited for these crops. The net result has been a dramatic increase in water stress and nutrient cycling through soil and water in much of the world. Dietary patterns have correspondingly changed, with the traditional diet relegated to ceremonial settings in many places. Food choices influence the water regimen, which in turn influences food productivity. Projections of future water- and climate-induced stress on food security typically rely on scenarios that extrapolate recent trends. The growing competition for water from the industrial and energy production sectors is likewise important to keep in mind, since these will likely be considered higher economic value uses than irrigation, at least in terms of what these industries are willing to pay for the resource.

Perhaps the most publicized global water issue through the United Nations Millennium Development Goals is that nearly a billion people lack access to safe drinking water. Drinking water links to food security, viewed broadly as the ability to provide an adequate quantity of the essential nutrients to all humans. The total quantity of water needed for direct human consumption is an insignificant fraction of the quantity used for food production and other economic uses, yet this water has to be high quality and free of biological pathogens and chemical contaminants. Unfortunately, over much of the world, agricultural pollution of water resources makes safe drinking water for all a much harder goal to achieve. Oddly enough, in some regions—for instance, in South and East Asia, which are nominally classified as having met the safe drinking water targets—very high quality groundwater is pumped in large quantities from depth for irrigation, and shallow, polluted groundwater is used as a drinking water source. A nutrition security goal that encompasses water for both drinking and food production may stimulate synergy in these areas, translating into better outcomes on both.

WATER AVAILABILITY AND INFRASTRUCTURE

The large spatial variability observed in mean annual precipitation across the world prescribes a variable upper bound on the local renewable water supply. In general, the spatial distribution of global population matches that of mean annual precipitation, revealing the important role that water endowment plays on human habitation. Interannual variation in precipitation is higher in the tropics and sub-tropics than in the mid to high latitudes. The Middle East and North Africa region, Central, West, and South Asia, northern China, western South and North America, northeast Brazil, East and southern Africa, and Australia stand out as areas with high climatic variability that potentially needs buffering irrigation.

By and large, the spatial distribution of groundwater recharge rates parallels the location of major river basins in the world and the areas with higher precipitation. Despite having relatively high mean annual rainfall, however, South Asia and northern China have dramatically lower renewable groundwater endowments relative to their population. In conjunction with the high variability in precipitation, these regions are of particular concern for regional water stress and food security.[1]

On the basis of global averages of water stocks and flows and groundwater recharge, and the recognition that 85 to 90 percent of the global water consumption and 70 percent of the total withdrawals of water are estimated to be for agriculture, water does not emerge as a significant constraint on present food production—or even on a 30 percent higher future food demand. Irrigated water is 16 percent of the renewable river and groundwater discharge, and rainfed agriculture consumes 33 percent of the annual precipitation less forest evapotranspiration. The fact that there is a current concern with water scarcity consequently reflects, not the total annual flux or storage available for water, but the ability to access this water. The question is how water's variability across regions and seasons can be best managed globally to provide reliable access for food production and other uses, particularly when population growth will be heavily concentrated in today's low-income tropics and subtropics, especially in Africa and Asia.

Experts predict only modest climate change influenced by human activity over the next two decades. In the absence of effective climate change mitigation, however, the variability of precipitation will increase globally, compounding the water stress agriculture faces in a warmer climate. Most climate model projections suggest a general drying in the subtropics and tropics and increased rainfall in higher latitudes. This will exacerbate the already severe situation in the arid and monsoonal regions, leading to increased pressure on groundwater resources. The temporal variability of climate is an important concern driving water and food security regionally and globally. One question is whether regional water storage and use strategies are robust enough in the event of recurrent or multiyear droughts. Could these adaptation strategies be improved if the climatic exigency could be anticipated through forecasts?

The importance of reducing volatility in water availability to secure food supplies has been appreciated for millennia. The history of surface water storage through the construction of dams and reservoirs, and the subsequent distribution by canals of the stored water, goes back to 3000 BC. Evidence of the earliest such structures has been found in Jordan and in the Tigris-Euphrates basin in the Middle East. Development of groundwater for agricultural use through qanats and wells dates back to approximately the same period.

Lehner et al. (2011) provide the most complete inventory of contemporary dams and reservoirs, including extrapolated estimates for the total storage volume in smaller undocumented sites. They estimate that there are over 16 million reservoirs in the world, with a total storage capacity in excess of 8,000 cubic kilometers, or about 16 percent of the total annual river flow. The most striking aspect of the spatial distribution of dams is the relatively high density of reservoir storage (and hence irrigation potential) in Europe, North America, and parts of South America relative to Africa, given that these regions are comparatively well endowed with water with lower variability.

Lempérière (2006) indicates that 95 percent of the global investment in dams occurred after 1950, but the construction of dams—except very large dams—has decreased dramatically in the last 20 years. Fully half of the world's reservoir storage is associated with just 100 very large dams. Given that most dams have multipurpose reservoirs, he argues, the total storage dedicated to irrigation may only be of the order of 1,000 cubic kilometers, but since multipurpose dams provide additional benefits, storage capacity for irrigation may double globally by 2050 (Lempérière 2006). With the prospect of rising food prices and climate variability, dam construction is likely to proceed, at least in areas where construction costs are low. Often there are economies of scale with dam construction, yet significant expenses and delays are incurred in environmental and social impact assessments, and a strong social movement opposes dams. At the same time, the lack of effectiveness of canal systems to deliver reliable supply in many developing countries has been raised as a governance issue. Thus, continued public investments in surface water storage and delivery may not increase at these projected rates.

In contrast to large dam and canal projects that are publicly financed, private investment has focused on groundwater development and small on-farm storage mechanisms. Shah (2009) estimated that the capital investment by individuals in groundwater irrigation in India since 1980 is equal to nearly three-quarters of the total public investment in surface water irrigation in the country since 1950. Indeed, more than one million new wells a year have been added in India since 1990. Given the proliferation of microfinance schemes and nongovernment organizations that support small-scale water development, one can expect similar rapid growth in both small surface storage development and groundwater development in Asia, Africa, and South America.

Finally, in areas where groundwater depletion is an issue, there are efforts to store water by increasing groundwater recharge to shallow and deep aquifers by artificial means. The clear advantages are that no water is lost to evaporation from a groundwater store and a certain amount of filtration is achieved in the recharge process. Nonpumping solutions can be low cost and may be particularly effective in shallow aquifers where topography is available to develop recharge. On the other hand, clogging due to biological activity and chemical reactivity is a problem, and given that groundwater movement is slow, recharge rates in many settings may be too slow. Further, energy must be expended to recover the water, and those who inject it in the ground may have little control over who recovers it. Thus, there are governance challenges with this storage approach if done at scale.

BLUE WATER

Some 20 percent of the world's arable cropland is under irrigation, with irrigated crops accounting for 40 percent of the global harvest (Bruinsma 2003). Nonrenewable groundwater extraction is estimated to have tripled between 1960 and 2000. Rice, wheat, and maize use the most water (Siebert and Döll 2010). Blue water use as percentage of total crop water use was highest for date palms (85 percent), cotton (39 percent), citrus fruits (33 percent), rice (33 percent), and sugar beets (32 percent). More than two-thirds of the world's irrigated land is located in Asia, with hardly any irrigation in sub-Saharan Africa.

Soil, crop types, and regional economic and labor markets determine the methods used for irrigation in different parts of the world. Estimates of total irrigation consumption and withdrawals vary widely, reflecting a lack of direct measurement. The uncertainties are particularly large for groundwater withdrawals, since there is little accessible data on usage or even on groundwater levels that would provide a coarse check on total withdrawal. On average, 25 percent of the water withdrawn for irrigation is used by the crop, 56 percent likely infiltrates back into the soil or runs off and may be available to others to use, and 19 percent is unproductively lost or evaporated from the soil (Sauer et al. 2010). The 56 percent of return flow may not be reliably or economically available to other users. Addressing these huge inefficiencies in irrigation promises a significant opportunity for productivity increases.

Sauer et al. (2010) found significant regional variations in the deployment of different irrigation technologies and their water use efficiency. The most common method of irrigation is basin and furrow—flooding the field—followed by sprinkler, and then by drip. The major cereal crops that cover the largest land areas are predominantly grown using basin and furrow. Water application efficiency ranges from 20 to 55 percent for basin and furrow, 60 to 86 percent for sprinkler, and 80 to 93 percent for drip. The lowest water use efficiencies for any method are in the Middle East, North Africa, and South Asia, and the highest are in Europe and North America.

Siebert and Döll (2010) estimate that globally the average yield of irrigated cereals is 442 metric tons per square kilometer, and of rainfed cereals is 266 metric tons per square kilometer. In their model they explore how total crop production would change if irrigation was removed. Cereal production on irrigated land would decrease by 47 percent, leading to a 20 percent loss of total cereal production. The largest cereal production losses are projected for northern Africa (66 percent) and South Asia (45 percent). The model projects very low losses for northern Europe (0.001 percent), western Europe (1.2 percent), eastern Europe (1.5 percent), and Central Africa (1.6 percent). This reflects the need for and distribution of irrigation for these crops—a significant regional food security risk.

A near doubling in irrigated area in the last 50 years has coincided with an increase in crop yields at a rate faster than population growth, leading to a net decline in real food prices over the period (Molden 2007). However, the rate of growth in irrigated areas has slowed to near zero in the last decade, and the growth rate in yields in many regions has consequently also slowed. As yield increases, so does water productivity, suggesting that improved practices lead to higher production for the same water applied. They argue that the greatest opportunity for water productivity lies in improving the yields at the low end, which are typically in poorer countries. However, it is interesting to note from their figures that water productivity for the low-performing settings (500 kilograms per hectare to 2,500 kilogram per hectare) ranges from 0.1 to 1 kilogram per cubic meter, whereas, for the high-yield regions (5,000 to 8,000 kilogram per hectare) the productivity range is 0.5 to 2.5 kilogram per cubic meter. This suggests that in terms of water productivity there is considerable room for improvement at both the high- and the low-yield ends. Rice typically has the lowest water productivity, and hence one opportunity is to shift to other cereals or potato to obtain the same calories. If we consider today's food security acceptable, then even if the population grows by 30 percent to nine billion and per capita consumption increases by 50 percent, one could in principle meet these demands by doubling productivity in the areas that currently have low yields and water productivity. Bridging the dry spells in the rainy seasons in Africa, Asia, and South America, and introducing efficient irrigation systems is the contribution needed from the water sector.

Hoekstra (2012) estimates that using developed country productivity levels and consumption of 3,400 kilocalories per day per person, a diet that includes meat and vegetables leads to a water footprint of 3,600 liters per day, versus 2,300 liters per day for one that uses dairy and vegetables. This indicates a saving of 36 percent of the water by shifting between these two diets. This idea can be extended to a consideration of the specific types of water use that make up a water footprint and their implications for regional water stress. Specifically, the implications of a regional water footprint might be best judged in terms of green versus blue water use and the timing of the water use.

Mekonnen and Hoekstra (2011) use a grid-based water balance model to compute the water footprint associated with many crops and derived products. Countries with the largest water footprints are India, China, and the United States. The Indo-Gangetic plains in South Asia account for 25 percent of the blue water footprint related to crop production. They experience a large blue water footprint throughout the year, since multiple crops are planted and irrigated over the year. The blue water footprint is large for only part of the year in Tigris-Euphrates, Huang He (Yellow River), Murray-Darling, Guadiana, Colorado, and Krishna river basins (Hoekstra et al. 2012). The significance of the temporal variation in the water footprint is that a seasonal footprint could conceivably be met by rain or river flow stored within the season or from the prior season, whereas in monsoonal settings such as the Indo-Gangetic plains, which have a year round high blue water footprint and a short rainy season, groundwater extraction is a likely consequence.

Irrigation imposes greater environmental stress and costs more than rainfed agriculture. Since the majority of the world's croplands are still classified as rainfed, improving water productivity in rainfed agriculture would make a significant contribution to water and food sustainability, and is needed as advocated by Rockström et al. (2009). However, irrigation and supplemental irrigation have enormous value for buffering climate variability and boosting crop yields. There has and continues to be significant public and private spending on irrigation systems. Unfortunately, a corresponding effort to improve irrigation application systems for higher water efficiency has been lacking. Government efforts often go towards subsidies or incentives to promote drip irrigation, a technology that is applicable to certain cropping systems and can deliver high water use efficiency. However, drip irrigation requires a pressurized delivery system and has higher capital, operation, and maintenance costs. Relatively inexpensive soil moisture monitoring systems (Perveen et al. 2012), and irrigation practices such as mulching, alternate furrow irrigation, facilitated by appropriate government incentives and regulation could deliver 30 percent savings in water applied for rice irrigation, which is a major water consumer. A key aspect of achieving better water use and crop yields is an effective agricultural extension program. Those who cannot afford to implement better technologies or lack access to knowledge through effective extension programs are locked in a productivity and poverty trap, even if inexpensive water is available to irrigate the crop. For instance, in India, neighboring areas can have crop yields that differ by a factor of two to five, largely due to differences in farmer practices mediated by extension programs.

WATER STRESS, TRADE, AND CONFLICT

Estimated water requirements for meeting future food needs vary dramatically, depending on assumptions about dietary trends, population growth, climate change, global cropping patterns, yields, and water-use technologies. Yet one message that always comes through is that significant changes with respect to agricultural water productivity will be needed if future food needs are to be met reliably. Most recent model-based estimates agree that, unless there are significant changes in where different crops are grown and in the water productivity of both green and blue water systems, water stress will likely be magnified significantly in the main agricultural regions of the world that are already water stressed. The range of cereal yields alone across the world varies by a factor of ten, and for a given level of yield, the water use varies by a factor of five (Molden 2007)—so the good news is that, from a technological or biophysical perspective, there is quite a bit of room for adaptation to the situation.

Given the disparities in water availability and population around the world, a question that arises is whether trade can address regional water and food stress. Dalin et al. (2012) estimate that the total volume of water traded in 2007 was 567 cubic kilometers per year, or about 22 percent of their estimate of the total global freshwater withdrawal for agriculture. This represents global water savings of 9 percent over the amount of water that would have been used if the crops were grown in the region of consumption. Dramatic increases in water savings associated with wheat, corn, and soy are noted. Increasing virtual water trade could increase opportunities for regions that are more efficient producers to export to less efficient regions. However, this trade may not help the importers improve their productivity, so water stress in the importing country may not improve (Allan 1998; D'Odorico et al. 2010).

Virtual water trade nearly doubled between 1986 and 2008, exceeding the rate of population growth. About 40 percent of the global virtual water exports now come from just three countries: Brazil, Argentina, and the United States (Carr et al. 2012). These countries are exporters of meat and cereal, both of which represent large water footprints. Interestingly, many of the virtual water exporters are also water-stressed countries, such as India. Even within India, the semi-arid northwest provides much of the rice that is consumed in the eastern part of the country that has nearly five times as much mean annual precipitation. Productivity, rather than water stress, appears to drive virtual water trade, and in this sense, the trade may exacerbate regional water stress in the near term.

Scholars and policy makers have expressed concern that growing water shortages induced by climate change or by a regional imbalance between supply and demand may trigger wars or transboundary conflict. In reality, nearly 80 percent of the incidents related to water reported in the International Crisis Behavior database pertained to government rhetoric rather than to actions. There is much stronger evidence for cooperation than conflict on transboundary water issues, with more than two-thirds of the water-related events in the Basins at Risk database falling on the cooperative scale (Yoffe et al. 2003). At subnational scales the link between water scarcity and conflict may be more complex. Allouche (2011) cites studies that indicate that countries with high population density and environmental degradation, including deterioration of fresh water supplies, may have an increased likelihood of civil wars. In summary, the way water is used or allocated, rather than absolute scarcity, is undoubtedly the most direct progenitor of conflict. Consequently, concerns regarding conflict induced by water stress could be addressed pre-emptively through the institution of appropriate water allocation and sharing principles that are agreed to, and the development of physical and institutional infrastructure to help implement these principles. This would include monitoring systems for water supply (quantity and quality) and use; the public disclosure of the resulting information; and pipes, pumps, and control systems that can assure delivery to the designated users.

STRATEGIES FOR SUSTAINABILITY: TECHNOLOGICAL SOLUTIONS

Many regions are struggling with serious water scarcity due to an imbalance between supply and demand. Renewable water supplies are subject to climate-induced variability. A traditional approach to addressing this situation is to increase water storage. The twentieth-century emphasis on developing large dam and canal systems led to significant ecological and social impacts and is now in disfavor. Dam and canal systems are a mature technology, and there are successful examples of how to construct and operate them to limit negative impacts.

Rockström et al. (2009) and other authors argue that, because the water fluxes and land areas associated with agriculture continue to be dominated by green water use, strategies to use green water more effectively through on-farm rainwater harvesting and well-timed supplemental irrigation is likely to pay bigger dividends than reservoir and canal development. This argument is consistent with ecological and social activists' "small is beautiful" mantra. Van der Zaag and Gupta (2008) discuss scale issues associated with water storage projects from a physical and social perspective. They compare a collection of small reservoirs versus a large centralized reservoir in an idealized setting: the large reservoir has a residence time of two years compared to 0.2 years for the small reservoirs, thus providing substantially longer carryover storage that is effective for addressing multiseason cropping or drought needs. Despite a higher capital cost per beneficiary, the larger system delivers higher returns per beneficiary. These relative benefits could be even higher if ancillary benefits for hydropower production, fisheries, recreation, and other uses from the larger reservoir are considered. In addition, losses of capacity due to siltation, salinization, and evaporation are likely to be much higher over time for the smaller reservoirs.

On the other hand, large dams lead to inequitable distribution of water resources and income and pose a much more significant governance and ecological challenge with differential impacts on upstream and downstream residents. Upstream residents may be displaced by dam construction, while downstream residents enjoy irrigation benefits, but may lose fisheries. Given the importance of storage, how should the issue of the appropriate scale of storage, and whether it should be surface or groundwater, be addressed?

In my view, the traditional debate of small vs. big storage is not a meaningful one. Small, on-farm storage is helpful for meeting supplemental irrigation needs in periods between rainfall events in the wet season. Larger storage may be needed to address within- or across-season deficits in water availability. However, contrary to the traditional development strategy of selecting a crop and an associated area to be irrigated and developing a storage and canal system, a simultaneous optimization of areal crop allocation and river basin level water storage and allocation strategy may be more useful.

Lall and Kaheil (2009) considered the economically optimal selection of regional crops and a mix of large dams and small, on-farm rainwater harvesting or storage systems in a river basin in south India. Irrigation was the only use considered for all the water-storage elements. The model solution was to select one large dam and some rainwater harvesting storage at all the farms. Canals provided water to regions downstream of the large dam. Crops whose yields were robust to reduction in water supply were allocated to the upstream section, while crops such as rice that have higher water requirements were allocated to the downstream section. The large reservoir was operated to buffer water shortages beyond those that could be handled by the smaller on-farm storage, with irrigation water for supplemental irrigation rather than as a primary resource for cropping. Thus, regional storage development and operation that is cognizant of the centralized and distributed elements and a cropping strategy optimized for such a system, rather than the traditional strategy of developing a dam and canal system with a planned target area to be fully irrigated by the system, emerges as economically sound.

Asia, Africa, and Latin America will require additional surface water storage in coming decades as part of a strategy for climate change adaptation and water and food security. The per capita water storage in these regions is currently well below that in Europe and North America, despite facing a higher variability in water availability.

With the rapid development of groundwater resources in areas that are increasingly water stressed and support intensive agriculture, mechanisms for inducing artificial groundwater recharge to increase subsurface storage have been receiving scrutiny. Such storage has the advantage of not being subject to direct evaporation. The techniques for artificial groundwater recharge include surface infiltration, if large land areas with highly permeable surface soils are available; vadose zone infiltration through recharge trenches that are backfilled with sand and gravel; and wells, where the water is pumped in after capture and surface detention and may be pumped back for subsequent use. Soil clogging and biofouling is a potential issue with all three technologies, and energy costs can be high for wells. Despite considerable interest in these technologies since the 1980s, large-scale application in agriculture has not emerged. Examples of the surface and vadose zone infiltration for subsequent irrigation application are actually plentiful in South Asia and Africa, where such systems have been shown to be locally quite successful, but a topographic gradient is needed for these methods to be effective, and hence their success depends on a suitable location. Wells for aquifer recharge and recovery are in use in several places, but the economics are not always favorable relative to the current prices for agricultural water. All of the techniques for artificial groundwater recharge raise the issue of who pays for the recharge mechanism and who can actually recover the water. The recharged groundwater can exfiltrate into a river system and be pumped by groups distant from the recharge location, so there are additional governance challenges that emerge as the scale of recharge increases.

Improvements in agricultural water productivity may be a more important contributor to global sustainability than further efforts at blue water development. Switching from furrow to sprinkler to drip are steps that offer the possibility of dramatic reductions (30 to 70 percent) in irrigation water demand. Sprinkler and drip irrigation require pressurization of the irrigation delivery system, which in turn requires an energy source—an additional operational expense. Subsurface drip, a method that delivers higher water efficiencies, is so far limited to certain types of crops, and crop rotation can require removal and reinstallation, adding to labor costs. However, even with basin and furrow irrigation of cereal crops that account for the dominant use of irrigation water, savings of as much as 30 percent are reported through better timing of irrigation and soil moisture control. Soil moisture sensors and daily precipitation and temperature forecasts can be used effectively to achieve such gains in water usage. Crop yields also increase, since the reduced humidity near the surface leads to lower incidence of several plant diseases. Overapplication of irrigation water, fertilizers, and pesticides causes leaching and pollution of surface and ground water bodies. Recent directions in precision agriculture suggest the possibility of ubiquitous low-cost sensors for soil water and chemicals that could transform application strategies for irrigation water, nutrients, and chemicals to address this situation (Adamchuk et al. 2004). The use of sensors to guide the frequency and duration of basin and furrow irrigation may provide the most economical strategy for water productivity improvement, rather than a transition to drip irrigation—contrary to many existing policy directives that offer high subsidies and incentives for drip.

In addition to advances in the adoption of the irrigation technologies, there are also advances in irrigation practices. The system of rice intensification (SRI) and direct seeding of rice (DSR) are examples of changed practices that in certain situations lead to 10 to 40 percent reduction in water use, better soil organics, and comparable or higher yields (Dobermann 2004; McDonald et al. 2006; Farooq et al. 2011; Kumar and Ladha 2011). Increased expenses in weed control are a negative impact. Both systems are gaining in popularity and experience in Asia.

Rockström et al. (2011) argue for a "triply green revolution" through dramatic improvements in green water management. The primary strategies they consider are vapor shift—reduction of bare soil evaporation by up to half through mulching and different tillage systems—and local rainwater harvesting or retention systems to meet water needs during breaks in the rainy season. Using vapor shift, global crop productivity could be improved by 2 to 25 percent, with potential for greater than 20 percent improvements in semi-arid regions in the midwestern United States, parts of South America, the Sahel, southern Africa, Central Asia, and southeastern Australia (Rost et al. 2009). Rost et al. estimate the corresponding boost for rainwater harvesting to be from 4 to 30 percent as the level of storage is increased, leading to a potential combined increase of 7 to 53 percent. The wide range in these estimates results from modeling uncertainties. The key to field success at any one of these ranges will need to be very aggressive extension programs. These would include significant improvements at all levels, from soil testing and chemistry monitoring to education to marketing and skilled labor for development. Farmers tend to be risk averse, and unless new practices that add labor and capital costs are successfully demonstrated in their setting over a number of years, the prospects for adoption are low.

Identifying traits that promote drought or salt tolerance and breeding or genetically engineering crops using these traits is an active area of research. Modifications to cotton, rice, and a variety of other crops to promote drought and salt tolerance are under way, with water efficiency gains of 20 to 25 percent in research fields, but there is still limited field data to support these gains. Salt tolerance has been demonstrated in genetically modified versions of rice, tomato, and tobacco, but the yields are typically lower. There are also traditional cultivars of rice that are tolerant of salt water in coastal margins and yield one to two tons per hectare. These are promising areas of research that could significantly change the food–water stress situation in the future.

STRATEGIES FOR SUSTAINABILITY: POLICY MEASURES

Policies that determine access to water and energy play a significant role in determining whether farms adopt technologies and conservation strategies to improve water use efficiency. Extensive public- and private-sector supply chains have emerged in the past fifty years for farm-level procurement and subsequent processing and marketing of food products. These provide several opportunities for policy signals that lead to improvements in on-farm water productivity.

Optimizing where particular crops are grown, taking into account climate, access to water, and soils, can lead to dramatic reduction in total regional and global crop water use (Sauer et al. 2010; Candela et al. 2012). Spatial optimization can be shaped as part of a sourcing strategy for a private corporation that is a major food processor, such as Proctor and Gamble or PepsiCo; by a nation, like India, that has an active food procurement and distribution system; or by aid organizations, such as the World Food Programme. Procurement price guarantees through a contract farming model (Huh et al. 2012) or insurance against price or yield failure can help to facilitate the transition, especially if the farmer is being asked to try a new practice or technology. Farmers can also be offered performance incentives for water productivity as part of a corporate social responsibility program or as part of the sourcing strategy. Verification and associated transaction costs can be a factor for such schemes. Insurance premiums to be paid by farmers or sourcing firms could be indexed to climate forecasts and to the associated likelihood of crop failure to stimulate better choices.

Appropriate pricing schemes, water rights and their regulated trading, metering of use, and enforcement are factors that are routinely cited in discussions of agricultural water economics. These concepts are usually applied to irrigation water—most commonly to surface water resources. Prices for irrigation water are usually lower than the cost of resource development, operation, and maintenance, and in much of the world, irrigation water or the energy to pump the water from the ground is highly subsidized. Such subsidies are controversial, especially in India, China, and other developing countries that suffer from physical water scarcity and competition. In some of these countries, informal water markets emerged in the 1980s and have become pervasive (Shah et al. 2012). Timely and reliable provision of energy and water are key ingredients in the success of formal pricing and allocation systems. If these can be assured, then a “lifeline” pricing structure that provides a certain amount of water or energy per unit area that is associated with highly efficient use at a fixed cost, and then charges a unit price for higher usage, could promote efficient use and stimulate conservation. Corresponding incentives or subsidies for conservation techniques could then facilitate their adoption.

Significant water reforms—including canal monitoring, irrigation infrastructure improvement, short- and long-term water trading, pricing, and targets for 30 percent use reduction to improve ecological functioning—were put in place over the last ten years in the Murray-Darling Basin in Australia. The reforms’ economic impact is controversial (Randall 2012), but they have clearly ushered in a significant transition in water use and productivity. Sankarasubramanian et al. (2009) present an example of how forward water contracts based on season ahead streamflow forecasts can be used effectively to improve multisectoral water allocation and provide signals for the selection of crops and irrigated area in northeast Brazil. Brown et al. (2006) discuss how dynamic pricing of energy in conjunction with seasonal rainfall forecasts can be used to promote groundwater sustainability. As the indications of regional water scarcity increase, discussion and use of such innovations is likely to increase and help improve efficiency of resource use and allocation. A safety net to ensure that inequities do not develop in the process is essential. Further innovation in the application of such ideas would help bring the experience and confidence to apply them in different settings. I expect that these two examples are some of many that are rapidly evolving today and are a harbinger for practices we can expect to be more commonplace in the next 10–20 years.

Virtual water trade can help to alleviate local water stress, and its liberalization could be a factor in global food and water management. However, to be most beneficial, it may be useful to consider it as one element of the spatial optimization of crops.

TOWARD MORE EFFICIENT AGRICULTURAL WATER USE

Water and food security are inexorably linked. The twentieth century was marked by dramatic increases in the yields per unit land area of selected crops that led to improvements in food security. The pathway toward sustainability in the twenty-first century requires similar gains in yield per unit consumption of water and other agricultural inputs. Human societies are increasingly global, so water use efficiency measures need to be considered at multiple scales from on-farm application to watershed to global. Governments need to develop and implement appropriate management structures needed to facilitate addressing increasing food demands and climate variability over the next two decades. The following actions should be top priorities if we hope to achieve sustainability in the next 10 to 20 years:

Regulation through measurement and pricing: Irrigation water use needs to be measured and appropriately priced to signal the benefits of conservation. Reliable provision of water is crucial for such measures to succeed politically. Subsidy and incentive programs need to be structured to promote water conservation rather than consumption. Water quality and quantity impacts of agriculture need to be targeted. Local, state, and national government policies need to be developed in each setting.

Extension: Robust and effective public- and private-sector extension programs are needed to promote the exposure to and adoption of better irrigation, soil moisture management practices, and cultivar selection. These need to be built as part of current supply chain development programs.

Risk management: Farmers are inherently risk averse. Their perception of climate, price, cultivar, and technology risks can be addressed through extension-based demonstrations, cooperatives, and insurance. Integrating these into public and private supply chain development programs is needed.

Procurement systems: Public and private supply chain designs need to consider the spatial optimization of crop selection to better address the regional water and climate constraints and integrate appropriate food and water storage systems. Rethinking international and intra-national trade barriers is a critical part of the success of global spatial optimization.

Irrigation development: The majority of global agriculture is rainfed, yet irrigation dramatically increases food productivity. A transition to reliable, efficient hybrid irrigation systems that exploit the economies of scale offered by large storage systems and better point-of-use control offered by localized rainwater harvesting and storage is critical. Contrary to the arguments in some of the literature, rather than looking for a choice between green and blue water strategies, and between small and large systems for supplemental irrigation, simultaneous improvements in these areas are needed and possible. These would need to be executed by state and national planners as well as private-sector sourcing specialists.

Irrigation and soil management practices: Significant improvements are possible in water use and quality impacts through active management using low-cost sensors and adapted tillage practices. Translational research and extension is needed to make these accessible as part of a policy package of subsidy and incentive reform. This may be best conducted by regional agricultural universities and through private-sector supply chain development.

Genetic improvements of cultivars: Breeding efforts to improve drought and salt tolerance, and to improve productivity of dryland crops hold significant potential for helping meet food and water security goals. This is an active area of public- and private-sector research.

The potential for improved effectiveness of water systems in agriculture is high and achievable if the policy and technology initiatives are appropriately packaged. With even modest improvements over the next ten years, we stand to meet global water–food goals. Such improvements are crucial to long-term food security and deserve to be the focus of a major global initiative.

ACKNOWLEDGMENTS

I would like to thank Chris Barrett of Cornell University for his helpful comments on earlier drafts of this chapter, as well as his team of students and editors for their feedback and editorial assistance.

NOTES

1. Shiklomanov (2000) and Oki and Kanae (2006) provide estimates of global water storage and fluxes. Bouwer (2002) and Döll and Fiedler (2008) provide estimates of groundwater recharge rates. A global groundwater map is available at http://www.whymap.org (accessed March 24, 2013).

REFERENCES

Adamchuk, V. I., J. W. Hummel, M. T. Morgan, S. K. Upadhyaya. 2004. On-the-go soil sensors for precision agriculture. *Computers and Electronics in Agriculture 44*: 71–91.

Allan, J. A. 1998. Virtual water: A strategic resource global solutions to regional deficits. *Ground Water 36*(4): 545–6.

Allouche, J. 2011. The sustainability and resilience of global water and food systems: Political analysis of the interplay between security, resource scarcity, political systems and global trade. *Food Policy 36*: 53–8.

Bouwer, H. 2002. Artificial recharge of groundwater: Hydrogeology and engineering. *Hydrogeology Journal 10*(1): 121–142.

Brown, C., P. Rogers, and U. Lall. 2006. Demand management of groundwater with monsoon forecasting. *Agricultural Systems 90*(1): 293–311.

Bruinsma, J., ed. 2003. *World agriculture: Towards 2015/2030–An FAO Perspective.* Rome: Food and Agriculture Organization.

Candela, L., F. J. Elorza, J. Jiménez-Martínez, W. Von Igel. 2012. Global change and agricultural management options for groundwater sustainability. *Computers and Electronics in Agriculture 86*: 120–130.

Carr, J., P. D'Odorico, F. Laio, L. Ridolfi, and D. Seekell. 2012. Inequalities in the networks of Virtual Water flow. *EOS, Transactions American Geophysical Union 93*(32): 309–10.

Dalin, C., M. Konar, N. Hanasaki, A. Rinaldo, and I. Rodriguez-Iturbe. 2012. Evolution of the global virtual water trade network. *Proceedings of the National Academy of Science USA 109*: 5989–99.

Dobermann, A. 2004. A critical assessment of the system of rice intensification (SRI). *Agricultural Systems 79*(3): 261–81.

D'Odorico, P., F. Laio, and L. Ridolfi. 2010. Does globalization of water reduce societal resilience to drought? *Geophysical Research Letters 37*: L13403.

Döll, P., and K. Fiedler. 2008. Global-scale modeling of groundwater recharge, *Hydrology and Earth System Sciences 12*: 863–885.

Farooq, M., K. H. Siddique, H. Rehman, T. Aziz, D. J. Lee, and A. Wahid. 2011. Rice direct seeding: Experiences, challenges and opportunities. *Soil and Tillage Research 111*(2): 87–98.

Hoekstra, A. Y. 2012. The hidden water resource use behind meat and dairy. *Animal Frontiers 2*: 3–8.

———, M. M. Mekonnen, A. K. Chapagain, R. E. Mathews, and B. D. Richter. 2012. Global monthly water scarcity: Blue water footprints versus blue water availability. *PLoS ONE 7*(2): e32688.

Huh, W. T., S. Athanassoglou, and U. Lall. 2012. Contract farming in a developing country with possible reneging: Can it work? *IIMB Management Review 24*(4): 187–202.

Konikow, L. F., and E. Kendy. 2005. Groundwater depletion: A global problem. *Hydrogeology Journal 13*: 317–20.

Kumar, V., and J. K. Ladha. 2011. Direct seeding of rice: Recent developments and future research needs. *Advances in Agronomy 111*: 297.

Lall, U., and Y. Kaheil. 2009. Basin scale water infrastructure investment evaluation and crop selection guidance considering climate risk. Project report to the Asian Development Bank. Columbia Water Center.

Lehner, B., C. R. Liermann, C. Revenga, C. Vörösmarty, B. Fekete, P. Crouzet, P. Döll et al. 2011. High-resolution mapping of the world's reservoirs and dams for sustainable river-flow management. *Frontiers in Ecology and the Environment 9*: 494–502.

Lempérière, F. 2006. The role of dams in the XXI century: Achieving a sustainable development target. *Hydropower and Dams 3*: 99–108.

McDonald, A. J., P. R. Hobbs, S. J. Riha. 2006. Does the system of rice intensification outperform conventional best management? A synopsis of the empirical record. *Field Crops Research 96*(1): 31–6.

Mekonnen, M. M., and A. Y. Hoekstra. 2011. The green, blue and grey water footprint of cropsand derived crop products. *Hydrology and Earth System Sciences 15*: 1577–1600.

Molden, D. 2007. *Water for food, water for life: A comprehensive assessment of water management in agriculture.* London: Earthscan, and Colombo: International Water Management Institute.

Oki, Taikan, and Shinjiro Kanae. 2006. Global hydrological cycles and world water resources. *Science 313*(5790): 1068–1072.

Perveen, S., C. K. Krishnamurthy, R. S. Sidhu, K. Vatta, B. Kaur, V. Modi, R. Fishman, L. Polycarpou and U. Lall, 2012. Restoring Groundwater in Punjab, India's Breadbasket: Finding Agricultural Solutions for Water Sustainability. Columbia Water Center White Paper.

Randall, A. 2012. Property entitlements and pricing policies for a maturing water economy. *Australian Journal of Agricultural and Resource Economics 25*(3): 195–220.

Rockström, J., M. Falkenmark, and L. Karlberg. 2011. Global food production in a water constrained world: Exploring "green" and "blue" challenges and solutions. In *Water resources planning and management,* ed. R. Q. Grafton and K. Hussey, 131–51. Cambridge: Cambridge University Press.

Rockström, J., M. Falkenmark, L. Karlberg, H. Hoff, and S. Rost. 2009. The potential of green water for increasing resilience to global change. *Water Resources Research 45*: W00A12.

Rost, S., D. Gerten, H. Hoff, W. Lucht, M. Falkenmark, and J. Rockström. 2009. Global potential to increase crop production through water management in rainfed agriculture. *Environmental Research Letters 4*: 044002.

Sankarasubramanian, A., U. Lall, F. A. Souza Filho, and A. Sharma. 2009. Improved water allocation utilizing probabilistic climate forecasts: Short-term water contracts in a risk management framework. *Water Resources Research 45*(11): W11409.

Sauer, T., P. Havlík, U. A. Schneider, E. Schmid, G. Kindermann, and M. Obersteiner. 2010. Agriculture and resource availability in a changing world: The role of irrigation. *Water Resources Research 46*: W06503.

Shah, T. 2009. Taming the anarchy: Groundwater governance in South Asia. Washington, DC: Resources for the Future, Colombo, Sri Lanka: International Water Management Institute.

——, M. Giordano, and A. Mukherji. 2012. Political economy of the energy-groundwater nexus in India: Exploring issues and assessing policy options. *Hydrogeology Journal 20*(5): 993–1006.

Shiklomanov, Igor, A. 2000, Appraisal and assessment of world water resources. *Water International 25*(1): 11–32.

Siebert, S., and P. Döll. 2010. Quantifying blue and green virtual water contents in global crop production as well as potential production losses without irrigation. *Journal of Hydrology 384*: 198–207.

van der Zaag, P., and J. Gupta. 2008. Scale issues in the governance of water storage projects. *Water Resources Research 44*: W10417.

Vörösmarty, C. J., P. J. Green, J. Salisbury, R. B. Lammers. 2000. Water resources vulnerability from climate change and population growth. *Science 289*: 284–8.

Wada, Y., L. P. H. van Beek, C. M. van Kempen, J. W. T. M. Reckman, S. Vasak, and M. F. P. Bierkens. 2010. Global depletion of groundwater resources. *Geophysical Research Letters 37* L20402: 1–5.

Yoffe, S., A. T. Wolf, M. Giordano. 2003. Conflict and cooperation over international freshwater resources: Indicators of basins at risk. *Journal of the American Water Resources Association 10*: 1109–26.

6

Managing Marine Resources for Food and Human Security

TIMOTHY R. MCCLANAHAN,
EDDIE H. ALLISON, AND JOSHUA E. CINNER

Fisheries and aquaculture employ 260 million people, contribute a global mean of 17 kilograms per person per year of micronutrient-rich animal food, and contribute about $100 billion annually to global trade.[1] Fish is among the most traded of food commodities, with 38 percent of all recorded fishery production traded across national borders in 2010. Developing countries supplied just over 50 percent of global fishery and aquaculture exports by value and 60 percent by weight; 67 percent of these exports were to developed countries (Food and Agriculture Organization 2012). These are minimum estimates based largely on transcontinental trade. Much unreported cross-border trade takes place, particularly of fish from inland waters, where the waterways themselves often form the national boundaries. Developing countries tend to export more higher-value species, such as shrimp and tuna, and import lower-value species like herring, sardine, and mackerel (Smith et al. 2010).

The fisheries sector is complex, with fishing operations ranging from hand-gathering shellfish on mudflats to fishing the open ocean with factory trawlers, and farming ranging from growing fish in farm ponds to intensive land-based and sea-cage operations. The sector supplies thousands of different products to highly segmented markets, from small sun-dried fish available to the poor to luxury sashimi for the wealthy. Aquatic organisms are also harvested and traded for medicinal products, both traditional and pharmaceutical, and for ornamentation and decoration. Fishing is also a major leisure pursuit among people of all social strata, with the line between recreation and subsistence frequently blurred among poorer people. Increases in marine fishing yields in the second half of the twentieth century were almost entirely from expansion of fishing efforts offshore and into the tropics. Yield increases have leveled off in the last 20 years. The rapid growth of aquaculture—both marine and freshwater—over the last 30 years explains the continuing increase in per capita global supply of fish, despite increasing global population.

The fisheries sector faces a number of food and human security challenges, chiefly governance, including perverse subsidies; weak or inappropriate access and property rights; uncontrolled and illegal fishing and other maritime criminal activity; climate change threats in sensitive regions; human settlement and infrastructure in densely populated, low-lying areas; development and trade policies that can conflict with the poor's access to resources; and an increasingly affluent population, especially among the emerging middle classes of developing and transitional economies, demanding more seafood. These are the areas where conflicts over access and resource declines can trigger wider unrest or "fish wars" (Pomeroy et al. 2006). Individuals and communities regularly engage in violent conflicts over access to marine resources. These local conflicts have led to international political incidents, such as the famed Cod Wars between Iceland and the UK in the 1950s and 1970s, the 1920s to 1930s Yellow Croaker dispute between China and Japan, and the 1995 Turbot War between Canada and Spain. These conflicts over marine resources have the potential to lead to wider instability, particularly where food insecurity is high, people are vulnerable, and governance is weak or autocratic. The current scale of fisheries is vast and there are many opportunities for large-scale failure and collapse. Consequently, after outlining the key issues we conclude with suggestions for maintaining environmental and social equity and food security.

FOOD SECURITY AND VULNERABILITY

The vulnerability of countries, communities, or populations of consumers dependent on fish for income, revenue, or nutrition is based on the exposure or degree to which the fishery system is stressed by direct human use of the resources and the range of indirect environmental stressors on fishery production, the population's sensitivity to this exposure, and the social adaptive capacity of the system that can reduce this potential impact. According to the Intergovernmental Panel on Climate Change (Watson and Albritton 2001):

Vulnerability = (Exposure + Sensitivity) − Adaptive Capacity

Exposure is characterized by the magnitude, frequency, duration, and spatial extent of climatic and human disturbances. The degree to which stress actually modifies the response of a system is known as sensitivity; in this case, the sensitivity of the food and livelihood system is measured by the degree of dependence on marine resources for food, revenue, and income. Finally, vulnerability is influenced by the capacity of people to adapt to changes associated with the exposure and sensitivity (dependency). Although more commonly applied to climate change vulnerability analysis, this analytical framework is also applicable to assessment of food insecurity (Hughes et al. 2012). The framework will thus form the basis for our evaluation of the threats to marine fisheries, and to food and nutrition security, and potential approaches to reducing these threats.

Globally, fishing effort has expanded massively since the 1950s. A 2.4 times increase in yield has been achieved by a fourfold expansion in the fishing area (Swartz et al. 2010). This expansion extended from the coastal waters of the North Atlantic and West Pacific to the open ocean, southern hemisphere, and tropics. Expansion was greatest from 1980 to the early 1990s, at a rate of almost a degree of latitude per year. By the mid 1990s, one-third of the world's ocean and two-thirds of the continental shelves were exploited at levels where more than 10 percent of marine primary production was appropriated to support fisheries. Given that nearly 90 percent of primary production is used in supporting ecosystems, this is near the maximum allowable by thermodynamic and food web constraints (Odum 1988). The final frontiers were the unproductive open oceans and the inaccessible poles, and the number of areas potentially open to new exploitation declined after the 1990s. Consequently, unlike agriculture where a doubling of agriculture production was associated with only a 10 percent increase in the area in cultivation over 35 years (Tilman 1999), this expansion of catch from 34 to 83 million tons is associated with a nearly 400 percent increase in the area fished during the same period (Swartz et al. 2010).

The expansion of fisheries is associated with a decline in the biomass of fishes, both target and incidental catch, and with subsequent ecological and biodiversity changes (Worm et al. 2006). Declines in biomass are a necessary part of fisheries exploitation but reducing the indirect effects on ecosystems and biodiversity is an increasing concern for modern fisheries management and decision making (Worm et al. 2009; Salomon et al. 2011). Reducing biomass to 25 to 50 percent of unexploited levels typically maximizes their yields, while going beyond this level can result in losses of diversity and other ecological processes (McClanahan et al. 2011). Most of the fisheries biomass levels in developed countries had reached this level since the 1980s while less developed regions have approached this value since the mid 2000s (Worm and Branch 2012). Fishing effort continues to rise even though yields have stabilized or potentially declined slightly since the mid 1990s. Watson et al. (2012) document a ten-fold increase in the power used in offshore fishing and the catch per unit power in 2006 was half of what it was in the 1950s.

Global assessments use average values and there is a wide variation around these values, with 25 to 50 percent of the total fished stocks reduced to levels below 10 percent of unexploited abundance, and thus considered as "collapsed" (Worm et al. 2006). While some stocks may recover at this level, many will not and this loss of stocks is of long-term concern for global food security.

The highest frequencies of collapsed stocks are in species-poor environments, typically in high latitudes (Worm et al. 2006). One effect of the collapse of major stocks of fish in north-temperate waters is an increased demand for fish from tropical waters to make up the shortfall in supply in developed countries. This is contributing to the net flow of fish from the waters of developing countries to developed countries (Smith et al. 2010).

This net loss of fish from developing countries might be seen negatively—as taking fish from the mouths of the poor and putting them on the tables of the rich. Trade may, however, contribute positively to food security by stimulating both export-orientated and domestic production and market development, particularly in aquaculture; by creating employment; and by supporting economic growth (Jaunky 2011). Improving trade balances may enable the import of food staples; for example, in Senegal during the 1980s and 1990s, the value of exported fish approximately equaled the value of imported food staples (FAO 2007). These positive economic contributions may lead to net improvement in food security, poverty reduction, and economic growth in fish-exporting developing countries, though such benefits have been difficult to demonstrate (Kurien 2004; Bene et al. 2010).

In the developed world, observers are increasingly recognizing a need to reduce fishing effort and increase conservation measures. An estimated 63 percent of the better-studied fished stocks require rebuilding, and this process is being implemented at various rates in Europe and North America (Worm et al. 2009). Key trade-offs associated with changing employment numbers, catch, and biodiversity often limit rebuilding programs (Froese and Proelb 2010; Hilborn et al. 2012). Furthermore, one of the indirect effects is to move the excess fishing capital and effort to less developed countries in Africa (Worm et al. 2009) where the fisheries may be poorly understood and estimates of potential yields overestimated (Le Manach et al. 2012). There are incentives to overestimate yields from both developed and developing countries, which can lead to short-term profits but undermines long-term sustainability.

Sub-Saharan Africa has seen the largest decline—about 15 percent—in per capita consumption of fish between 1990 and 2002 (Bene and Allison 2011), yet this is an area of increasing fishing contracts and exports of fish to Europe and Asia. In many cases, governments subsidize fisheries, particularly in developed countries, and this often works to make developed countries' fisheries more competitive with countries that do not subsidize them and allows them to potentially exploit the fisheries of countries without subsidies. For example, average fuel subsidies for fisheries in developed countries were $5.02 billion, but only $1.35 billion per year in developing countries in the early 2000s (Sumaila et al. 2006). Declining fish consumption in Africa is not, however, simply caused by increased access by foreign fishing fleets and associated exports: slow development of aquaculture, fully or over-exploited wild fisheries, and increasing populations are also key explanatory factors (Allison 2011).

FISHERIES CONFLICT AND HUMAN SECURITY

The relationship between fisheries and political insecurity is complex: wars affect fisheries and conversely fisheries conflicts can spark wider local or regional political insecurity. Wars tend to reduce fishing pressure in areas of conflict as it becomes unsafe for fishermen to venture out to sea. This respite allows stocks to rebuild and results in bonanza catches after conflicts, as was seen in the North Atlantic after the six years of World War II (Jennings et al. 2001). Nevertheless, the need to feed troops can lead to increased fishing in grounds far away from conflict areas. For example, wartime fishing for sardines led to collapse of stocks in Monterey Bay, California, in the 1940s (Palumbi and Sotka 2010).

Redeployment of labor, population displacement, counter-insurgencies, and opportunistic encroachment on the resource can, however, considerably delay any fishery recovery despite recovery of fish populations. In some instances, conflict may simply displace fishing effort spatially—fishing moves to adjacent areas unaffected by conflict which results in the heavier exploitation and serial depletion of neighboring fisheries. In a study of 123 civil conflicts between 1952 and 2004, Hendrix and Glaser (2011) found that civil conflicts reduced the catch by 16 percent compared to prewar levels and recovery was slow, taking nine years to return to prewar levels. This effect is 13 times larger than that of a strong climate anomaly, such as the El Niño and the Southern Oscillation (ENSO) events that drive down fish catches in the Pacific (Watson and Pauly 2001) and, therefore, represent a considerable food security loss in countries with long and repeated conflicts.

Sometimes, fisheries conflicts can escalate into military confrontations, a classic example being the cod wars of the 1950s to 1970s between Iceland and the United Kingdom. Iceland, seeking to gain control over its resources after gaining its independence from Denmark in 1944, used evolving international ocean law to extend its territorial waters successively from the internationally recognized 3 to 200 nautical miles by 1975. The early extensions sparked conflict with British distant water fleets that had been fishing for cod in Icelandic waters since the fifteenth century. As well as threatening tens of thousands of livelihoods in the UK's fishing ports and the supply of cod for the British working-class staple of fish and chips, this was a challenge to wider British interests in the "freedom of the seas" (Johannesson 2004).

In the succeeding decades, most nations extended and maximized their territorial limits to 200 nautical miles, or to the edge of their continental shelves, establishing the present ocean governance regime of "exclusive economic zones," leaving the usually much less productive open oceans as international territory or "the high seas." Fishing rights in these areas were thus nationalized and nation states have powers to limit or exclude other nations, bearing in mind historical rights of access. Such rights have been painstakingly negotiated in international courts and a plethora of regional fishery management organizations govern access to particular resources. Developing countries get fishing license revenues and other benefits from allowing other nations' fleets to fish their waters, though these fees are low compared to the value of the resources (Le Manach et al. 2012).

Conflicts over access to resources persist where weak, autocratic, or failing states are unable to govern their exclusive economic zones and where states are unwilling or unable to defend the interests of their own fishing fleets. Piracy off the Horn of Africa is a contemporary example, and a cross-country analysis found that increased piracy between 1995 and 2007 was related to state weakness and reduced opportunities in the domestic fishing sector (Daxecker and Prins 2012). Piracy costs the shipping industry $25 billion per year. This is small compared to the $11.8 trillion value of global maritime trade in 2008 but still significant (Mildner and Gross 2011). As the principle of territorial limits has become accepted and codified in international law, fishery conflicts of the type witnessed in the 1970s to 1990s are increasingly rare, but where maritime boundaries remain disputed, fishery access rights can be a flashpoint. Typically, however, such conflicts are associated with claims on mineral rights, particularly subsea oil and gas, and sometimes with militarily strategic location.

Fishery governance analysts have classified fishery conflicts into five types. There is no quantitative data available on the frequency of each type, but all are common and may be ubiquitous in small-scale fisheries in developing countries. The first and second types—struggles for resource control and for the mechanisms of control—are found in the everyday acts of resistance by many fisherfolk to central state-imposed management of resources, and have sometimes led to the transition toward more participatory forms of resource governance. Conflicts over who controls the fishery often manifest as interstate conflicts over territorial waters, village boundary disputes, and intra-community struggles for power and authority. Conflicts over how a fishery is controlled often relate to how management systems are implemented, how quotas or territories are allocated, the use of seasonal or area closures, and technical restrictions. Devolving governance to the local level doesn't always reduce these conflicts, particularly where local interests clash with broader national policy. Nevertheless, state–community partnerships are thought to reduce the incidence of minor resource conflicts (Muawanah et al. 2012).

The third type of conflict—conflict among users—is also very frequent. Fisheries are diverse industries and conflicts between different user groups are inevitable. Conflicts often arise out of relations between user groups, such as artisanal and industrial fishers, static and passive gear users, migrants and residents, men and women, and among different ethnic and religious groups. Conflicts arising from overlapping use of fish stocks by small-scale coastal and larger-scale industrial fishing are particularly significant to food security and can involve physical violence at sea (DuBois and Zografos 2012). Most aquaculture-related conflicts fall into the fourth category—conflict among different users of the same resource system. Conflicting uses of marine resources include aquaculture, coastal or floodplain farming, water abstraction, hydropower, tourism, leisure, conservation, residential or industrial development, and waste disposal. The conflicts between common property mangrove forest users and commercial shrimp farmers are a common example. This conversion and privatization of common property has impacted coastal livelihoods and led to civil society campaigns and boycotts of farmed shrimp (Veuthy and Gerber 2011). Fishing, farming, tourism and leisure, industrial and urban development, power generation, nature conservation, and waste disposal all compete for space in coastal and riparian areas and can be a source of conflict between interest groups.

Fishing communities, particularly artisanal or small-scale ones, are often portrayed as rebellious and ungovernable by government officials. Conflicts related to the fifth type—wider governance, economic, and social change contexts—occur when fishers feel marginalized by state authority and land-based elites (Allison et al. 2012). Conflictual issues external to fisheries but which affect fisheries nonetheless include rule of law, corruption, elite capture, demographic change (e.g., resettlement), economic policy change, and political discrimination. While the fishery sector is thus often rife with conflict, there is little direct evidence that food insecurity results from loss of fisheries access or that decreases in productivity drive wider conflict. There are certainly conflicts over resource access and loss of livelihood, but the contribution of fish to food security is through the "hidden hunger" of micronutrient deficiency. Food riots, such as in 2007, are usually sparked by rising prices of staples (Bush 2010) and not the micro-nutrient deficiencies that may result from fish shortages.

CLIMATE CHANGE PROJECTIONS

Climate change will alter fisheries production through changes in winds, water temperature, dissolved oxygen, and increasing ocean acidity. Based on projections of temperature changes, species distributions, and temperature-related growth rates, fisheries production has and should continue to increase in some high-latitude regions because of warming and decreased ice cover (MacNeil et al. 2010). Low-latitude regions are, however, governed by different processes, and production is predicted to decline as a result of reduced mixing of the water column and nutrients, and increased temperature and acid-induced damage to key habitats, especially coral reefs (Brander 2007; Pratchett et al. 2008). Behrenfeld et al. (2006) used an

ENSO-related warming period to evaluate changes in ocean productivity and found declines of 74 percent, mostly associated with stratified conditions in the low-latitudes. Given the net primary production consumed by harvested species (Swartz et al. 2010), this lost production of the plankton will also reduce fish catch rates.

Cheung et al. (2008, 2009, 2010) has made climate change projections for more than one thousand species based on existing theories linking thermal tolerance limits, feeding energetics, body size metabolisms, geographic ranges, and primary production. These models suggest a 60 percent turnover in present biodiversity by 2055, with numerous local extinctions of fishes in the subpolar regions, the tropics, and semi-enclosed seas. The most intense species invasions are projected for the Arctic and the Southern Oceans. These models predict a large-scale redistribution of global catch potential, with a 30 to 70 percent increase in high-latitude regions and a 40 percent drop in the tropics. The greatest catch potential declines are projected for the southern coasts of semi-enclosed seas. Norway, Greenland, Alaska, and Russia are projected to benefit the most while Indonesia, Chile, China, and the continental United States are projected to lose the most catch potential. A similar study focused just on the western African region based on 128 exploited species projected a 21 percent drop in the landed value, a 50 percent decline in fisheries jobs, and a total annual loss of $311 million by 2050 (Lam et al. 2012).

The sophistication of the climate projection models has further increased to include other factors, such as pH, dissolved oxygen, and phytoplankton community structure (Cheung et al. 2011). A northeast Atlantic study of 120 exploited species found that including changes in acidity and oxygen content in seawater reduced growth performance, increased the rate of range shift, and reduced catch potential 20–30 percent by 2050 relative to simulations with only temperature changes. There were still increases in catch potential in most of the northern regions but, in the southern Atlantic, regional declines in catch were predicted if the species were sensitive to changes in oxygen and pH, factors that are not well understood for these species at present.

Habitat quality is also a considerable concern for some fisheries, particularly tropical coral reefs and other ecosystems where habitat is created by living organisms that are sensitive to climate and other anthropogenic disturbances (Pratchett et al. 2008). There have been large-scale and continuous declines in coral cover in both the Caribbean and Pacific (Gardner et al. 2003; Bruno and Selig 2007) and changes in the taxonomic composition in the Indian Ocean (McClanahan et al. 2011). These changes are projected to result in declines in coral reef fishes but improved management can override some of the negative effects (McClanahan 2010). High and sustained temperatures are one of the more important stresses to corals and the fish and fisheries associated with them, but high light level, still and stratified waters, sediments, and pollution can also contribute to or exacerbate the stress, while temperature variability and water movement by tides can reduce the impacts. Global maps of these stresses suggest regions such as the western Pacific Coral Reef Triangle, the northern Indian Ocean, and eastern Pacific are at the greatest risk of reduced productivity of coral reef fisheries (Maina et al. 2011).

Recent global surveys that included fisheries, economics, management, and conservation biology related issues have attempted to identify areas for future fisheries problems, or hotspots of problematic change (Alder et al. 2010; Worm and Branch 2012). The surveys used different methodologies and therefore identified somewhat different regions, but the overlaps are notable. Alder et al. (2010) evaluated 53 countries based on 14 indicators of biodiversity, social value, and jobs, and scaled them by the management priorities of markets, policy, security, or sustainability. The best-performing countries differed based on the above priorities but Argentina, Bangladesh, Faeroes, Iran, and Ukraine were found in the bottom for more than one management priority.

Sumalia et al. (2012) considered the economic opportunity costs or economic incentives to overexploit versus the intrinsic growth rates of fish, and identified the north-eastern coast of Canada, the Pacific coast of Mexico, the Peruvian coast, the southeastern coast of Africa, the Pacific coast of New Zealand, and the Antarctic region as future overfishing hotspots. In these regions, there is a negative balance between the incentive to exploit and the potential for fish production to accommodate this exploitation rate. Worm and Branch (2012) observe that some developed-country fisheries are improving, and that this improvement is diverting unsustainable fishing practices to poorer nations with less management capacity. They considered the factors of management effectiveness, numbers of species, and increasing and potentially unsustainable catch trends and identified the Pacific Coral Reef Triangle (Indonesia), the southeastern coast of Africa, the Red Sea, northwestern Africa, and the Gulf of California (Mexico) as future problem regions. The southeastern coast of Africa is identified by both of these studies and provides a basis for a detailed case study (McClanahan and Cinner 2012), which is briefly summarized below.

As a means to reduce exposure of fishery systems to stresses and pressures the principles of fisheries science recommend restrictions on fishing effort, gear use, fishing areas, and time, and on the species and sizes of caught individuals (Walters 1986). Environmental sciences would recommend reducing greenhouse gases and other pollutants to minimize climate warming and water quality issues. These recommendations, while apparently simple, are associated with costs or subsidies, such as management, and trade-offs, such as jobs and net income, that are difficult to resolve, optimize, or implement unless the human adaptive capacity is more fully considered.

FOOD SECURITY FROM MARINE FISHERIES: SCALING SENSITIVITY AND ADAPTIVE CAPACITY

Fisheries are overcapitalized and over-subsidized, leading to losses in gross food production and also larger losses in potential income (World Bank 2009). The sensitivity, or degree to which a system is likely to be modified by an event or stressor, of a nation or community to this suboptimal usage of resources is a reflection of the importance of fisheries to livelihood and nutrition security, and vulnerability will vary with the scale of the analysis and the severity of disturbance or exposure to pressures on the fishery system. In the context of food security, sensitivity may be affected by national-level dependence on trade, while at the local level dependence on natural resources, such as the relative production and profitability of local fisheries and agriculture, can be critical.

In fish-dependent communities, fish provide an important source of protein, and some species provide iron, iodine, vitamin A, calcium, and essential amino acids critical for child development, maternal health, and maintenance of good health in later life. Smaller fish, consumed whole and more affordable by lower-income consumers, often provide the most micronutrient-rich contributions to diets. In households that can replace beans and domestic animal protein with fish, considerable food security stability exists if local and national conditions are conducive to providing a choice (Kawarazuka and Bene 2011).

Kawarazuka and Bene (2011) scaled the 30 nations most dependent on fish consumption and found that between 34 and 76 percent of the populations' protein from animals comes from fish. Lower-income consumers in many of these countries still mostly rely on plant sources of protein, such as beans and pulses, but in some import-dependent countries, such as small-island developing states (SIDS) like the Maldives, Solomon Islands, Comoros, Seychelles, and Vanuatu, plant sources of protein may be less available and fish more accessible.

National-scale statistics about the contribution of fisheries to diet or GDP may not reflect the realities of fishing dependence among many communities. As an example of the potential mismatch between country-level data and community-level realities, a study of western Indian Ocean nations found that the fisheries sector does not appear to contribute greatly to the GDP or workforces at the national scale except in the Seychelles, and to a lesser extent the Maldives and Comoros (McClanahan and Cinner 2012). However, since many of the countries are large, have sizable inland populations, and artisanal coastal catches are not well quantified, national-level statistics do not properly reflect the sensitivity of coastal communities to changes in the fishery. Community-level data in certain communities reveal an extremely high dependence on fishing, either as a primary or secondary occupation, in contrast to the national-level statistics (Cinner and Bodin 2010).

Studies using network analysis as a means to visualize the relationships between fisheries and other occupational sectors, referred to as a "livelihood landscape" (Cinner and Bodin 2010; Cinner et al. 2012a), have allowed researchers to examine the level of participation in different occupational sectors, whether and how households participated in multiple sectors (i.e., the linkages between sectors), and whether and how specific occupations were consistently ranked as being more or less important than others. This technique shows how the local-level dependencies and interactions differ at the local level in national economies of various levels of poverty. The poorest nations appear more connected, and this is likely to provide them with a risk-spreading approach that can buffer disturbances (Cinner and Bodin 2010).

Capacity to deal with change or adaptive capacity requires assets and flexibility to help people weather disturbances, and this may increasingly involve having secondary occupations persist through severe global shocks, but these sectors provide some net gains or benefits to livelihood costs (Cinner et al. 2009a). In most tropical coastal communities, people regularly engage in a range of occupational sectors, such as agriculture, fishing, and informal economic activities, such as casual labor or entrepreneurial activities, including small shops or transportation, that are generally not monitored and taxed by the government (Allison and Ellis 2001; Barrett et al. 2001; Pomeroy et al. 2006). The ways that fishing is embedded in a landscape of other economic activities has critical implications for how food security influences political stability. Cinner et al. (2012a) showed that there is considerable variance between communities within countries that can mask the specific sensitivities and adaptive capacity and, therefore, the specific adaptation or development needs of communities (Fig. 6.1; Cinner et al. 2012a)

HOW TO BUILD ADAPTIVE CAPACITY FOR FOOD SECURITY: SOME KEY FACTORS

Adaptive capacity is critical to reducing food insecurity and refers to the conditions that enable people to anticipate and respond to changes; minimize, cope with, and recover from the consequences of change; and take advantage of new opportunities (Adger and Vincent 2005). People with high adaptive capacity are less likely to suffer from environmental and human disturbances and are therefore better able to take advantage of the opportunities to increase their food security. Progressive and successful policies and management are most likely to be specific to the social-ecological context and identify key factors limiting progress toward the desired social-ecological or food security state (Andrew et al. 2007; Ostrom 2007; McClanahan et al. 2008, 2009b). One-size-fits-all solutions, such as the creation of marine protected areas or

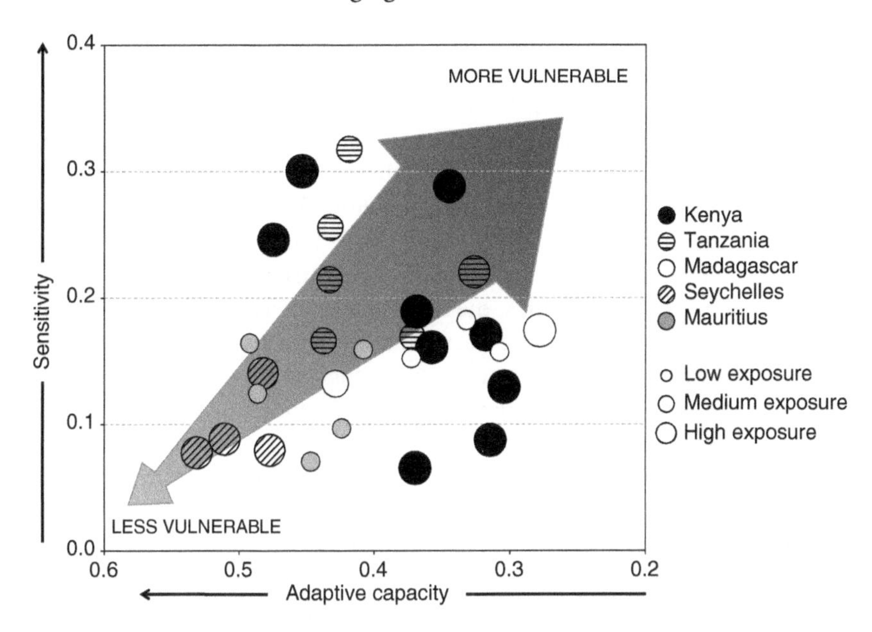

Fig. 6.1. Plot of the vulnerability of coastal communities to the impacts of coral bleaching on fisheries

Note: Adaptive capacity (x-axis; high adaptive capacity is on the left) is plotted against Sensitivity (y-axis). More vulnerable communities are in the top right of the graph and less vulnerable communities in the bottom left. Exposure is represented as the size of the bubble (larger = more exposure)

Source: Cinner et al. (2012a).

introducing private or community fishing rights, cannot solve the diversity of specific and immediate problems facing fisheries.

Flexibility

Flexibility of individuals and institutions is a critical component of adaptive capacity. Understanding where sources of flexibility already exist and where flexibility can be increased are key to building and managing the resilience of social-ecological systems. The above analysis of national-level vulnerability to food insecurity for coral reefs identified socio-economic equity as measured by the GINI coefficient and trade as a measure of flexibility at the national level. We are unaware of studies that have evaluated social equity on fisheries, but we predict that more even access to resources will result in more efficient usage. Comanagement, which is a form of social organization promoting equity has, however, been evaluated and found to increase fish biomass in a fishery (Cinner et al. 2012b) and reduce both local and international conflicts over access to territorial waters in Asian fisheries (Pomeroy et al. 2006).

Trade is another aspect of flexibility. Fish products are now the largest globally traded food commodity and often considered a key way for nations to improve their economies and food security. Nevertheless, the relationship between international trade in fish and food security is complex and appears to vary greatly from country to country (Allison 2011). A preliminary analysis found that some countries with large offshore fisheries, such as Namibia, greatly benefited from international trade, while those with coastal fisheries, such as Ghana, Philippines, and Kenya, did not (Kurien 2004).

In many coastal communities, the flexibility to switch between livelihood strategies is critical for resource users to cope with the high uncertainties and seasonal variability associated with fishing, farming, and other livelihoods (Allison and Ellis 2001). Flexibility can influence how people cope with change and avoid poverty traps where low assets and job options can prevent overcoming disturbances (Carter and Barrett 2006). In a hypothetical set of questions, Kenyan fishers were asked how they would respond to sustained declines in their catch, and those most inclined to remain in the fishery were poorer and had few livelihood alternatives (Cinner et al. 2009b). Nevertheless, when a similar analysis was conducted at a regional scale, these responses varied by country along a gradient of national wealth, and suggested the reverse response when moving across national scales in this region (Daw et al. 2012).

Daw et al. (2012) found that site-level factors had the greatest influence on readiness to exit, but contrary to the findings about household wealth, readiness to exit declined with increased community-level infrastructure development, economic vitality, and the biomass of the fish. Fishers in poor countries with more livelihoods and lower catch value were more willing to exit probably because they had a risk-spreading strategy that allowed them to easily move between alternatives as described above for livelihood landscapes. Consequently, adaptive responses are influenced by factors at multiple scales and social-ecological context.

Assets

Resources and economic assets help people to adapt to environmental changes by providing resources to draw on during sparse and changing conditions. Assets include various technologies that can increase production or provide monitoring and early warning systems and provide key infrastructure and sufficient biological resources that assist coping with environmental change. A key aspect of adaptive capacity is access to assets needed to mobilize resource stakeholders in the face of change (Adger 1999; Barnett 2001). The asset metrics used above included the GDP, sanitation infrastructure, and areas in marine protected areas, but other important aspects include the abundance and ecological state of the resource and alternative infrastructures, such as aquaculture.

Aquaculture production now accounts for nearly half of the current consumption of seafood (including freshwater fish) products and is expected to increase with demand for seafood and declining profitability of wild-caught fisheries. Global per capita food consumption of farmed fish has increased from 0.7 kilograms to 7.8 kilograms from 1970 to 2008, or a 6.6 percent annual rate of increase—higher than human population growth (FAO 2011). The markets, however, tend to be specialized and the production is very regional. By far the largest development is in Asia, mostly in China. This growth has been driven by rising demand from growing and urbanizing populations, stagnating supplies from capture fisheries, investment in education and technology research, a dynamic private sector, and high levels of public investment in infrastructure to support agricultural development.

China, Vietnam, Thailand, Indonesia, and the Philippines have developed a vibrant small and medium enterprise sector during the past 15 years, which targets both domestic and international markets (Beveridge et al. 2010). Aquaculture accounts for between 15–50 percent of the value of their agriculture in these countries (Scholtens and Badjeck 2010). Aquaculture developments frequently claim increased food security and reduced poverty benefits but this assumption is rarely tested and it is most likely that revenues accrue largely to a small number of relatively wealthy people in small and medium-enterprise production, and in processing and trade, and the extra production is sometimes directed toward exports to developed countries or for consumption by the growing middle classes in Asian cities. There is some evidence, however, that aquaculture development drives down the price of fish in domestic markets and makes it more affordable to the poor (Beveridge et al. 2010).

One surprising trend in aquaculture is the steady price over the past 15 years during a time when wild-caught fish prices are rising (FAO 2011). Given that fishmeal produced from wild-caught fisheries is commonly used in aquaculture, and conversion ratios from wild fish to farm fish production were typically 7:1, this is an unexpected outcome that arises during the early evolution of market processes. These conversion ratios are declining with various improvements and efficiencies that have continued over the past 30 years and are expected to continue (Tacon and Metian 2008).

This aquatic Green Revolution has, unfortunately, not resulted in many quantitative studies assessing the impact on poverty and food security (Allison 2011). Nevertheless, one clear message that emerges is that benefits of increased trade are dependent on existing relations of power, with those at the end of global value chains (small-scale producers, processing factory workers) sometimes remaining in poverty despite increases in trade revenue and domestic formal-sector employment (Geheb et al. 2008; Islam 2008). There are inevitable trade-offs between maintaining coastal commons and developing coastal aquaculture, between growing rice on coastal land and converting it to prawn farms.

Improved coastal, land-use, and wetland and water resource planning can help avoid disputes escalating into conflicts, or macro-economic choices leading to social injustices. In general, improved environmental and social governance in aquaculture is addressing some of the worst excesses of the past, such as large-scale mangrove conversion and displacement of small-scale fishers and farmers (Hall et al. 2011). Many studies have suggested that aquaculture at the smallest scale (farmers with ponds or coastal household-based enterprises) fails more frequently compared to small- and medium-scale enterprise approaches, and that aquaculture development therefore requires an intermediate level of social and economic capital to succeed (Allison 2011).

Projections of aquaculture production into the future, given reasonable IPCC emission scenarios and projected fishmeal and fish oil price estimations and technological development in aquaculture feed technology, predict that meeting the consumption rates is feasible if the wild fish resources, small pelagic fish from which fish feeds are derived, are managed sustainably and the animal feeds industry reduces its overall reliance on wild fish feed sources (Merino et al. 2012).

Specific policy advice for aquaculture from a global review by Allison (2011) recommends supporting the growth of the small- and medium-scale aquaculture enterprise sector, and particularly its role in increasing the availability of nutritious, affordable food in domestic markets. In countries with nascent aquaculture sectors, particularly in Africa, this can be achieved by support for innovation systems capable of contributing to the growth of the sub-sector supplying domestic and regional markets. The nutritional, equity, and environmental dimensions of aquaculture development also require policy attention to ensure that the sustainability and poverty-reduction benefits of aquaculture development are maximized.

Aquaculture is a potential asset for many countries, but again, it is most likely to succeed in regions with medium to high human capital and adaptive capacity, high domestic demand, or good trade links, and a poorer state of the wild-capture fisheries or dependence on imports, which will mean prices for fish are moderate to high. Ecological resources are likely to be a key asset for less developed and more food insecure countries where fishing capital and economies of scale have not developed sufficiently to access profitable offshore fishing and aquaculture. For example, fish biomass is one ecosystem asset that has proved useful in evaluating the state of coral reef fisheries resources in the western Indian Ocean, where considerable heterogeneity has been found and associated with different levels of human development and management (Cinner et al. 2009c; McClanahan et al. 2011). McClanahan et al. (2011) found that as the biomass of fish is reduced by fishing, the ecosystem also changes through a series of switch points where there are rapid changes in ecological characteristics, such as increased variability in benthic algal abundance, the ratio of algae/coral, the rates of predation on sea urchins, the numbers of fish species, sea urchins, calcifying algae, and amount of coral cover.

The suggested maximum yields of fish catch occur before a series of the switches, where the passage of each switch point is expected to make it more difficult for the ecosystem to maintain its full set of assets and recover to a state productive for fisheries. The biomass of fish varies greatly in the western Indian Ocean, where some countries have more or fewer ecological assets based on fishing pressure and fisheries management options, such as gear restrictions and fisheries closures. Overall, however, some fishing restrictions—often associated with national or co-management systems—tend to maintain the ecosystem for continuing fish production (Cinner et al. 2009c, 2012b; McClanahan et al. 2011). Stewardship of fish resource assets before the ecological switch points seems likely to offer a beneficial trade-off between fishery productivity and ecological resilience, allowing vulnerable populations to avoid social-ecological poverty traps and food insecurity.

Learning

Food security rests on the ability of individuals, institutions, and societies to assess potential response strategies and adapt to changes in causal factors that influence food production (Smit and Pilifosova 2003). If people do not perceive connections between human activities and the condition of resources, they are not likely to support management initiatives that restrict or manage resources. In some cases, the capacity to learn can be constrained by cultural or religious views that attribute all change to supernatural agents, natural processes, and decision makers too distant, immutable, or capricious to control. Low levels of education, poverty, and particularly weak governance can promote this vision (Norris and Ingelhart 2004), so investments in formal and informal education, scientific research on natural resources, and strengthening governance institutions to involve resource users in management decision-making are wise responses.

Adaptive management has been promoted as a useful way of learning in complex social-ecological systems (Walters 1986). This adaptive process can be quite sophisticated in taking advantage of scientific principles of replication, repeatability, and Before-After-Control-Impact (BACI) designs of management impacts on resources and human organization. The costs are, however, often too prohibitive to be replicated in all of the needed and poorest locations (Johannes 2002).

Social Organization

Collective action is critical for maintaining food security and responding to disturbances that threaten this security (Adger 2003). Vulnerability to food insecurity is influenced by the effectiveness of collective institutions, social networks, and demographic trends, such as population growth and migration (Thornton et al. 2008). Governance and services provided by national governments, societies, and informal organizations depend on levels of political conflicts, corruption, accountability, and overall trust (Barnett and Adger 2007; Vollan and Ostrom 2010). Aspects of social organization, such as leadership and effective communication, have proved to be critical to the success of fisheries (McClanahan et al. 2009a; Gutierrez et al. 2011; Cinner et al. 2012b, c).

One synthetic and useful approach to evaluating social organization is the institutional design principles that are part of successful management of common property, such as fisheries (Ostrom 1990). These principles were derived from the study of well-documented cases of long-enduring common-pool resource systems. These design principles interacting with local conditions can provide credible commitments that resource users will maintain and invest in their institutions over time. Recently, the usefulness of these design principles has been evaluated in the context of making water management institutions adaptive to climate change, resulting in some modifications and expansions of Ostrom's original concepts (Huntjens et al. 2011).

The design principles for climate change adaptation include: (1) clearly defined boundaries, such as geographic or institutional membership rights; (2) equitable distribution of risks, benefits, and costs; (3) congruence between rules and local conditions (i.e., scale and appropriateness); (4) collective choice arrangements that provide resource users with rights to make, enforce, and change the rules; (5) transparent monitoring and evaluating of both the resource and the management process; (6) effective arenas for discussion and agreement such as conflict resolution; (7) the degree to which they are nested within other institutions; (8) graduated sanctions (penalties that increase with the severity or number of infringements); (9) robust and flexible processes; and (10) the capacity for policies to be adaptive and change as learning occurs (adapted from Huntjens et al. 2011; Ostrom 1990; Cinner et al. 2012c). Importantly, many of these design principles are means to achieving adaptive and sustainable institutions, but they are not ends. In some contexts, there may be other, more locally appropriate, means to achieving similar ends.

Several key design principles have been investigated in fisheries to evaluate their current strengths and weaknesses (Cinner et al. 2009d, 2012b; Table 6.1). All fisheries have some of these elements, but it is common for some of them to be either missing or weak. Problems are typically evident in physical monitoring of the resources and informing local users, accountability mechanisms for those monitoring the rules, and sanctions systems. National governments often monitor the resource or catch, but in many cases this information is not fed back to the local level where it can influence decisions, but used farther up the management chain by national and international governments, leading to long delays and inappropriate recommendations for local management.

Table 6.1. A comparison of whether key institutional design principles are present in the marine resource co-management frameworks in Kenya, Madagascar, Mexico, Papua New Guinea (PNG), and Indonesia [a]

Design principle	Description	Indonesia	PNG	Mexico	Kenya	Madagascar
Clearly defined membership rights	Clear delineation of membership rights to co-managed area	Most sites	Most sites	Yes	Yes	Yes
Congruence	Whether scale and scope of rules are appropriate for the local conditions	All sites	All sites	Yes	Yes	Yes
Collective choice rights	Whether resource users have rights to make, enforce, and change the rules	All sites	All sites	Yes	Yes	Yes
Conflict resolution mechanisms	Rapid access to effective conflict-resolution forum	Some sites	Some sites	Yes	Yes	Yes
Nested enterprises	Nested within lead agencies or partner organizations at critical stages	Moderate	Weak	Very weak	Partially	Partially
Monitoring of process	Whether there are accountability mechanisms for those enforcing the rules	Not present in any site	Not present in any site	Unknown	Partially	Partially*
Clearly defined geographic boundaries	Clear delineation of co-management area (OS)	Most sites	Most sites	Yes	Partially	Partially
Graduated sanctions	Whether sanctions increase with numerous offences or the severity of the offense	Some sites	Some sites	Yes	Partially	Partially
Monitoring of resources	quantitative or qualitative monitoring of resource conditions	Few sites	Qualitative monitoring at all sites	Yes	Partially	No

[a] The study of Papua New Guinea, Mexico, and Indonesia (Cinner et al. 2012c) was a comparison of a number of specific indigenous fishery case studies, while the study of Kenya and Madagascar (Cinner et al. 2009d) was of the broader fisheries co-management framework. These different scales of analysis are reflected in the language used to describe the degree to which the design principles were present. Several of the design principles proposed by Huntjens et al. (2011), such as policy learning and robust flexible processes were not evaluated and difficult to quantify post hoc, so are not included.

Source: Cinner et al. 2009d; Cinner 2012c

Policy Coherence and Sequenced Development Intervention

Food security will rise when the above factors are fully considered and integrated into the various levels of development. We recommend that policymakers at the highest levels work more in partnership with development economists, planners, and practitioners to avoid a narrowly sectoral perspective and to ensure the sector's development goals fit with wider national economic development policy priorities; make poverty and food security goals and strategies explicit in fisheries and aquaculture sector policy; ensure coherence between major cross-sectoral development policies and programs; and identify the most promising pathways for positive food security impact through a cross-sectoral diagnosis. Inclusion of these activities is expected to reduce some of the conflicts that have plagued past fisheries development programs.

When it is possible to plan fisheries developments, we recommend a sequence of activities where the first priority is securing basic human rights of vulnerable resource users, so that people can effectively defend exclusive fishing rights designed to sustain ecosystems and economically viable fisheries. These fishing rights could be privately owned individual transferable quotas, community rights, or state-controlled licensing systems, according to context. These first and second priorities of strengthening both human rights and fishing rights broadly align with ongoing attempts to implement the 1995 FAO Code of Conduct for Responsible Fisheries (Allison et al. 2012). Only when poor people's basic human rights and exclusive fishing rights are relatively secure are they likely to benefit from the current trend toward stronger links with global markets. Promotion of greater global market integration should also promote the development of a "green economy" in the fisheries sector through the use of existing instruments governing global value chains, such as Marine Stewardship Council certification ("ecolabelling") and fair trade principles (Hatanaka et al. 2005). This sequence will counter some of the problems that arise from unregulated market-driven developments that can undermine natural resources and the livelihoods and food security of dependent people. When markets are poorly coordinated with local social-ecological interactions and institutions they can become globalized faster than sustainable management can be implemented (Cinner and McClanahan 2006).

Conflict Prevention and Resolution

Improved implementation of the FAO Code of Conduct for Responsible Fisheries (FAO 1995)—which, for example, urges states to set aside areas for exclusive use by small-scale fisheries—is likely to reduce local conflicts. Additionally, stakeholder-based comanagement that creates forums for public discussion among different types of fishers can assist conflict resolution.

Promoting the rights and inclusion of small-scale fishers in wider development processes will help reduce their marginalization and associated conflicts. National and marine spatial planning of developments such as dam construction, coastal land reclamation, tourism, and conservation planning that includes fisher and common property users' opinions is expected to reduce conflicts (Douvere 2008). Improving the mechanisms, such as judicial services, for resolving disputes that are fair and transparent within various decision-making and use contexts can help ensure fair governance (Ratner and Allison 2012). While the above factors are important, the wider governance and economic reforms that reduce marginalization will be critical to conflict resolution, as fisheries are part of this wider context.

TOWARD BETTER FISHERIES GOVERNANCE

If not resolved, current trends—declining net resources, increased costs and sensitivity by high resource dependence, inescapable poverty traps through lack of resources and alternatives, and increasing competition for access—predict more conflicts between poor and wealthy nations over equitable trade, and potentially less food security in the future. Future conflicts between nation states over fishery resource access are likely to be found where there are weak forums and societal dialogue to handle local and international disputes. Disputed territories like the Spratley Islands are potential international flash points where mineral rights, fishing rights, and strategic location could drive escalating disputes. Within national boundaries, weak governance and autocracies that ignore democratic processes and reforms are likely to combine with other non-fishery issues to create internal conflicts. Examples include the Horn of Africa, West Africa, Sri Lanka, Cambodia, Philippines, and Indonesia, where popular resistance has equalized, or has attempted to equalize, disparities between local elites, outsiders, and criminal or military elements. Avoiding these resource inequities will require coherence and cooperation between policies for economic growth, resource management, poverty reduction, and food security.

Fish may become a luxury, inaccessible to those who rely on common property. Policies, access rights, and markets and governance systems that function effectively to distribute the benefits of local and global trade while strengthening common property rights will help to avoid the marginalization and food insecurity typical of poor people dependent on aquatic common property (Gutierrez et al. 2011; Cinner et al. 2012c). Improving local and regional markets and small- and medium-scale aquaculture and animal husbandry offer counterweights to the possibility of rising global exports and inequity of access (Allison 2011).

Some hopeful developments now emerging, such as fewer subsidies, reduced capital investment, precautionary management to minimize risks of ecosystem collapse, conservation of remaining resources, diversified portfolios of production and markets, and greater equity in contracts and distribution, predict a future where these conflicts can potentially be resolved. The diversity of issues and circumstances requires a diagnostic process to assess and guide policy formulation. We recommend a contextual diagnostic and environmental justice framework to assess a range of options for fishery governance reform in terms of their likely impact on the most dependent, food-insecure people (Ratner and Allison 2012). In general, addressing illegal fishing and strengthening property rights (whether state, private, or communal); ensuring that wider aquatic ecosystem service functions are preserved via habitat management; inclusion of fishing-dependent regions in climate change adaptation planning; aligning fishery sector policy with wider environmentally aware development planning; and support for more inclusive, transparent, and accountable forms of decision-making in fish value chains will all help to secure the future contribution of aquaculture and fisheries to human food security and well-being.

ACKNOWLEDGMENTS

We are grateful to the Wildlife Conservation Society, James Cook University, Australian Research Council, the University of East Anglia, and the Rockefeller and John D. and Catherine T. MacArthur Foundations for providing the capacity to complete this chapter.

NOTES

1. A longer version of this paper appeared as McClanahan T. R., E. H. Allison, and J. E. Cinner. 2013. Managing fisheries for human and food security. *Fish and Fisheries* (forthcoming).

REFERENCES

Adger, W. N. 1999. Social vulnerability to climate change and extremes in coastal Vietnam. *World Development 27*: 249–69.

———. 2003. Social capital, collective action, and adaptation to climate change. *Economic Geography 79*: 387–404.

———, and K. Vincent. 2005. Uncertainty in adaptive capacity. *Comptes Rendus Geoscience 337*: 399–410.

Alder, J., S. Cullis-Suzuki, V. Karpouzi, K. Kaschner, S. Mondoux, W. Swartz, P. Trujillo, R. Watson, and D. Pauly. 2010. Aggregate performance in managing marine ecosystems of 53 maritime countries. *Marine Policy 34*: 468–76.

Allison, E. H. 2011. Aquaculture, fisheries, poverty and food security. Working Paper, 2011-65. Penang, Malaysia: World Fish Center.

——, and F. Ellis. 2001. The livelihood approach and management of small-scale fisheries. *Marine Policy 25*: 377–88.

Allison, E. H., B. D. Ratner, B. A. Asgard, R. Willmann, R. D. Pomeroy, and J. Kurien. 2012. Rights-based fisheries governance: From fishing rights to human rights. *Fish and Fisheries 13*: 14–29.

Andrew, N. L., C. Béné, S. J. Hall, E. H. Allison, S. Heck, and B. D. Ratner. 2007. Diagnosis and management of small-scale fisheries in developing countries. *Fish and Fisheries 8*: 227–40.

Barnett, J. 2001. *The meaning of environmental security: Environmental politics and policy in the new security era.* New York: Zed Books.

——, and W. N. Adger. 2007. Climate change, human security and violent conflict. *Political Geography 26*: 639.

Barrett, C. B., T. Reardon, and P. Webb. 2001. Nonfarm income diversification and household livelihood strategies in rural Africa: Concepts, dynamics and policy implications. *Food Policy 26*: 315–31.

Behrenfeld, M. J., R. T. O'Malley, D. A. Siegel, C. R. McClain, J. L. Sarmiento, G. C. Feldman, A. J. Milligan, P. G. Falkowski, R. M. Letelier, and E. S. Boss. 2006. Climate-driven trends in contemporary ocean productivity. *Nature 444*: 752–55.

Béné, C., R. Lawton, and E. H. Allison. 2010. Trade matters in the fight against poverty: Narratives, perceptions, and (lack of) evidence in the case of fish trade in Africa. *World Development 38*: 933–54

——, M. Phillips, and E. H. Allison. 2011. The forgotten service: Food as an ecosystem service from estuarine and coastal zones. In *Treatise on Estuarine and Coastal Science*, ed. E. Wolanski and D. S. McLusky, 147–80. Waltham, MA: Academic Press.

Beveridge, M. C. M., M. Phillips, P. Dugan, and R. Brummet. 2010. Barriers to aquaculture development as a pathway to poverty alleviation and food security. In *Advancing the Aquaculture Agenda*, 345–59. Paris: Organization for Economic Cooperation and Development.

Brander, K. 2007. Global fish production and climate change. *Proceedings of the National Academy of Sciences 104*: 19709–14.

Bruno, J. F., and E. R. Selig. 2007. Regional decline of coral cover in the Indo-Pacific: Timing, extent, and subregional comparisons. *PLOS ONE 2*: e711.

Bush, R. 2010. Food riots: Poverty, power and protest. *Journal of Agrarian Change 10*: 119–129.

Carter, M. R., and C. B. Barrett. 2006. The economics of poverty traps and persistent poverty: An asset-based approach. *Journal of Development Studies 42*: 178–99.

Cheung, W. W. L., C. Close, V. Lam, R. Watson, and D. Pauly. 2008. Application of macroecological theory to predict effects of climate change on global fisheries potential. *Marine Ecology Progress Series 365*: 187–97.

——, J. Dunne, J. L. Sarmiento, and D. Pauly. 2011. Integrating ecophysiology and plankton dynamics into projected maximum fisheries catch potential under climate change in the Northeast Atlantic. *ICES Journal of Marine Science 68*: 1008–18.

——, V. W. Y. Lam, J. L. Sarmiento, K. Kearney, R. E. G. Watson, D. Zeller, and D. Pauly. 2010. Large-scale redistribution of maximum fisheries catch potential in the global ocean under climate change. *Global Change Biology 16*: 24–35.

——, V. W. Y. Lam, J. L. Sarmiento, K. Kearney, R. Watson, and D. Pauly. 2009. Projecting global marine biodiversity impacts under climate change scenarios. *Fish and Fisheries 10*: 235–51.

Cinner, J. E. and Ö. Bodin. 2010. Livelihood diversification in tropical coastal communities: A network-based approach to analyzing "livelihood landscapes." *PLOS ONE* 5: e11999.

——, T. Daw, and T. R. McClanahan. 2009b. Socioeconomic factors that affect artisanal fishers' readiness to exit a declining fishery. *Conservation Biology 23*: 124–30.

——, N. Muthiga, C. Abunge, S. Hamed, B. Mwaka et al. 2012c. Transitions toward co-management: The process of marine resource management devolution in three east African countries. *Global Environmental Change 22*: 651–58.

——, M. Fuentes, and H. Randriamahazo. 2009a. Exploring social resilience in Madagascar's marine protected areas. *Ecology and Society 14*: 41.

——, and T. R. McClanahan. 2006. Socioeconomic factors that lead to overfishing in small-scale coral reef fisheries of Papua New Guinea. *Environmental Conservation 33*: 73–80.

——, T. M. Daw, N. A. J. Graham, J. Maina, S. K. Wilson, and T. P. Hughes. 2009c. Linking social and ecological systems to sustain coral reef fisheries. *Current Biology 19*: 206–12.

——, T. R. McClanahan, M. A. MacNeil, N. A. J. Graham, T. M. Daw, A. Mukminin, D. A. Feary et al. 2012b. Comanagement of coral reef social-ecological systems. *Proceedings of the National Academy of Sciences 109*: 5219–22.

——, T. R. McClanahan, S. Stead, N. A. J. Graham, T. Daw, J. Maina, A. Wamukota, O. Bodin, and K. Brown. 2012a. Vulnerability of coastal communities to key impacts of climate change on coral reef fisheries. *Global Environmental Change 22*: 12–20.

——, A. Wamukota, H. Randriamahazo, and A. Rabearisoa. 2009d. Toward institutions for community-based management of inshore marine resources in the Western Indian Ocean. *Marine Policy 33*: 489–96.

Daw, T. M., J. E. Cinner, T. R. McClanahan, K. Brown, S. Stead, N. A. J. Graham, and J. Maina. 2012. To fish or not to fish: Factors at multiple scales affecting artisanal fishers' readiness to exit a declining fishery. *PLOS ONE 7*: e31460.

Daxecker, U., and B. Prins. (forthcoming). Insurgents of the sea: Institutional and economic opportunities for maritime piracy. *Journal of Conflict Resolution.*

Douvere, F. 2008. The importance of marine spatial planning in advancing ecosystem-based sea use management. *Marine Policy 32*: 762–71.

DuBois C. and C. Zografos. 2012. Conflicts at sea between artisanal and industrial fishers: Inter-sectoral interactions and dispute resolution in Senegal. *Marine Policy 36*: 1211–20.

FAO. 1995. *The Code of Conduct for responsible fisheries.* Rome: United Nations Food and Agriculture Organization.

——. 2007. Making fish trade work for development and livelihoods in West and Central Africa. New directions in fisheries: A series of policy briefs on development issues. No. 10. Rome: United Nations Food and Agriculture Organization.

——. 2011. *The State of Fisheries and Aquaculture.* Rome: United Nations Food and Agriculture Organization.

Fazey, I., J. A. Fazey, J. Fischer, K. Sherren, J. Warren, R. F. Noss, and S. R. Dovers, 2007. Adaptive capacity and learning to learn as leverage for social-ecological resilience. *Frontiers in Ecology and Environment 5*: 375–380.

Froese, R., and A. Proelb. 2010. Rebuilding fish stocks no later than 2015: Will Europe meet the deadline? *Fish and Fisheries 11*: 194–202.

Gardner, T. A., I. M. Cote, J. A. Gill, A. Grant, and A. R. Watkinson. 2003. Long-term region-wide declines in Caribbean corals. *Science 301*: 958–60.

Geheb, K., S. Kalloch, S. Medard, A.-T. Nyapendi, C. Lwenya, and M. Kyangwa. 2008. Nile perch and the hungry of Lake Victoria: Gender, status and food in an East African fishery. *Food Policy 33*: 85–98

Gutierrez, N. L., R. Hilborn, and O. Defeo. 2011. Leadership, social capital and incentives promote successful fisheries. *Nature 470*: 386–9.

Hall, S. J., A. Delaporte, M. Phillips, M. C. M. Beveridge and M. O'Keefe. 2011. *Blue frontiers: Managing the environmental costs of aquaculture.* Penang, Malaysia: The World Fish Center.

Hatanaka, M., C. Bain, and L. Busch. 2005. Third-party certification in the global agro-food system. *Food Policy 30*: 354–69.

Hendrix, C. S., and S. M. Glaser. 2011. Civil conflict and world fisheries, 1952–2004. *Journal of Peace Research 48*: 481–95.

Hilborn R., I. J. Stewart, T. A. Branch, and O. P. Jensen. 2012. Defining trade-offs among conservation, profitability, and food security in the California current bottom-trawl fishery. *Conservation Biology: The Journal of the Society for Conservation Biology 26*: 257–66.

Hughes, S., A. Yau, L. Max, N. Petrovicc, F. Davenport, M. Marshall, T. McClanahan, E. Allison, and J. E. Cinner. 2012. A framework to assess national level vulnerability from the perspective of food security: The case of coral reef fisheries. *Environmental Science and Policy 23*: 95–108.

Huntjens, P., L. Lebel, C. Pahl-Wostl, R. Schulze, J. Camkin, and N. Kranz. 2011. Institutional design propositions for the governance of adaptation to climate change in the water sector. *Global Environmental Change 22*: 67–81.

Islam, M. D. S. 2008. From sea to shrimp processing factories in Bangladesh: Gender and employment at the bottom of a global commodity chain. *Journal of South Asian Development 3*: 211–36

Jaunky, V. 2011. Fish exports and economic growth: The case of SIDS. *Coastal Management 39*: 377–95.

Jennings, S., M. J. Kaiser, and J. E. Reynolds. 2001. *Marine Fisheries Ecology*. Oxford, UK: Blackswell Science.

Johannes, R. E. 2002. The rennaissance of community-based resource management in Oceania. *Annual Review of Ecology and Systematics 33*: 317–40.

Johannesson G. T. 2004. How "cod war" came: The origins of the Anglo-Icelandic fisheries dispute, 1958–61. *Historical Research 77*: 543–74

Kawarazuka, N. and C. Béné. 2011. The potential role of small fish species in improving micronutrient deficiencies in developing countries: Building the evidence. *Public Health Nutrition 14*: 1927–38.

Kurien, J. 2004. *Fish trade for the people: Toward understanding the relationship between international fish trade and food security.* Rome: Food and Agriculture Organization of the United Nations and the Royal Norwegian Ministry of Foreign Affairs.

Lam, V. W. Y., W. W. L. Cheung, W. Swartz, and U. R. Sumaila. 2012. Climate change impacts on fisheries in West Africa: Implications for economic, food and nutritional security. *African Journal of Marine Science 34*: 103–17.

Le Manach F., C. Gough, A. Harris, F. Humber, S. Harper, and D. Zeller. 2012. Unreported fishing, hungry people and political turmoil: The recipe for a food security crisis in Madagascar. *Marine Policy 36*: 218–25

McClanahan, T. R. 2010. Effects of fisheries closures and gear restrictions on fishing income in a Kenyan coral reef. *Conservation Biology 24*: 1519–28.

———, J. C. Castilla, A. T. White, and O. Defeo. 2009a. Healing small-scale fisheries by facilitating complex socio-ecological systems. *Reviews in Fish Biology and Fisheries 19*: 33–47.

———, and J. E. Cinner. 2012. *Adapting to a changing environment: Confronting the consequences of climate change.* New York: Oxford University Press.

———, J. E. Cinner, N. A. J. Graham, T. M. Daw, J. Maina, S. M. Stead, A. Wamukota, K. Brown, V. Venus, and N. V. C. Polunin. 2009b. Identifying reefs of hope and hopeful actions: Contextualizing environmental, ecological, and social parameters to respond effectively to climate change. *Conservation Biology 23*: 662–71.

———, Cinner, J. Maina, N. A. J. Graham, T. M. Daw, S. M. Stead, A. Wamukota, K. Brown, M. Ateweberhan, and V. Venus. 2008. Conservation action in a changing climate. *Conservation Letters 1*: 53–59.

———, J. M. Maina, and N. A. Muthiga. 2011. Associations between climate stress and coral reef diversity in the Western Indian Ocean. *Global Change Biology 17*: 2023–32.

MacNeil, A. M., N. A. J. Graham, J. E. Cinner, N. K. Dulvy, P. A. Loring, S. Jenning, N. V. C. Polunin, A. T. Fisk, and T. R. McClanahan. 2010. Transitional states in marine fisheries: Adapting to predicted global change. *Philosophical Transactions of the Royal Society B: Biological Sciences 365*: 3753–63.

Maina, J., T. R. McClanahan, V. Venus, M. Ateweberhan, and J. Madin. 2011. Global gradients of coral exposure to environmental stresses and implications for local management. *PLOS ONE 6*: e23064.

Merino, G., M. Barange, J. L. Blanchard, J. Harle, R. Holmes, I. Allen, E. H. Allison et al,. 2012. Can marine fisheries and aquaculture meet fish demand from a growing human population in a changing climate? *Global Environmental Change 22*: 795–806.

Muawanah U., Pomeroy R. S., and Marlessy C. 2012. Revisiting fish wars: Conflict and collaboration over fisheries in Indonesia. *Coastal Management 40*: 279–88.

Norris, P., and R. Inglehart. 2004. *Sacred and secular: Religion and politics worldwide.* Cambridge: Cambride University Press.

Odum, H. T. 1988. Self-organization, transformity, and information. *Science 242*: 1132–9.

Ostrom, E. 1990. *Governing the Commons: The Evolution of Institutions for Collective Action.* Cambridge: Cambridge University Press.

——— 2007. A diagnostic approach for going beyond panaceas. *Proceedings of the National Academy of Sciences 104*: 15181–7.

Palumbi, S., and K. Sotka. 2010. *The Death and Life of Monterey Bay: A story of revival.* Washington, DC: Washington Island Press.

Pomeroy, R. S., B. D. Ratner, S. J. Hall, J. Pimoljinda, and V. Vivekanandan. 2006. Coping with disaster: Rehabilitating coastal livelihoods and communities. *Marine Policy 30*: 786–93.

Pratchett, M. S., P. L. Munday, S. K. Wilson, N. A. Graham, J. E. Cinner, D. R. Bellwood, G. P. Jones, N. V. C. Polunin, and T. R. McClanahan. 2008. Effects of climate-induced coral bleaching on coral-reef fishes: Ecological and economic consequences. *Oceanography and Marine Biology: An Annual Review 46*: 251–96.

Ratner, B. D. and E. H. Allison. 2012. Wealth, rights, and resilience: An agenda for governance reform in small-scale fisheries. *Development Policy Review 30*: 371–98.

Salomon, A. K., S. K. Gaichas, O. P. Jensen, V. N. Agostini, N. A. Sloan, J. Rice, T. R. McClanahan et al. 2011. Bridging the divide between fisheries and marine conservation science. *Bulletin of Marine Science 87*: 251–74.

Scholtens, J. and M-C. Badjeck. 2010. Dollars, work and food: Understanding dependency on the fisheries and aquaculture sector. IIFET 2010: Economics of Fish Resources and Aquatic Ecosystems: Balancing Uses, Balancing Costs. July 13–16, Montpellier, France.

Smit B, and O. Pilifosova. 2003. From adaptation to adaptive capacity and vulnerability reduction. In *Climate change, adaptive capacity and development*, ed. J. Smith, R. Klein, and S. Huq, 9–28. London: Imperial College Press.

Smith, M. D., C. A. Roheim, L. B. Crowder, B. S. Halpern, M. Turnipseed, J. L. Anderson, F. Asche et al. 2010. Sustainability and global seafood. *Science 327*: 784–86.

Swartz, W., E. Sala, S. Tracey, R. Watson, D. Pauly, and S. A. Sandin. 2010. The spatial expansion and ecological footprint of fisheries (1950 to present). *PLOS ONE 5*: e15143.

Tacon, A. G. J., and M. Metian. 2008. Global overview of the use of fish meal and fish oil in industrially compounded aquafeds: Trends and future propects. *Aquaculture* 285: 146–158.

Thornton, P., T. Jones, T. Owiyo, R. Kruska, M. Herrero, V. Orindi, S. Bhadwal et al. 2008. Climate change and poverty in Africa: Mapping hotspots of vulnerability. *African Journal of Agricultural and Resource Economics 2*: 24–44.

Tilman, D. 1999. Global environmental impacts of agricultural expansion: The need for sustainable and efficient practices. *Proceedings of the National Academy of Sciences 96*: 5995–6000.

Veuthy, S. and J. F. Gerber. 2011. Valuation contests over the commoditisation of the moabi tree in south-eastern Cameroon. *Environmental Values 20*: 239–64.

Vollan, B. and E. Ostrom. 2010. Cooperation and the commons. *Science 330*: 923.

Walters, C. 1986. *Adaptive Management of Renewable Resources.* New York: Mamillan Publishing Co.

Watson, R. T. and D. L. Albritton. 2001. *Climate change 2001: Synthesis report.* Cambridge: Cambridge University Press.

Watson, R. A., W. W. L. Cheung, J. A. Anticara, U. R. Sumaila, D. Zeller, and D. Pauly. 2012. Global marine yield halved as fishing intensity redoubles. *Fish and Fisheries*. doi: 10.1111/j.1467-2979.2012.00483.x.

Watson, R. and D. Pauly. 2001. Systematic distortions in world fisheries catch trends. *Nature 414*: 534–6.

World Bank. 2009. *The sunken billions: The economic case for fisheries reform.* Washington, DC and Rome: The World Bank and FAO.

Worm, B., E. B. Barbier, N. Beaumont, J. E. Duffy, C. Folke, B. S. Halpern, J. B. C. Jackson et al. 2006. Impacts of biodiversity loss on ocean ecosystem services. *Science 314*: 789–90.

——, and T. A. Branch. 2012. The future of fish. *Trends in Ecology and Evolution 27*: 594–599.

——, R. Hilborn, J. K. Baum, T. A. Branch, J. S. Collie, C. Costello, M. J. Fogarty et al. 2009. Rebuilding global fisheries. *Science 325*: 578–85.

7

Crop Technologies for the Coming Decade

SUSAN MCCOUCH AND SAMUEL CROWELL

Improvements in agricultural productivity over the last half-century have largely been achieved through technological change enabled by investments in agricultural research, development, extension, and education. While the world's population increased from 3.1 billion in 1960 to over seven billion in 2012, total cereal production grew faster than the population, from 643 million metric tons in 1960 to more than 2,280 million metric tons in 2010 (FAO 2010). This boost in global crop productivity was brought about by a combination of crop genetics and breeding, drastically altered agronomic practices, greater access to credit and information by farmers, and enhanced real-time communication networks. These developments were supported by investments in agricultural research and development in both the public and private sectors.

Despite significant global increases in aggregate output since 1961, yield gains in staple food crops have slowed during 1990 to 2007 compared to 1961 to 1990 (Alston et al. 2010). With population and income growing rapidly around the world and climate change making extreme weather events more likely, slowing yields raise serious concerns about our ability to sustain the increases in agricultural supply needed to meet future demands for food, feed, fiber, and fuel. Demand is growing fastest in Asia and Africa, where there is little spare land or water available for agricultural expansion. Thus, increasing crop productivity through technological change represents the best hope for improving food security in these regions.

Although large-holder farmers are crucial for maintaining affordable food supplies in urban centers, an estimated 2.3 billion people (one-third of humankind) currently depend on income derived from small farms of less than two hectares in Asia and Africa (Barrett et al. 2001). Farm yields in key crops still vary significantly between farming regions, however, and the cost–benefit ratios of different agricultural practices often remain far below their potential, which results in a significant "yield and sustainability gap" (Byerlee et al. 2009).

There are many reasons for the shortfalls, including fundamental resource constraints (poverty); insufficient information or education about improved agronomic practices; lack of timely access to inputs, such as fertilizer, pesticides, herbicides, or water; use of crop varieties that are susceptible to pests, diseases, and abiotic stresses; post-harvest losses due to poor storage facilities; and ineffective market integration. Existing technologies and management systems can sometimes address these disparities when coupled with investments in extension and outreach. In other cases, new crop technologies are key ingredients that can boost the productivity and profitability of a farm operation.

SCIENTIFIC HAVES AND HAVE-NOTS

Over the past two decades, publicly funded agricultural R&D programs have been drastically reduced, while the private agricultural sector has experienced unprecedented growth—a trend that is particularly noticeable in wealthy countries since the 1990s (Pardey et al. 2006). This shift has brought about a suite of changes that affects every aspect of agriculture in both the developed and developing world. Mechanical and chemical technologies that generate greater returns on investments have become R&D's major focus, while development of new inbred varieties in staple food crops and environmental management strategies that enhance on-farm productivity has dwindled. With a strengthening of intellectual property rights (IPR); restrictions on the exchange of germplasm; introduction of costly biosafety regulations; and a shift toward endeavors that are more likely to maximize profit, such as improving food quality and developing novel medical, energy, and industrial applications of agricultural products (Pardey and Pingali 2010), basic research and extension focused on increasing crop productivity has been critically underfunded.

As wealthy countries reorient their agricultural R&D away from the types of technologies that are most easily adopted by farmers in developing countries, they limit low-income producers' rapid uptake of new technology. A significant geopolitical divide in ownership of scientific technology worldwide is leading to a growing gap between the scientific haves and have-nots (Pardey et al. 2006). Gene grabs, land grabs, and water wars undertaken by governments and private-sector institutions seeking to control scarce resources are manifestations of this divide, threatening political stability as poorer or weaker nations and populations are outcompeted for access to essential natural resources. Obstacles that continue to hinder the flow of improved germplasm, technology, and information deepen the gap between rich and poor, leading to greater vulnerability in resource-poor populations threatened by climate change and global economic instability.

Public–private partnerships that are explicitly pro-poor and have a commitment to R&D intended to bridge the technology gap and address the needs of smallholder farmers offer some hope of addressing the root causes of social unrest and political upheaval. Where the public sector typically lacks resources to make these investments—and the private sector often lacks the motivation to do so—foundations can step in and foster these partnerships. Complex institutional arrangements are often required to negotiate access to technology, establish biosafety protocols, and engage farmers and farming communities in their own development. If well designed, public–private partnerships can maximize gains for both investors and small-scale farmers (Ferroni and Castle 2011).

GENETICS AND AGRONOMICS: TWO SIDES OF THE SAME COIN

We use the term "crop technologies" to refer to the combination of genetics (new crop varieties) and agronomic practices (management of soil, water, fertilizer, pests, diseases, and so on). Crop technologies are multicomponent systems developed by multiple individuals or institutions, and they continue to evolve as farmers deploy them. Successful execution of crop technologies depends on farmers' ability to adapt basic ideas and innovations to new situations using locally available materials to integrate different components in different combinations—the old and the new, the low tech and the high tech. Ingenious combinations that optimize on-farm profitability, productivity, and input use efficiency are often developed by resource-poor farmers who need them the most, sometimes with the help of extension agents, nonprofit institutions, or private-sector agricultural advisers. Involving farmers and community members in disseminating and adopting new crop technologies helps the investments to take root and invites local innovation that is needed to drive development.

Over the last 50 to 60 years, yield gains in the world's major staple crops were achieved through a combination of intensive plant breeding and large-scale environmental engineering. Farmers optimized the performance of their cropping systems by combining improved seed, new forms of mechanization, and intensive management of water, fertilizer, pesticide, and herbicide—doubling or tripling yields per unit area. From the 1960s to the 1980s, the Green Revolution increased agricultural production around the world by fitting into this paradigm: international crop improvement efforts led by the International Maize and Wheat Center (CIMMYT) and the International Rice Research Institute (IRRI) provided high-yielding, semi-dwarf varieties of wheat and rice to farmers (with no intellectual property protection) and dramatically increased harvestable yields, but only if farmers could supply irrigation, fertilizer, and pest and disease management. A supportive policy environment also provided access to credit and expanded markets, both critical in driving the productivity gains made possible by improved genetics. Today, this yield optimization paradigm has given way to new realities, with more attention to renewable resource management and long-term environmental impact. Slowing productivity gains, shifting global climate trends, and public outcry against inefficiencies in the use of agricultural inputs leading to contamination of food, soil, and waterways highlight the need for a fresh approach to variety development, pest management, and soil and water resource conservation.

Conservation agriculture evolved as a way for smallholder farmers to enhance soil resources and manage expensive chemical and water inputs (Hobbs 2007), but these practices quickly permeated all levels of agriculture, as they represent a sustainable and economic way of managing farm resources at any level of income or production. Conservation agriculture involves three interlinked principles: (1) minimal mechanical soil disturbance (no-tillage), (2) permanent organic soil cover (living cover crop or green mulch), and (3) diversified crop rotations (FAO 2012). Zero tilling maintains a permanent or semipermanent organic soil cover, which physically protects the soil from sun, rain, and wind. Soil microorganisms and fauna are encouraged to perform functions, such as aeration of soil and nutrient release, that tillage used to play in conventional farming systems. A varied crop rotation scheme involving more than two crops enhances the diversity of the system both above and below ground, reducing disease and pest problems.

Conservation agriculture represents one approach towards developing more sustainable farming systems. However, there are still many questions concerning how basic principles and practices should be adapted and integrated into local farm conditions. Plant breeders are working to develop new varieties that fit into more sustainable cropping systems. They face questions about which combinations of crops will be most productive, which traits are most urgently needed, what breeding technologies are most appropriate, and how to breed for a future when pests, diseases, and climate are extremely difficult to predict.

ACCESS TO PLANT GENETIC RESOURCES

Plant breeding's success depends on a breeder's ability to access, utilize, and exchange natural forms of genetic variation found in wild and domesticated plant species. Traditionally, farmers selectively maintained favorable plants and eliminated weak, diseased, or unattractive plants from wild or early-domesticated populations. Over time, this form of breeding led to the development of landraces of crops (or farmer varieties) and ornamental plants that provided the basis for virtually all of modern plant breeding. These early plant varieties and other genetic resources were historically considered part of the global genetic commons and could be freely collected and shared across the world. This notion of a shared genetic heritage led to the establishment of large botanical collections and the wide dissemination of valued crop plants. During the twentieth century, the Consultative Group on International Agriculture (CGIAR) and the United Nations Food and Agriculture Organization (FAO) established international seed banks to collect, maintain, and disseminate genetically diverse strains of economically important plant species.

Ownership of genetic resources is now a major point of discussion affecting both public- and private-sector breeding programs. Traditional forms of collecting and exchanging germplasm are giving way to new realities. Intellectual property rights are being asserted over plant varieties at a rapid pace and countries are closing their borders to exchange of indigenous genetic resources. The global genetic commons is shrinking at a very rapid pace, heightening tensions about access to both genetic resources and the technology needed to harness their potential—and deepening the divide between scientific haves and have-nots.

Plant Variety Protection and Utility Patents: Closing the Commons

Treating the products of agricultural research as private goods is not a new concept. The US Plant Patent Act (1930) allowed patent protection for asexually reproduced plants, focusing mostly on horticultural and ornamental species and specifically excluding potato. A second wave of appropriation started in Europe during the 1940s and 1950s with the adoption of a system of plant breeders' rights. These rights were revised and extended to other countries with the establishment of the International Convention for the Protection of New Varieties of Plants (UPOV) in 1961. The system granted plant breeders a form of IPR referred to as Plant Variety Protection (PVP) that is currently in place in about 70 countries. PVP allows plant breeders to protect varieties that they develop from appropriation by others, but it guarantees the breeders' exemption, allowing other breeders to use PVP-protected varieties to generate further varietal improvement. It also guarantees farmers' privilege, allowing farmers to save seed of PVP varieties for their own reuse. These provisions are designed to protect plant breeders' rights while ensuring continued access to plant genetic resources and sharing of benefits derived from their use.

While in many parts of the world crop genetic resources are still managed as public goods, the situation changed dramatically in the United States with the 1980 *Diamond v. Chakrabarty* Supreme Court decision. This decision allowed utility patents to be issued on plant varieties for the first time. Utility patents confer the patent holder exclusive rights to exploit the "invention" and to exclude others from making, reproducing, using, selling, or importing the patented variety for a period of 20 years. There is neither breeders' exemption nor farmers' privilege, and unlike industrial utility patents, there is no requirement for the plant variety patent holder to disclose the technical process by which a new plant variety was developed. This contributes to a "closing of the commons," because the private innovator's rights supplant the rights of both the public breeding and traditional farming sectors, preventing access to technological innovation and enhanced germplasm that are the basis of innovation. Today, plants are eligible for utility patents in the US, Australia, and the EU, although in the EU a plant variety must be documented to carry a "particular gene" to be eligible for patent protection.

The 1995 Agreement on Trade-Related Aspects of Intellectual Property (TRIPS) inextricably ties trade with patent protection and is a requirement for members of the World Trade Organization (WTO) (Pardey et al. 2004). Countries participating in TRIPS are required to provide a system for IPR on plant varieties, involving either the use of patents or PVP as implemented under UPOV. Alternatively, a *sui generis* system of plant variety protection may be implemented, as in the 2001 Indian Protection of Plant Varieties and Farmers' Rights Act; it contains provisions for "benefit sharing," whereby local communities are acknowledged as contributors of landraces and farmer varieties in the breeding of new plant varieties. The Philippines and Thailand also implement *sui generis* protection systems. In some cases, countries have gotten together to develop regional IP regimes, such as the Andean Community (Bolivia, Colombia, Ecuador, Peru, and Venezuela), which expressly prohibits patents on plants and animals.

The Convention on Biological Diversity: Redefining the Commons

In 1993, a decade after *Diamond v. Chakrabarty*, the United Nations ratified the Convention on Biological Diversity (CBD), which states that a country has sovereign rights to its indigenous genetic resources. The CBD sought to establish a mechanism by which countries could assign value to genetic resources and derive benefit from making them available as inputs for breeding new varieties and other forms of agricultural R&D, institutionalizing a dramatic change in the concept of ownership in the global genetic commons.

Most countries interpreted the CBD to mean that they could require payment for access to germplasm. Subsequently, the 2004 International Treaty on Plant Genetic Resources for Food and Agriculture was developed to determine how benefits created using these resources would be equitably shared and to establish a multilateral system to facilitate access to genetic resources. The treaty currently governs the international exchange of germplasm for 35 crop species, including major cereals such as rice, wheat, and maize, but excluding soybean, peanut, and other major crops. It is vague on a number of points, however, including how monetary benefits from commercial products developed from gene bank materials are to flow. Farmers who conserve indigenous genetic resources are supposed to benefit, especially those in developing and transitional economies, but the particulars of how this will happen have yet to be clarified.

The lack of clarity in the CBD has taken a toll on international germplasm exchange and has created an environment in which commercial breeders often look for a work-around to access genetic materials that were distributed prior to 1993 when the CBD went into effect (Hammond 2010). For example, US gene banks maintain a major portion of the materials from the international plant germplasm collections funded by the CGIAR. Most of these materials were freely shared prior to 1993, and because the United States has signed but not ratified the treaty, it allows these materials to be accessed without applying the provisions of the treaty; that is, without the mandate for sharing the benefits of commercialization stipulated by the CBD. Many of these materials are now being conserved in private germplasm collections with no requirement to distribute them further. It is ironic that the CBD, aiming for access and benefit sharing, appears for the moment to have accomplished the opposite (Hammond 2010).

Paying for Gene Banks: Who Owns the Commons?

Following *Diamond v. Chakrabarty*, commercial breeders and some governments began to amass large collections of improved germplasm for their own use. Despite the fact that many of these resources come directly from the public sector, genetic materials maintained in private collections are generally not shared. They guarantee the owners access to the large reservoirs of genetic variation needed to breed and patent new varieties. Yet private-sector breeders still look to the public sector to collect, characterize, and perform prebreeding enhancement of landrace and wild relatives (Duvick 1991). This dichotomy in the way germplasm resources are valued and paid for lies at the root of an uneasy relationship between public and private plant breeders today. Varieties developed by the private sector are protected by formal IPR, in contrast to wild relatives and landraces that are still considered a public good (Kronstad 1996).

Private breeding enterprises and seed companies benefit greatly from access to publicly available genetic resources, but have so far been unwilling to contribute financially to long-term international germplasm conservation efforts. Rather, the strategy seems to be for major multinational breeding companies and some governments to try to appropriate entire collections of germplasm from international organizations such as the IRRI or the CIMMYT, or other countries that still support free exchange of germplasm. This gene grab or genetics arms race is taking place across the globe. As the strategic importance of germplasm diversity is obvious, the closure of the commons creates urgency in the face of potential scarcity.

Many breeding systems in the developing world are poorly equipped to deal with the rapid changes that are occurring, because they have depended so extensively on international free exchange of germplasm. As plant research in the industrialized world has come to be dominated by private companies, the uptake of new technologies and the spillover effects of global investments in agricultural R&D have slowed. Additionally, there is currently no mechanism for ensuring that the benefits of modern breeding will reach breeders and farmers in the developing world. Many argue that farmers in developing countries made an essential contribution to plant breeding and modern crop variety development by developing landraces, but received no payment or recognition for that contribution (Day 1997). The financial incentives and rewards of the patent system that have been put in place to drive innovation in modern plant breeding are often in conflict with the essential "public goods" nature of the germplasm resources and crop varieties that underpin those innovations. New mechanisms for funding international, public-sector agricultural R&D are urgently needed to ensure that the benefits of modern plant breeding translate into appropriate products and information systems that are made available to those in need of them in the developing world.

CLASSICAL PLANT-BREEDING TECHNOLOGY

Plant breeding has been practiced for thousands of years—in many cases inadvertently. Domestication of landraces from wild relatives represents the simplest form of plant breeding: selection of desirable individuals from existing variation, with no conscious crossing (hybridization) to generate new combinations of genes. Classical breeding exerts selection on diverse populations of plants in similar ways to nature, by imposing non-random forms of mating between closely or distantly related individuals (cross-pollination) and allowing some plants to reproduce at the expense of others every generation. Humans typically impose more intense selection pressure than nature does, speeding the process. Sexual crosses underpin the entire business of plant breeding in both the public and private sectors and are the foundation upon which all other breeding technologies are built. Newer approaches simply increase the efficiency, precision, or profitability of classical plant breeding—none has yet replaced it.

The level of technical knowledge required to manage a classical breeding program is usually relatively low, but a high level of decision making and management expertise is required to coordinate a successful plant improvement program. It takes years of experience to become familiar with a crop, its germplasm, and the range of environments in which it is grown. A breeder may make and evaluate dozens to thousands of crosses per year, and these crosses may be evaluated by teams of professional breeders or done in on-farm trials in consultation with farmers. In each generation the breeder selects the lines with the most desirable traits and continues the breeding cycle. This cyclical process recombines and recycles genes that, for the most part, have been available in the cultivated gene pool for thousands of years. It is important to note that there is no requirement to describe the underlying genetic composition of a new variety that has been bred using classical crossing and selection. In most cases, the exact genetic make-up is not known.

The time it takes from the first cross to final release of a new variety is generally between 10 and 15 years for annual crop species and may be considerable longer for perennials. Breeding populations must be evaluated in different years and locations to provide insight into performance stability in different target environments; after identification, several more years will pass before the release of successful varieties. With few exceptions, the public sector breeds self-pollinating, inbred species at government-supported National Agricultural Research and Extension Systems (NARES) and universities. Because self-pollinators are genetically uniform and perform stably when seed is saved from one generation to the next, and are thus difficult to protect with IPR, the private sector has little interest in inbreds. International crop improvement centers have played and continue to play a major role in supporting breeding activities in inbreeding crops by collaborating with NARES partners—tailoring the majority of varieties that currently meet smallholder needs.

HYBRID BREEDING TECHNOLOGY

Hybrid varieties were the primary driver of private investment in the plant breeding and seed business before the 1990s. Hybrid breeding began in 1909 when Shull and colleagues demonstrated that by carefully mating genetically distinct, inbred parents, vigorous corn hybrids could be produced that are genetically uniform and show superior performance to either parent. Parental lines are maintained by inbreeding and their identity is protected as a "trade secret," while the offspring generated from crossing parental lines (F1 hybrid seed) is generated annually and sold every season to farmers. The genetic mechanisms controlling this phenomenon, known as heterosis or hybrid vigor, are still not entirely understood today.

The time required to develop a new hybrid variety is also 10 to 15 years. In terms of breeding, however, hybrid varieties are considerably more complicated and expensive to develop, due to integration of elaborate male sterility systems and rigorous pollination control at each step in the process. Fresh seed must also be generated every year to supply market demand, and seed delivery systems must be reliable. Due to these constraints and the business opportunity that they represent, most hybrid varieties are delivered by the private sector. Hybrid technology also has an added advantage—it is easily protected as a "trade secret" by simply limiting access to the inbred parents. Due to gene segregation, seeds saved from F1 hybrids have very poor ability to reproduce the traits of interest in the next growing season. This inexpensive form of intellectual property protection is not subject to any legal requirements or restrictions, making hybrid technology valuable and easily controllable. Because development of hybrid seed represents a significant upfront investment, it is often more expensive to purchase and is typically used where farmers have greater access to credit and other kinds of agricultural inputs. However, companies still must maintain competitive pricing in order to attract farmers to purchase their products. Both large and smallholder farmers around the world have widely adopted hybrid varieties, attesting to the success of hybrid breeding.

GENOMICS-ASSISTED BREEDING TECHNOLOGY

New DNA sequencing technologies have expanded modern breeding into the field of genomics, the study of entire genomes of organisms. With the availability of high-resolution genomics platforms, there is growing interest in using DNA markers to enhance the efficiency and precision of classical breeding to provide a more comprehensive understanding of a breeding line's potential. DNA sequence differences (polymorphisms) mark particular regions of a chromosome and act as an individual's genetic fingerprint. Marker-assisted selection (MAS) uses DNA polymorphisms to predict performance, often before genes or traits are even expressed (Eathington et al. 2007). Breeders are beginning to use information about thousands or millions of DNA polymorphisms across the genome simultaneously to make genomic predictions about plant performance (Heffner et al. 2009). Scientists also use genomics to search for rare, valuable alleles in large germplasm collections in a process known as "allele mining" (Bhullar et al. 2009). These applications of genomics represent a major competitive advantage in commercial plant improvement for at least three reasons: they increase the efficiency of selection, saving time and lowering the cost of field evaluation; they facilitate the identification of valuable accessions in gene banks; and they provide a competitive advantage in securing IP.

Using high throughput genomics data, breeders can predict which traits a plant carries and infer how well it is likely to perform in the field. This ability represents a major competitive advantage in the context of commercial plant improvement but puts small programs at a disadvantage because of the cost of genotyping large numbers of plants. Expenses are incurred with every genotyping run, and qualified computational and bioinformatics specialists are needed to analyze the data. While large private-sector breeding companies usually do their genotyping in-house, public-sector breeding operations often contract out to commercial genotyping laboratories that were originally developed to service the health and pharmaceutical industries. Linking with the human health community to optimize access to genotyping services at prices that take advantage of economies of scale, agricultural research programs in developing countries can now evaluate varieties utilizing up-to-date technology platforms adapted for a wide range of plant species. This capacity allows many research institutions to avoid sending plant or DNA samples out of the country, quelling concerns about patenting and ownership of resources. Commercial genotyping laboratories profit by staying abreast of new technologies, maintaining strict quality control standards, keeping costs competitive, and respecting the proprietary nature of data. Ensuring access to reliable, state-of-the-art technical support provides an invaluable service to a global community of publicly funded research institutions—dissemination of technological innovation without creating dependencies on private seed companies.

Genomics technology and MAS have also opened up many new opportunities to utilize rare genes from wild species and landraces located in the world's gene banks, which could improve pest and disease resistance, abiotic stress tolerance, nutritional quality, or yield (Brar and Khush 2003; Septiningsih et al. 2008; Yan et al. 2010). MAS is also widely used to move transgenes into new varieties (Gao et al. 2011; Nishiguchi and Mori 1998; ProVitaMinRice Consortium 2008).

GENETIC ENGINEERING

Genetic engineering involves the use of recombinant DNA technology, rather than sexual crossing, to expand the gene pool available within a given crop species. All transgenic crops on the market today have been developed using a species of bacteria, *Agrobacterium tumefaciens*, which has the natural ability to insert DNA randomly into the chromosomes of a plant that it infects—a process known as transformation (Klee et al. 1987). It is a common misconception that crop biotechnology involves a simple cut and paste procedure (Gepts 2002). Because *Agrobacterium* randomly inserts a gene of interest into a plant's genome, transformation events must be screened to confirm that new traits are expressed correctly and no other characteristics have been disrupted. Sometimes multiple transgenes are put into a crop variety, a process known as gene stacking. This can be used to introduce multiple independent traits (Bt and Roundup Ready, for example) or multigenic forms of a single trait (different forms of herbicide tolerance) (Que et al. 2010). Patented traits developed at different times or by different companies can be combined in the same variety, although complex licensing agreements must be established to ensure proper revenue flow.

Crops have varying degrees of transformation efficiency, and often transformation is only possible in a few varieties of a given species (van Wordragen and Dons 1992). As a result, transgenes are regularly introduced into one variety and later transferred between varieties using sexual crosses and MAS. In asexually propagated crops like potato, sweet potato, and cassava, however, it is difficult to use crosses to improve existing varieties, let alone to move transgenes between varieties. Most yield gains in potato over the past 150 years are not the result of genetic improvement, but have instead come from improved agronomic practices that result in healthier tubers (Douches et al. 1996). Genetic engineering can speed varietal development in asexually propagated crops by bypassing sexual barriers altogether, although transformation protocols must be established for every variety within these species—a hurdle that can be difficult to overcome (Prakash and Varadarajan 1992).

The Precautionary Principle and Regulation of Transgenic Crops

Public apprehension about genetically modified organisms (GMOs) is intertwined with moral, religious, political, economic, and scientific concerns that play out in complex ways. Because biotechnology facilitates transfer of traits between totally unrelated species and essentially leapfrogs millions of years of evolution, many liken genetic engineering to "playing God" or tampering with the balance of nature. Others in the anti-GMO movement see biotechnology as an example of wealthy nations, companies, or institutions using patents and trade secrets to block access to technologies, crop varieties, and essential knowledge that are needed in the developing world. Scientific apprehension about genetically modified crops is rooted in awareness of how little we understand the complex ecosystems that sustain life on Earth. Many ecologists and evolutionary biologists believe that formal scientific investigation is warranted to better understand the long-term consequences of what appear to be small changes brought about by genetic engineering. Novel and potentially potent new trait combinations may have emergent properties in complex biological or ecological systems that are not predictable using the reductionist principles of molecular biology (Regal 1996).

All of these concerns have led to the application of the "precautionary principle" and the development of biosafety regulations designed to protect the health of the consumer, the environment, the agricultural system, and the ecosystem as a whole. Each transgenic event must be registered, evaluated, and officially released in the country where the crop is to be grown. The identity and copy number of the individual transgene or genes in the plant genome are carefully documented, and the GMO is evaluated in field trials over years and environments to confirm the identity and stability of the new trait. Multiple government agencies are generally involved in the regulation of biotech crops, often creating numerous technical and bureaucratic hurdles that impose significant financial costs and delays on the approval process. Regulatory bodies are also required to successfully manage growth, sale, distribution, and trade of GMO seed and produce. Many developing regions of the world have difficulty establishing policies due to lack of trained personnel, political instability, or confusion surrounding the safety of GMOs. Regional efforts are emerging in order to overcome these hurdles, including the East African Community, Common Market for Eastern and Southern Africa, Southern African Development Community, and the Economic Community of West African States. It is interesting to note that varieties generated using traditional plant breeding techniques undergo much less stringent testing for food and environmental safety than varieties produced using genetic engineering.

How Have GM Crops Impacted Farmers?

GMOs have only been in commercial production in the United States since 1996. Since then, global biotech production has increased 94-fold, from 1.7 million to 160 million hectares in 2011—representing the fastest adoption of any crop technology in history. Today, biotech crops account for roughly 11 percent of the world's agriculturally productive 1.5 billion hectares and are expected to reach 13.3 percent of total production by 2015. The United States continues to have the largest acreage of GMOs for a single country (69 million hectares in 2011), but 44 percent of the world's biotech crops were planted in Brazil, India, China, Argentina, and South Africa—developing countries that represent about 40 percent of worldwide population. Biotechnology has had well-documented positive impacts globally: of the estimated 16.7 million farmers growing GMOs in 2011, 90 percent reside in developing countries and subsist on less than two hectares of land (ISAAA 2011).

Nearly all biotech crops in production today are soybean, maize, cotton, and canola varieties engineered for only two traits—insect resistance (Bt) and herbicide tolerance. These traits were designed to reduce manual labor, eliminate the need for highly toxic chemicals, and facilitate conservation agriculture practices. Globally, there has been little incentive to develop GMOs for human consumption because of public opposition to biotechnology. As a result, only a few GM crops in the United States have reached consumers as produce in the grocery store; examples include virus-resistant papaya (Tennant et al. 1994), virus-resistant squash (ISAAA 2011), and Bt sweet corn (Shelton 2012). While these crops have little impact on global food security, they represent an emerging trend that is likely to pick up in the future.

Bt crystal (cry) toxins are derived from the soil bacterium *Bacillus thuringiensis* and have been used for decades in organic agriculture, due to their safety for both human consumption and the environment. Different cry genes have been engineered into several major crop plants to specifically target certain insects, such as the cotton bollworm. Bt crops control insect pests, reduce pesticide usage, and have dramatically increased on-farm yields; from 1996 to 2008, Bt crops reduced chemical insecticide use in corn by 35.3 percent and in cotton by 21.9 percent (Brookes and Barfoot 2010). These changes increased income for millions of smallholder farmers. Kathage and Qaim (2012) tracked the economic impact of Bt cotton in India from 2002 to 2008 and showed that smallholder adopters had a 50 percent gain in cotton profit and an 18 percent increase in monetary living standards. However, these changes did not occur without considerable controversy. In the early 2000s, local farmers introduced Bt genes into Indian cotton cultivars from an unknown source of germplasm. The use of Bt cotton became so widespread that it led to the subsequent approval of its cultivation by the Indian government (Herring 2007). In this case, local farmers were more successful than multinational corporations in battling anti-GMO activists, who were too busy organizing widespread protests, burning Monsanto Bt cotton fields, and spreading unsupported rumors concerning farmer suicides (Herring 2007; Herring and Rao 2012). Today, seven million smallholder farmers in India and seven million more in China cultivate Bt cotton. India accounts for one-third of global cotton hectarage, 88 percent of which has one or several Bt genes (ISAAA 2011).

Herbicide tolerant crops have had a similar success story. Glyphosate (Roundup) was discovered in the 1970s and has been heralded as the least toxic and most environmentally benign chemical herbicide. It has a low tendency to bioaccumulate, is relatively immobile in the soil, and is not a major polluter of waterways; it is quickly bound by organic matter and subsequently degraded by microbes, a process that only takes several weeks (Sprankle et al. 1975; Shuette 1998). Glyphosate inhibits an enzyme that is absent from mammals but that is essential for synthesizing key aromatic amino acids in plants (Amrhein et al. 1980; Steinrucken and Amrhein 1980); Roundup Ready crops contain a modified version of this enzyme that is resistant to the chemical, so that when fields are sprayed only the crop survives. Glufosinate (Liberty) herbicides contain the active ingredient phosphinothricin, which kills plants by blocking an enzyme responsible for nitrogen metabolism and ammonia detoxification—a by-product of plant metabolism. Liberty-link crops are engineered to express an enzyme that blocks the effects of phosphonothricin. Herbicides have become important in the adoption of conservation agriculture (CA) in mechanized farming systems. In 2010 alone, it is estimated that the use of herbicide-tolerant crops saved 17.6 billion kg of carbon dioxide from entering the atmosphere by facilitating CA practices, equivalent to taking about eight million cars off of the road (ISAAA 2011).

Biotechnology: A Private Endeavor

Although genetic engineering is not a magic bullet for the majority of challenges faced by farmers in resource poor regions of the world, it will continue to offer many opportunities for enhancing food security when integrated into sustainable crop production systems. Large companies are willing to invest in biotechnology because patents and use-conditions ensure a constant revenue stream, as utility patents in the United States, the European Union, and Australia force farmers to repurchase seed every year if they want to use a biotech variety. Roundup Ready crops require farmers to purchase herbicide as well as the GM seed, ensuring a double revenue stream. However, farmers are only willing to purchase GM crops and extra inputs as long as the benefits of the GM trait outweigh additional costs incurred. By competitively pricing GM seed, private companies have successfully introduced a technology into the marketplace that is easily identifiable, proprietary, and demonstrably profitable for both industry and farmers.

Development of biotechnology is not a risk-free enterprise. The average cost of researching, developing, and gaining approval to release a single transgenic crop variety is estimated at $135 million (ISAAA 2011), with no guaranteed payoff in the form of market or production success. This is a significant barrier for both small companies and public-sector breeders that want to enter the biotech market, and often only large, well-financed companies have the capacity to take innovative R&D and successfully bring it to the market as a product. Thus, many multinational seed companies buy out smaller start-up companies that do groundbreaking research and incorporate their R&D into existing biotech development pipelines. This concentration of biotechnology patents within the hands of a few large corporations is no different than the germplasm arms race, and it continues to impede R&D in the public sector from moving forward—particularly when developing crop varieties for humanitarian efforts where no profit is to be gained.

By establishing a presence within the farming community, the private sector has arguably begun to sway public opinion concerning the value in biotech crops—a movement that is visible all over the world. Parts of Europe are now growing biotech maize, and Japan imports transgenic soy and papaya from the United States, although it still prevents production of GM crops within the country. As developing regions of the world continue to accept the use of biotech traits as a major means of combatting biotic stressors such as pests and weeds, anti-GM regions that are dependent on importing food will be increasingly likely to encounter transgenics as the major available commodity. Additionally, countries such as China, India, Brazil, and South Korea are generating their own native biotechnology industries and collections of patents. Many of these countries will seek to export technology, while continuing to be major importers of food, fiber, and fuel crops. This new set of players will alter the supply and demand chains all over the world. As observed in cotton, biotech crops may even become the preferred way to meet the global demand for certain agricultural commodities.

CROP TECHNOLOGIES IN THE COMING DECADE

Next-generation crop varieties targeted to enter the market in the coming decade will be bred for novel traits, including disease and pest resistance, abiotic stress tolerance, and in some cases, nutritional quality. In order to deliver improved varieties in a timely manner, many of these traits will have to be bred using combinations of several technologies, which will require new institutional collaborations, both public and private, on a much greater scale than ever before.

Biotechnology is expected to bring significant improvements to numerous crop species in the coming decade, largely by addressing abiotic and biotic stressors that are currently difficult to manage using traditional means. However, despite increasing adoption rates of biotechnology globally, there are still considerable political and social barriers surrounding the adoption of transgenics that will have to be overcome—especially if transgenics are to reach their maximum potential in addressing issues of food stability.

New Forms of Biotic Stress Resistance

Recent discoveries in plant pathology offer exciting new opportunities to develop targeted forms of disease resistance that do not depend on the introduction of transgenes. Scientists have discovered that the pathogenic bacteria *Xanthomonas* uses a class of proteins known as transcription activator-like (TAL) effectors to alter gene expression and cause disease in rice, pepper, and other crops (Boch et al. 2009). By deleting DNA sequences targeted by TAL effectors, it is possible to confer immunity to disease (Li et al. 2012). This exciting development is completely heritable, precisely targeted, and easily bred into subsequent generations of plant varieties. Furthermore, it differs from standard genetic engineering because it carries no footprint of human intervention and is identical to many natural mutations causing small deletions in the genome. Whether this approach will be subjected to the same biosafety regulations as other forms of genetic engineering has yet to be determined, and public acceptance cannot be predicted. Nonetheless, this new strategy of intercepting the signaling between a pathogen and its host is opening new doors in plant breeding.

Cisgenics

In polyploid, asexually propagated, or perennial crops, such as potato, sweet potato, or apple respectively, genetic engineering will speed up and diversify the deployment of resistance genes. Transforming plants with genes from wild or domesticated relatives is being rebranded as "cisgenics" rather than transgenics (Schubert and Williams 2006), because genes are coming from sexually compatible species and there is no introduction of "foreign DNA"—though genetic engineering is still involved. Cisgenic transformation is being used to develop apple varieties resistant to apple scab caused by *Venturia inaequalis* (Vanblaere et al. 2011), and potato varieties resistant to late blight caused by the fungus *Phytophthora infestans,* the same pathogen that was responsible for the Irish potato famine in the late 1800s (Boonekamp et al. 2008; Jacobsen and Schouten 2008). Stacking multiple disease resistance genes from cultivated or wild relatives in modern varieties using cisgenics will provide durable resistance to diseases without the long and arduous process of conventional approaches to breeding.

Biotech Vegetables

Some of the largest gains in yield and nutritional quality in the coming decade will likely come from introducing transgenes into vegetable crops. The amount of arable land devoted to cultivation of vegetable crops is expanding 2.8 percent annually, more than any other type of crop globally. As of 2010, vegetable crops demanded the use of more pesticides than any other major crop—twice what was used in rice, and more than in corn and cotton combined (Shelton 2012). The relatively high use of pesticides is due to expansion of cultivation and the fact that market success of a vegetable crop is directly dependent on cosmetic appearance. Chemical control of pests and diseases is currently the most common strategy for meeting consumer demands.

Pest-resistant Bt potatoes, crucifers, and eggplant have already been developed, but have not yet reached the market due to high costs of regulation, absence of regulatory systems, confusion surrounding the safety of GM crops, and political lobbying. The Genetic Engineering Approval Committee (GEAC) of India approved the cultivation of Bt eggplant in 2010, after rigorous testing showed that it met the "laws of substantial equivalence" and was just as safe to cultivate and consume as non-engineered varieties (Shelton 2012). Non-engineered eggplant currently experiences 60 to 70 percent losses due to pests and is sprayed an average of 27 times during a growing season. Bt eggplant produces twice the yield of modern hybrids and could provide economic gains of $108 million annually, due to increased availability of produce to consumers and significantly reduced pesticide usage (Krishna and Qaim 2008). However, political lobbying by major nonprofits such as Greenpeace led to an indefinite moratorium on Bt eggplant's release. Ironically, the moratorium was invoked citing the precautionary principle, yet has resulted in the continued overuse of deadly pesticides (Kolady et al. 2010).

There are similar stories surrounding the release of Monsanto's NewLeaf potato during the 1990s, which was stacked with Bt genes, a potato Y virus resistance gene, and a resistance gene for potato leaf roll virus. Monsanto ultimately closed its potato breeding program in 2001, after intense political lobbying and boycotts from major US food chains like McDonalds. Farmers also contributed to NewLeaf's lack of success because they opposed the requirement to plant refuge areas and purchase new tubers every year (Shelton 2012). It is noteworthy that the significant public opposition to the use of GM vegetables, even in the United States, keeps them from reaching the market—despite the benefits they offer farmers and consumers alike.

Although debates surrounding biotechnology in the coming decade are inevitable, transgenic variety development is continuing at a faster pace than ever—in hopes that the benefits of genetic engineering will ultimately outweigh concerns. Insect-, fungal-, and viral-resistant varieties of cassava, pigeon pea, and sweet potato are being developed for Africa (ISAAA 2011), and Bt rice is expected to be released in China in the coming decade (Choudhary and Gaur 2009). Bt eggplant may also be approved in the Philippines and Bangladesh, regardless of its status in India (Shelton 2012).

Biofortification

Nutritional enhancement is a target of both traditional breeding and genetic engineering that requires different strategies for execution, depending on the nutrient of interest. For micronutrients present in the soil, breeding efforts focus on more efficient uptake of nutrients by the roots and partitioning to the plant organs that we ultimately consume (White and Broadley 2009). Traditional breeding for vitamin A-enriched maize, sorghum, and sweet potatoes, and high iron beans is well documented. Because many crop varieties do not contain the genes that are responsible for partitioning and accumulating micronutrients and vitamins, biotechnology represents a way of introducing novel nutrient pathways into a crop variety (Naqvi et al. 2009). Iron, zinc, vitamin A, and vitamin E-enhancement are currently being genetically engineered into cassava, banana, sorghum, and rice (ProVitaMinRice Consortium 2008), with vitamin A-enriched golden rice expected to reach production in the Philippines and other Asian countries by 2013 or 2014 (ISAAA 2011). While the best long-term solution for addressing micronutrient disorders is clearly diet diversification, delivering critical micronutrients in a food crop for which delivery channels are already well established has support from many sectors.

CONCERNS AND OPPORTUNITIES FOR THE COMING DECADE

There are many promising new crop technologies in the pipeline that can offer farmers relief from devastating diseases, insects, and environmental stresses, but continued investments in traditional breeding are still a high priority. Although biotechnology has facilitated successful delivery of defensive traits into multiple species, it has yet to develop a variety from scratch or improve yield per se—and thus it is still dependent on traditional breeding. Adapted plant varieties are integrated biological systems whose underlying genetic circuitries are replete with built-in redundancies, feedback loops, and finely tuned genes. While inserting or modifying a small number of carefully selected genes can alter plant physiological processes and response to plant disease and environmental cues, we need to continue to invest in germplasm enhancement and traditional forms of plant breeding, including genomic-assisted breeding, to maintain a steady flow of improved germplasm that provides the basis for both biotechnology and essential crop production.

As a complement to plant breeding, there is a growing global awareness of the need to manage natural resources more sustainably in agricultural production systems. The major challenges for the coming decade revolve around how new crop technologies will be deployed, who will benefit from them, and what political, economic, and environmental consequences will ensue. National and international policy support will be critical for balancing competing interests as people, institutions, and governments vie for the resources needed to meet the rising demand for food, fuel, and fiber, and cope with the social and ecological disruption associated with climate change and the displacement of human populations. This will require complex, system-level thinking and constant attention to changing realities on the ground.

Genetic Uniformity and Vulnerability

Wide-scale adoption of hybrids and other modern varieties has been criticized for contributing to the uniformity of crop genetics in farmers' fields

(Rubenstein et al. 2005). Genetic uniformity imposes extremely high selection pressure on pests, weeds, and diseases, which quickly evolve to overcome barriers to their survival due to relatively short reproductive cycles and prolific numbers of offspring (Delp 1980). In some cases, resistance breakdown can occur within just a few years (McDonald and Linde 2002a, 2002b; Palloix et al. 2009). A restricted crop repertoire also narrows the nutritional options available to humans, especially when poor infrastructure, geographic barriers, or political pressures restrict trade.

Geographically, crop genetic diversity can be quantified based on the number of varieties in the field at any point in time, the distribution of varieties within an area, and the genetic distance between varieties (number of species) within a given area or period of time (Rubenstein et al. 2005). In traditional farming systems, spatial diversity helped lower the risk of crop loss due to pests, diseases, and variable environmental stress. In modern production systems, temporal diversity represents an alternative risk-mitigating strategy, but requires a pipeline of new, genetically diverse varieties that can be steadily released over time (Duvick 1991). While genetic variation is a key component of ecosystem stability, some forms of genetic variation are more resilient than others. Some crop varieties and some mixtures of varieties are more resistant to pests, diseases, or abiotic stresses than others (Wolfe 1985; Mundt 2002). Thus, modern varieties, including transgenics, that are bred to provide multiple and complex forms of pest resistance may provide more durable resistance to pests, pathogens, and weeds than landraces.

The Irish Potato Famine of the 1840s represents an extreme example of low genetic diversity leading to famine due to widespread susceptibility to a single pathogen—*P. infestans* (Donnelly 2001). Similarly, the US Southern corn leaf blight epidemic in 1970 occurred due to the use of a single source of male sterility (T-cytoplasm) in hybrid maize in the United States, which caused susceptibility to the fungus *Helminthosporium maydis*—a disease which was never before a serious problem. Widespread use of this form of male sterility resulted in a 15 percent loss of the US maize crop in a single season, despite abundant variation in the nuclear genome of maize varieties (Tatum 1971; Ullstrup 1972). Today, widespread use of a single source of cytoplasmic male sterility in rice hybrids warrants caution (Cheng et al. 2007). Looking to the unpredictability of extreme weather events, particularly temperature and rainfall, it is not difficult to imagine scenarios where widespread deployment of a single variety of any crop could result in serious loss, with potential to disrupt the entire global food supply.

Similar concerns surround transgenic crops. The Roundup Ready trait has been incorporated into numerous species, and all are dependent on a single chemical herbicide, glyphosate. With the spraying of glyphosate on millions of hectares globally year after year, it is inevitable that herbicide-tolerant weeds will emerge; indeed, many have already been reported (Powles 2008).

Glufosinate-tolerant crops warrant similar concern. Even Bt-resistant crops are dependent on just a handful of cry genes conferring insect resistance. Insect-resistance to Bt has also been reported (Tabashnik et al. 2009), although the rate of resistance has been far slower than with most conventional insecticides (Bates et al. 2005), thanks in part to the mandated use of refuges.

Using genetically diverse germplasm in breeding can greatly reduce uniformity in modern crop varieties. Genomics can facilitate this process by allowing breeders to selectively introgress traits from diverse sources without disrupting the adapted gene complexes and quality traits that have been painstakingly developed over many decades. Placing value on enhancing genetic diversity within a breeding program and deploying diversity at the landscape level introduces a long-term perspective that is key to mitigating the effects of climate change and coping with new pest and pathogen populations.

At the landscape level, the release of new hybrid varieties can be orchestrated to maximize genetic variation over both space and time. Because farmers must purchase F1 hybrid seed every year or every other year, if competitive varieties carrying diverse forms of resistance to insects, weeds, and diseases are developed using numerous hybrid genetic backgrounds, they can be strategically released into the market to ensure that farmers plant a mosaic of different hybrid varieties every year. Diversity in the marketplace is the surest way to avoid uniformity in the field.

CA also contributes to crop diversity by emphasizing the use of crop rotation, which ensures different species are planted in succession over the years in the same field. Although seed companies and smallholder farmers do not reap any short-term benefits from maximizing diversity, and thus cannot be expected to prioritize this when deciding what to put in the field, it may be possible for policymakers to introduce incentives supporting crop diversity. Policies and interventions that encourage the use of spatial and temporal variation in cropping systems should be investigated as a critical and cost-effective policy contribution to food security.

To slow development of super weeds and super pests, modern transgenics has stacked transgenes. Gene stacking is often combined with planting non-transgenic, pest-susceptible refuge areas, practices that are required by law in most countries growing transgenics. It is important to maintain refuge areas, even in the presence of gene stacking, as non-GM areas help to lower the pressure on populations of susceptible insects and weeds to develop resistance (Bates et al. 2005). Durable pest- and disease-resistant varieties are also in the pipeline.

The Technology Gap and Agricultural R&D

Most of the crop technologies discussed in this chapter have been around for decades and could be easily applied to numerous crop species and scenarios with obvious advantages for smallholder farmers in the developing world. Why is this not happening? What is needed to catalyze more effective integration of crop technologies with the farming practices of smallholder farmers?

More than 90 percent of global agricultural research is conducted in developed countries and is focused on the crops important in those economies. The private sector accounts for more than half of those R&D expenditures, influencing the kind of research and IP protection that emerges from investments (Pardey et al. 2006). For every $100 of agricultural output, developed countries invest eight to ten times the money on research and development—and they do so with an expectation of return on that investment. In developing countries, public funds are still the major source of support, and the private sector accounts for just 6 percent of R&D (Pardey and Pingali 2010). These figures underscore the different phases and philosophies of agricultural R&D that separate developing and developed countries. There is enormous potential for agronomic and genetic improvement and rapid market development for many orphaned indigenous food crops, such as those that feed a large percentage of the African population: yams, pigeon pea, millet, sorghum, and cassava. Currently, the private sector is the major source of multinational investment in agricultural R&D, but because these crops are deemed to hold little commercial promise, they remain a low priority for investment.

Yet industry is motivated to invest when there are markets that offer a reasonable return on investment. Technology alone cannot motivate development: a supportive environment is a prerequisite. In most developing countries today, the level of public investment in agricultural R&D is too low to foster and maintain a skilled workforce, making private investment unlikely to come to the rescue. There has to be enough public investment in place to support local infrastructure, as well as skilled people who can adapt new technologies to local needs. If GM crops are involved, the existence of appropriate biosafety legislation and administration is essential. Potential market size and an economic and intellectual infrastructure that can support and motivate a high level of innovation and entrepreneurship are also crucial.

Beyond Patent Protection

Historically, open sharing of germplasm and technology has been the first step to developing a vibrant agricultural R&D sector (Smolders 2005). The explosion of patenting in the developed world, the pace of discoveries, and the high level of investment in the biological sciences that underpin new agricultural technologies have created barriers to innovation that continually marginalize those most in need. New forms of public–private partnerships that protect assets accumulated with public funding are urgently needed, along with reconsideration of the utility patent system as it relates to plant variety development.

Adopting Plant Variety Protection as a form of IPR in most developing countries is more appropriate than adopting a utility patent system for crop varieties, at least during the early phases of agricultural development. Ensuring breeders' exemption and farmers' privilege can facilitate access to germplasm and reward local breeders for sharing and utilizing efficient breeding technologies, rather than seeking to protect or own them. Allowing utility patents to enter the scene too early is likely to impede crop technology development that is essential to addressing smallholder needs (Smolders 2005). The history of corn breeding in the United States suggests that the absence of strong IPR played a critical role in facilitating the development of the hybrid corn industry. Initially, all hybrid corn companies had access to the same public inbreds, and each company self-pollinated superior hybrids of their competitors to develop new inbreds (Troyer 2004). Thus, hybrid varieties and inbred lines are essentially derived from varieties developed by public-sector breeders in multiple countries. Gouache (2004) argued that if the intellectual property practices of today had been in place 50 years ago, it is very unlikely that US corn yields would have reached today's level.

Indeed, crop improvement was and largely remains a cumulative process. New varieties are built upon past selection, and the progressive nature of this process means that past discoveries and related research are an integral part of contemporary innovations. Current intellectual property laws created to protect the linear process of developing a single crop technology do not reflect this cyclical need to return to "old" germplasm resources or technologies when developing new varieties.

Although patents ultimately exist to motivate innovation and disclosure, too many competing patent rights surrounding an invention can lead to less innovation and may actually prevent useful and affordable products from reaching the marketplace. This situation, referred to as the "tragedy of the anti-commons" (Heller and Eisenberg 1998), has become pervasive in the world of biotech crops. It is often challenging to cut through the dense web of overlapping IP rights surrounding an invention. Given the pace of technology development in genetic research, there are often dozens or hundreds of patents held by multiple companies relating to a single variety. These "patent thickets" require users and other innovators to make licensing deals with multiple patent holders to commercialize new products that build on prior inventions. The transactions add significant costs to the licensing process that are passed on to consumers and frequently prevent entry into new markets. These unintended consequences of the current patent system stifle competition and are especially insidious when they prevent the benefits of technology that address the needs of smallholder farmers from trickling down to the public breeders.

Collaborative and Open-Source Licensing

In cases where utility patents are already in place or cannot be avoided, as is often the case for biotech crops, cooperative licensing can help newcomers navigate through the patent thicket, lower transaction costs, hasten innovation, and commercialize new technologies (Shapiro 2001). Patent pools, clearinghouses, and open-source licensing are examples of cooperative licensing strategies that have been used in the international plant breeding arena to facilitate access to new genetic technologies for humanitarian purposes.

A patent pool is an agreement between two or more patent owners to license one or more of their patents as a package to one another or to third parties that are willing to pay the associated royalties (Merges 2001). Patent pools could clear the patent thickets and lower the barrier to entry for new technology users. An example is the Golden Rice pool, where six key patent holders agreed to grant licenses for more than 70 patents, free of charge, to developing countries. This pool was established by a public–private partnership that formed a single licensing authority to speed the development of biofortified Golden Rice varieties for dissemination to farmers in Africa and Asia. Market segmentation may also be negotiated as part of a patent pool. The poorest smallholder farmers can obtain improved biofortified rice varieties at minimum cost, while those who can afford to pay shoulder higher costs.

A clearinghouse is a mechanism whereby providers and users of goods, services, information, or patents are matched with each other to improve access to patented technologies (Hope 2008, 2009). An example of a clearinghouse is the Public Intellectual Property Resource for Agriculture (PIPRA). PIPRA's goal is to mobilize patented technologies from a wide range of public or private technology providers to address specific projects for the improvement of subsistence and specialty crops. They seek to make agricultural biotechnology inventions more accessible to emerging countries by providing one-stop shopping for IP information, supporting the development of IP management best practices, enhancing IP management capacity in developing countries, facilitating access to public-sector patented technologies, and developing gene transfer and gene-based trait technologies that have maximum legal freedom to operate.

Open-source licensing means anyone, anywhere, is allowed to copy, modify, and distribute the technology for any purpose, without having to pay royalties to the owner of the license, with the stipulation that any improvements made to the technology must be shared under the same open-source license. A relevant example is the Biological Innovation for Open Society (BiOS) License from the Centre for Applications of Molecular Biology in International Agriculture (CAMBIA) (www.bios.net/daisy/bios/mta/license-intro.html (accessed March 26, 2013)). The CAMBIA-BiOS initiative has developed a public-access database of patents in the life sciences to help users analyze the IP landscape, navigate IP thickets, acquire freedom to operate, and forecast trends and new technology; created new genetic technologies and delivered them through open-access licenses; and developed new mechanisms for licensing and contract agreements to influence national and international policy and encourage democratized problem solving.

Open-source initiatives occupy an important niche in the complex IP landscape today. Open-source business models pose different advantages, opportunities, and trade-offs than do proprietary models; the co-existence of both can drive competition and promote efficiency and innovation in the marketplace (Lerner and Schankerman 2010). While the two approaches are historically perceived as competitors, the same organization often engages in both forms of IP simultaneously, depending on which is best suited to reaching the target audience or user group. This suggests that a dynamic combination of the two strategies could be highly synergistic in developing and delivering crop technologies to small-scale farmers in the developing world. Acknowledging and valuing this combined approach opens up new opportunities for public–private partnerships.

Open-Source Breeding

One example would be an open-source collaborative breeding program, whereby a central research body, such as a CGIAR center, collaborates with national program partners and/or small seed and breeding companies. In an era where hundreds or thousands of varieties can be genotyped more quickly and economically than they can be phenotyped in the field, CGIAR centers can provide genotyping services while taking advantage of partners' investments in phenotyping across years and environments. In exchange for information about phenotypic performance, the CGIAR center integrates genotypic and phenotypic information and provides predictions about the performance of new lines and decision support for the selection of future parents and crosses. The collective information is much greater than the sum of the parts: the genomic information can be leveraged to provide more accurate predictions about performance than any single breeding program could hope to produce. Open-source information sharing would be compatible with other forms of IP protection allowable on the seeds of improved varieties emerging from the program.

The breeding efforts of the CGIAR center would focus on validating the predictions, developing new breeding lines, and improving source populations, to address deficiencies identified broadly across its client breeding programs. Open-source breeding could be particularly valuable when breeding for stress-prone environments. Lines evaluated in specific mega-environments share information on a target environment with other members of a cohort (breeding population). Through data pooling, a number of small but cooperating breeding programs could leverage each other's efforts to increase breeding efficiency. IP on successful varieties derived from the program could include a negotiated royalty flow back to the CGIAR to support its activities.

Capacity Building and Institutional Change

Several new players, including Brazil, China, India, and South Korea, have developed productive, innovative, self-sustaining agricultural research potential over the last 30 years and are poised to provide new contributions to agricultural R&D. International activities funded by foundations and other donors are facilitating access to high throughput genotyping, phenotyping, and genetic engineering technologies on a wide range of crop species (Generation Challenge Program 2012), but technical advances have outpaced the availability of adequately trained personnel in all areas of plant breeding, creating a large gap in the ability to integrate and use these technologies. Most of the expertise needed is currently found in the private sector, where cost-effective integration of new and older breeding technologies is farthest advanced. Universities are having a hard time keeping up with the demand for new breeders, and most qualified graduates are snatched up immediately by the private sector. New training initiatives are urgently needed to prepare breeders, managers, and extension personnel for the complex world they will inhabit. Leveraging the comparative advantage of the private sector to help train people is worth exploring. However, technical expertise is not enough: these professionals must also be grounded in the realities of the developing world and have a passion and commitment to translate technical potential into crop technologies that will contribute to the well-being of smallholder farmers in that world.

Any decision to integrate modern crop technologies into breeding programs within developing countries will take time and require the evolution of an entirely new institutional culture. Genomics can help breeders more efficiently utilize the raw genetic resources that abound in many developing countries. To be successful, however, open-source breeding programs will require reorganization of management structures and reporting protocols (Eathington et al. 2007). The technology itself imposes transparency and rigor, requiring goals and objectives to be explicitly spelled out, time frames set for each activity, both progress and impediments tracked and monitored, and information kept transparent at all levels.

Even as governmental spending on agricultural R&D is outstripped in many countries by investments in health, education, and other social services, rational policy decisions by national governments could encourage the emergence of regional centers that create economies of scale to help attract transnational agricultural R&D investment (Pardey and Pingali 2010). Regional centers would increase access to and dissemination of technical expertise, harmonize the regulatory environment for release of new crop varieties and management practices, and prevent redundancy in agricultural research among neighboring countries. From a political perspective, regional centers would attract larger R&D investments and could help address food security problems before they become politically destabilizing.

TOWARD PRODUCTIVITY AND SUSTAINABILITY

Accelerating the development and deployment of appropriate crop technologies and information-sharing systems aimed at closing the agricultural productivity and sustainability gap for millions of small landholders in Asia and Africa represents the most rational long-term strategy for avoiding major disruptions in the global food system and addressing inequalities that exacerbate political and economic turmoil. In order to meet this challenge, new kinds of public–private partnerships and international collaborations are needed to bring together complementary skills and resources from the public sector, the private sector, government organizations, NGOs, international organizations, and foundations. The emerging economies of Brazil, China, India, and South Korea represent new sources of crop technologies and new engines of growth for international development. These nations are potential investors and should have a stake in global stability, but will compete for natural resources to address concerns about national food security.

The world's international germplasm repositories are essential to plant breeding in the United States and around the world, and are critical to sustaining the world's food supply. They are irreplaceable and must be supported. Communities of skilled local breeders, agronomists, extensionists, and entrepreneurs that can contribute to closing the yield and sustainability gap in their own countries are essential to economic growth and political stability. In many cases, encouraging the use of Plant Variety Protection rather than utility patents is warranted to promote sharing of breeding technologies and germplasm and to drive innovation. The creation of regional centers for training and research in agricultural R&D can help attract new people and expand the talent pool, harmonize biosafety standards, facilitate the introduction of new crop technologies, and develop new kinds of partnerships and licensing options that will open the door for greater international investment and economic growth.

ACKNOWLEDGMENTS

We are grateful for helpful comments and suggestions that greatly improved an earlier draft of this paper from Tony Shelton, Ron Herring, Ellen McCullough, Joanna Upton, and Chris Barrett. We acknowledge financial support from NSF Plant Genome Research Program grant #1026555 (to SMc) and NSF Graduate Research Fellowship Program (PhD fellowship to SC). We thank Cheryl Utter, Cynthia Mathys, and the many undergraduate students who helped with formatting and editing.

REFERENCES

Alston, J. M., P. G. Pardey, and J. M. Beddow. 2010. Global patterns of crop yields and other partial productivity measures and prices. In *The shifting patterns of agricultural production and productivity worldwide*, ed. J. M. Alston, B. Babcock and P. G. Pardey, Chapter 3. Ames: Iowa State University.

Amrhein, N., B. Deus, P. Gehrke, and H. C. Steinrücken. 1980. The site of the inhibition of the shikimate pathway by glyphosate: II. Interference of glyphosate with *chorismate* formation *in vivo* and *in vitro*. *Plant Physiology 66*(5): 830–4.

Barrett, C., T. Reardon, P. Webb, eds. 2001. Income diversification and livelihoods in rural Africa: Cause and consequence of change. *Food Policy 26*(4) (Special issue).

Bates, S. L., J-Z. Zhao, R. T. Roush, and A. M. Shelton. 2005. Insect resistance management in GM crops: Past, present and future. *Nature biotechnology 23*(1): 57–62.

Bhullar, N. K., K. Street, M. Mackay, N. Yahiaoui, and B. Keller. 2009. Unlocking wheat genetic resources for the molecular identification of previously undescribed functional alleles at the *Pm3* resistance locus. *PNAS 106*(23): 9519–24.

Boch, J., H. Scholze, S. Schornack, A. Landgraf, S. Hahn, S. Kay, T. Lahaye, A. Nickstadt, and U. Bonas. 2009. Breaking the code of DNA-binding specificity of TAL-type III effectors. *Science 326*: 1509–12.

Boonekamp, P., A. Haverkort, R. Hutten, E. Jacobsen, B. Lotz, G. Kessel, R. Visser, and E. van der Vossen. 2008. Sustainable resistance against *Phytophthora* in potato through cisgenic marker-free modification. Wageningen, The Netherlands: Project Group DuRPh, Plant Research International.

Brar, D. S., and G. S. Khush. 2003. Utilization of wild species of genus *Oryza* in rice improvement. In *Monograph on genus Oryza*, ed. J. S. Nanda and S. D. Sharma, 283–309. Enfield, New Hampshire and Plymouth, UK: Science Publishers.

Brookes, G., and P. Barfoot. 2010. Global impact of biotech crops: Environmental effects, 1996–2008. *AgBioForum 13*(1): 76–94.

Byerlee, D., A. de Janvry, and E. Sadoulet. 2009. Agriculture for development: Toward a new paradigm. *Annual Review of Resource Economics 7*(1): 15–33.

Cheng, S. H., J. Y. Zhuang, Y. Y. Fan, J. H. Du, and L. Y. Cao. 2007. Research and development on hybrid rice: A super-domesticate in China. *Annals of Botany 100*: 959–66.

Choudhary, B., and K. Gaur. 2009. The development and regulation of Bt brinjal in India (eggplant/aubergine). ISAAA Brief No. 38–2009. http://www.isaaa.org/resources/publications/briefs/38/default.asp (accessed June 6, 2012).

Day, P. R. 1997. Biodiversity and the equitable use of the world's genetic resources. In *Global genetic resources: Access, ownership and intellectual property rights,* ed. K. E. Hoagland and A. Y. Rossman. *Beltsville symposia in Agricultural Research.* Washington, DC: The Association of Systematics Collections.

Delp, C. J. 1980. Coping with resistance to plant disease. *Plant Disease 64*: 652–7.

Donnelly, J. S. 2001. *The great Irish potato famine.* Gloucestershire, UK: Sutton Publishing.

Douches, D. S., D. Maas, K. Jastrzebski, and R. W. Chase. 1996. Assessment of potato breeding progress in the USA over the last century. *Crop Science 36*: 1544–52.

Duvick, D. N. 1991. Industry and its role in plant diversity. *Forum for Applied Research and Public Policy 6*(3): 90–94.

Eathington, S. R., T. M. Crosbie, M. D. Edwards, R. S. Reiter, and J. K. Bull. 2007. Molecular markers in a commercial breeding program. *Crop Science 47*: S154–63.

FAO. 2010. Food outlook. Rome: Trade and Markets Division, FAO.

FAO. 2012. Conservation agriculture. Rome: FAO. http://www.fao.org/ag/ca/1a.html (accessed June 5, 2012).

Ferroni, M., and P. Castle. 2011. Public-private partnerships and sustainable agriculture development. *Sustainability 3*: 1064–73.

Gao, S., B. Yu, H. Zhai, H., S. Z. He, and Q. C. Liu. 2011. Enhanced stem nematode resistance of transgenic sweetpotato plants expressing *Oryzacystatin-I* gene. *Agricultural Sciences in China 10*(4): 519–25.

Generation Challenge Program. 2012. The GCP molecular marker toolkit. See http://www.grandchallenges.org/ImproveNutrition/Challenges/NutrientRichPlants/Pages/default.aspx (accessed May 3, 2013).

Gepts, P. 2002. A comparison between crop domestication, classical plant breeding, and genetic engineering. *Crop Science 42*(6): 1780–90.

Gouache, J. C. 2004. Balancing access and protection: Lessons from the past to build the future. ISF International Seminar on the Protection of Intellectual Property and Access to Plant Genetic Resources, Berlin, Germany May 27–28, 2004.

Hammond, E. 2010. The sorghum gene grab. A briefing paper by the African Centre for Biosafety. June 2010. http://www.acbio.org.za/index.php/publications/biopiracy/314-the-sorghum-gene-grab (accessed August 13, 2012).

Heffner, E. L., M. E. Sorrells, and J. L. Jannink. 2009. Genomic selection for crop improvement. *Crop Science 49*: 1–12.

Heller, M. A., and R. S. Eisenberg. 1998. Can patents deter innovation? The anticommons in biomedical research. *Science 280*: 698–701.

Herring, R. J. 2007. Stealth seeds: Bioproperty, biosafety, biopolitics. *Journal of Development Studies 43*(1): 130–157.

——— and. N. C. Rao. 2012. On the "failure of Bt cotton": Analysing a decade of experience. *Economic & Political Weekly 47*(18): 45–53.

Hobbs, P. R. 2007. Conservation agriculture: What is it and why is it important for future sustainable food production? *Journal of Agriculture Science 145*: 127–37.

Hope, J. 2008. *Biobazaar: The open source revolution and biotechnology.* Cambridge, MA: Harvard University Press.

——— 2009. Patent pools, clearinghouses, open source models and liability regimes. In *Gene patents and collaborative licensing models*, ed. G. Van Overwalle. Cambridge, MA: Cambridge University Press.

ISAAA. 2011. Global status of commercialized biotech/GM crops: Brief No.43–2011. http://www.isaaa.org/resources/publications/briefs/43/ (accessed June 5, 2012).

Jacobsen, E., and H. J. Schouten. 2008. Cisgenesis, a new tool for traditional plant breeding, should be exempted from the regulation on genetically modified organisms in a step by step approach. *Potato Research 51*(1): 75–88.

Kathage, J., and M. Qaim. 2012. Economic impacts and impact dynamics of Bt cotton in India. *PNAS. 109*(29): 11652–6.

Klee, H., R. Horsch, and S. Rogers. 1987. Agrobacterium-mediated plant transformation and its further applications to plant biology. *Annual Review of Plant Physiology 38*(1): 467–86.

Kolady, D., D. J. Spielman, and A. Cavalieri. 2010. *Intellectual property rights, private investment in research, and productivity growth in Indian agriculture: A review of evidence and options.* IFPRI Discussion Paper 1031. Washington, DC: International Food Policy Research Institute.

Krishna, V. V., and M. Qaim. 2008. Potential impacts of Bt eggplant on economic surplus and farmers' health in India. *Agricultural Economics 38*: 167–80.

Kronstad, W. E. 1996. Genetic diversity and the free exchange of germplasm in breaking yield barriers. In *Increasing yield potential in wheat: Breaking the barriers*, ed. M. P. Reynolds, S. Rajaram, and A. McNab. Mexico City: CIMMYT 19–27.

Lerner, J., and M. Schankerman. 2010. *The conmingled code: Open source and economic development.* Cambridge, MA: MIT Press.

Li, T., B. Liu, M. H. Spaulding, D. P. Weeks, and B. Yang. 2012. High-efficiency TALEN-based gene editing produces disease-resistant rice. *Nature biotechnology. 30(5):* 390–2.

Merges, R. 2001. Institutions for intellectual property transactions: The case of patent pools. In *Expanding the boundaries of intellectual property*, ed. R. Dreyfuss, D. Leenheer Zimmerman, and H. First, 123–66. Oxford: Oxford University Press.

McDonald, B. A., and C. Linde. 2002a. Pathogen population genetics, evolutionary potential, and durable resistance. *Annual Review of Phytopathology 40*: 349–79.

——— 2002b. The population genetics of plant pathogens and breeding strategies for durable resistance. *Euphytica 124*(2): 163–80.

Mundt, C. C. 2002 Use of multiline cultivars and cultivar mixtures for disease management. *Annual Review of Phytopathology 40:* 381–410.

Naqvi, S, C. Zhu, G. Farre, K. Ramessar, L. Bassie, J. Breitenbach, D. P. Conesa et al. 2009. Transgenic multivitamin corn through biofortification of endosperm with three vitamins representing three distinct metabolic pathways. *PNAS 106*: 7762–7.

Nishiguchi, M. and M. Mori. 1998. Virus resistant transgenic sweet potato with the CP gene: Current challenge and perspective of its use. *Phytoprotection 79*: 112–16.

Palloix, A., V. Ayme, and B. Moury. 2009. Durability of plant major resistance genes to pathogens depends on the genetic background, experimental evidence and consequences for breeding strategies. *New Phytologist 183*(1): 190–9.

Pardey, P. G., N. Beintema, S. Dehmer, S. Wood. 2006. *Agricultural research: A growing global divide?* Washington, DC: International Food Policy Research Institute.

Pardey, P. G., B. Koo, and C. Nottenburg. 2004. Creating, protecting and using crop biotechnologies worldwide in an era of intellectual property. In *WIPO/UPOV Symposium on Intellectual Property Rights in plant biotechnology.* Geneva, October 10, 2003. Geneva: WIPO.

Pardey, P. G., and P. L. Pingali. 2010. Reassessing international agricultural research for food and agriculture. Report prepared for the Global Conference on Agricultural Research for Development (GCARD), Montpellier, France March 2010, 28–31.

Powles, S. B. 2008. Evolved glyphosate-resistant weeds around the world: Lessons to be learnt. *Pest Management Science 64*(4): 360–5.

Prakash, C. and U. Varadarajan. 1992. Genetic transformation of sweet potato by particle bombardment. *Plant Cell Reports 11*(2): 53–7.

ProVitaMinRice Consortium. 2008. Grand challenges in global health. http://www.goldenrice.org/Content5-GCGH/GCGH1.html (accessed June 5, 2012).

Que, Q., M-D. M. Chilton, C. M. de Fontes, C. He, M. Nuccio, T. Zhu, Y. Wu, J. S. Chen, and L. Shi. 2010. Trait stacking in transgenic crops: Challenges and opportunities. *GM Crops 1*(4): 220–9.

Regal, P. J. 1996. Metaphysics in genetic engineering: Cryptic philosophy and ideology in the "science of risk assessment." In *Coping with deliberate release: The limits of risk assessment*, ed. Ad Van Dommelen, 15–32. Tilburg/Buenos Aires, International Centre for Human and Public Affairs,.

Rubenstein, K. D., P. Heisey, R. Shoemaker, J. Sullivan, and G. Frisvold. 2005. Crop genetic resources: An economic appraisal. Economic Research Service/USDA Economic Information Bulletin No. 2.

Schubert, D. and D. Williams. 2006. Cisgenic as a product designation. *Nature Biotechnology 24*(11): 1329.

Septiningsih, E. M., A. M. Pamplona, D. L. Sanchez, C. N. Neeraja, G. Vergara, S. Heuer, A. M. Ismail, D. J. and Mackill. 2008. Development of submergence-tolerant rice cultivars: The Sub1 locus and beyond. *Annals of Botany 103*(2): 151–60.

Shapiro, C. 2001. Navigating the patent thicket: Cross licenses, patent pools, and standard-setting. In *Innovation Policy and the Economy.* ed. I. Jaffe and B. Adam et al., 119–50. Cambridge: MIT Press.

Shelton, A. M. 2012. Genetically engineered vegetables expressing proteins from Bacillus thuringiensis for insect resistance: Successes, disappointments, challenges and ways to move forward. *GM Crops and Food: Biotechnology in Agriculture and the Food Chain 3*(3): 1–9.

Shuette, J. 1998. Environmental fate of glyphosate. Environmental Monitoring & Pest Management Department of Pesticide Regulation Sacramento, CA. www.cdpr.ca.gov/docs/emon/pubs/fatememo/glyphos.pdf (accessed March 27, 2013).

Smolders, W. 2005. Plant genetic resources for food and agriculture: Facilitated access or utility patents on plant varieties? IP Strategy Today No. 13-2005. pp. 1–17.

Sprankle, P., W. Meggitt, and D. Penner. 1975. Adsorption, mobility, and microbial degradation of glyphosate in the soil. *Weed Science 23*(3): 229–34.

Steinrücken, H. C., and N. Amrhein. 1980. The herbicide glyphosate is a potent inhibitor of 5-enolpyruvyl-shikimic acid-3-phosphate synthase. *Biochemical and Biophysical Research Communications 94*(4): 1207–12.

Tabashnik, B. E., J. B. J. Van Rensburg, and Y. Carrière. 2009. Field-evolved insect resistance to Bt crops: Definition, theory, and data. *Journal of Economic Entomology 102*: 2011–25.

Tatum, L. A. 1971. The southern corn leaf blight epidemic. *Science 171*: 1113–16.

Tennant, P., C. Gonsalves, K. Ling, M. Fitch, R. Manshardt, J. Slightom, and D. Gonsalves. 1994. Differential protection against papaya ringspot virus isolates in coat protein gene transgenic papaya and classically cross-protected papaya. *Phytopathology 84*(11): 1359–65.

Troyer, A. F. 2004. Background of U.S. hybrid corn. *Crop Science 44*: 601–26.

Ullstrup, A. J. 1972. The impacts of the Southern Corn Leaf Blight epidemics of 1970–1971. *Annual Review of Phytopathology 10*: 37–50.

International Union for the Protection of New Varieties of Plants (UPOV). 2002. General introduction to the examination of distinctness, uniformity and stability and the development of harmonized descriptions of new varieties of plants. Geneva, April 19. Version TG/1/3.

Vanblaere, T., I. Szankowski, J. Schaart, H. Schouten, H. Flachowsky, G. A. L. Broggini, C. and Gessler. 2011. The development of a cisgenic apple plant. *Journal of Biotechnology 154*(4): 304–11.

van Wordragen, M. F., and H. J. M. Dons. 1992. Agrobacterium tumefaciens-mediated transformation of recalcitrant crops. *Plant Molecular Biology Reporter 10*(1): 12–36.

White, P. J., and M. R. Broadley. 2009. Biofortification of crops with seven mineral elements often lacking in human diets—iron, zinc, copper, calcium, magnesium, selenium and iodine. *New Phytologist 182*(1): 49–84.

Wolfe, M.S. 1985. The current status and prospects of multiline cultivars and variety mixtures for disease resistance. *Annual Review of Phytopathology 23:* 251–73.

Yan, J., C. B. Kandianis, C. E. Harjes, L. Bai, E. H. Kim, X. Yang, D. J. Skinner. 2010. Rare genetic variation at Zea mays crtRB1 ß-carotene in maize grain. *Nature Genetics 42*(4): 322–27.

8

Livestock Futures to 2020: How Will They Shape Food, Environmental, Health, and Global Security?

JOHN MCDERMOTT, DOLAPO ENAHORO, AND MARIO HERRERO

Livestock is of central importance to overall agricultural and economic performance and to food and other security issues. The greatest changes in livestock production are occurring in low- and middle-income countries, where livestock production can be broadly grouped into three production systems: smallholder mixed crop–livestock, semipastoral and pastoral, and larger-scale commercial livestock systems. Smallholder mixed crop–livestock systems are household farming units that vary from less than one hectare to 10–20 hectares, depending on rainfall and temperature. Smallholder farming dominates tropical Africa and Asia, producing more than 75 percent of these regions' milk, 65 percent of their beef, and 50 percent of their cereal (Herrero et al. 2010). Pastoral and semipastoral systems contribute less than 10 percent of livestock, but these livestock are essential where crops cannot reliably grow. Livestock represent between 40 and 80 percent of the income for pastoral families. Industrial systems, primarily raising poultry but also pigs, are the fastest growing sector in low- and middle-income countries. These are sited to supply urban areas and take advantage of widely available inputs, management, and infrastructure developed by large private-sector organizations.

Livestock provide important and widely available benefits. For rural people in low- and middle-income countries, livestock are a critical mechanism for saving income for times of ill-health or other household emergency or need. They provide more regular income as well as employment opportunities, particularly from milk production. This potential has been most realized in the smallholder dairy sector in South Asia and East Africa. India, through smallholder dairy, is the world's largest dairy producer. Very large economic benefits for the smallholder dairy sector, particularly in terms of income and employment of poor people, have been documented (Kaitibie et al. 2008). In many smallholder systems, animal manure is the main source of fertilizer and animals are the main source of traction and transport. Milk, meat, and eggs are critical in nourishing poor populations, particularly pregnant women and young children. In many livestock systems, particularly pastoral and semipastoral, livestock play a central role in culture and socioeconomic transactions.

The large benefits from livestock come at a cost. Livestock production activities cover approximately 25 percent of ice-free land for grazing (Steinfeld et al. 2006). This large land footprint also extends to water resources: livestock account for 32 percent of global water use, according to Lannerstad et al. (2012). Feed and biomass demands from the livestock sector are also significant—33 percent of global crop area and increasing rapidly, in competition with human food and biofuels (Steinfeld et al. 2006). Globally, livestock are major contributors to greenhouse gases, with global estimates that vary between 10 and 18 percent, depending on attribution and other methodological issues. Greenhouse gas efficiencies of the livestock sector can differ more than 20-fold depending on the degree of industrialization (Herrero et al. 2011; FAO 2009).

Growth in demand for livestock and livestock products due to population and income growth is dramatic and is a major contributor to the relative tightening of agricultural supply relative to demand (Rosegrant et al. 2012) and the increasing susceptibility of agri-food systems to shocks. The growth is almost exclusively in low- and middle-income countries, greatest in China and increasingly other Asian countries, but also in Latin America and urbanizing areas of Africa. Increases in production have to a large extent mirrored demand, but with the emergence of Brazil and Argentina as major exporters of livestock products and feeds.

The commercial livestock sector has provided the most rapid response to the escalating demand. This has been primarily in poultry and secondarily in pig production, growing at rates between 5 and 7 percent per annum (FAO 2009). The inputs and organizational arrangements for poultry and to a lesser extent pig production have been easily transferred globally and are important contributors to effectively and efficiently supplying urban consumers with high quality and competitively priced meat and eggs.

Mixed smallholder systems of Asia and Africa dominate in terms of overall production and producers involved. Despite increasing urbanization in both these regions, rural populations are likely to remain very large and smallholder farming systems dominant for the foreseeable future. Livestock play a critical role in the smallholder mixed farming systems. The key issue is how these smallholder systems can intensify to provide greater income and food, often under circumstances of small and still shrinking farm sizes. To date, progress has not been rapid (World Bank 2007). Livestock can play a critical role in intensifying overall production through enhancing nutrient flows and providing high-value products for income. The critical issue is how to accelerate sustainable livestock intensification by combining available feed, breeding, and health technologies with institutional arrangements for more efficient input supply, credit, and output markets.

This chapter focuses on livestock production and how it influences security. Three main livestock security pathways are described. The first focuses on the influence of livestock in overall food availability, accessibility, and utilization. The second pathway looks at the role that livestock plays in the competition for scarce natural resources. The third pathway looks at the intensification of livestock production and marketing systems and infectious disease risks both for animal and human populations. Given the fast-growing demand for livestock products and the expansion of livestock production to meet this demand in low- and middle-income countries, governments and key livestock actors will need to pay increasing attention to the responsible growth of the livestock sector and its broader implications for the development and security of societies.

MAJOR SECURITY CONCERNS

Food security depends on the availability of food, whether it can be accessed, and how it is utilized in diets. Stability of food availability and access across seasons is also critical. Given the importance and expansion of livestock in low- and middle-income countries, livestock has a great impact on food security. The most positive benefits relate to food utilization, dietary diversity, and food accessibility. Animal source foods (milk, meat, eggs, and fish) are arguably the most important source of nutrient rich foods in diets globally. These foods make a critical contribution to nourishing children in undernourished populations. This is particularly true in the internationally recognized 1,000 days' target from conception to two years of age. Young children, who have much higher nutrient requirements than adults, are most easily nourished by nutrient-dense animal source foods (Neumann et al. 2002; Randolph et al. 2006).

Livestock also play a major role in food accessibility and stability. Livestock are the main mechanism by which poor rural people, both smallholder and pastoral, build assets during good times, and provide regular incomes throughout the year. Livestock also increase food availability. This is most critical in pastoral and semipastoral communities, where meat and milk can provide the majority of food, particularly during hunger periods.

In smallholder systems, the important future challenge is to sustainably intensify production. Overall in smallholder systems, the intensification of livestock production has lagged relative to crop production (Erenstein et al.

2010) in this huge sector of global agriculture. Thus, there is lots of room for improvement from a low base. Sustainably improving livestock production offers a lot of potential for increasing the productivity from, and income generated by, the smallholder sector, with huge implications for food and nutritional security. Given available technologies, major increases in livestock production may be possible. The challenge will be to ensure that the necessary feed, breeding, and health technologies are appropriately combined and linked to improved infrastructure and institutional arrangements for input supply, finance, and saving, and access to markets (McDermott et al. 2010).

Beyond this good news story of livestock's positive role in food security in pastoral and smallholder systems, the growing demand for meat, milk, and eggs, particularly in urbanizing low- and middle-income countries, contributes to the overall tightening of agricultural supply relative to demand (Rosegrant et al. 2012). In the short term, the continuing and even accelerating growth of demand for poultry (meat and eggs) and pork will continue. Some short-term feed efficiency gains in commercial production, particularly in China and Brazil, to match efficiency in high-income countries can be expected, but growing livestock feed demand, particularly from corn (maize) and soybeans, will continue to directly and indirectly compete with both human food and biofuel demand for these crops. The growing feed demand will also place increasing pressure on scarce land and water resources.

Conflicts involving livestock production have largely been rural and linked to competition for pasture land and water. Such conflicts have been traditional in pastoral and semipastoral areas for millennia. However, the nature of such conflicts and their intensity have evolved and increased over the past two to three decades. One of the major evolutionary changes in conflicts in the last century has been the increasing conflict between mobile pastoralists and settled smallholder farmers as traditional dry season grazing areas for mobile pastoralists have been taken over by settled smallholders. These types of conflicts between different mobile pastoral groups among themselves and with settled smallholders will likely increase in importance, compounded by climate change, particularly across the Sahelian zone of Africa. The particularly worrying trend is that such conflicts have evolved from relative low-intensity conflict to higher-intensity conflicts with the increasing availability of sophisticated weapons, including high-intensity conflicts in the Horn of Africa and more recently, in West Africa (Shettima and Tar 2008).

In parts of southern and eastern Africa, there are also natural resource conflicts between both pastoral and smallholder livestock keepers and wildlife conservation (see Pica-Ciamarra et al. 2007). These conflicts have traditionally been of low intensity and have often been managed successfully by more sophisticated and inclusive arrangements, allowing pastoral people to co-manage and share benefits from wildlife conservation as part of broader pastoral area development and diversification programs (Reid et al. 2009; Osofsky 2005).

Natural resource conflicts around livestock also occur at broader national, regional, and global levels given the large demands for water and land for livestock feed production, particularly for larger commercial livestock production. For example, in the Nile basin of northeastern Africa, upstream countries such as Ethiopia and downstream countries such as Sudan and Egypt may conflict over water allocation. In the Nile basin, livestock are important users of water and the smarter positioning of livestock and feed production within different ecologies of the basin is an important potential response to managing water demands (Peden et al. 2007). The reduction of the Amazon rainforest for livestock and feed (particularly soybean) production has allowed Brazil to advance as one of the global leaders in livestock and feed exports. Given the growing population and livestock demand in Africa, it is easy to foresee a similar competition for Central African forests accelerating in the next few years.

BIOSECURITY AT RISK

A final and worrying security threat is to biosecurity. Animals may act as infectious agents, transmitting disease to people through food and water or direct and indirect contact. There is an age-old history of infectious agents crossing species barriers, with many examples of animal pathogens becoming important in people, and some infections moving in the other direction (Grace and McDermott 2011). Some of these zoonoses, such as Rift Valley fever, are already endemic in parts of the developing world and could easily be spread by available vectors in North America and Europe. Others may emerge as zoonoses, such as Nipah or Hendra virus.

A second biosecurity threat can be caused by strict or relatively strict animal pathogens, such as foot and mouth disease virus, that are endemic in many developing countries but could become economy-wrecking epidemics in rich countries. The 2001 outbreak of foot and mouth disease in the United Kingdom brought estimated losses of $30 billion, mostly indirect due to non-agricultural economic disruptions. Estimates of losses from a global influenza pandemic would be in the order of $1 trillion.

For biosecurity threats, we need the capacity to predict, prevent, and respond. Prediction, particularly in low- and middle-income countries, is very difficult, and the best we can do is to have some basic biomedical surveillance capacity, focus surveillance capacity on highest risk systems, and build innovative systems based on crowd sourcing and new knowledge and information technologies, such as mobile phones and social media. Given resource constraints, surveillance systems for pathogens should be well coordinated between human and animal health sectors. Encouragingly, collaboration among the World Organization for Animal Health (OIE), Food and Agriculture Organization (FAO), and World Health Organization (WHO) has been greatly enhanced through their common Global Early Warning Systems and, specifically for animal and human pathogens, the Global Livestock Early Warning Systems. Regular surveillance data and crowd-sourcing information have been put to good use at the international level in the Global Outbreak Alert and Response Network. This kind of data is also used in many countries from newspaper and Internet sources (e.g., UK). This type of sophisticated surveillance, however, has not trickled down to the national or regional level in low- and middle-income countries.

More targeted pathogen surveillance is also needed for prevention or early response. Epidemics usually require a large population of hosts with sufficient contacts so that new susceptible individuals are available to fuel the epidemic. In human populations, only cities with populations of at least 250,000 can usually support epidemics. For livestock, the dense poultry and pig populations in Asia are likely targets. In the case of the recent H5N1 avian influenza, Indonesia was a newly affected country that had a sufficient population to maintain virus transmission, while African countries did not. Increasing intensification of livestock production and marketing operations mean that the number of populations and their risk of pathogen spread are increasing rapidly.

Many emerging diseases of humans come from wildlife. While some potential emerging viruses pass directly from wildlife, particularly from primates, other wildlife pathogens with epidemic potential for people, such as bat viruses, may need a livestock population to amplify them sufficiently to create a human disease outbreak. A number of bat viruses have the potential to be amplified in pig populations—and some already have been (Nipah virus). Ebola-Reston, a non-pathogenic virus for people, has done this in the Philippines, pointing to what could happen to highly pathogenic Ebola and other viruses in rapidly growing pig populations in Uganda, Nigeria, and some other West and central African countries.

Food and water are also potentially important pathogen spreading mechanisms at different stages. SARS is the modern example of a pathogen that started as a food-transmitted agent and then adapted to be directly transmitted between people. For surveillance and prevention of food-borne pathogens, appropriate technologies, incentives, and regulations are needed in the dynamic world of meat, milk, and egg value chains, from informal to supermarket supply chains. Two potential threats are important here. The first is moving potential novel viruses through the meat trade, especially the illegal bush meat trade. The exact amount is unknown, but the estimate for bush meat harvested from the Congo Basin every year is five million tons. Five tons per week is estimated to be smuggled through Paris Charles de Gaulle airport (Laurence 2012). The second threat is the challenge of increased diarrheal diseases from more common pathogens, such as coliforms, in long food chains with poor refrigeration.

In most low-income countries, the response to identified pathogen hazard emergence is extremely weak to nonexistent. At a minimum, these countries will require a national capacity for disease detection and response at a basic level. For emerging disease outbreaks, support for sequencing and other rapid diagnostic tools from regional or international response teams and diagnostic labs will be required. Surveillance and sequencing for new influenza strains is currently inadequate (Butler 2012). Technologies make viral surveillance for potential pandemic pathogens much easier and cheaper, but there are important intellectual property and sovereignty issues that need to be negotiated to safeguard the interests of poor countries and global public goods while encouraging vaccine development.

Given the need for rapid decisions and actions, government agencies responsible for disease control often employ careful contingency plans and exercises so that they have some advance preparation for crises. For endemic diseases that can become epidemic, such as Rift Valley fever, decision support tools are critical in phasing and guiding decision making. Kenya has developed a decision support tool for responding to increased risk of Rift Valley fever. Such plans often need new thinking to enhance response and control time. For example, during the 2001 foot and mouth disease outbreak in the United Kingdom, the Netherlands experienced an associated outbreak. To limit the spatial spread of the virus, a rapid ring vaccination campaign was conducted, stopping transmission and allowing time for slaughter and disinfection procedures to be completed.

There are real opportunities for win-win situations in controlling infectious diseases in developing countries. International cooperation can reduce disease burdens in endemic areas, while the reduction in transmission risk lowers the potential for accidental or deliberate global spillover.

GLOBAL AND REGIONAL TRENDS AND PLAUSIBLE FUTURE SCENARIOS

Much has been written about the growing demand for livestock products in developing countries (Delgado et al. 1999; FAO 2009). This dietary transition is linked largely to income growth (see Figure 8.1) and urbanization in middle- and even low-income countries and, with population growth, is expected to continue for several decades to come.

In general the growing demand has been supplied by domestic production growth but with a modest growth in livestock trade, currently varying from 11–17 percent of agricultural exports, still dominated by staple crops for food

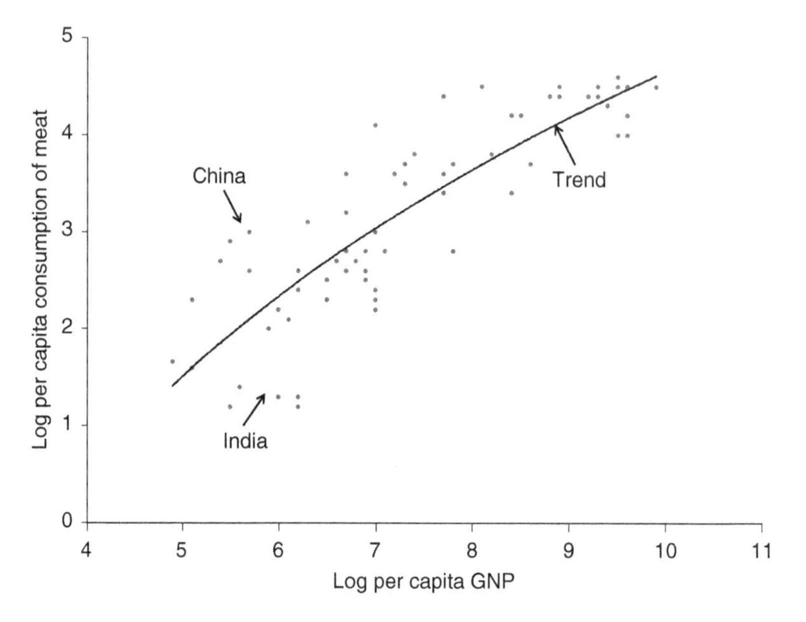

Fig. 8.1. Relationship between per capita consumption of meat and per capita GNP
Source: Livestock to 2020: The next food revolution, a joint IFPRI, FAO, ILRI study

and feed. The supply response has been dominated by commercial poultry production with a tenfold increase between 1961 and 2007. This is followed by pig meat production (fourfold increase), followed by a doubling of ruminant meat production.

Milk production has also increased dramatically in South and East Asia, and to a lesser extent East Africa. Regionally, increased meat supply has been dominated by industrial poultry production systems in Asia. Relatively cheap feed grain and the ability to import complete commercial packages of improved breeding stock, feeds, vaccines, and other inputs have greatly aided production increases. In the past, smallholder livestock meat production has not intensified and livestock yields have largely lagged behind crop yields in low-income countries. In some urban markets in Africa, people eat imported Brazilian chicken meat, eggs, and pork. While we expect commercial production to continue to grow fastest over the next 10 to 20 years, production from agro-pastoral and smallholder systems will also increase, albeit from a lower base (Figure 8.2)

While increased demand in middle-income countries and urban areas of low-income countries will continue, there are some important supply side challenges. The major issue is increased competition for natural resources. In addition to a growing relative scarcity of land and water in some regions, plant sources, particularly corn, for food and feed are also now being used

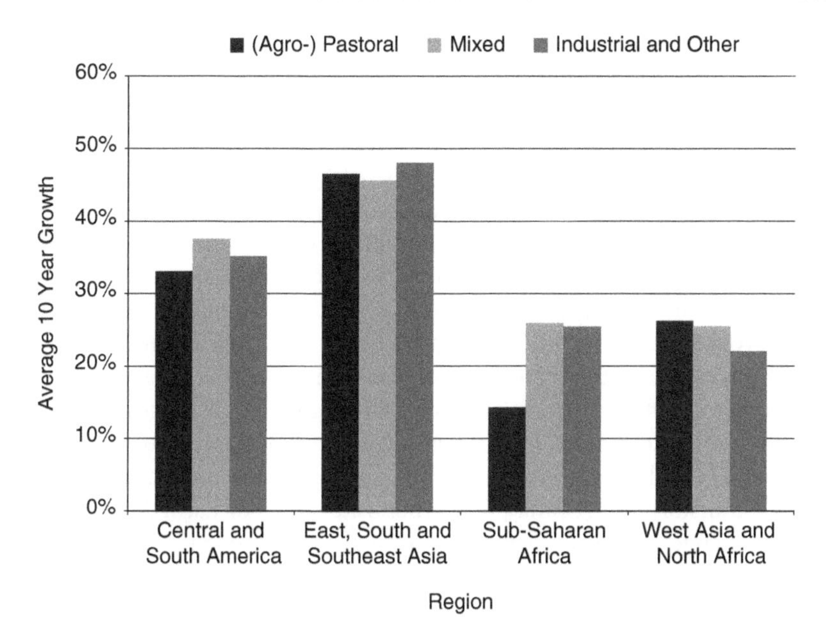

Fig. 8.2. Projections of average 10-year growth in meat production in different production systems in the developing world, 2000–2030
Source: Data from Herrero et al. (2009)

for energy through biofuels. This has led to dramatic changes in what cereals are grown, with a marked increase in corn production in Asia relative to other cereals, largely for livestock feed, as much American corn goes for biofuels. As described by Rosegrant and colleagues (2012), this is contributing to a much greater tightening of food supplies and a change of historical patterns. Beyond these more global drivers, relative regional scarcities of water will become increasingly important (Comprehensive Assessment 2007). Livestock and feed production can be very water intensive and will need to be carefully managed in water-scarce environments. Peden et al. (2007) provide specific suggestions for increasing water efficiency in livestock systems, including better feed management and upstream sourcing, feed improvement and productivity efficiency, control of stocking rates and other range management methods, and implementing water-harvesting methods for livestock.

Increases in overall livestock production have come about through a combination of increasing livestock numbers and increasing yields. The dramatic increases in poultry meat production in East and South Asia from 1980 to 2007 are largely the result of a massive increase in numbers of already relatively high-yielding birds. The fourfold increases in pork production in East Asia have been due to an expansion of numbers and, to a lesser extent, yield.

For beef, milk, and eggs, most regions apart from East Asia have relied on modest increases in numbers and yield (FAO 2009, 17).

Productivity gaps, as measured by differences between low- and high-yielding groups in the same smallholder farming systems, are large (two to six times) for indigenous, crossbred, and improved livestock species in smallholder farming systems (see Staal et al. 2010). Combinations of improved feeding, genetics, health, and other inputs are needed to improve yields. Individual inputs such as feed or improved breed on their own will not suffice. Thus, there is great potential for increasing livestock productivity in smallholder systems, with major progress requiring bundled inputs, finance, marketing, and know-how.

In the livestock sector, there are dramatic variations in the type and nature of production in different countries, largely varying with their stage of economic development. The World Development Report (2008) described agricultural development in three major types of countries: agrarian (much of sub-Saharan Africa), transforming (China), and urbanized (much of Latin America). The greatest changes to date in demand, and supply response to that demand, have been from large commercial systems in transforming or urbanized countries. In the coming decades, much greater demand is almost certain to be in countries that are currently described as agrarian, particularly in South Asia and Africa. These regions are likely to remain dominated by smallholders. The big question is how much smallholder systems can provide an effective supply response to this demand opportunity.

The potential for sustainable intensification of mixed crop–livestock production systems has been described by McDermott et al. (2010). Smallholder systems with livestock are very important in the world, affecting the livelihoods of approximately 700 million people in the largest smallholder regions of South Asia and sub-Saharan Africa. In these regions, transitions from smallholder farming to other activities have been relatively slow and it is likely that smallholder systems will dominate for the next generation or more. Intensification in smallholder livestock systems has lagged beyond intensification in crops, and increased production has been a function of modest increases in numbers and yield. One area in which smallholders have been very competitive is dairy production. India has become the world's largest dairy producer for the past few years, almost exclusively based on smallholder production. This has been achieved through various institutional models, including cooperatives, less formal farmer organizations, and with input and output markets organized by the private sector. In Africa, smallholder dairy has also been quite successful, especially in Kenya. The smallholder dairy success has not been readily transferred to other livestock species/enterprises (McDermott et al. 2010).

In terms of land area, extensive pastoral and semipastoral systems are important, particularly in Africa and Central Asia. However, their populations are much smaller than in smallholder systems. A key issue in extensive livestock systems is their susceptibility to drought, disease, and market shocks.

These systems are increasingly fragile, with sometimes severe socioeconomic disruptions.

One of the important issues for livestock production globally is greenhouse gas production by the livestock sector. This is a complex issue that will greatly influence livestock production in the future. Most estimates of livestock's global contribution range between 10 and 18 percent (Herrero et al. 2010; Steinfeld et al. 2006). Particularly in Africa, important currently arable areas are likely to become unsuitable for agriculture in the future, and people will need to revert to livestock production or other livelihood strategies to survive (Jones and Thornton 2009). Fortunately, beyond this adaptation role, ruminant livestock can play a significant role in mitigation. With modest gains in already low productivity, greenhouse gases per unit of output can decrease dramatically (Herrero et al. 2011). Given the low contribution of greenhouse gases by Africa relative to other regions, Africa should not have to reduce emissions. However, investments to reduce greenhouse gases in Africa will be relatively cost effective, so that high greenhouse gas emitters from other regions could finance improved productivity and decreasing greenhouse gases in African livestock systems.

MODELING POTENTIAL LIVESTOCK CHANGES TO 2020

In the late 1990s, impressive work began on assessing and modeling future livestock supply and demand globally. A seminal publication, *Livestock to 2020: The next food revolution* (Delgado et al. 1999) summarized this work and highlighted the dramatically increasing importance of developing countries in the livestock sector. This work built upon a broader agricultural modeling base developed by Rosegrant and colleagues called the International Model for Policy Analysis and Trade (IMPACT). This broader framework, with some adaptations, can be very effective in assessing cross-cutting trade-offs related to plant biomass utilization between food, feed, and fuel.[1]

Livestock modeling work with IMPACT has two major components. The first is factoring livestock into the larger agricultural picture. The second, started two years ago, is building in more specific livestock elements into the IMPACT model. From the major global and regional trends presented, we focus on three important scenarios that might have important implications for food security to 2020, globally, regionally, and nationally.

The first scenario looks at the relative supply and demand of meat versus feeds from major staple crops. Most of the IMPACT modeling of meat and cereal supply and demand has been from the perspective of changing meat and milk demand. The earliest work was done by Rosegrant and colleagues in 2001. The scenario investigated was of markedly increased demand for milk and meat in India, from relatively low levels at that time. It was estimated that for a projected increase of 18 kilograms per capita of meat demand from 1997 to 2020, domestic productivity yields would need to double and 1.8 million tons of meat imports would be required. The expansion of domestic production would require an increase from 181 million to 281 million tons over the period leading to cereal imports of 31.4 million tons rather than 24 million. Price changes for both meat and cereals for food were relatively modest, from 2–7 percent for meat, 2 percent for milk, and 2 percent for cereals. The only major price change was an increase of 28 percent in formulated feed concentrates.

In a recent analysis (Msangi and Rosegrant 2011), various scenarios of lowered meat consumption were tested for high-income countries, and Brazil and China. The scenario of halved meat consumption in the selected countries led to marked decreases in grain prices, including estimated price reductions of 18 percent for maize and 20 percent for coarse grains. Rice prices were relatively unaffected but wheat prices declined by 7 percent. This certainly illustrates the importance of livestock demand on some cereal prices. A halving of meat consumption in China and significant slowing of meat demand growth in low-income countries is an unlikely scenario. Thus it is hard to see much long-term easing of the livestock demand contribution to tightening agricultural markets, especially given the increase in population over the coming decades from current low-income, low-consumption countries.

One interesting scenario that deserves further exploration is how increases in cereal and other staple crop prices will influence livestock production. Increases in the relative prices of cereals versus meat should shift some meat production from monogastrics fed cereals to ruminants grazing on pasture or crop byproducts. This would also favor production in regions with natural pastures and forage production, such as Latin America and Africa. These and other questions are important in trying to rebalance the role of livestock in overall food security.

A second scenario investigates likely future impacts of livestock productivity improvements. In many smallholder regions of Africa and South Asia, it is quite feasible to double ruminant milk and meat productivity increases over a 10-year period. These increases would follow the successful introduction of a package of technological improvements (improved breeds, feed, health, and production management) combined with enabling policy and institutional arrangements. The implications of improvements to per unit production of ruminant livestock products (beef, milk, and lamb) were tested for selected countries in South Asia (India), eastern Africa (Ethiopia, Kenya, Tanzania, and Uganda), and southern Africa (Mozambique). The selected countries in this analysis are representative of developing countries for which research investment to enhance livestock productivity is considered of very high importance. In these countries, the necessary technologies for improving livestock productivity have already been developed, or are in the early development phase. Further, uptake and successful adoption of specific livestock yield-enhancing technologies and management practices may be considered plausible within the next decade. Following expert calculations on what production systems could expect to gain from livestock yield improvement interventions in 10 years, annual growth rates of milk, beef, and lamb products were estimated for the target countries.

Milk yields were assumed to grow by an additional 2.5 percent each year (an approximate doubling of the current baseline), beef by 2 percent per year, and lamb (referring to sheep and goat meat production) by 1.5 percent yearly. Sustained in the long run, these growth rates could lead to substantial livestock yield improvements in the target countries. If fully achieved, milk production in target systems, for example, would increase by 5 percent per annum or a total of approximately 60 percent in a ten-year period. From now to 2020, however, projected total productivity gains are more modest. These increases did not lead to significant changes in estimated prices.

In summary, intensifying ruminant meat and milk productivity for smallholders shows considerable promise and can certainly help improve food and nutritional security for individual households and communities in some countries by 2020, but big aggregate changes in ruminant meat and milk supply from the smallholder sector will take longer to achieve.

A third scenario looks at a drought in the Horn of Africa and other countries in eastern Africa. Eight countries were assumed to be affected: Djibouti, Eritrea, Ethiopia, Kenya, Somalia, Sudan, Tanzania, and Uganda. The researchers used sub-national data on animal numbers and livestock production recently developed by the International Livestock Research Institute (ILRI) and the International Institute for Applied Systems Analysis to estimate the country-level effect in terms of losses to total animal populations (Havlik et al. 2012).[2] In particular, standard assumptions on drought-related livestock losses in arid and semiarid rangelands and extensive livestock systems in selected countries were extrapolated to the general livestock population in the countries using the proportions of livestock to be found in the extensive systems.

The impacts of drought varied dramatically by country, depending on the proportion of livestock sector output in dry and drought-vulnerable areas. Djibouti and Somalia were almost universally affected, with lower overall impacts in Kenya, Tanzania, Ethiopia, and Uganda. For example, milk production fell by 50 percent in Djibouti but only 7 percent in Ethiopia. For lamb and beef production, losses in Ethiopia were approximately 10 percent of total production versus 50 percent in Somalia. As expected, drought impact in Djibouti and Somalia was massive. The other Horn of Africa countries with sizeable highland areas had some scope for covering supply decreases through imports and alternative production, with only modest estimated increases in prices.

To 2020, the impacts of a Horn of Africa drought would be devastating for food security and humanitarian conditions in dry areas. However, the generalized impact on the larger East African countries in terms of meat and milk supply and products would be modest. Other more general security concerns may not be so localized.

PLAUSIBLE LIVESTOCK FUTURES

Projections based on past trends are instructive but limited, as livestock futures will be shaped by dynamic forces, some of which relate to major trend drivers such as demography and urbanization and some to processes and decisions on how countries work together and manage global resources.[3] In 2006, ILRI and FAO convened a diverse group of thinkers around developing livestock future scenarios. Four scenarios were developed, framed by two important global decision-making axes relevant to livestock and other agriculture and natural resource sectors. The first is the level of global and regional cooperation and the second is the level to which natural resource challenges are handled, either proactively or reactively. These axes are consistent with those used in scenario development for the 2005 Millennium Ecosystem Assessment (MEA).

From these scenarios, a number of common thematic areas emerged, but with different narratives, projections, and forecasts. Common livestock themes included in the narratives are changes in key livestock production systems; demand for meat and milk; market institutions and arrangements and trade; livelihoods, equity, and food security; natural resource management and sustainability; climate change; food safety, emerging diseases, and biosecurity; and science and technology advances and options. Key predictions from the different livestock future scenarios are summarized in Table 8.1.

How do these trends influence thinking about livestock and global security concerns? For overall livestock production, scenarios are generally optimistic about capacity to manage larger-scale industrial and commercial monogastric production to supply growing demand for meat, milk, and eggs at relatively stable prices. The main variations in scenarios over the next ten years are the degree to which these expansions are integrated globally or in regional blocks. Optimism is more varied across scenarios for increased production and productivity efficiency in the mixed crop–livestock smallholder sector. In the short term, because of the size and importance of smallholder farming, this will be an important safety valve in agrarian and transforming countries for reducing the overall tightening of agriculture supply and demand.

In terms of food production, livestock production in arid and semi-arid areas is important locally, but is not a major influence globally. However, in a regional context, diversifying and building the capacity of these systems is critical in

Table 8.1. Summary of key livestock future predictions under four different scenarios to 2020

Key Dimensions	Global Orchestration	Order from Strength	Adapting Mosaic	Techno-garden
Changes in key livestock production systems	Smallholder multi-purpose systems shift to more vertically integrated arrangements (contract farming and out-growers). Faster pace of farm consolidation as more successful neighbors take over. Increased feed prices shift livestock productivity to large land-based countries, such as those in Latin America.	Trend of more vertical integration of smallholders but with more variations between regions and countries. Major expansion of livestock production in most countries—led by commercial poultry but also improved smallholder ruminant (meat and milk) and pig production.	Local and regional livestock production systems with significant variation. Expansion of livestock production in most areas. Many innovative local approaches are working but many are not.	Important shifts in livestock production linked to environmental services. Relative decline in intensive industrial livestock systems, banned in many peri-urban areas and increase in mid-size and free-range production.
Demand for meat and milk, market institutions, and arrangements and trade	Greater external trade among many countries but important trade laggards. Africa fails to hold Middle East markets. Expanded imports to African cities from BRIC countries. Dominance of huge global firms that set their own standards.	Export trade in livestock becomes increasingly regulated and declines except for a few niche markets. Decentralized multinationals control markets and adapt to national and local conditions.	Trade in livestock becomes increasingly regulated by "free-trade" rules. Multinationals important in livestock food chains. Value chains are more vertically integrated and concentrated and overall trade increases and stabilizes.	Improved linkages between rural and urban areas. Increased demand and prices for organic and free-range products.
Livelihoods, equity, and food security	Exclusion of smallholders and systems in poorer countries. Instability and unrest in SSA and regions of South Asia landless and urban poor as well as pastoral communities. Some localized successes in smallholder farming communities.	Smallholders integrated well in some countries and excluded in others. Better coordination of pastoralists and extensive systems in the Sahel and southern Africa but increased conflicts in the Horn of Africa and South Asia.	Overall improvement in production and income not equitably spread. Important disparities lead to decline in livelihoods, especially in rural areas. Significant migration from rural to urban areas.	Rural livelihoods improve due to better utilization and preservation of natural resources. Pro-active investments and greater opportunities in smaller towns and rural areas.

(Continued)

Key Dimensions	Global Orchestration	Order from Strength	Adapting Mosaic	Techno-garden
Natural resource management and sustainability	Continued expansion of livestock into forests in LAC and expansion in Africa. Rapid deterioration of marginal lands in Horn of Africa, Sahel and South and Central Asia.	In general, environments degrade much faster in poorer countries and marginal lands. Increasing nationalization of water, forests, oil, and mining.	Continued expansion of livestock into new areas. In more favorable areas, successful intensification. In more marginal areas, land degradation and failure of traditional systems to manage.	Moderate intensification in more favorable developing country settings and increased demand for free-range animals leading to more sustainable livestock systems in many rural areas.
Climate change	Better early warning systems and capacity of vertically integrated systems. Marginal systems have less adaptive capacity and suffer.	Little emphasis on mitigation or adaptation investments. Impacts of climate changes exacerbate conflicts in Horn of Africa and South and Central Asia.	Weak capacity to adapt to climate change. Regions with extreme climate variability do not cope well and livestock production often disrupted.	Some improvements in local adaptability but overall climate change negatively impacts production and health of livestock systems.
Food safety, emerging diseases and biosecurity	Strengthened capacity for food safety and disease control but sometimes spectacular failures. More coordinated international actions improve.	Increasing frequency of disease outbreaks in peri-urban areas and between livestock-and wildlife. Food safety improved in some countries but worse in others.	After initial failures, local capacity for epidemic disease control and food safety improves. Improving standards allow increased regional trade in West Africa, Southern Africa, Horn of Africa, and Middle East.	Improved food safety and disease control systems correlated with improvements in livestock production systems. Biosecurity largely managed with occasional failures.
Science and technology advances and options	IP regulations more strictly enforced, limiting benefits to more able countries participating in trade. Technology adoption increases productivity in China, SE Asia, and LAC.	Most livestock S&T linked to private sector. Emergence of locally based livestock research where private sector not active.	S&T controlled by multi-nationals. Technologies are available and adopted by successful farmers but not more widely.	Emphasis on S&T is in green technologies. Combining livestock production with environmental services provides better livelihoods in many rural areas.

BRIC (Brazil, Russia, India, and China), SSA (sub-Saharan Africa), SE (South-East), LAC (Latin America and the Caribbean), S&T (Science and technology).

Source: Adapted from ILRI and FAO (2006)

conflict avoidance (for example in the Sahel and the Horn of Africa). All scenarios are relatively pessimistic about the prospects for sustainable development of pastoral systems. All scenarios are also consistently pessimistic that pastoral systems will be able to adapt to anticipated climate shocks. This sustainable pastoral development challenge is much greater than the livestock production and marketing challenge, and will require investments in infrastructure and human resource development and a more aggressive agricultural and nonagricultural diversification strategy to enhance resilience and provide alternative livelihoods.

To varying degrees, most scenarios are pessimistic about the ability to sustainably manage livestock sector growth without natural resource degradation and the expansion of livestock and feed production into natural areas such as wetlands and forests. Given the importance of sustainably managing natural resources and the large impact of the livestock sector, these scenarios identify an important gap for global and regional action and for research and investment into new approaches. The emergence of new pathogens and infectious disease outbreaks are the greatest risks in the different scenarios, both globally and locally. Experts felt that these risks will be difficult to manage given the current weakness in pathogen surveillance and response capacities, and the scenarios in which these were managed in the longer term were invariably built on short-term failures.

LIVESTOCK TECHNOLOGIES AND THEIR LIKELY IMPACTS TO 2020

We reviewed a diversity of current and future livestock technologies that might be adopted in different regions and livestock production and marketing systems. In addition, we considered new bioscience advances, including bio-informatics, molecular biology, cell biology, genetics, nanotechnology, vaccine and biological development technologies, and plant biology and biomass manipulation, as well as new information technology advances and their applications for knowledge management, risk management, and improved policies, regulations, and investment. Technology options that are currently available and in the process of adoption are likely to have an impact to 2020. Individual technologies are unlikely to have much impact by themselves, but will need to be combined in packages.

Production and Productivity-Enhancing Technologies

A premise for sustainably intensifying the smallholder livestock sector is that technology packages that will increase overall systems productivity with reduced environmental degradation can affect technological change. A large part of the investment focus in livestock research and technology has been directed toward these kinds of technologies. Technology packages with associated management have been very successful in increasing the production and productivity efficiency of commercial poultry and pig production. Greater challenges lie in developing and combining technologies for sustainable productivity increases in smallholder livestock systems. Potential productivity-enhancing technology packages include technologies to improve livestock feed resources and animal genetics.

Potential livestock feeds in developing countries include pastures, crop residues, cultivated forages, concentrate feeds (agro-industrial by-products, grains, and feed supplements), and household wastes. The relative importance of these resources varies across production systems and their availability is influenced by agro-ecology, seasonality, land tenure, and management practices at the farm level (Table 8.2).

For smallholder livestock systems, the most important in the short term will be (Blummel et al. 2005) crop residues (food–feed crops) for mixed crop–livestock systems; feed grains for smallholder poultry and pigs, and as supplements for ruminants; cultivated forages and trees for mixed and grassland-based systems of relatively high productive potential; and rangelands in pastoral and agro-pastoral systems.

Inadequate feed resources constrain the productivity of mixed crop–livestock systems throughout the tropics. A particularly promising approach to increasing feed resources (Hall et al. 2004) is to improve the nutritive quality of crop residues through plant breeding and selection. Much research progress has been made in this over two decades, and methods are available for widespread adoption of dual-purpose food–feed crops through seed distribution systems for most major crops (Blummel et al. 2005; Lenne et al. 2003).

Next to crop residues, planted forages are generally the most important source of biomass for livestock feed, especially in crop livestock systems in

Table 8.2. Relative importance and trends in use of feed resources in main production systems

Systems	*Rangelands*	*Crop residues*	*Cultivated forages and trees*	*Grain and concentrate feed*
Pastoral and agro-pastoral	High	Low	n.a.	Low
Mixed crop-livestock	Low	High	Low	Low-Medium
Industrial	None	Low	Medium	High
Trends in importance and use	Decreasing	Increasing rapidly	Increasing slowly	Increasing rapidly

Source: Author

areas with limited land availability, where grazing is limited and livestock are increasingly maintained in cut-and-carry systems. These types of technologies offer high quality feeds for livestock from small areas in the farm. Currently about five million hectares of forage legumes and more than 42 million hectares of forage grasses are grown in the tropics in a range of production systems from smallholder crop–livestock systems to more extensive grassland/grazing based systems (Shelton et al. 2005). These technologies can increase animal and land productivity significantly, doubling milk and meat production per animal in some cases.

Forage technologies have had variable adoption in the livestock systems of developing countries (Shelton et al. 2005), and it has been shown that the successful integration of sown forages depends on there being a genuine need for improved feed by farmers. This is clearly demonstrated in the adoption of African forage grasses for improving grasslands in the extensive farming systems of Latin America in support of market-oriented beef production. By 1996, over 40 million hectares were sown to *Brachiaria* pastures in Brazil (Miles et al. 1996). This adoption was supported by public–private partnerships for forage seed production and driven by strong demand for livestock products in the region (Peters and Lascano 2003). Forage adoption has also been successful in some parts of Asia, with adoption of *Stylosanthes* in both India (Ramesh et al. 2005) and Thailand (Phaikaew et al. 2004; Phaikaew and Hare 2005). In contrast to Latin America, forage adoption by smallholder farmers in Africa has been slow, in large part because of smaller farm sizes and a poor fit with the farming systems. Beyond their feed attributes, forages have important environmental benefits including ground cover, steep slope stabilization, as wind breaks, as well as carbon sequestration and enhancing the water productivity of the system (Peters and Lascano 2003).

The meat and milk production yield differences for livestock within smallholder systems in the developing world are large, and genetic improvement of livestock is a key ingredient for bridging those (Staal et al. 2010). The genotypes observed in current livestock herds in developing countries are mostly indigenous, with some crossbreeding or the use of high-performance breeds as the systems become more intensive.

In developing countries, within-breed selection of desirable traits and crossbreeding are the standard approaches to genetic improvement. Much of our discussion on breed enhancement for smallholder systems is taken from the reviews by Enahoro et al. (2011a, b). They outline promising livestock genotype improvement interventions applicable to smallholder systems, including laboratory-based and community-based approaches to livestock genotype selection and a number of institutional approaches to livestock genotype enhancement.

Marker-based selection is a generic term that has been used to represent all such technologies as marker-assisted selection, gene-assisted selection, marker-assisted introgression, and genomic selection.[4] In marker-based selection, biological, biochemical, morphological, cytological, or DNA-based identifiers are used for indirect selection of the genetic determinants of traits such as productivity and disease resistance that are the goal of the genetic improvement program. While the benefits of these technologies for genetic improvement have long been recognized, resource and organization constraints in smallholder and developing country settings have limited their successful application. Specific constraints that genetic improvement programs face in smallholder environments include small herd sizes, poor management, low inputs, uncontrolled mating, indiscriminate crossbreeding, large proportions of non-genetic variations, and poor records of animal performance and pedigree.

Progeny testing and performance-recording mechanisms are critical for success in standard breeding programs. However, they have been proved impractical by most smallholders. Livestock breeding programs that facilitate collective access of local farmers to superior genetic material adapted for their use could provide alternatives to traditional breeding strategies. Community-based breeding, sire rotations, and loan schemes are a few such mechanisms.

Reproductive technologies are useful in livestock breed improvement schemes, for the creation of genetic enhancements, and for dissemination of the new genetic material within livestock populations. In smallholder livestock systems, reproductive mechanisms could be applied to increase fecundity of the parent animals, potentially reducing the need for a large parent stock. Juvenile *in vitro* fertilization and egg transfer are used for early selection, reducing the generation intervals. Artificial insemination is used by farmers to facilitate access to exotic genetic material. Another reproductive technology with potential use in genotype improvement in smallholder systems is semen sexing—the ability to sort semen cells to distinguish the sexes. In particular, developments in fluorescence-activated cell sorting have made sexed semen in dairy cattle commercially available, making it possible to increase the numbers of offspring of one sex in a closed population. This technology could allow the farmer to select a larger number of replacement heifers from her own herd, for example, or to select for heifer replacements using superior genetic material only—potentially reducing the lag for improving the genetic level of the entire herd.

Other breeding alternatives include the conservation and use of indigenous breeds that may have important roles as the climate changes, and also species substitution (say, from cows to camels) to ensure adequate provision of livestock products.

Health Interventions

The impacts of diseases on the reduced productivity and mortality of livestock are a serious constraint to production. The type of disease determines the type of intervention needed in the system. There are many diseases, such as mastitis, that can be limited through better management or infrastructure. Vaccines for livestock diseases have been very successful in some cases, such as the eradication of rinderpest. In other cases, long development times or impractical deployment requirements have created difficulties. Vaccines often have high investment costs, both in development and implementation. Periods of 15–20 years to develop some of these are not uncommon in the field.

Disease interventions are also hampered by poor veterinary surveillance and monitoring services. New opportunities for linking with human health investments for vaccine development (Global Alliance for Vaccines Initiative) and deployment, point-of-care diagnostics (Foundation for Innovative New Diagnostics, Grand Challenges Canada and others), and high throughput molecular sequencing and discovery sequencing are transforming diagnostic possibilities. Many of these human health interventions will become more widely available in the next 10–15 years and can be adapted to animal health.

Risk Reduction Technologies

In some livestock systems, such as pastoral systems, securing the existing livestock assets is more important than increasing productivity. Several technologies are available for attaining this. They include insurance-based schemes, such as the Index Insurance Innovation Initiative (Chantarat et al. 2007); improved market information for timely selling and buying of animals; disease surveillance and monitoring; early warning systems for drought, such as the Famine and Early Warning System Network (FEWS NET); and feed supplementation during the dry season.

Success in implementing these technologies has often been hampered by low investment, leaving pastoralists without access to key services and infrastructure. As a result—and especially during drought periods—these communities experience very large losses of animals.

Knowledge Management and Information Technologies

Perhaps the greatest current innovation that could lead to widespread change is in communication and knowledge management technologies. Impressive gains have been made in low-income countries in the update of mobile phones for information seeking, financial transactions, and other uses. The potential of mobile devices as information, knowledge management, and social media and mobilization tools is enormous. Applications, such as disease surveillance and reporting and social media, such as HealthMap and Riff, are greatly enhancing the speed at which information can be collected and responses provided. Given their accessibility, information and knowledge management tools have the capability to allow low- and middle-income countries to transform their approach to diagnosis, analysis, action, and information exchange for innovation.

LIVESTOCK ACTIONS FOR IMPROVING SECURITY

The central concerns for secure food, feed, and energy supplies are improved productivity and functional characteristics of plant biomass and the trade-offs in biomass's utilization. Two broad actions can help with the sustainable intensification of smallholder systems, which will allow livestock production in mixed smallholder systems to better complement crop agriculture. The first is specific investments in scaling up the application of dual-purpose food–feed crops, particularly for ruminant and small-scale pig production. The second is a more general investment in sustainable intensification of smallholder animal productivity improvements linked to strong institutions and smarter policies. Given the current large productivity efficiency gaps, doubling or tripling smallholder production per unit of input should be easily attained with significant natural resource efficiency use and reduction in greenhouse gas emissions. Benefits may take two to three decades to attain, but there will be short-term benefits as well as longer term ones.

Another set of actions is required to reduce the "hoof-print" of the livestock sector in the use of scarce water, land, and forest resources for food production. An initial step would be to develop, through broad consensus, guidelines and practices for responsible livestock development. This can include information and knowledge resources to support policymakers and civil society in different countries and with regional bodies in the case of watershed and transboundary land and forest concerns.

While food security is important, it does not always directly provide nutritional security. The elimination of stunting in young children has correctly risen to the top of the international agenda. Children's nutritional demands are much greater than adults', and meeting their needs is not just a question of expanding caloric intakes and food availability and access. To promote dietary diversity and better utilization, broad-based incorporation of nutrient-rich animal source foods in the diets of small children will be critical. Some of this will be through the expansion of smallholder dairy in favorable regions. As part of funding for improved nutrition in the first 1,000 days, maternal and young child health and development programs need to support innovative structured demand programs for supplying milk, meat, and eggs (and fish) to weaned and preschool children.

In tracking progress for these livestock actions, a number of more general food and nutritional security indicators will be needed. These include market information monitoring systems on prices and marketed volumes for main agricultural commodities—including milk, eggs, beef, lamb, pork, and poultry—to monitor volatility. Donors and governments should continue to provide support for FEWS NET, which provides good information on emerging food security changes and enhanced tracking of standard malnutrition indicators for young children, including monitoring prices of a food basket that could feed one to two-year-old children.

Many of these livestock actions are closely integrated into broad agricultural improvements, but will require better information on livestock productivity trends as part of broader agricultural and economic surveys for tracking performance. Crucial information will include mapping livestock within water basins and forests, tracking associated land degradation, and identifying hot spots with additional analysis. A critical requirement is to develop in-country and regional capacity for analysis and decision-making that can tap into broader global knowledge and link to local decision-making.

Livestock have been the main livelihood strategy in semiarid and arid areas, but increasing population and competition for scarce grazing and water resources have complicated pastoral livelihood strategies. In some key regions, low-level conflicts have escalated to high-intensity ones. Clearly specific security and conflict resolution actions will be important in these regions, particularly in empowering the peace-keeping perspectives and capacities of local communities.

These conflicts over livestock, however, arise from fundamental grievances linked to disempowerment and lack of livestock options. Proposing general livestock and other development actions in this context is a daunting challenge. While the security problem is large and worthy of investment, it is more difficult to advise on what strategies will work, as pastoral areas are the graveyards of many failed interventions and investments. Given the primacy of local knowledge and action, long-term support of the priorities of communities and local leaders is critical. In the short term, funders need to dramatically increase investments in supporting local communities and leaders on livelihood diversification strategies and support these plans with investments. Diverse strategies could include livestock production and marketing, wildlife or natural area conservation, education, and potential for nonagricultural activities. A lot of this support will be about sharing knowledge and information and connecting leaders to other groups, as well as building their capacity for analysis, decision making, and advocacy.

Much of the short-term investment will be in food assistance. These investments need to be designed so that food assistance and safety net programs are planned for at least ten years and linked to development initiatives and risk reduction programs such as low transaction cost insurance programs. One of the serious failures in pastoral development has been the disconnect between pastoralist needs and desires, and overall government policies and plans. There is often a serious mismatch between land tenure and other basic laws that disadvantage pastoral communities. Much progress could be made by recognizing these disconnects and discussing solutions with community leaders.

Information and indicators will be important as part of this process. These can build on the very successful FEWS NET. These systems can be supplemented with crowd sourcing information on drought, diseases, and conflicts linked to dissemination of maps and other decision aids for local communities. An important element will be monitoring of youth unemployment and underemployment as a warning of future conflict.

Biosecurity is widely viewed as an important livestock risk with widespread security implications. To mitigate these risks, an obvious short-term effort should be to share existing national and regional diagnostic capacities more effectively, linking them to international support. To increase efficiency, joint animal, human, and even plant disease surveillance should be linked. New technologies should be employed to improve sampling for molecular surveillance of pathogens linked to environmental changes, and for access and use of crowd-sourcing information through mobile devices and social media. This will need to be supplemented by specific capacity development, particularly on the response side. Capacity development for improved contingency planning and response decision making for animal and human health services and civil authorities is critical.

At the international level, we need agreements on access and benefit sharing—similar to the International Treaty for Plant Genetic Resources for Food and Agriculture—to enable freer exchange of pathogens for research and vaccine development. This initiative to hasten development of new vaccines for emerging pathogens should be part of a broader investment program in vaccines for animal diseases and zoonoses that can be linked to a variety of exciting innovations in the human vaccine field.

There are a number of opportunities for enhancing monitoring and response to biosecurity threats. Important building blocks already exist for molecular surveillance and crowd sourcing of events. Integrating these effectively through existing networks will be a useful short-term investment. In addition, targeted surveillance data for high-density livestock, tracking such factors as wildlife interfaces, primate bush meat hunting, and intensifying pig populations in contact with bats, would be a good investment to track the most likely hot spots for disease and pathogen emergence.

Livestock is a large and often forgotten sector of agriculture. The livestock sector is intimately linked with overall food production, food and nutritional security, environmental concerns, and infectious disease risks. This chapter has described a number of short-term actions that can enhance food, environmental, and health security, and reduce conflicts. Such actions will have multiple benefits in addition to managing these key security-related risks.

ACKNOWLEDGMENTS

We would like to thank Mark Rosegrant and Gerald Nelson for discussions on the section on IMPACT modeling and Steve Kemp for discussions on genetic and animal health technologies.

NOTES

1. The IMPACT model and its assumptions are described by Rosegrant et al. (2008). Nelson et al. (2010) describe some additional refinements and considerations relevant to the modeling reported here.
2. With the aid of spatial modeling and herd dynamics modeling tools, Havlik et al. (2012), disaggregated FAOstat country-level information on animal numbers and production into sub-national data representing distinct agro-ecological and management-system-based livestock production systems (following Sere and Steinfeld 1996; Wint and Robinson 2007).
3. The material in this section draws heavily on ILRI/FAO (2006) and Freeman et al. (2007).
4. See Marshall et al. (2011) and Rege et al. (2011) for further details.

REFERENCES

Butler, D. 2012. Flu surveillance lacking. *Nature 483*: 520–522.

Chantarat S., C. B. Barrett, A. G. Mude, and C. G. Turvey. 2007. Using weather index insurance to improve drought response for famine prevention. *American Journal of Agricultural Economics 89* (5): 1262–68.

Delgado, C., M. Rosegrant, H. Steinfeld, S. Ehui, and C. Courbois. 1999. Livestock to 2020: The next food revolution. Food, Agriculture and the Environment Discussion Paper 28. Washington, DC.

Enahoro, D., M. Herrero, and N. Johnson. 2011a. Promising technologies for improving livestock production and productivity in smallholder systems in developing countries I: Livestock breeds, vaccines and feeds. Global futures project report. Nairobi: ILRI.

———. 2011b. Promising technologies for improving livestock production and productivity in smallholder systems in developing countries II: Assets, markets, institutions and management. Global futures project report. Nairobi: ILRI.

FAO. 2009. State of food and agriculture: Livestock in the balance. Rome.

Freeman, A. H., P. K. Thornton, J. A. van de Steeg, and A. McLeod. 2007. Future scenarios of livestock systems in developing countries. In *Animal production and animal science worldwide*, 219–32. Rome: World Association for Animal Production.

Grace D., and J. McDermott. 2011. Livestock epidemic. In *The Routledge handbook of hazards and disaster risk reduction*, ed. B. Gaillard J. C, Kelman, 348–59. London: Routledge.

Hall A., M. Blümmel, W. I. Thorpe., and C. T. Bidinger. 2004. Sorghum and pearl millet as food-feed-crops in India. *Animal Nutrition and Feed Technology 4*: 1–15.

Havlik, P., H. Valin, A. Mosnier, M. Obersteiner, J. S. Baker, M. Herrero, M.C. Rufino, and E. Schmid. 2013. Crop productivity and the global livestock sector: Implications for land use change and greenhouse gas emissions. *American Journal of Agricultural Economics,* 95(2): 442–448 (January) (published online December 6, 2012).

Herrero, M., P. K. Thornton, A. Notenbaert, S. Msangi, S. Wood, R. Kruska, J. Dixon et al. 2009. Drivers of change in crop–livestock systems and their potential impacts on agro-ecosystems services and human well being to 2030: A study commissioned by the CGIAR System-wide Livestock Programme. International Livestock Research Institute. Nairobi, Kenya.

Herrero, M., P. K. Thornton, A. M. Notenbaert, S. Wood, S. Msangi, H. A. Freeman, D. Bossio et al. 2010. Smart investments in sustainable food production: Revisiting mixed crop–livestock systems. *Science 327*: 822–25.

Herrero, M., P. K. Thornton, P. Havlík, and M. Rufino. 2011. Livestock and greenhouse gas emissions: Mitigation options and trade-offs. In *Climate change mitigation and agriculture,* ed. E. Wollenberg, A. Nihart, M. L. Tapio-Bistrom, and C. Seeberg-Elverfeldt, 316–32. London: Earthscan.

ILRI–FAO. 2006. The Future of Livestock in Developing Countries to 2030, 13–15 February, Nairobi, Kenya. Meeting Report.

Jones, P. G., and P. K. Thornton. 2009. Croppers to livestock keepers: Livelihood transitions to 2050 in Africa due to climate change. *Environmental Science and Policy 12*: 427–37.

Kaitibie, S., A. Omore, K. Rich, B. Salasya, N. Hooton, D. Mwero, and P. Kristjanson. 2008. Influence pathways and economic impacts of policy change in the Kenyan dairy sector: The role of smallholder dairy project. Research report for the CGIAR Standing Panel on Impact Assessment, 64. Nairobi: ILRI.

Lannerstad, M., J. Heinke, M. Herrero and P. Hávlik. 2012. Livestock production systems—green, green, blue consumptive water use. In Water and livestock: Interactions, tradeoffs and opportunities symposium. World Water Week: Water and food security. August 26, 2012. Stockholm, Sweden.

Laurence, J. 2012. African monkey meat that could be behind the next HIV. *Independent*, May 25. http://www.independent.co.uk/life-style/health-and-families/health-news/african-monkey-meat-that-could-be-behind-the-next-hiv-7786152.html.

Lenne, J., S. Fernández-Rivera, and M. Blümmel. 2003. Approaches to improve the utilization of food–feed crops. *Field Crops Research 64* (1 and 2): 1–227.

McDermott, J. J., S. J. Staal, H. A. Freeman, M. Herrero, and J. Van de Steeg. 2010. Sustaining intensification of livestock production systems in the tropics. *Livestock Science 130*: 95–109.

Marshall, K., C. Quiros-Campos, J. H. J. van der Werf, and B. Kinghorn. 2011. Marker-based selection within smallholder production systems in developing countries. *Livestock Science 136:* 1–54.

Miles, J. W., B. L. Maass, and C. B. do Valle, eds. 1996. *Brachiaria: Biology, agronomy and improvement.* Cali, Colombia: CIAT.

Millennium Ecosystem Assessment. 2005. *Ecosystems and human well-being.* Washington: Island Press.

Msangi S., and M. Rosegrant. 2011. Feeding the future's changing diets: Implications for agriculture, markets, nutrition and policy. 2020 Conference Paper. Washington DC: IFPRI.

Neumann, C., D. M. Harris and L. M. Rogers. 2002. Contribution of animal source foods in improving diet quality and function in children in the developing world. *Nutrition Research 22* (1): 193–220.

Osofsky, S.A. 2005. Conservation and development interventions at the wildlife/livestock interface. Occasional Paper of the IUCN Species Survival Commission No. 30. Gland, Switzerland and Cambridge, UK: IUCN.

Peden, D., G. Tadesse, A. K. Misra, F. A. Ahmed, A. Astatke, W. Ayalneh, M. Herrero et al.. 2007. Water and livestock for human development. In *Water for food, water for life: A comprehensive assessment of water management in agriculture*, ed. D. Molden, 485–514. London, UK and Colombo, Sri Lanka: Earthscan and IWMI.

Peters M., and C. Lascano. 2003. Forage technology adoption: Linking research with participatory methods. *Tropical Grasslands 37*: 197–203.

Phaikaew, C., and M. D. Hare. 2005. Stylo adoption in Thailand: Three decades of progress. In XX *International Grassland Congress*, eds. F. P. O'Mara, R. J. Wilkins, L. t'Mannetje, D. K. Lovett, P. A. M. Rogers, and T. M. Boland, 323. Netherlands: Wageningen Academic Publishers.

Phaikaew, C., C. R. Ramesh, Yi Kexian, W. Stür. 2004. Utilisation of Stylosanthes as a forage crop in Asia. In *High-yielding anthracnose-resistant Stylosanthes for agricultural systems*, ed. S. Chakraborty, 65–73. Canberra: Australian Centre for International Agricultural Research, Monograph No. 111.

Pica-Ciamarra, U., J. Otte and P. Chilonda. 2007. Livestock policies, land and rural conflicts in sub-Saharan Africa. Pro-poor livestock policy initiative (PPLPI) research report, 07–04. Rome.

Ramesh, C. R., S. Chakraborty, P. S. Pathak, N. Biradar and P. Bhat. 2005. Stylo in India—much more than a plant for the revegetation of wasteland. In XX *International Grassland Congress,* eds. F. P. O'Mara, R. J. Wilkins, L. t'Mannetje, D. K. Lovett, P. A. M. Rogers, and T. M. Boland, 320. Netherlands: Wageningen Academic Publishers.

Randolph, T. F., E. Schelling, D. Grace, C. F. Nicholson, J. L. Leroy, D. C. Cole, M. W. Demment, A. Omore, J. Zinsstag, and M. Ruel. 2007. Role of livestock in human nutrition and health for poverty reduction in developing countries. *Journal of Animal Science 85*: 2788–800.

Rege, J. E. O., K. Marshall, A. Notenbaert, J. H. K. Ojango, and A. M. Okeyo. 2011. Pro-poor animal improvement and breeding—What can science do? *Livestock Science 136*: 15–28.

Reid, R. S., D. Nkedianye, M. Y. Said, D. Kaelo, M. Neselle, O. Makui, L. Onetu et al. 2009. Evolution of models to support community and policy actions with science: Balancing pastoral livelihoods and wildlife conservations in savannas of East Africa. *Proceedings of the National Academy of Science.* doi: 10.1073/pnas.0900313106.

Rosegrant, M., M. Paisner, S. Meijer, and J. Witcover. 2001. *Global food projections to 2020: Emerging trends and alternative futures.* Washington, DC: International Food Policy Research Institute.

Rosegrant, M., C. Ringler, S. Msangi, T. Sulser, T. Zhu, and S. Cline. 2008. *International Model for Policy Analysis of Agricultural Commodities and Trade (IMPACT): Model Description.* Washington, DC: International Food Policy Research Institute.

Rosegrant, M., S. Tokgoz, and P. Bhandary. 2012. The new normal? A tighter global agricultural supply and demand relations and its implications for food security. *American Journal of Agricultural Economics,* 1–7. doi: 10.1093/ajae/aas041.

Shelton, H.M., S. Franzel, and M. Peters. 2005. Adoption of tropical legume technology around the world: Analysis of success. In: XX *International Grassland Congress,* eds. F. P. O'Mara, R. J. Wilkins, L. t'Mannetje, D. K. Lovett, P. A. M. Rogers, and T. M. Boland, 149–168. Netherlands: Wageningen Academic Publishers.

Shettima, A. G., and U. A. Tar. 2008. Farmer–pastoralist conflict in West Africa: Exploring the causes and consequences. *Information, Society and Justice 1*(2): 163–84.

Staal, S., J. Poole, I. Baltenweck, J. Mwacharo, A. Notenbaert, T. Randolph, W. Thorpe, J. Nzuma, and M. Herrero. 2010. Targeting strategic investment in livestock development as a vehicle for rural livelihoods. Nairobi: International Livestock Research Institute.

Steinfeld, H., P. Gerber, T. Wassenaar, V. Castel, M. Rosales, and C. de Haan. 2006. *Livestock's long shadow: Environmental issues and options.* Rome: FAO.

Wint, W., and T. Robinson. 2007. *Gridded livestock of the world—2007.* Rome: Food and Agriculture Organization of the United Nations.

World Bank. 2007. *Agriculture for development: World Development Report, 2008.* Washington, DC: World Bank.

9

Labor Migration and Food Security in a Changing Climate

ROBERT MCLEMAN

When migration and climate change have been linked in the international security literature, it has often been done in the context of larger scenarios in which food and water resources become scarcer, natural disasters become more frequent, intergroup violence erupts, and environmental refugees flow out of troubled regions. The present example of Darfur is held up as an example of what to expect more of in a changing climate (UNEP 2007). This sort of scenario may become more common in the future, but it is only one possibility among many, for the relationship between climate, food security, political instability, and migration is complex. Security scholarship tends to focus on how migration can be a cause of instability or an outcome of it, but less attention is paid to the fact that migration can also offset or avert food insecurity and political instability. While climatic events and conditions can and do have a significant influence on migration in many parts of the world, cases like Darfur are, thankfully, infrequent. Most climate-related migration events, even those that seem chaotic or undesirable from an outside perspective—like a pulse of migrants out of a storm-damaged or drought-stricken area—are, from the point of view of affected households, rational responses to challenging circumstances.

This chapter examines the important relationship between labor-related migration and food security in the context of climate change, and considers the circumstances under which it may lead to political instability. Labor migration forms a major part of global migration flows, and is used by households in many low-income countries (LICs) as an integrated part of ongoing strategies to maintain food security and enhance the family's collective well-being. In the face of a sudden-onset food security crisis, climate-related or otherwise, labor migration can be a critical household coping strategy when institutional responses are inadequate or absent. In some LICs, labor migration is so essential to maintaining the food security of rural and urban households, that any external factors that interfere with its normal functioning should be viewed as having the potential to trigger socioeconomic disruptions and political instability. Such factors include anthropogenic climate change, which is the main focus of this volume, as well as others that will be discussed to varying degree in this chapter, such as demographic change, rapid urbanization, growing global demand for food, and state restrictions on mobility.

GENERAL LINKAGES BETWEEN LABOR MIGRATION AND FOOD SECURITY

Although there are many explanations as to why people participate in voluntary migration, it has long been held that economic or employment-related reasons are among the most common (Ravenstein 1889). The traditional microeconomics-based view has been that labor migration occurs opportunistically, with individuals making rational choices about where their skills, training, and aptitudes have the greatest wage income potential, and then choosing their place of residence accordingly (Todaro 1969). This assumes individuals have freedom of mobility and access to information about labor markets; if so, the individual's human capital and financial means to relocate become the key constraints on labor migration. This is a point worth emphasizing: having the desire to migrate and having the necessary assets to do so are two separate things.

International labor migration is much more heavily constrained by structural factors than is internal migration within countries (Samers 2010).[1] Governments and international trade agreements often deliberately favor and facilitate the movement of higher skilled and higher paid workers across borders, while restricting the legal movement of lower skilled workers (Lalonde and Topel 1997). This does not mean to say that lower paid workers do not migrate internationally—they do so in great numbers—but that they must often do so through irregular channels as opposed to formal ones (Gross and Schmitt 2012). The result is a segmented global labor market within which specialized groups enjoy a greater opportunity to maximize their incomes through migration (Grogger and Hanson 2011), while a much larger population of labor migrants get channeled into low-paying, seasonal, or dangerous occupations unattractive to nonmigrants (Samers 2010).

Income maximization is not, however, the only reason people participate in labor migration. In regions with persistent threats to household security—such as challenging environmental conditions, uncertain local income opportunities, or lack of access to lending institutions and financial markets—households may depend heavily on remittances from members who have migrated elsewhere to obtain employment (Adger et al. 2002; Schneider and Knerr 2000). International remittances are estimated in the hundreds of millions of US dollars annually, and dwarf official development assistance transfers to LICs (Gammeltoft 2002; World Bank 2011). In absolute dollar value, the greatest flow of remittances is to large migrant source countries like India, China, and Mexico, but in less populous LICs like Lesotho, Moldova, Nepal, Tajikistan, and Tonga, migrant remittances can form as much as a third of gross domestic product (World Bank 2011).

Labor migration can enhance food security for rural and urban households alike, with tangible benefits flowing back and forth between migrant sending and receiving areas. In Namibia, for example, urban migrants often receive food transfers from their rural relatives, and extended families will act collaboratively to take advantage of price discrepancies between rural and urban locations (Frayne 2005; Greiner 2011). Smit (1998) has observed similar food transfers along rural–urban networks in South Africa. These rural–urban linkages can be long-lasting, with evidence from Kenya showing that migrants to Nairobi often maintain connections to their rural home regions for many decades (Mberu et al. 2012). Remittances received from urban relatives are often invested by rural households to expand and improve their food-production capacity and household income (Tiffen 2003). Labor migration and household food security are linked in regions other than Africa, as well. In Vietnam, poorer households use short-term labor migration earnings to enhance their caloric intake and diversity of food sources (Nguyen and Winters 2011). Remittances help households cope with sudden shocks to food security, with de Brauw (2011) observing that during a 2008 food price crisis in El Salvador, children living in households receiving remittances from abroad showed less indication of diet-related setbacks in physical development.

The connection between labor migration and household food security can be organized into a simple hierarchy (Figure 9.1). Figure 9.1 suggests that the expected returns from migration ascend from short-term crisis management to longer-term security with respect to basic needs, to household betterment through greater savings, housing quality, education, and finally to opportunity seeking. Movement from lower to higher levels of the hierarchy implies that labor migration becomes less directly connected with food security, and is more likely pursued for purposes of socio-economic advancement as it is in in higher income countries.

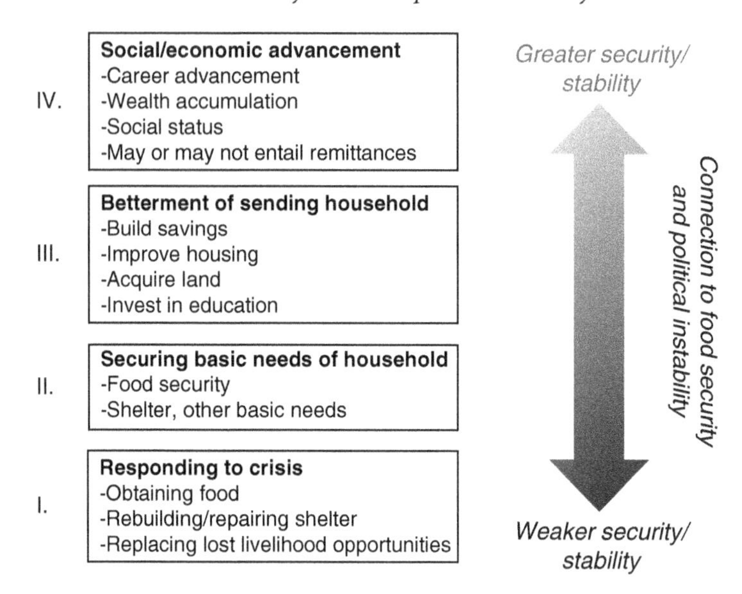

Fig. 9.1. Simple hierarchy of expected returns from labor migration
Source: Author

PATTERNS OF LABOR MIGRATION IN THE PURSUIT OF FOOD SECURITY

A regular pattern of ongoing, intranational, or regional seasonal migration timed to coincide with natural fluctuations in environmental conditions often emerges in regions where precipitation is highly seasonal or cyclical in nature. In sub-Saharan Africa, off-farm work and labor migration have long been key components of rural livelihood strategies (Reardon 1997). In the semi-arid Sudano-Sahelian region of West Africa, rural populations have adopted a regular cycle of seasonal migration described as "eating the dry season" (Rain 1999). During the dry season, when their labor is not needed at home and when household food reserves risk becoming depleted, young men and, depending on the cultural group, young women, travel to seek wage employment in regional urban centers or in other rural areas with longer productive seasons. Their departure allows the household to reduce the number of people dependent on stored food, which in some areas may only last for seven months out of the year (Quaye 2008) and use remitted funds to purchase additional food. The migrants return to the countryside when the next crop production season is ready to begin and their labor is once again needed at home. When drought conditions strike—that is, extended periods of little precipitation much more severe than usual levels of dryness—a surge in migration of rural people to local urban centers may ensue, with some families sending children to live with relatives outside the drought-stricken zone (Roncoli et al. 2001). There tends not to be a commensurate surge in long-distance migration out of the region during droughts (Henry et al. 2004), likely due to the costs associated with sending a migrant abroad. Instead, labor migration abroad is more likely to be undertaken under favorable environmental conditions, when rural incomes are strong and the household can afford both the expense of migration and the loss of the migrant's labor.

Seasonal labor migration is also a common strategy used by the rural poor in many Asian countries to enhance food security, including Bangladesh, India, the Philippines, Thailand, and Vietnam. In Nepal, where 10 percent of households suffer from chronic food insecurity (Maharjan and Joshi 2011), members of poor rural households have long engaged in seasonal migration to lowland areas and/or to Indian cities during periods of annual food shortage. In parts of India characterized by poor soils and chronic water scarcity, up to one-half, and sometimes three-quarters of the rural population may be engaged in regular seasonal migration (Deshingkar 2006).

Sudden surges in labor migration may emerge in the wake of events that disrupt household food security directly, through direct damage to crops, livestock, or housing, or indirectly, through loss of livelihood, loss of wage income opportunities, or price shocks. The poor are disproportionately exposed to natural disasters, and such events often stimulate pulses of labor migrants seeking to replace lost livelihood income and, in the case of floods and extreme storms, raise money to repair damage to homes and other fixed assets. In the months following 1998's Hurricane Mitch, spikes were observed in the numbers of Nicaraguan agricultural workers seeking employment in Costa Rica, Guatemalan workers seeking farm work in Belize, and young Honduran males seeking to enter the United States clandestinely (Carr 2008; Murrugarra and Herrera 2011). Flood events in China have led to pulses of labor migration to urban areas, as occurred in the Beijiang basin in Guangdong province in 1994 and following the great Yangtze River floods of 1998 (Wong and Zhao 2001). As a reminder that these phenomena are not exclusive to lower- and middle-income countries, it is worth recalling the short-term food security crisis poor families experienced following Hurricane Katrina, and the surge of evacuees seeking employment in cities like Baton Rouge and Houston (Pyles et al. 2008).

Droughts also have impacts on food security, and the size of resulting migration can be considerable, as witnessed in Depression-era drought migrations on the North American Great Plains (McLeman 2006) and in East Africa in more recent decades (Ezra and Kiros 2001). Migration to the United States from Mexico spikes during droughts, with Feng et al. (2010) suggesting that a 10 percent decrease in Mexican crop yields due to drought can be expected to result in a 2 percent increase in migrant numbers. Indian drought events stimulate considerable temporary migration, as seen in a 2001 drought in Orissa state where an estimated 300,000 people undertook short-term migration (Deshingkar 2006).

In Nepal, environmental degradation and food insecurity can influence internal migration patterns, but seems to have no observable effect on longer-duration labor migration to more distant destinations such as the Middle East (Massey et al. 2010; Poertner et al. 2011). Referring back to Figure 9.1 above, the Nepal experience suggests short-distance internal migration is associated with the lower levels of the hierarchy—closely tied to basic needs and food security—while international migration emerges from higher-level, opportunity-seeking motives.

Institutionally led actions that deliberately exclude people's access to resources, such as conservation or development projects and the construction of large dams, can have tremendous impacts on both voluntary and involuntary migration patterns as households adjust. Often, those displaced find themselves in worse economic circumstances as a result of inadequate or incompetent government resettlement assistance (Scudder 2012). For example, an estimated 1.3 million people have been displaced since 1992 by the development of China's Three Gorges Dam project (Xi and Hwang 2011). Official resettlement programs that sought to relocate farmers to nearby farmlands proved to be unpopular, with the new land being less productive and more easily eroded, leaving many farmers wanting to resettle in nearby urban centers where employment opportunities, wages, and services were better (Tan et al. 2003). Other programs that channeled people to more distant locations have also often proven unsuccessful, as many migrants have been unable to integrate into new settings and hold on to government-arranged jobs (Tan et al. 2005). In response, large numbers of working-age displacees have of their own accord migrated to coastal cities and the Pearl River delta, where they work in factories, shops, and restaurants and remit significant amounts of money to relatives remaining behind (Jim and Yang 2006).

Where food and other critical resources are scarce, economic decline may stimulate further changes in migration patterns. In rural western India, the effects of severe ecological pressure of a large rural population on the land, combined with heavy household debt loads, lead the rural poor to use labor migration as a defensive strategy whenever economic conditions turn bad (Mosse et al. 2002). In the 1990s, Tanzania, Uganda, and second-tier cities in Ghana saw urban-to-rural migration increase as households struggled with food security during economic crises (Potts 1995). In the last decade, inflation, unemployment and lack of housing have led significant numbers of urbanites in Cote d'Ivoire, Zambia, and Zimbabwe to migrate to the countryside to pursue subsistence livelihoods (IRIN 2006; Potts 2005; Beauchemin 2011).

HOW CLIMATE CHANGE INFLUENCES LABOR MIGRATION AND FOOD SECURITY

Many scholars view migration as part of a larger range of possible actions by which individuals and households adjust or adapt to the effects of climatic variability and change (McLeman and Smit 2006). Because food production is shaped by climate, it is logical to expect anthropogenic climate change to have significant implications for food availability at all scales from the local to the global (Schmidhuber and Tubiello 2007) and, as a potential consequence, on labor migration patterns as households adapt and adjust. This logic applies to households that are actively engaged in food production and are thus directly exposed to climatic impacts, as well as to those households that purchase their food and are sensitive to changes in food prices triggered by climatic events locally or elsewhere.

Figures 9.2 and 9.3 provide a simple, generalized representation of how climate change might be expected to influence labor migration and household food security dynamics at a given location. The representation has its origins in the "pressure-and-release" model of vulnerability to environmental change first developed by Blaikie et al. (1994) and is consistent with subsequent conceptualizations of food security (e.g., Gregory et al. 2005) and climate-related migration (e.g., McLeman and Smit 2006). Figure 9.2 assumes

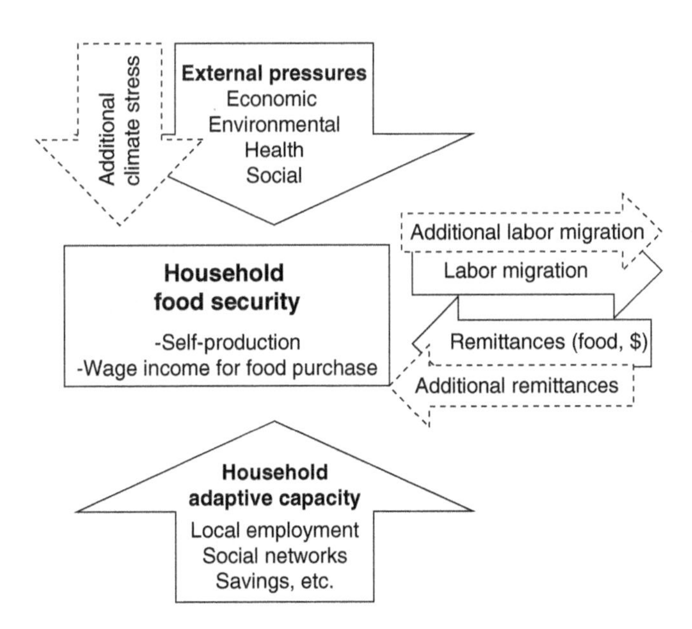

Fig. 9.2. Labor migration in the context of food security
Source: Author

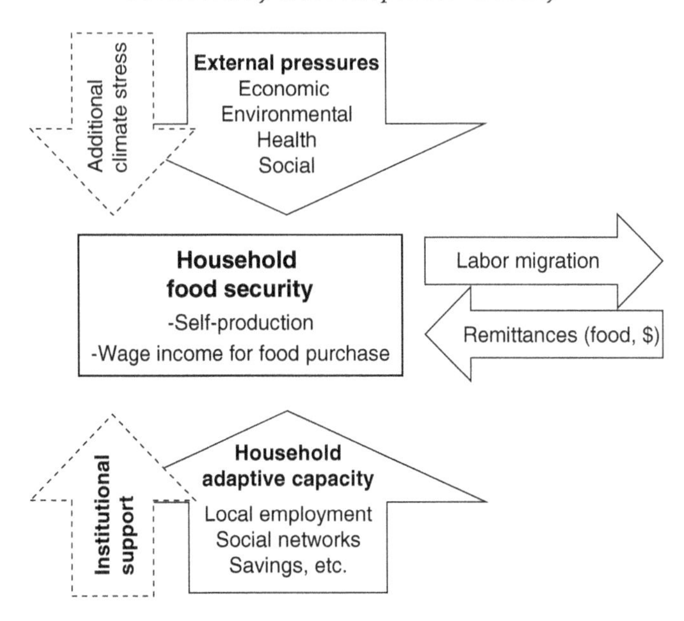

Fig. 9.3. Labor migration, food security, and climate change in the context of institutional capacity-building
Source: Author

that the food security of a given household—whose access to food may be through self-production, food purchases, or a combination of both—is subject to an ongoing mix of pressures of various origins. These pressures might be economic (wage or market price fluctuations), environmental (weather or pest outbreaks), social (access to education and training), or of some other nature (the health of household members). How a household adapts or adjusts to these ongoing pressures will depend on its members' access to economic capital, social networks, skills, and similar qualities. Households that have access to significant amounts of these various forms of capital may be able to create successful livelihoods and achieve an acceptable standard of living without any members having to engage in labor migration (though some may do so by choice, in order to seek out opportunities). Other households may engage in labor migration as part of short- or longer-term strategies to achieve food security and longer-term stability.

Climate change can be expected to introduce additional pressures on household food security, as suggested by the dashed arrow in the upper left. These additional pressures could be experienced as direct impacts on local food production, or indirectly, such as adverse impacts elsewhere that shock market prices for food. Households that have already been relying on remittances to meet their basic needs may be obliged to send additional members in search of remittances (or send them away for longer periods), and households that were previously able to succeed without migration may find themselves needing to engage in migration. As a consequence, participation in labor migration may rise beyond the normal background levels, a dynamic witnessed in the wake of many extreme events. In the short run, the additional migration would likely include many short distance and temporary migrants, but should the climatic regime shift permanently, the duration and destination of migration may become more extensive. In an ideal situation, the out-migration and subsequent remittances would offset the degree of harm caused by additional stressors, and enable the population to regain a state of equilibrium, which is the state shown in Figure 9.2. Should labor migration be restricted or remittances fall short or fail, however, equilibrium is not achieved and the risk of instability within the population will rise.

This raises the important question of institutional behavior. It may be the case that, in either planning for or reacting to the impacts of climate change, governments and other institutional actors might initiate strategies, programs, or policies that support households and enhance their *in situ* adaptive capacity. Should this be done effectively, significant changes in the background level of labor migration might be avoided, at least in terms of risk minimization behavior, and a state of equilibrium maintained. This is illustrated in Figure 9.3, and foreshadows conclusions and recommendations made below.

CLIMATE CHANGE RISKS AND LABOR MIGRATION IN THE NEXT 20 YEARS

Evidence from the recent past suggests three particular climate phenomena have the greatest direct influence on migration patterns: tropical cyclones, droughts, and river valley flooding (McLeman and Hunter 2010). In some instances governments have been actively involved in facilitating relocations, but in most cases, the resulting migration has been initiated at the household level, consistent with the patterns and processes described above. Across various regions, each of these three types of events is expected to occur more frequently or severely over coming decades as a result of anthropogenic climate change (IPCC 2007). In addition, two new types of change are emerging: rapid loss of Arctic land and sea ice and changes in mean sea levels. There are no historical precedents for these types of change, but there is recent evidence emerging that they are already beginning to stimulate changes in labor migration patterns. Using expected climate change impacts identified by the IPCC (2007), McLeman and Hunter (2010) summarized the likely implications for climate change on broad-scale migration patterns, and identified four broad conditions likely to affect labor migration patterns in the next two decades (Table 9.1). Each of these is now elaborated in turn.

Table 9.1. Likely influences of climate change on labor migration, by region

Affected region	*Expected climatic change*	*Likely impact on migration*
Mid- and low-latitude dry-land and seasonally dry regions (e.g., South Asia, western China, western North America, western South America, Sudano-Sahelian Africa)	Lower annual river runoff, greater frequency and/or spatial extent of drought events; decreasing crop productivity and/or greater risk of seasonal crop damage/losses.	Greater frequency of drought-related migration, especially seasonal and temporary regional migration; increased levels of indefinite rural–urban migration in areas where conditions become persistent; greater dependence on remittances in affected rural areas.
Low-lying coastal plains, deltas (especially heavily populated Asian deltas) and small-island states	Increased risk of erosion, flooding and storm surge damage	Hurricane Katrina/Mitch-style displacements and distress migration; greater labor migration out of exposed islands; need for organized relocation from highly exposed areas.
Mid- to high-latitude river valleys and some wet tropical areas	Increased average annual runoff	Increased risk of flood-related displacements; regional labor migration in response
Arctic	Loss of ice cover on sea and land	Opening of shipping lanes, accessibility of seabed resources to create influx of labor migrants

Source: Adapted by author from McLeman & Hunter (2010); climate change projections from IPCC (2007); with implications for labor migration derived from McLeman and Hunter (2010) and additional sources cited in this chapter.

Increased Droughts

Drought events are expected to occur more frequently and with greater duration and intensity across many of the world's arid and semiarid regions in coming decades, to be accompanied by corresponding downturns in food production (Li et al. 2009). The world's agriculturally productive drylands can be loosely categorized into those that support large rural populations that are directly dependent on small-scale agricultural livelihoods (Pakistan, India, Sudano-Sahelian Africa), those that host a small rural population base and a commercially oriented agricultural system (e.g. North American Great Plains, Australia, Russia), and those in transition between the two (Brazil, China, Mexico). As observed in examples given previously in this chapter, droughts already stimulate responses in labor migration patterns in the first category of regions and in those that are in transition to smaller rural populations. Future droughts can be expected to accelerate participation in ongoing rural–urban migration and rural–rural migration (from dry to less-dry areas), trigger new pulses of distress migration, and encourage further growth of remittance economies within these regions. Local and regional declines in food production will put pressure on household food security, which would be magnified should ever these coincide with droughts in any of the large export-producing regions producing regions that trigger price shocks.

By contrast, given their relatively small agricultural population bases, food-exporting dryland regions are unlikely to see significant rural population movements during future droughts, such as used to occur, for example, on the North American Great Plains in the last century (McLeman 2006). However, there are urban centers on the Great Plains whose surface water supplies are showing worrying downward trends (Schindler and Donahue 2006), and should severe droughts occur, households and communities could experience localized water rationing and temporary relocations. The sizable contribution of Great Plains agriculture to global food supplies should give cause for concern in terms of future food price stability. Long-term climatological evidence suggests that the twentieth century was an unusually wet one, in spite of the severe drought events of the 1930s, 1950s, and 1970s. Indeed, in past centuries multi-decade droughts were not uncommon (Sauchyn et al. 2003). Any sudden declines in Great Plains food production due to drought are likely to stimulate price shocks in Asia and other export markets. Recent droughts in Russia and Australia led to short-term drops in grain exports from those countries, which in turn contributed to upward spikes in global grain prices (Piesse and Thirtle 2009; Wegren 2011).[2] A severe, multi-year drought on the Great Plains can be expected to have even greater impacts on global food prices and generate ripple effects on labor migration patterns in LICs as poorer households struggle with higher food prices. These could include new movements of urban poor back to the land and/or from rural subsistence households to urban areas.

Sea-level Change and Storm Surges

Average sea levels are rising from the combined effects of thermal expansion and increasing input from glacier and snowpack decline, with the rate of change being highest in the western Pacific (Nicholls and Cazenave 2010). The IPCC (2007) estimate of 0.6 meters mean sea-level rise by 2100 based on current emissions trends is increasingly viewed as being overly conservative, with estimates of 0.8 meters or greater being reasonable (Nicholls and Cazenave 2010). Even at faster rates, the greatest impacts of mean sea-level rise on migration will be experienced beyond the next two decades. Given the scale of the potential disruptions by mid-century, anticipatory adaptation will need to begin sooner.

Despite popular reports in the media, there is no evidence that anthropogenically induced mean sea-level rise is presently causing migration or displacements, even in Pacific atoll nations that have the highest degree of physical exposure (Mortreux and Barnett 2009). There are examples of coastal settlements in various regions having to be relocated because of erosion, but conclusive links to anthropogenic climate change have not been demonstrated. That said, many small islands exposed to mean sea-level rise already experience high rates of labor migration, such as the many Pacific Island states whose nationals migrate to Australia and New Zealand (Barnett and Campbell 2010). Out-migration rates from such areas should be expected to accelerate in coming decades as the impacts of climate change take hold. The absolute number of additional global labor migrants due to mean sea-level rise over the next two decades is not likely to be great; for example, the total population of all Pacific Island states combined is only about ten million people (Barnett and Campbell 2010). Nonetheless, from a regional geopolitical perspective, the effects could be politically disruptive.

Densely populated, low-lying Asian and African river deltas, currently home to more than 400 million people and urbanizing fast, are also highly exposed to mean sea-level rise, but these areas are net receivers of labor migrants (Seto 2011). Even with mean sea-level rise, it is unlikely there will be any changes in their attractiveness as labor migration destinations in the absence of catastrophic events. At the same time, Asian production of rice—a key staple crop for populations living in Asia's mega-deltas—could decline and global rice prices rise as a result of changing sea-levels (Chen et al. 2012). Vafeidis et al. (2011) estimate that by 2030, anywhere between 930 million and 1.3 billion people could be living in low-elevation coastal zones, up from their estimate of 630 million in the year 2000. Overall, then, the outlook for the next two decades is a rapid increase in the global population exposed to coastal hazards and the risk of rice market price shocks—a potentially destabilizing combination for coastal South and Southeast Asia and parts of coastal Africa.

The preceding discussion does not take into account the effects of climate change on tropical cyclones. Some existing research suggests that the intensity of tropical cyclones in the North Atlantic is increasing (Webster et al. 2005), but overall there is no scientific consensus in terms of trends in the frequency, intensity, and spatial extent of tropical cyclones over the next two decades (Knutson et al. 2010). Even in the absence of a change in trends, tropical cyclones are, and will continue to be, events with the potential to trigger large-scale distress migration and short-term political instability.

River Valley Flooding

The risk of flooding due to heavy precipitation events and snowmelt is very sensitive to changes in climatic conditions, with anthropogenic climate change expected to increase the risk of flooding in many coastal regions and river valleys in coming decades (Knox 2000; Kundzewicz et al. 2010). An estimated 85 percent of nations have experienced significant flood events in recent decades (Bakker 2009), and the Millennium Ecosystem Assessment (2005) suggests as many as two billion people worldwide may live in areas exposed to flooding risk. Four countries—the United States, China, India, and Indonesia—account for the largest number of reported floods in recent years,[3] most of which were associated with heavy rains (Adhikari et al. 2010). The broad geographical regions with the greatest exposure to flooding are Southeast Asia, China, Central America, and the Caribbean, regions that have large populations living in exposed locations and where monsoon precipitation patterns and tropical cyclones are common. Asia in particular, with its many large, densely-populated river valleys and deltas, hosts the greatest number of people at risk (Jonkman 2005).

The scale of human displacement, fatalities, and damage to property and infrastructure during flood events can be tremendous. A 1998 flood in China's densely populated Yangtze Valley affected more than 200 million people, destroyed almost ten million homes, and caused an estimated $420 billion in damage (Yin and Li 2001; Zong and Chen 2000). That same year, over 60 percent of Bangladesh was inundated for more than two months, affecting 30 million people, damaging five million homes, and requiring the evacuation of over a million people (Kundzewicz et al. 2010; Kunii et al. 2002). In LICs, the poor typically lack access to insurance plans or ad hoc government compensation schemes that might assist them in overcoming the double impact of lost shelter and income, making them particularly predisposed to migration when flood events do occur. Floods have a further impact on the food security of rural populations when they destroy stored food reserves and seeds being saved for future planting.

A common response to floods is labor migration, and this should be expected to continue to be the case in the next two decades, especially in Asia. In parts of Asia where flooding is an annual event and rice production predominates, seasonal labor migration is embedded in rural livelihood strategies (Ellis 2000). In the case of extreme flood events, the migration's distance and duration is highly dependent on local socioeconomic dynamics. Evidence suggests that where permanent relocation does occur, it often takes place over relatively short distances within the same rural area or to the nearest regional center. Such responses can be expected to continue in coming decades. In some particularly flood-prone locales, it may be necessary to initiate organized settlement relocations with the assistance or instigation of government authorities.

Arctic Sea Ice Decline

Opportunity-seeking labor migrants can be expected to be drawn to northern communities as the rapid decline in sea ice leads to more regular seasonal open-water conditions in the Arctic, allowing for greater trans-Arctic shipping, offshore resource development, and marine-based tourism. The most recent evidence suggests that Arctic waters will regularly be ice-free during the months of September and October within the next two to three decades, although there is no clear evidence of a year-round ice-free period this century (Jones 2011). Ho (2010) suggests that marine navigation in the Arctic will remain regional at first and dominated by gas, oil, and other resource shipments, with the volume and extent of shipping to accelerate quickly. The Chinese government is already planning for the possibility of trans-Arctic shipping, and Russian officials are already promoting the benefits of the Northeastern Passage (Jakobson 2010; McLeman and Smit 2012).

The high costs of hydrocarbon extraction in the Arctic mean that the pace of future offshore resource development will be driven as much by market demand as by warming trends, with state-controlled Russian firms likely being the first to take advantage of the changing climate (Harsem et al. 2011). Consequently, it can be expected that labor migration stimulated by infrastructure construction for shipping and resource development will emerge first in northern Russia and Scandinavia in no more than a decade or so, and spread to Alaska and Canada fairly soon afterwards. The number of new Arctic settlers will not be great in a global sense—likely in the tens of thousands, or perhaps low hundreds of thousands by 2030—but their presence will feed into renewed geostrategic rivalries among the Arctic states and China. Large Arctic settlements will face food supply challenges and occasional logistical emergencies; for example, food needed to be airlifted to residents of the Canadian territory of the Yukon in the summer of 2012 when landslides closed the only highway. In-migration will also be of concern for the fast-growing indigenous populations of the north, whose traditional subsistence livelihoods are already under threat and whose long-term food security will depend on how successfully they can be incorporated into an expanding wage-based economy (Ford and Pearce 2010).

FUTURE PATHWAYS TO INSTABILITY AND INDICATORS FOR MONITORING

Labor migration can be an outcome of instability, a moderator of it, and/or a contributor to it, depending upon the situation. This section identifies more precisely how political instability can emerge from and feed into labor migration dynamics, suggests particular sets of dynamics that should be watched for in the next two decades, and identifies possible indicators for these where possible. Although the analogy is not perfect, imagine that the flow of migrants and remittances between a source area and the destination is like a flow of water or air, and that sociopolitical instability is akin to turbulence within that flow. Turbulence is created when that flow encounters obstacles that create resistance, the greater the resistance, the greater the turbulence. Figure 9.4 illustrates the flow of migrants and remittances passing over a field of potential turbulence-inducing obstacles, examples of which are contained among the examples given in preceding sections and are summarized in Table 9.2.

Table 9.2 is not an exhaustive list, and it is worth noting that none of the listed factors operates independently. Some have effects on sociopolitical stability in the migrant sending area, some at the receiving area, and some at both. The weight of the effect of each factor relative to the others will vary depending on local conditions. With sufficient local-level data—on wage differentials, costs of migration, housing stocks, and so on—it may be possible to estimate the impact of each factor to create a set of metrics for monitoring labor migration movements for changes that might feed sociopolitical instability.

The top left corner of Figure 9.4 contains an arrow that is a placeholder for emergent drivers of concern in the next two decades. Of the many possibilities, several stand out as having potential for generating regional or global instability:

Table 9.2. Factors in generating sociopolitical instability

Factor	Considerations
Government barriers to mobility	Formal migration regulations, internal and international, are typically known; in some cases additional informal barriers exist, like the need to bribe government officials. Where regulations prohibit legal labor migration (in total or of certain types of individuals), migrants may take on additional costs to migrate (e.g., engaging smugglers or undertaking dangerous employment) or become trapped, unable to migrate
Financial costs of migration	For regular migrants these are known and can be budgeted for. For irregular migrants, these can be highly variable, unpredictable, and require pooling of resources or taking on debt
Labor market entry/access at destination	Includes such considerations as the availability of wage labor opportunities, being legally permitted to work at destination, ease of accessing jobs (e.g., are source-area credentials recognized at destination)
Destination area wages	Pre-existing wage differentials can factor into choice of destination; does migrant influx have impact on average wages at destination?
Impact of remittances	Are safe transfer mechanisms in place? Do migrants earn enough to remit? Are social connections sufficiently strong to ensure migrant does not lose interest in remitting? Do remittances increase socio-economic stratification in migrant source area?
Other migrant impacts on destination	Includes impacts on prices/availability of housing, food and other basic needs/services; does migration stimulate environmental degradation at destination?

Source: Author

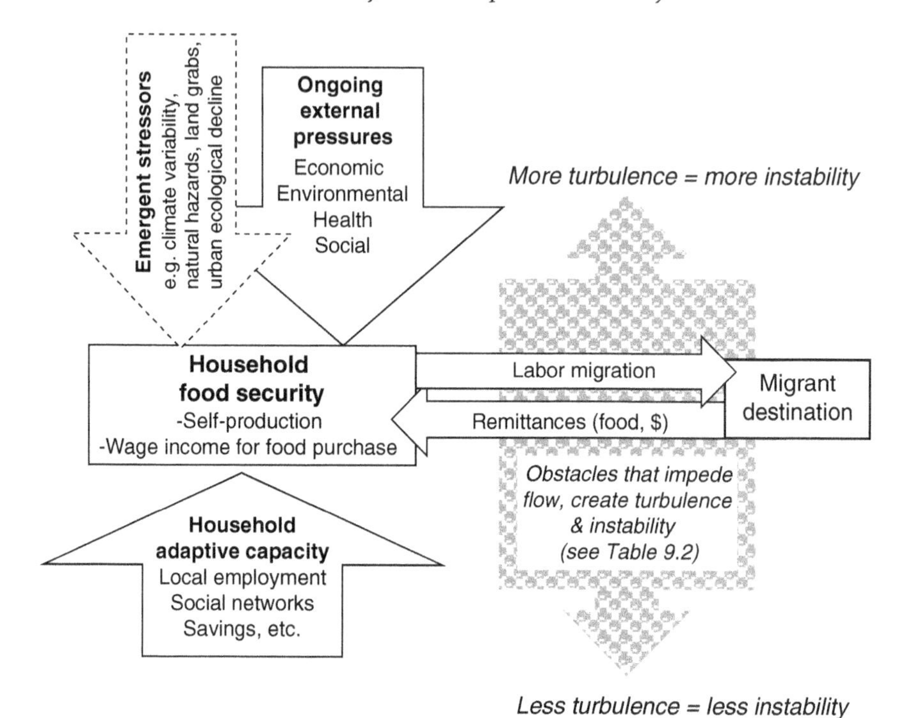

Fig. 9.4. Turbulent migrant and remittance flows contributing to instability
Source: Author

Migration pulses from natural hazard events. The potential for sudden population movements from climate-related events is expected to increase in coming decades (Black et al. 2011). As seen with hurricanes Mitch and Katrina, sudden-onset events can trigger spikes in local labor market participation and in labor migration away from the affected area. While neither of these events led to violent conflict, natural hazards and environmentally driven scarcity have the potential to exacerbate preexisting sociopolitical tensions (Brown et al. 2007). When areas receiving migrants from hazard-affected areas are able to incorporate a pulse of in-migrants, the potential for instability or disruption is low. By contrast, when governments and institutions are unwilling or unable to allow mobility, to provide assistance and support for affected areas, or are—as in the cases of Sudan and Zimbabwe—actively contributing to unrest, situations will present an increased potential for political instability and violence (McLeman 2011).

Sudden-onset natural hazards are notoriously difficult to predict in advance. Advance notice of severe storms ranges from minutes to a few days. Seasonal flood forecasts are sometimes possible, while monitoring of drought can be done with greater accuracy. However, event forecasting is an insufficient indicator, since not all events stimulate pulses of labor migration. There are, however, indicators of those that do. Degree of damage to housing has a strong correlation with displacement and distress migration, and this can be quickly measured after events. While there is no universal metric for effectiveness of emergency response, it is possible to assess the success of institutional response to past events and look to see if improvements have been made. This provides a general indication of institutional capacity to adapt to future hazard events and, as shown above, strong institutional capacity reduces pressure on households. GIS models that incorporate crop-performance-linked climate data and soil quality may be used to identify areas where rural populations might have high rates of migration in response to droughts (McLeman et al. 2010; McLeman and Ploeger 2012).

"Land-grabs" and population displacements. Population displacements, food security, and political instability arising from "land grabs" are a growing concern. Lands in LICs—most commonly in Africa, but also in parts of South and Southeast Asia—are being acquired by outside private investors for conversion to commercial export agriculture. The nature of these acquisitions varies considerably from one example to the next in their size, scale, purpose, and legal arrangements, as well as the promises made to those displaced (Cotula et al. 2011). While the general merits, equitable considerations, broad-based economic benefits, and best practices of these types of transactions have been vigorously discussed, there exists little reliable evidence or research on how many people have been displaced by these transactions, where they resettle, their welfare after they are resettled, and the implications for local and regional labor movements. The types of crops being produced on these lands typically require modest amounts of labor, thereby providing few local employment opportunities for those displaced (Li 2011).

Some express concerns that the conversion of land from subsistence to export production may trigger local food security problems (Daniel 2011). There are examples of local and national-level resistance to land grabs, the best known one being in Madagascar, where the government proposed to lease half the country's arable land to a Korean multi-national (Daniel 2011). There consequently remains an ongoing need to better study and monitor the consequences for displaced populations and the potential for instability. In addition, Woodhouse (2012) notes quite presciently that many African lands being acquired through these processes are in water-scarce areas, and so the longer-term risks of conflict over water rights may be as great or greater than the land acquisition itself. Finally, de Sherbinin et al. (2011) note that some programs and regimes being envisaged to reduce global greenhouse gas emissions, such as those under the United Nations program for Reducing Emissions from Deforestation and Forest Degradation in developing countries, may also lead to involuntary displacements in LICs, creating an additional driver for population movements and instability. Given the relative newness of the land-grab phenomenon and lack of data on displacement outcomes, those seeking indicators of the potential for sociopolitical instability to rise would likely look to measurements of population change in informal urban/peri-urban settlements near to the land-grab site, and changes in housing prices, food prices, and wages (formal and informal). The availability of such data in LICs is highly variable, and often requires targeted field collection.

Populations trapped in unemployment, poverty, or conflict. Households may become trapped and unable to migrate for a number of reasons.[4] Examples from recent decades include those suffering famine events in East Africa (Ezra and Kiros 2001) and rural displacements in Mugabe's Zimbabwe (Kinsey 2010). The combined effects of climate variability, population growth, and variability in global food production and prices raise the likelihood of more trap situations emerging in LICs in the next two decades. Rural households can get stuck in poverty traps that leave them inadequate resources to participate in labor migration, while labor migrants arriving in urban/peri-urban slums do not always find hoped-for economic opportunities, leaving the migrant and the sending household less well-off than before. Poverty traps, regardless of type of location, provide ripe conditions for political dissent and instability to emerge. The International Labour Organization (ILO 2012) warns that unrest associated with growing rates of economic instability and unemployment is currently increasing social instability in sub-Saharan Africa, the Middle East, and northern Africa/Maghreb. To make such projections, ILO uses a range of indicators across countries related to unemployment and inactivity rates (including youth employment), change in the quality of employment, and changes in poverty rates and income inequalities, as well as a weighted index of Gallup-collected poll data regarding perceptions of future economic performance, government effectiveness, and so forth. These provide a template for identification of potential regional hotspots for poverty traps, but would need to be complemented by collection of similar indicator data at more local and regional scales to more precisely identify specific poverty traps that could influence labor migration dynamics.

Violent conflicts typically have impacts on local and regional migration patterns, most obviously by driving involuntary migration, but also by shaping labor migration patterns and affecting the economic returns of migration to households. Those not directly affected by violence may nonetheless weigh the risks and decide to escape the trap by relocating elsewhere, even if the wages and other economic benefits thereby gained are less than desirable. All situations of violent conflict should be regarded as traps that affect labor migration, and so no particular indicators are necessary.

Declining urban ecological security. Rapid urbanization in low-income countries—including the conversion of agricultural land, a growing ratio of urban residents to food producers, and changes in food demand—has significant impacts on food security (Satterthwaite et al. 2010). Urbanization puts increased pressure on formal food systems and on informal rural–urban linkages. In addition to food security challenges, much of the population growth in LICs' urban centers takes place in peri-urban areas that lack basic services, utilities, housing, and formal sector employment opportunities (Davis 2006). For those urban settlements situated in low-lying coastal zones or in areas where water is scarce, climate change will have direct effects on the safety and quality of life of the urban poor, along with indirect effects on food prices (Hunt and Watkiss 2011).

Where local governance structures and capacity are weak, economic growth poor, or rural–urban linkages eroded, the potential for tensions and instability can be expected to increase (Homer-Dixon and Deligiannis 2009). Urban centers in already-fragile states are especially at risk of becoming more volatile. As with natural hazards, not all changes in urban ecology are potentially destabilizing. The most important indicators of urban sociopolitical stability are those linked to availability of and access to potable water and food availability and price on local markets.

Regional rural labor shortages. In Africa, some low-income countries already experience localized rural labor shortages in areas of high rural out-migration, as do parts of Southeast Asia and southwestern China. At present, rural wage labor opportunities and domestic migration to take advantage of them are still widespread across much of South Asia, even in the face of high rates of urbanization (Ellis 2000; Li 2010). China presently has shortages of skilled labor in some coastal cities and a surplus of unskilled rural laborers. However, because of declining birth rates and high urbanization rates, the next two decades will likely see this dynamic shift, with rural labor becoming scarcer in China, as well as India and many African and Asian LICs. When persistent declines in rural labor supply are offset by improved gains in agricultural productivity, the adverse impacts on food production may be minimal or localized, but there is the risk that local food production may suffer in areas where small-scale farming continues to predominate. Indicators for monitoring these situations will again need to be regional or local in scale, with a focus on measuring rates of agricultural intensification, investment, mechanization, and productivity per unit labor.

HOW TO AVOID PATHS LEADING TO INSTABILITY

Let me conclude with some brief policy recommendations, the most important being to view environmentally related migration less as a potential threat to be mitigated and more as a potential opportunity to be managed. The era of labor surpluses and unlimited numbers of international migrants willing to fill labor market gaps in wealthier countries will end in the next two to three decades.

When properly managed, migration provides short- and long-term benefits to sending and receiving communities, as well as to migrants. The challenge for policymakers is to craft domestic and international migration programs and policies that facilitate the movement of labor migrants, allow them legal and legitimate status in the receiving area, and ensure simple mechanisms for the safe transfer of remittance money.

This does not require open borders that allow indiscriminate movements without any controls, but simply a shift from policies that seek to restrict migration to ones that manage it in an orderly fashion. The current immigration policies of many countries—including the United States and Europe—offer few legal and legitimate ways for labor migrants to enter, save for small numbers of highly skilled workers and those with close family ties. These systems do not eliminate international labor migration, they simply force it into irregular channels that benefit neither the migrant nor the receiving population as a whole. They do, however, benefit organized crime groups, which should prompt more attention from the security community than it does.

By providing legal outlets for labor migrants, governments will allow households in vulnerable locations better options for adapting to climate change and threats to food security. Long-term development assistance and short-term emergency financial assistance could then be dedicated to (1) assisting trapped populations that lack the means to use migration to escape difficult conditions, and to (2) building adaptive capacity in locations, such as cities in water-scarce areas and coastal plains, that will be at greater risk in coming decades. The reality is that urban centers in highly exposed locations will continue to grow in population, and so greater investment in improving urban ecological security in cities must become an international policy priority. Inaction simply increases the potential for distress migration events that are undesirable and holds the potential for generating political instability.

Greater research is needed into the effects of the land-grab phenomenon, as well as the consequences of programs that seek to reduce greenhouse gas emissions from deforestation and desertification. While these may in principle offer net global benefits for food supply and avoiding dangerous climate change, past development-induced displacements that led to bad outcomes for local populations were also a product of good intentions. Except for very specific locations, such as erosion-prone settlements in river valleys or coastal plains, there should be few instances in the next 20 years where the impacts of climate change will require organized relocations of people. People who live in those rare areas that may need to be abandoned later this century, however, may use labor migration to improve their household well-being in advance, and it makes good sense to facilitate successful migration when possible.

A final recommendation relates to the old adage that an ounce of prevention is worth a pound of cure. In the early 1990s, security scholars, military analysts, and policymakers began worrying that global warming would become a destabilizing force, a threat multiplier, a trigger for environmental refugees—however one wishes to describe it. Now, we see global warming being labeled as a causal factor in events like the Darfur refugee crisis and the Arab Spring uprisings (UNEP 2007; Johnstone and Mazo 2011). The merits of such claims can be argued indefinitely. The heart of the matter is this: if we believe that the impacts of climate change pose a serious threat to global security—and many serious people do—then the security community needs to become the leading champion for policies that address the underlying causes of it, build adaptive capacity in the most vulnerable places, and allow orderly international movements of labor migrants. So long as leadership on these issues is left to environment ministries, labor secretaries, and others not operating at the highest tier of political influence, meaningful and durable solutions will not be found.

ACKNOWLEDGMENTS

This chapter benefitted from suggestions by Chris Barrett, Sheri Englund, Cynthia Mathys, Marc Rockmore, Joanna Upton, and Lindy Williams. The organizers of the workshop at Cornell University where this paper was first presented are also thanked. The author would like to acknowledge the ongoing financial support he has received from the Social Science and Humanities Research Council of Canada, which has made his research program possible.

NOTES

1. An important exception being China, where internal migration is regulated under the *hukou* system; those who decide to move without state permission join what is typically described as the "floating population" of workers without full legal rights, protections, and social benefits (see, for example, Zhu 2007).
2. Australia's wheat exports in the dry 2009–2010 crop year were roughly 10.5 million metric tonnes, as compared with over 15 million metric tonnes the following year (Wheat Exports Australia 2011). Russia halted grain exports in 2010 in the midst of a severe drought, but is expected in 2012 to resume large-scale exports with forecasts at time of writing of 25 million tonnes for 2012 (RT Online 2012). Even in very productive years, Russian and Australian wheat exports combined do not match North American exports.
3. Each experienced more than 100 significant flood events between 1998 and 2008 according to Adhikari et al. 2010.
4. The final report of the Foresight Project (2011) provides a detailed overview of the perils of trapped populations in particular, and on understanding the relationship between environment and migration more generally. It is highly recommended as further reading.

REFERENCES

Adger, W. N., P. M. Kelly, A. Winkels, L. Q. Huy, and C. Locke. 2002. Migration, remittances, livelihood trajectories and social resilience. *Ambio 31*(4): 358–66.

Adhikari, P., Y. Hong, K. R. Douglas, D. B. Kirschbaum, J. Gourley, R. Adler, and G. R. Brakenridge. 2010. A digitized global flood inventory (1998–2008): Compilation and preliminary results. *Natural Hazards 55*(2): 405–22.

Bakker, M. 2009. Transboundary river floods and institutional capacity. *Journal of the American Water Association 45*: 5 53–566.

Barnett, J., and J. Campbell. 2010. *Climate change and small island states: Power, knowledge and the South Pacific.* London: Earthscan.

Beauchemin, C. 2011. Rural–urban migration in West Africa: Towards a reversal? Migration trends and economic situation in Burkina Faso and Côte d'Ivoire. *Population, Space and Place 17*(1): 47–72.

Black, R., D. Kniveton, and K. Schmidt-Verkerk. 2011. Migration and climate change: Towards an integrated assessment of sensitivity. *Environment and Planning A 43*: 431–50.

Blaikie, P., T. Cannon, I. Davis, and B. Wisner. 1994. *At risk: Natural hazards, people's vulnerability, and disasters.* New York: Routledge.

Brown, O., A. Hammill, and R. McLeman. 2007. Climate change as the "new" security threat: Implications for Africa. *International Affairs 83*(6): 1141–54.

Carr, D. L. 2008. Migration to the Maya Biosphere Reserve, Guatemala: Why place matters. *Human Organization 67*(1): 37–48.

Chen, C.-C., B. McCarl, and C.-C. Chang. 2012. Climate change, sea level rise and rice: Global market implications. *Climatic Change 110*(3–4): 543–60.

Cotula, L., S. Vermeulen, P. Mathieu, and C. Toulmin. 2011. Agricultural investment and international land deals: Evidence from a multi-country study in Africa. *Food Security 3*(1): 99–113.

Daniel, S. 2011. Land grabbing and potential implications for world food security. *Sustainable Agricultural Development 1*: 25–42.

Davis, M. 2006. Planet of slums. *New Perspectives Quarterly 23*(2): 6–11.

de Brauw, A. 2011. Migration and child development during the food price crisis in El Salvador. *Food Policy 36*(1): 28–40.

de Sherbinin, A., M. Castro, F. Gemenne, M. M. Cernea, S. Adamo, P. M. Fearnside, G. Krieger. 2011. Preparing for resettlement associated with climate change. *Science 334*(6055): 456–57.

Deshingkar, P. 2006. Improved livelihoods in improved watersheds in India: Can migration be mitigated? In *Earthscan Reader in Rural–Urban Linkages,* ed. C. Tacoli, 215–228. London: Earthscan.

Ellis, F. 2000. The determinants of rural livelihood diversification in developing countries. *Journal of Agricultural Economics 51*(2): 289–302.

Ezra, M., and G.-E. Kiros. 2001. Rural out-migration in the drought prone areas of Ethiopia: A multilevel analysis. *International Migration Review 35*(3): 749–71.

Feng, S. F., A. B. Krueger, and M. Oppenheimer. 2010. Linkages among climate change, crop yields and Mexico–US cross-border migration. *Proceedings of the National Academy of Science 107*(32): 14257–62.

Ford, J. D., and T. Pearce. 2010. What we know, do not know, and need to know about climate change vulnerability in the western Canadian Arctic: A systematic literature review. *Environmental Research Letters 5*(1): 14008.

Foresight: Migration and Global Environmental Change. 2011. *Final project report.* London. http://www.bis.gov.uk/assets/bispartners/foresight/docs/migration/11-1116-migration-and-global-environmental-change.pdf (accessed March 29, 2013).

Frayne, B. 2005. Survival of the poorest: Migration and food security in Namibia. In *Agropolis: The social, political, and environmental dimensions of urban agriculture,* ed. L. J. A. Mougeot, 31–50. London: IDRC/Earthscan.

Gammeltoft, P. 2002. Remittances and other financial flows to developing countries. *International Migration 40*(5): 181–211.

Greiner, C. 2011. Migration, translocal networks, and socio-economic stratification in Namibia. *Africa 81*(4): 606–27.

Gregory, P. J., J. S. I. Ingram, and M. Brklacich. 2005. Climate change and food security. *Philisophical transactions of the Royal Society London, Biological sciences, Series B 360*(1463): 2139–48.

Grogger, J., and G. H. Hanson. 2011. Income maximization and the selection and sorting of international migrants. *Journal of Development Economics 95*(1): 42–57.

Gross D. M., and N. Schmitt. 2012. Low- and high-skill migration flows: Free mobility versus other determinants. *Applied Economics 44*: 1–20.

Harsem, Ø., A. Eide, and K. Heen. 2011. Factors influencing future oil and gas prospects in the Arctic. *Energy Policy 39*(12): 8037–45.

Henry, S., V. Piché, D. Ouédraogo, and E. F. Lambin. 2004. Descriptive analysis of the individual migratory pathways according to environmental typologies. *Population and Environment 25*(5): 397–422.

Ho, J. 2010. The implications of Arctic sea ice decline on shipping. *Marine Policy 34*: 713–715.

Homer-Dixon, T., and T. Deligiannis. 2009. Environmental scarcities and civil violence. In *Facing Global Environmental Change,* ed. H. G. Brauch, Ú. O. Spring, J. Grin, C. Mesjasz, P. Kameri-Mbote, N. C. Behera, B. Chourou, and H. Krummenacher, 309–23. Berlin: Springer.

Hunt, A., and P. Watkiss. 2011. Climate change impacts and adaptation in cities: A review of the literature. *Climatic Change 104*(1): 13–49.

ILO. 2012. World of work report 2012: Better jobs for a better economy. Rome: International Labour Organization.

IPCC. 2007. Summary for policymakers. In *Climate Change 2007: Synthesis Report: Contribution of Working Groups I, II and III to the Fourth Assessment Report of the Intergovernmental Panel on Climate Change,* ed. B. Metz, O. R. Davidson, P. R. Bosch, R. Dave, and L. A. Meyer. Cambridge: Cambridge University Press.

IRIN. 2006. *Zimbabwe: Ruralisation is the new trend.* Nairobi: Integrated Regional Information Networks, UN Office for the Coordination of Humanitarian Affairs. http://www.irinnews.org/Report/59751/ZIMBABWE-Ruralisation-is-the-new-trend (accessed March 29, 2013).

Jakobson, L. 2010. *China prepares for an ice-free Arctic.* Stockholm: SIPRI. See http://books.sipri.org/files/insight/SIPRIInsight1002.pdf (accessed 2 May 2013).

Jim, C. Y., and F. Y. Yang. 2006. Local responses to inundation and de-farming in the reservoir region of the Three Gorges project (China). *Environmental Management 38*(4): 618–37.

Johnstone, S., and J. Mazo. 2011. Global warming and the Arab Spring. *Survival 53*(2): 11–17.

Jones, N. 2011. Towards an ice-free Arctic. *Nature Climate Change 1* :381.

Jonkman, S. N. 2005. Global perspectives on loss of human life caused by floods. *Natural Hazards 34*(2): 151–75

Kinsey, B. H. 2010. Who went where . . . and why: Patterns and consequences of displacement in rural Zimbabwe after February 2000. *Journal of Southern African Studies 36*(2): 339–60.

Knox, J. C. 2000. Sensitivity of modern and Holocene floods to climate change. *Quaternary Science Reviews 19*(1–5): 439–57.

Knutson, T. R., J. L. McBride, J. Chan, K. Emanuel, G. Holland, C. Landsea, I. Held, J. P. Kossin, A. K. Srivastava, and M. Sugi. 2010. Tropical cyclones and climate change. *Nature Geoscience 3*: 157–63.

Kundzewicz, Z. W., Y. Hirabayashi, and S. Kanae. 2010. River floods in the changing climate—Observations and projections. *Water Resources Management 24*(11): 2633–46.

Kunii, O., S. Nakamura, R. Abdur, and S. Wakai. 2002. The impact on health and risk factors of the diarrhoea epidemics in the 1998 Bangladesh floods. *Public Health 116*(2): 68–74.

Lalonde, R. J., and R. H. Topel. 1997. Economic impact of international migration and the economic performance of migrants. In *Handbook of population and family economics vol. 1B,* ed. M. R. Rosenzweig and O. Stark, 799–887. Amsterdam: Elsevier.

Li, T.M. 2010. To make live or let die? Rural dispossession and the protection of surplus populations. *Antipode 41*(S1): 66–93.

——— 2011. Centering labor in the land grab debate. *Journal of Peasant Studies 38*(2): 281–98.

Li, Y., W. Ye, M. Wang, and X. Yan. 2009. Climate change and drought: A risk assessment of crop-yield impacts. *Journal of Climate Research 39*(1): 31–46.

Maharjan, K., and N. Joshi. 2011. Determinants of household food security in Nepal: A binary logistic regression analysis. *Journal of Mountain Research 8*(3): 403–13.

Massey, D. S., Axinn, W. G., and Ghimire, D. J. 2010. Environmental change and out-migration: Evidence from Nepal. *Population and Environment, 32*(2–3): 109–136.

Mberu, B. U., A. C. Ezeh, G. Chepngeno-Langat, J. Kimani, S. Oti, and D. Beguy. 2013. Family ties and urban–rural linkages among older migrants in Nairobi informal settlements. *Population, Space and Place* 19(3): 275–293.

McLeman, R. 2006. Migration out of 1930s rural eastern Oklahoma: Insights for climate change research. *Great Plains Quarterly 26*(1): 27–40.

——— 2011. *Climate change, migration, and critical international security considerations.* IOM: Geneva. http://publications.iom.int/bookstore/index.php?main_page=product_info&cPath=2_3&products_id=688 (accessed March 29, 2013).

———, S. Herold, Z. Reljic, M. Sawada, and D. McKenney. 2010. GIS-based modeling of drought and historical population change on the Canadian Prairies. *Journal of Historical Geography 36*: 43–55.

——— and L. M. Hunter. 2010. Migration in the context of vulnerability and adaptation to climate change: Insights from analogues. *Wiley Interdisciplinary Reviews, Climate Change 1*(3): 450–61.

——— and S. K. Ploeger. 2012. Soil and its influence on rural drought migration: Insights from depression-era southwestern Saskatchewan, Canada. *Population and Environment 33*(4): 304–32.

——— and B. Smit. 2006. Migration as an adaptation to climate change. *Climatic Change 76*(1–2): 31–53.

——— and B. Smit. 2012. Climate change and Canadian security. In *Evolving transnational threats and border security: A new research agenda*, ed. C. Leuprecht, T. Hataley, and K. Nossal, 97–108. Kingston, ON: Centre for International Defence and Policy, Martello Papers.

Millennium Ecosystem Assessment. 2005 *Ecosystems and human well-being: Synthesis.* Island Press, Washington DC.

Mortreux, C., and J. Barnett. 2009. Climate change, migration and adaptation in Funafuti, Tuvalu. *Global Environmental Change 19*(1): 105–12.

Mosse, D., S. Gupta, M. Mehta, V. Shah, and J. Rees. 2002. Brokered livelihoods: Debt, labour migration and development in tribal Western India. *Journal of Development Studies 38*(5): 59–88.

Murrugarra, E., and C. Herrera. 2011. Migration choices, inequality of opportunities and poverty reduction in Nicaragua. In *Migration and poverty: Towards better opportunities for the poor*, ed. E. Murrugarra, J. Larrison, and M. Sasin, 101–124. Washington DC: International Bank for Reconstruction and Development.

Nicholls, R. J., and A. Cazenave. 2010. Sea-level rise and its impact on coastal zones. *Science 328*(5985): 1517–20.

Nguyen, M. C., and P. Winters. 2011. The impact of migration on food consumption patterns: The case of Vietnam. *Food Policy 36*(1): 71–87.

Piesse, J., and C. Thirtle. 2009. Three bubbles and a panic: An explanatory review of recent food commodity price events. *Food Policy 34*(2): 119–29.

Poertner, E., M. Junginger, and U. Muller-Boker. 2011. Migration in Far West Nepal. *Critical Asian Studies 43*(1): 23–47.

Potts, D. 1995. Shall we go home? Increasing urban poverty in African cities and migration processes. *The Geographical Journal 161*(3): 245–64.

——— 2005. Counter-urbanisation on the Zambian copperbelt? Interpretations and implications. *Urban Studies 42*(4): 583–609.

Pyles, L., S. Kulkarni, and L. Lein. 2008. Economic survival strategies and food insecurity: The case of hurricane Katrina in New Orleans. *Journal of Social Service Research 34*(3): 43–53.

Quaye, W. 2008. Food security situation in northern Ghana, coping strategies and related constraints. *African Journal of Agricultural Research 3*(5): 334–42.

Rain, D. 1999. *Eaters of the dry season: Circular labor migration in the West African Sahel.* Boulder, CO: Westview Press.

Ravenstein, E. G. 1889. The laws of migration (second paper). *Journal of the Royal Statistical Society 52*(2): 241–305.

Reardon, T. 1997. Using evidence of household income diversification to inform study of the rural nonfarm labor market in Africa. *World Development 25*(5): 735–47.

Roncoli, C., K. Ingram, and P. Kirshen. 2001. The costs and risks of coping with drought: Livelihood impacts and farmers' responses in Burkina Faso. *Climate Research 19*: 119–32.

RT Online. 2012. Russia is expected to increase its grain exports in 2012. http://rt.com/business/news/russia-grain-export-growth-691/ (accessed March 29, 2013).

Samers, M. 2010. *Migration*. London: Routledge.

Satterthwaite, D., G. McGranahan, and C. Tacoli. 2010. Urbanization and its implications for food and farming. *Philosophical Transactions of the Royal Society B*, *365*(1554): 2809–2820.

Sauchyn, D. J., J. Stroich, and A. Beriault. 2003. A paleoclimatic context for the drought of 1999–2001 in the northern Great Plains of North America. *The Geographical Journal, 169*(2): 158–167.

Schindler, D. W., and W. F. Donahue. 2006. An impending water crisis in Canada's western prairie provinces. *Proceedings of the National Academy of Sciences 103*(19): 7210–16.

Schmidhuber, J., and F. Tubiello. 2007. Global food security under climate change. *Proceedings of the National Academy of Sciences 104*(50): 19703–708.

Schneider, G., and B. Knerr. 2000. Labour migration as a social security mechanism for smallholder households in sub-Saharan Africa: The case of Cameroon. *Oxford Development Studies 28*(2): 223–36.

Scudder, T. 2012. Resettlement outcomes of large dams. In *Impacts of large dams: A global assessment*, ed. C. Tortajada, D. Altinbilek, and A. K. Biswas, 37–67. Berlin: Springer.

Seto, K. C. 2011. Exploring the dynamics of migration to mega-delta cities in Asia and Africa: Contemporary drivers and future scenarios. *Global Environmental Change 21*(S1): S94–107.

Smit, W. 1998. The rural linkages of urban households in Durban, South Africa. *Environment and Urbanization 10*(1): 77–87.

Tan, Y., G. Hugo, and L. Potter. 2003. Government-organized distant resettlement and the Three Gorges project, China. *Asia Pacific Population Journal 18*(3): 5–26.

Tan, Y., B. Bryan, and G. Hugo. 2005. Development, land-use change and rural resettlement capacity: A case study of the Three Gorges project, China. *Australian Geographer 36*(2): 201–20.

Tiffen, M. 2003. Transitions in sub-Saharan Africa: Agriculture, urbanization and income growth. *World Development 31*: 1343–66.

Todaro, M. P. 1969. A model of labor migration and urban unemployment in less developed countries. *American Economic Review 59*(1): 138–48.

UNEP. 2007. Sudan: Post-conflict environmental assessment. Nairobi: UNEP. See http://postconflict.unep.ch/sudanreport/sudan_website/index.php (accessed 2 May 2013).

Vafeidis, A., Neumann, B., Zimmermann, J., and Nicholls, R. J. 2011. *Analysis of land area and population in the low-elevation coastal zone*. London, Foresight, Government Office for Science. http://www.bis.gov.uk/assets/foresight/docs/migration/modelling/11-1169-mr9-land-and-population-in-the-low-elevation-coastal-zone.pdf (accessed March 29, 2013).

Webster, P. J., G. J. Holland, J. A. Curry, and H. R. Chang. 2005. Changes in tropical cyclone number, duration, and intensity in a warming environment. *Science 309*(5742): 1844–46.

Wegren, S. K. 2011. Food security and Russia's 2010 drought. *Eurasian Geography and Economics 52*(1): 140–56.

Wheat Exports Australia. 2011. *Annual report 2010–2011: Deakin Act.* http://www.wea.gov.au/PDF/WEA_Annual%20Report%202010-11%20-%20Online%20version.pdf (accessed March 29, 2013).

Wong, K.-K., and X. Zhao. 2001. Living with floods: Victims' perceptions in Beijiang, Guangdong, China. *Area 33*(2): 190–201.

Woodhouse, P. 2012. New investment, old challenges: Land deals and the water constraint in African agriculture. *Journal of Peasant Studies 39*(3–4): 777–794.

World Bank. 2011. *Migration and remittances factbook 2011, 2nd ed.* Washington DC: World Bank.

Xi, J., and S.-S. Hwang. 2011. Relocation stress, coping, and sense of control among resettlers resulting from China's Three Gorges dam project. *Social Indicators Research 104*(3): 507–22.

Yin, H., and C. Li. 2001. Human impact on floods and flood disasters on the Yangtze River. *Geomorphology 41*(2): 105–109.

Zhu, Y. 2007. China's floating population and their settlement intention in the cities: Beyond the Hukou reform. *Habitat International 31*(1): 65–76.

Zong, Y., and X. Chen. 2000. The 1998 flood on the Yangtze, China. *Natural Hazards 22*: 165–84.

10

Trade Policies and Global Food Security

KYM ANDERSON

Between 2004 and 2008, real food prices in international markets rose by more than 50 percent on average, with grain prices spiking in mid-2008. Prices began to drop late in 2008, only to rise steeply again in 2010 to 2011, along with the prices of energy raw materials. Those price spikes caused consumer panic among buyers of staples such as rice and wheat, and it raised the cost of living dramatically for poor households in developing countries. Not surprisingly it triggered urban food riots and other forms of sociopolitical instability in a number of developing countries and may well have contributed to the overthrow of governments in 2008 in Haiti and Madagascar and in 2011 in Egypt and Tunisia.

This combination of high and volatile food prices understandably raises global food security concerns. One consequence has been a call for emergency physical grain reserves (Fan et al. 2011), with discussions in East Asia focusing on coordinating national rice reserves (Briones 2011). Governments in high-income countries also have expressed concern, and food price volatility has been high on the agenda of the G-8 and G-20 meetings (FAO et al. 2011).

However, the factors influencing the long-run trend level of food prices are not the same as those affecting the short-run volatility of food prices around that trend. The distributional and poverty effects of food price spikes—and of policy responses to them—differ from the effects associated with changes in the trend price level, as well. Unless societies and governments clarify what concerns them most, and understand the underlying causes, they will not be able to identify the most appropriate and cost-effective policy actions or reforms to ease those concerns and reduce sociopolitical instability.

Over most of the twentieth century, the real price of food in international markets fluctuated around a long-run downward trend of more than 0.5 percent per year. The conventional wisdom was that increasingly protectionist farm trade policies contributed to both the declining mean and the continuing volatility of prices. It was presumed that those policies also added to global poverty, because the vast majority of the world's poor have depended directly or indirectly on agriculture for their livelihood, as net sellers of food and other farm products. This line of thinking led to calls for agricultural trade liberalization on the grounds that such reform would boost not only aggregate world economic welfare, but also global food security, by expanding world food supplies and reducing poverty. The higher mean and variance of international food prices of the past few years, however, are projected to persist for the foreseeable future (Rosegrant et al. 2013). What does this fact suggest about the role that trade policy reform could play over the next decade in boosting food security and sociopolitical stability?

Answering the question requires taking into account the rapid industrialization of numerous emerging economies and the out-migration of rural people to urban slums. In some developing countries, a rise in the consumer price of food still may reduce poverty by boosting real incomes of more households in rural areas than it harms in urban areas, as a rise in farm prices boosts the demand for unskilled labor on farms, which could spill over to a rise in non-farm unskilled wages. But, in settings where many poor farmers are focusing on cash crops and are net buyers of basic foods, the reverse may be true. When food prices change, both rural and urban areas have winners and losers, depending on different groups' earning and spending profiles and assets and their access to or provision of remittances. Different groups may also pressure the national government to alter the price of food, with poverty being only one (and possibly not a major) driver of such influence peddling. In short, the links between high and volatile food prices, poverty, food insecurity, and sociopolitical instability are complex and not necessarily unidirectional.

Despite a reversal in the recent and prospective international food price trend, national governments' current trade-related policies remain important contributors to the volatility of international food prices and hence to global food insecurity. Thanks to the information technology revolution, the time is ripe for reform. National trade policies can now be replaced by domestic policy measures that will boost the food security of vulnerable individuals and households more efficiently and more equitably than trade-related measures that distort food prices. This chapter reassesses the role that trade policy reform over the present decade could play in reducing global poverty, food insecurity, and sociopolitical instability, and recommends some empirical indicators for monitoring the evolving impact of trade policy on food security and sociopolitical stability. It closes with some policy actions, both unilateral and multilateral, that could contribute to global food security and sociopolitical stability over the decade to come.

HOW HAVE PAST POLICIES ADDRESSED INTERNATIONAL FOOD PRICES?

Since the 1950s world agriculture has been characterized by the persistence of high agricultural protection in developed countries, by anti-agricultural and anti-trade policies in developing countries, and by the tendency for both groups of countries to vary their trade restrictions in an attempt to stabilize their domestic food market—thereby exacerbating price fluctuations in the international marketplace. The situation came to a head in the mid-1980s, with assistance to farmers in high-income countries peaking and international food prices plummeting in 1986, thanks in large part to an agricultural export subsidy war between North America and the European Community. Meanwhile, many developing countries had been holding down the incomes of their farmers, not only by heavily taxing agricultural exports (or grain production in-kind), but also by protecting manufacturers from import competition and overvaluing the national currency.

This disarray in world agriculture, as Johnson (1973) described it, meant there was overproduction of farm products in high-income countries and underproduction in needy developing countries. There was also less international trade in farm products than would be the case under free trade, thereby thinning the market for these weather-dependent products and making global commodity markets more volatile in the face of any global supply or demand shock. The impact of this pattern of policies on the trend level of international prices of farm products, however, is an empirical question: the outcome depends in part on whether the price-depressing effect in the international market of agricultural protection in higher-income countries was larger or smaller than the international price-raising effect of agricultural taxation in lower-income countries.

In the early 1980s Tyers and Anderson (1986, 1992) built a dynamic, stochastic model of the world's seven most important traded food products, and simulated numerous policy reform scenarios. They found that the international price-depressing effect of agricultural protection dominated the international price-raising effect of developing countries' direct agricultural taxes, although only slightly. Industrial country farm protection policies as of the early 1980s lowered the weighted average of international prices of those key foods by 20 percent, while the agricultural export taxes imposed by developing countries offset all but one of those 20 percentage points.

Yet both sets of policies shrunk international food trade: by an estimated 25 percent because of industrial countries' policies, but by 56 percent when developing countries' policies were added (Tyers and Anderson 1992, Table 6.9). To estimate the extent to which this thinning of the international food market added to its price volatility, the analysts ran 200 repeated simulations with random weather shocks. They concluded that the coefficient of variation of international food prices would have fallen by two-thirds, from 34 to 10 percent, if all countries had agreed in 1990 to cease their domestic price-insulating practice of varying their trade restrictions, and instead maintain constant percentage trade tax rates. In most of the 16 developing economies they considered individually, the coefficient of variation for domestic food prices would have fallen substantially if all countries had agreed multilaterally to refrain from using price-insulating measures (Tyers and Anderson 1992, Table 6.14).

As early as 1958, the Haberler report to GATT (General Agreement on Tariffs and Trade) Contracting Parties forewarned that agricultural protectionism might grow as economies industrialized, and indeed it did between the 1950s and the early 1980s in East Asia, North America, and Western Europe (Anderson et al. 1986). While that development was captured in the Tyers and Anderson partial equilibrium modeling study, the model was unable to capture fully the price-distorting effects of government policies of many developing countries that indirectly taxed their farmers by overvaluing the national currency and pursuing an import-substituting industrialization strategy by restricting imports of manufacturers. Together those latter measures taxed farmers indirectly more than export taxes taxed them directly.

Since the mid-1980s, numerous countries have been reforming their agricultural price and trade policies. High-income countries have gradually lowered their assistance to farmers following the export subsidy war of 1986 to 1987 and have decoupled some of that support from production, while developing countries have steadily lowered their farm export taxes and moved toward market-determined exchange rates.

To examine the extent of those reforms, the World Bank recently undertook another multicountry empirical study, summarized in Anderson (2009). These new estimates complement and extend both the regular estimates since 1986 for high-income countries by the OECD (2011) and the historical developing country estimates to the mid-1980s by Krueger et al. (1988). The World Bank study provides estimates for many additional developing and transition economies, estimates new and more comprehensive policy indicators, and adds measures of price distortions for nonagricultural tradables. These estimates have since been further updated to 2009–10 (Anderson and Nelgen 2012b).

These latest World Bank estimates cover 82 countries, which together account for more than 92 percent of the world's population, farmers, agricultural GDP, and total GDP. The sample countries also account for more than 85 percent of agricultural production and employment in each of Africa, Asia, Latin America, and the transition-economy region of Eastern Europe and Central Asia, as well as for all agricultural production and employment in OECD countries. The study computes nominal rates of assistance (NRAs) and consumer tax equivalents (CTEs) for 75 different farm products, with an average of almost 11 products per country. This product coverage represents about

70 percent of the gross value of agricultural production in each of the focus countries and thus just under two-thirds of global agricultural production when valued at undistorted prices over the period covered. Not all countries had data for the entire period from 1955 through 2010, particularly the transition economies of Europe and Asia, but the average number of years covered is 45 per country. Such comprehensive coverage of countries, products, and years offers the prospect of generating a reliable picture of long-term trends in policy indicators and annual fluctuations around those trends for individual countries and commodities, as well as for country groups, regions, and the world as a whole.

The updated estimates provide several new indicators of price distortions. The simplest one suitable for present purposes is the NRA, which is computed for each product as the percentage by which government policies have raised gross returns to farmers above what they would have been had the government not intervened (or lowered them, if NRA<0). This rate includes any product-specific input subsidies or taxes, plus the price-depressing effect for each product of multiple and overvalued exchange rates. A weighted-average NRA for all available products is derived using the value of production at undistorted prices as product weights. To this NRA for available (covered) products is added the country author's "guesstimates" of the NRA for non-covered exportables, import-competing goods, and nontradable farm products (on average, about 30 percent of the total in value terms), along with an estimate of the NRA from non-product-specific forms of assistance to (or taxation of) farmers.[1] Each commodity's industry is classified either as import-competing, or as producing exportables, or as producing a nontradable, so as to generate for each year the weighted-average NRAs for these three subsets of covered products. Note that a given industry's trade status can change over time.

The World Bank's NRA estimates are summarized in Figures 10.1a and b for the 50 years to 2010. Three points are worth emphasizing. First, assistance to farmers in high-income countries was on an upward trend until the mid-1980s, when the NRA peaked at an average of 60 percent during the food export subsidy war, before following a downward trend over the second half of this period to just one-quarter of that peak rate. Second, farmers in developing countries were taxed at more than 20 percent of their gross value of production on average for the first two decades of this period, but the extent of that disprotection gradually diminished, and by the mid-1990s had switched from negative to slightly positive assistance on average. Third, in both rich and poor countries the NRA for farmers fluctuated around the long-run trends, including the recent downward trend period for high-income countries. Rates spiked downwards in 1973 and 1974 and 2008 when international food prices spiked up, and did the opposite in 1986 and 1987 when international food prices slumped.

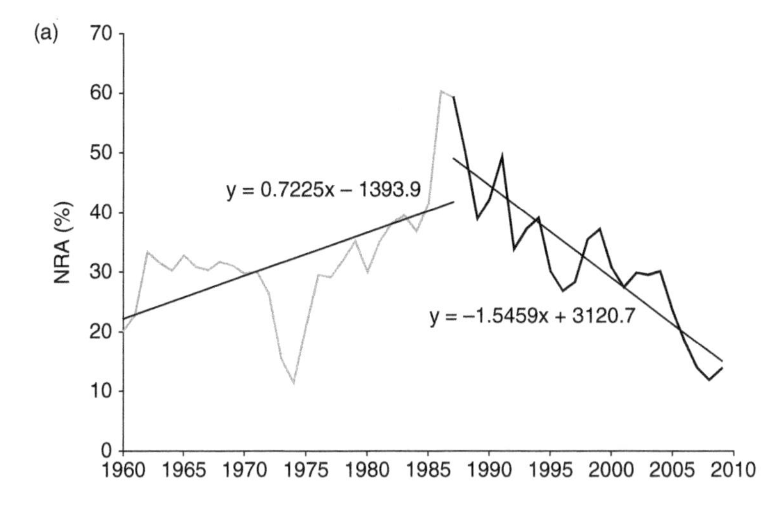

Fig. 10.1a. Nominal rates of assistance (NRAs) to farmers in high-income countries, 1960 to 2010 (percent)

Source: Based on the update of NRA estimates in Anderson and Nelgen (2012b)

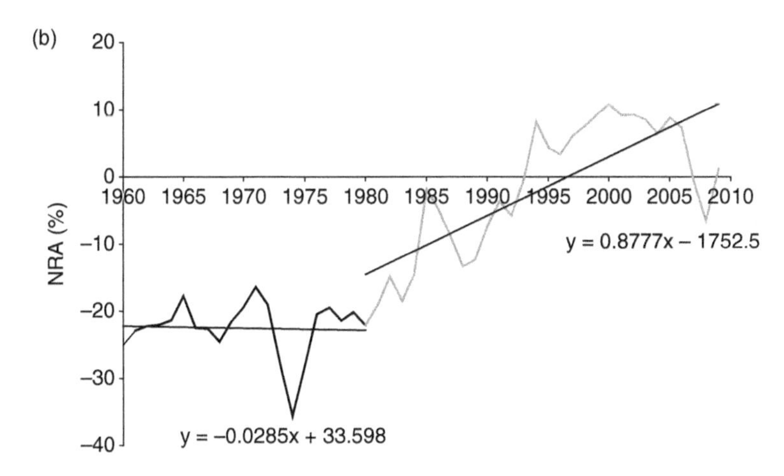

Fig. 10.1b. Nominal rates of assistance (NRAs) to farmers in developing countries, 1960 to 2010 (percent)

Source: Based on the update of NRA estimates in Anderson and Nelgen (2012b)

Two more pertinent findings from the recent World Bank study can be seen from the developing country indicators summarized in Figure 10.2. While the NRA in developing countries has been rising for farmers over the past few decades, it has been falling for producers of non-farm tradables (mostly manufactures). Indeed by the end of the twentieth century the former

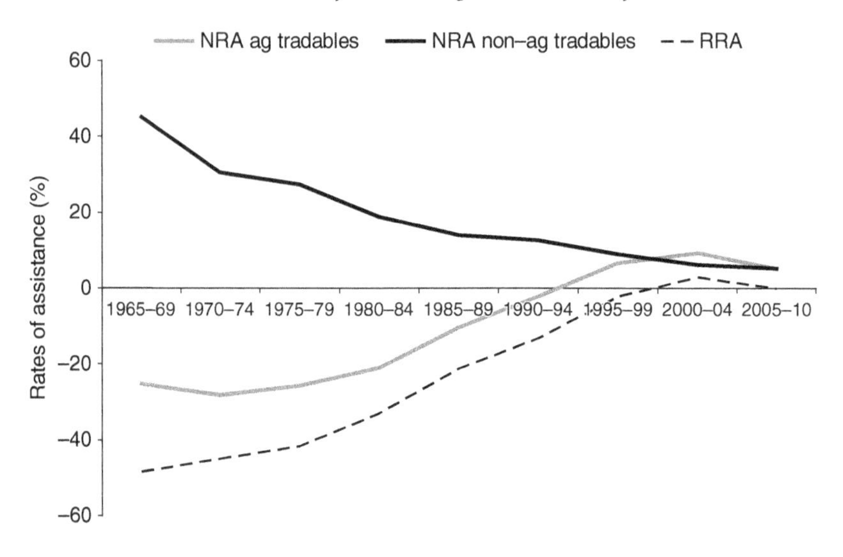

Fig. 10.2. Nominal rate of assistance (NRA) to farmers and producers of non-farm tradable goods, and relative rate of assistance, developing countries, 1965 to 2009 (percent, five-year averages)

Source: Based on the update of NRA and RRA estimates in Anderson and Nelgen (2012b)

exceeded the latter, so the estimated relative rate of farmer assistance (RRA)[2] rose above zero for the first time (Figure 10.2). This suggests that developing country governments on average are no longer discriminating against their farmers through their price-distorting policies. It needs to be kept in mind, however, that within the developing country group—not to mention among high-income countries—the spectrum of national RRA estimates as of 2000 through 2004 remained very wide (Figure 10.3), indicating great scope still for global economic welfare gains from further trade liberalization.

In addition, when the NRAs for developing countries' farmers are disaggregated into import-competing and exporting categories, it is evident that only the latter group has suffered negative NRAs. Indeed the import-competing subsector of farmers in developing countries enjoyed positive and rising protection over this long period, as seen by the regression trend line in the upper half of Figure 10.4—just as occurred in earlier decades for farmers in high-income countries. These findings suggest that export-focused farmers in developing countries are still discriminated against in two respects: by the anti-trade structure of assistance within their own agricultural sectors, and by the remaining protection afforded farmers in high-income countries.

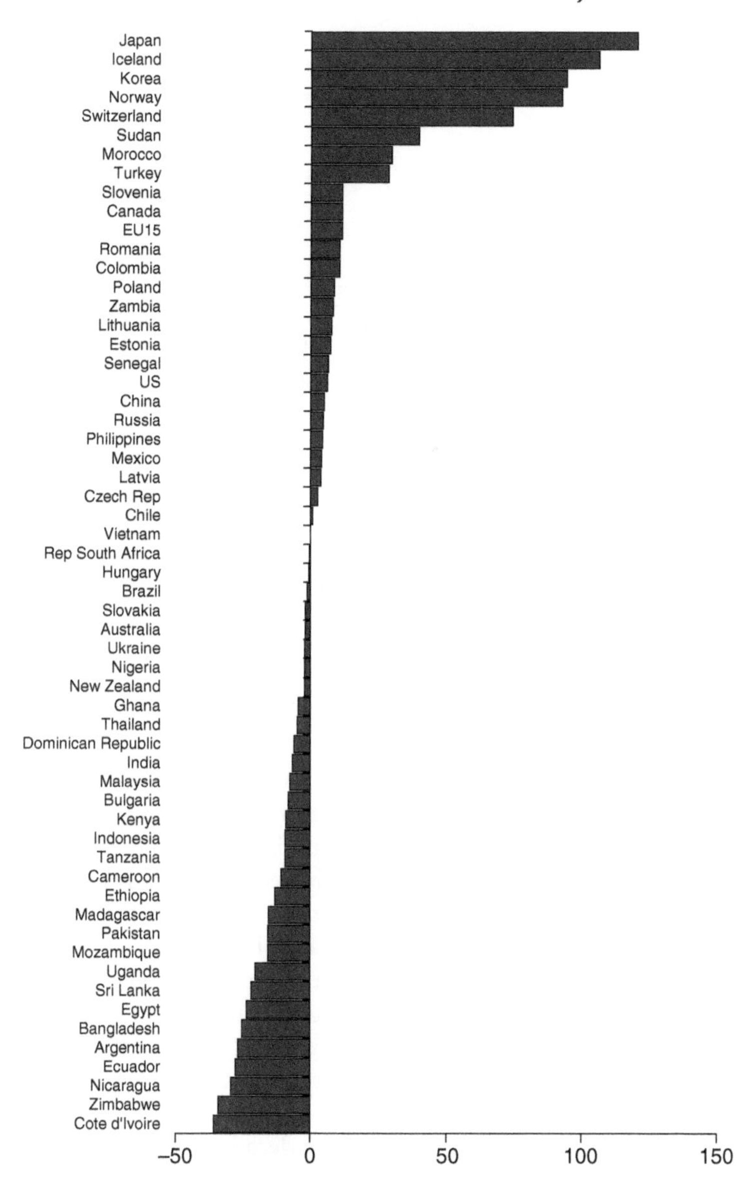

Fig. 10.3. Relative rate of assistance to farmers (RRA), by country, 2005-09 (percent)

Source: Based on the update of RRA estimates in Anderson and Nelgen (2012b)

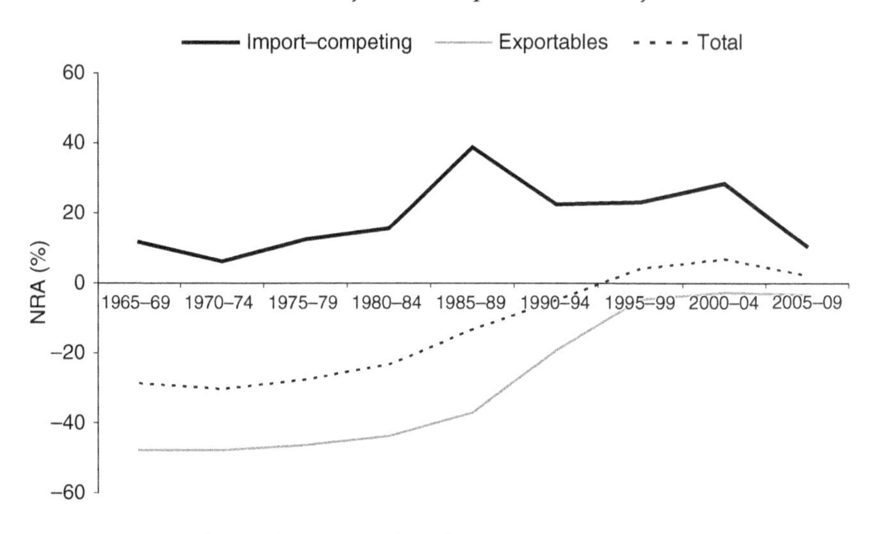

Fig. 10.4. Nominal rate of assistance (NRA) to exporting, import-competing, and all farmers in developing countries, 1965 to 2009 (percent, five-year averages)
Source: Based on the update of NRA and RRA estimates in Anderson and Nelgen (2012b)

FOOD PRICE TRANSMISSION

Each country tends to transmit less than fully any fluctuations in international food prices away from their trend. This widespread and systematic tendency means the estimated NRA for each product also fluctuates, and in the opposite direction to the international price. This can be seen clearly in Figure 10.5a for rice, and to a lesser extent for wheat in Figure 10.5b, despite the fact that the NRA is averaged over all 82 countries in the sample.

This propensity for national governments to alter individual product NRAs from year to year around their long-run trend does not appear to have diminished as part of the trade-related policy reforms that began in the mid-1980s. Table 10.1 focuses on the NRA's annual average deviation from trend in the two decades before versus after 1985. That average deviation from trend NRA is more than one-tenth higher in the latter two decades than in the earlier two decades in just as many cases as it is more than one-tenth lower. The deviations are significant: except for rice in high-income countries, the average deviation is well above the mean NRA for each product. According to Anderson and Nelgen (2012b), barely half the movement in international prices of primary food products is transmitted to domestic markets on average. In short, trade policies are still insulating domestic markets from the volatility of international food markets.[3]

Despite the comprehensiveness of trade reforms since the 1980s, reforms have raised only very slightly the extent to which farm products are traded

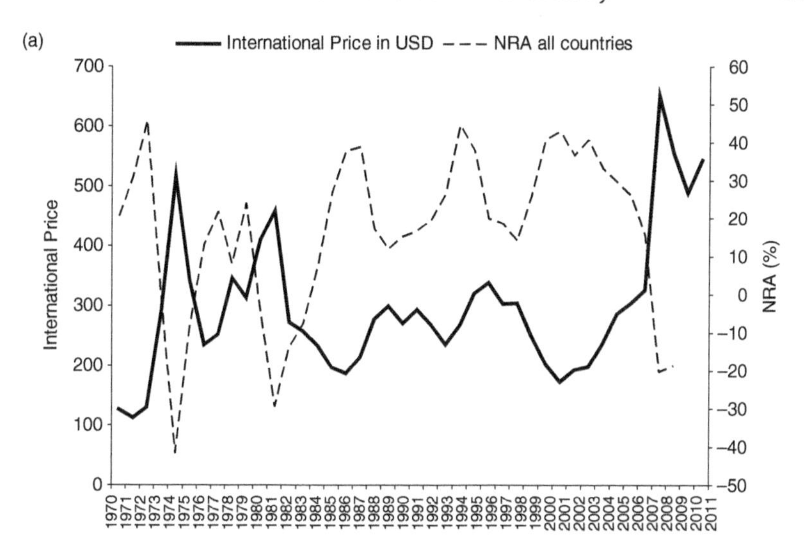

Fig. 10.5a. Rice NRAs and their international price, 82 countries, 1970 to 2010 (left axis is international price in current US$, right axis is weighted average NRA in percent)

Source: Based on the update of NRA estimates in Anderson and Valenzuela (2008) by Anderson and Nelgen (2012b)

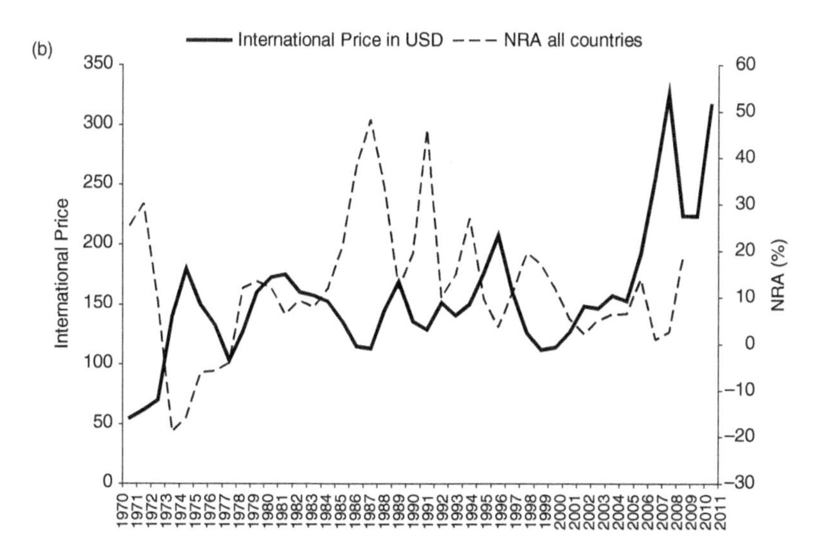

Fig. 10.5b. Wheat NRAs and their international price, 82 countries, 1970 to 2010 (left axis is international price in current US$, right axis is weighted average NRA in percent)

Source: Based on the update of NRA estimates in Anderson and Valenzuela (2008) by Anderson and Nelgen (2012b)

Table 10.1. Deviation of national NRA around its trend value,[a] key farm products,[b] developing and high-income countries, 1965–84 and 1985–2010 (percent)

	Deviation of national NRAs around trends				Weighted average of NRAs (%)			
	Developing countries		High-income countries		Developing countries		High-income countries	
	1965–84	*1985–2009*	*1965–84*	*1985–2010*	*1965–84*	*1985–2009*	*1965–84*	*1985–2010*
Rice	32	59	66	186	−20.1	0.9	136.8	351.8
Wheat	33	43	52	76	5.5	9.1	12.2	20.5
Maize	36	33	40	48	−3.4	2.3	6.9	11.9
Soybean	46	120	75	54	2.7	−2.1	0.1	5.2
Sugar	53	64	168	152	17.2	18	107.6	108.1
Cotton	38	32	42	30	−16	−2.7	21.3	10.4
Coconut	22	34	n.a.	n.a.	−11.5	1.2	n.a.	n.a.
Coffee	41	29	n.a.	n.a.	−37.3	−11.6	n.a.	n.a.
Beef	45	56	84	109	−12.4	2.6	22.7	37.9
Pork	81	58	73	69	23.6	−4.6	37.1	15
Poultry	109	69	91	175	26.3	11.8	24.5	25.4

Source: Updated from Anderson and Nelgen (2012b).
[a] Deviation, measured in NRA percentage points, is computed as the absolute value of (residual—trend NRA) where national trend NRA in each of the two sub-periods is obtained by ordinary least squares linear regression of the national NRA on time. [b] Estimates shown are an unweighted average of national NRA deviations each year, averaged over the number of years in each period.

across national boundaries: the share of primary agricultural production exported globally, including intra-European Union trade, rose only slightly in the two decades to 2000–04 (Sandri et al. 2007). One reason for this small aggregate global response is that high-income countries lowered not only their import restrictions, but also their export subsidies, with the latter offsetting somewhat the trade-expanding effect of cuts in import tariffs. A second reason is that while developing countries phased out their farm export taxes, they also raised tariffs on farm imports (Figure 10.4)—the latter again offsetting somewhat the trade-expanding effect of cuts in export taxes. As a result, international food markets are not much "thicker" now than they were a quarter-century ago.[4]

According to Valenzuela et al. (2009), liberalization of remaining trade barriers as of 2004 would raise the share of farm production exported globally from 8 to 13 percent, so plenty of scope still remains for trade reform to thicken international food markets (i.e., raise the shares of production exported and consumption imported) and thereby make them less volatile. That study also finds the developing countries' share of global exports of farm products would rise from 55 to 64 percent, suggesting more of those countries would become net food exporters and thus would be net beneficiaries of higher international food prices.

POVERTY EFFECTS OF TRADE POLICIES: 2004

Even if trade policies were not contributing to international food price volatility (of which more below), what can be said about their poverty effects as of 2004, before international food prices began rising steeply? A recent, detailed study summarized by Anderson et al. (2011) addresses that issue using numerous global and national economy-wide models, all calibrated to 2004 and incorporating the World Bank estimates of national price distortions.

In the World Bank's model, the incidence of poverty would be reduced under the full merchandise trade reform: there would be 2.7 percent (26 million) fewer people living on less than $1 a day in developing countries, and 3.4 percent (87 million) fewer on less than $2 per day. That result is based on a simple assumption that any increase in per capita income will have a poverty-reducing effect, but it is bolstered by results from ten individual country case studies. These case studies focus on price-distorting policies as of 2004, but they include more product disaggregation and make use of national survey data to consider multiple types of households and labor. When all merchandise trade is liberalized in each of these country case studies, the poverty reduction ranges from close to zero to about 3.5 percentage points, except for Pakistan where it is more than six points. On average nearly two-thirds of the alleviation is due to non-farm trade policy reform, and the contribution of own-country reforms to the reduction in poverty appears to be equally as important as rest-of-world reforms.

Those country case study estimates of national poverty alleviation are also subdivided into rural and urban sources. In every case rural poverty is reduced much more than urban poverty. This is true for both farm and non-farm trade policy reform, and for own-country as well as rest-of-world reform. Since the rural poor have been and are much poorer on average than the urban poor

(Ravallion et al. 2007), trade policy reform should also reduce inequality—and that is indeed what the results show for this sample of countries.

These poverty findings for permanent full trade liberalization of all goods are of course not expected to be the same as those resulting from a temporary alteration of trade restrictions in response to a spike in international prices for one or a few food items. Comprehensive empirical studies of the latter, as they relate to the recent food price spikes, are under way, but it is too early to be able to draw general conclusions. One study of the poverty impact in 28 developing countries found quite heterogeneous results, but concluded that the 2010 price spike for 38 agricultural commodities pushed an extra 68 million food-deficit people into poverty and raised only 24 million food-surplus people out of poverty at the extreme poverty line of $1.25 per day (Ivanic et al. 2012). A preliminary study of the rise in just grain and oilseed prices during 2006 through 2008 also finds a high proportion of developing countries in which poverty appears to have risen—but that study shows that the net poverty-reducing impact of insulating the domestic market from the international price rise is much less than its apparent impact once the combined effect of all governments' responses on amplifying the international price rise is taken into account (Anderson et al. 2012).

STANDING IN THE FOOTBALL STADIUM: VARIABLE NATIONAL TRADE RESTRICTIONS

National governments clearly dislike domestic food price volatility: net buyers of food complain about price rises, while net sellers (farmers) complain when their output prices fall. When some governments alter the restrictiveness of their food trade measures to insulate their domestic markets somewhat from international price fluctuations, however, the volatility faced by other countries is amplified. That reaction prompts more countries to follow suit. The irony is that when both food-exporting and food-importing countries respond, each undermines the other's attempts to stabilize domestic markets. This is a classic international public good problem.

The predicament can be seen by considering two country groups, food importers and food exporters. Suppose a severe weather shock at a time of low global stocks causes the international food price to suddenly rise. National governments wishing to avert losses for domestic food consumers may alter their food trade restriction so that only a fraction of the price rise is transmitted to their domestic market. For example, imposing or raising an export tax on food exports would ensure the domestic price in a food-surplus country rose less than the border price. Similarly, lowering any import tax on food would mean the domestic price in a food-deficit country would rise less than the price of an imported substitute. Hence it is not surprising that governments, in seeking to protect domestic consumers from an upward spike in international food prices, consider a variation in their degree of trade restriction as an appropriate response. That response raises the consumer subsidy equivalent or lowers the consumer tax equivalent of any such trade measure, and does the opposite to producer incentives (the NRA falls). Yet if such domestic market insulation using trade measures is practiced by both food-exporting and food-importing countries, it turns out to be not very effective in keeping domestic price volatility down.

Conversely, if the exogenous weather shock was of the opposite sort (a bumper global harvest) that depressed the international price even after purchases by stockholders, and if governments sought in that case to protect their farmers from the full force of the price fall, the international price fall would be accentuated to the benefit of food-importing countries. Clearly, both such attempts at domestic price insulation exacerbate international price volatility while doing little or possibly nothing to assist those most harmed by the initial exogenous weather shock.

More than that, this use of trade measures can be inefficient and possibly inequitable, and it may even *add to* global poverty despite a possible part of its motivation being to reduce the risk of a rise in national poverty. Note that an import tax is the equivalent of a consumer tax and a producer subsidy, hence lowering it also reduces the extent to which the measure assists producers of the product in question. Likewise, an export tax is the equivalent of a consumer subsidy and a producer tax, so raising it not only helps consumers but also harms farmers. If farming is thereby discouraged in both food-importing and food-exporting countries, the demand for labor on farms falls, and with it the wages of low-skilled workers not only in farm jobs but also in non-farm jobs—and the more agrarian the economy, the more low-skilled workers' wages will fall. Thus while poor households may benefit on the expenditure side from a measure that reduces the extent to which the cost of food consumption would otherwise rise, they could be harmed on the earnings side if they are sellers of food or suppliers of low-skilled labor. Such trade policy responses therefore could add to rather than reduce poverty (Ivanic and Martin 2008; Aksoy and Hoekman 2010). And even if there are some countries where poverty is reduced by the change in trade restrictions in response to a price spike (such as India, according to Do and Ravallion 2012), that response will have exacerbated the international price spike and hence worsened the incidence of poverty in those countries where the spike expanded the number of poor. The net effect could therefore be a worsening of global food insecurity, depending in part on whether the policy actions were more or less in food-importing countries than in food-exporting countries.

To see how effective were changes in trade restrictions in limiting the rise in domestic prices during 2006–08, Anderson and Nelgen (2012a) make use of a simple model developed by Martin and Anderson (2012). That model provides an estimate of the proportional rise in the international price of each grain had trade restrictions not been altered during those three years. That proportional rise, when multiplied by the international price rise during that period, is reported in the second column of Table 10.2, where it is compared with the proportional rises in the domestic price in the sample countries. The numbers for 2006 though 2008 suggest that, on average for all countries in the sample, domestic prices rose slightly more than the adjusted international price change for wheat, only slightly less for maize, and just one-sixth less for rice. The extent of insulation was greater in developing countries, especially for wheat and maize. These results suggest that the combined responses by governments of all countries have been sufficiently offsetting as to do very little on average to insulate domestic markets from the 2006–08 international food price spike. This is not to rule out the possibility that some developing countries' actions may have reduced the number of households that otherwise would have temporarily fallen into poverty, but rather to stress that when many countries take these actions in offsetting ways, it creates an international public "bad" that can adversely affect third countries.[5]

Table 10.2. Comparison of the domestic price rise with the rise in international grain prices net of the contribution of changed trade restrictions, rice, wheat and maize, 1972-74 and 2006-08 (percent, unweighted averages)

	International price rise		**Domestic price rise**		
	Including contribution of changed trade restrictions	Net of contribution of changed trade restrictions	All countries	Developing countries	High-income countries
1972-74					
Rice	300	220	59	72	27
Wheat	157	121	64	77	55
Maize	135	111	49	48	52
2006-08					
Rice	113	68	56	48	74
Wheat	70	56	77	65	81
Maize	83	75	73	62	82

Source: Based on Anderson and Nelgen (2012a)

HOW WILL TRADE POLICIES AFFECT GLOBAL FOOD SECURITY OVER THE NEXT DECADE?

The global financial crisis and the ongoing economic recession in Europe seem set to ensure that emerging economies will continue to grow faster than high-income countries. Led by China and India, developing economies have experienced rapid growth, doubling their share of global exports since the mid-1990s. This trend looks likely to continue, as does the growth in relative importance of South–South trade (ADB 2011; Anderson and Strutt 2012). Industrialization in emerging economies is deepening global production networks and contributing to greater trade in intermediate inputs, but it is also driving the strong demand for farm products, industrial raw materials, and energy. If this demand for energy, together with only slow increases in the taxing of carbon emissions globally, holds fossil fuel prices at current high levels as expected (IEA 2011), the United States and the European Union are likely to retain their biofuel subsidies and mandates for reasons of energy self-sufficiency. Consequently, prices for food and fuel will remain closely linked—in both height and volatility (Hertel and Beckman 2011). Food production variability is expected to increase, as well, thanks to climate change. Under this scenario, what effects of government assistance to agriculture should we expect?

If international food prices stay high, then high-income countries are unlikely to return to their former agricultural protection. However, people and governments in industrializing economies—especially large ones such as China, India, and Indonesia—may well feel more food-insecure if their farm sectors become less able to compete for mobile resources (despite higher international food prices) while domestic food and feed demands grow. Continuing growth in their agricultural protection cannot be ruled out, therefore, even if international food prices remain high (Anderson and Nelgen 2011). Ironically, greater agricultural protection will raise domestic prices of foods increasingly above those at their borders, thereby undermining food security for all their households except those that are net sellers of food and maybe some remaining farm laborers. The latter group will become an ever-smaller share of the population and workforce in the course of economic growth, but whether they become a smaller share of the poor in those countries is difficult to anticipate—so it is unclear what impact agricultural protection growth might have on the national poverty rate. Certainly per capita food consumption would grow less rapidly, and farm protection growth would dampen international food prices somewhat (Anderson and Strutt 2012). If average incomes in those emerging economies continue to grow rapidly, however, then domestic sociopolitical stability is more likely to be helped by assistance to farmers that reduces urban–rural income inequality, and the slight dampening of international food prices may well reduce the risk of sociopolitical instability in poorer developing countries. But this is not to say that rising agricultural protection is the best way for emerging economies to deal with these concerns.

As for fluctuations in NRAs around trend, past behavior suggests both high-income and developing countries' governments will continue to alter their food trade restrictions to insulate their domestic markets somewhat from international food price volatility. This behavior will continue to amplify price fluctuations in the international market, and if both exporting and importing countries continue to respond similarly, such interventions will keep being rather ineffective in preventing fluctuations in domestic food prices. How severe such volatility might be will depend on the size of any unanticipated exogenous shocks to world food markets and the global stocks-to-use ratios of the affected products at the time of any such exogenous shocks. If stocks were to be very low when harvests failed in significant regions, food price spikes of the magnitude experienced in 2008 and early 2011 could well be repeated if countries do not agree multilaterally to desist from altering their trade restrictions at such times.

POLICY FOR GLOBAL FOOD SECURITY

What policy actions, both unilateral and multilateral, could contribute to both global food security and sociopolitical stability over the remainder of the decade? The empirical evidence supports the view that national trade restrictions add to international food price volatility in at least two ways: by thinning international food markets, and by insulating domestic food markets from international price fluctuations. Together these policy attributes magnify the effect on international prices of any shock to global food supply or demand, especially when global food stocks are low.

The solution to the thinning problem is simple economically, if not politically: it is for countries to open further their markets to food trade. If food price spikes cause riots, reducing the risk of such sociopolitical instability is an additional benefit from global trade liberalization—on top of the usual welfare gains measured by economists. The political difficulty and the adjustment costs associated with such reform are minimized if countries can agree to liberalize their food and agricultural markets multilaterally, and to do so at the same time as nonagricultural markets are liberalized. That was what happened in the Uruguay Round, and it is what has been aspired to by members of the World Trade Organization in their Doha Development Agenda (DDA). After more than a decade of negotiating, the DDA is at a standstill, but there is still hope that the talks will be revived. Meanwhile, various plurilateral negotiations on options for regional integration and free-trade areas are under discussion, but the benefits from them are always far smaller than those of a multilateral agreement—and often agriculture is the sector liberalized least in preferential trade agreements (Anderson forthcoming).

The optimal solution to the problem of insulating domestic food markets also involves the WTO. In a world made up of many countries, the trade policy actions of individual countries can be offset by those of other countries—to the point that domestic policy interventions become ineffective in achieving their stated aim of reducing domestic food price volatility. This classic international public good problem could be solved by a multilateral agreement to restrain the variability of trade restrictions.

One of the original motivations for the contracting parties to sign the General Agreement on Tariffs and Trade (GATT), WTO's predecessor, was to bring stability and predictability to world trade. To that end, the membership adopted rules to encourage the use of trade taxes in place of quantitative restrictions on trade, and managed to obtain binding commitments on import tariffs and on production and export subsidies as part of the GATT's Uruguay Round Agreement on Agriculture. However, those commitments continue to be set well above applied rates by most countries, leaving plenty of scope for varying import restrictions without dishonoring the legal commitments under WTO. Meanwhile, there are no effective disciplines on export taxes, let alone bindings.

The current Doha Round of WTO negotiations includes proposals to phase out agricultural export subsidies, as well as to bring down import tariff bindings. Both proposals would contribute to global economic welfare, to the thickening of food and agricultural markets, and thereby to more stable international food prices and fewer food riots. However, proposals to broaden the Doha agenda to introduce disciplines on export restraints have struggled to gain traction. A proposal by Japan in 2000, for example, involved disciplines similar to those on the import side, with export restrictions to be replaced by taxes and those export taxes to be bound and gradually phased down. A year later Jordan proposed even stronger rules: a ban on export restrictions and, as proposed for export subsidies, the binding of all export taxes at zero. However, strong opposition to the inclusion of this export item on the Doha Development Agenda has come from several food-exporting developing countries, led by Argentina, where farm exports have been highly taxed since the country's large currency devaluation at the end of 2001. WTO negotiations have traditionally been dominated by interests seeking market access, yet the above analysis reveals the need for symmetry in the WTO's treatment of export and import disciplines.

If WTO member countries were to liberalize their food trade and bind their trade taxes on both exports and imports at low or zero levels, what alternative instruments could they use when international food prices spike in order to avert losses for significant groups in their societies that might otherwise lead to sociopolitical instability? A standard answer from economists is that food security for consumers, most notably food affordability for the poor, is best dealt with using generic social safety net measures that offset the adverse impacts of a wide range of different shocks on poor people—net sellers as well as net buyers of food—without imposing the costly by-product distortions that necessarily accompany the use of less effective trade policy instruments for social protection. These social safety net measures might take the form of targeted income supplements to only the most vulnerable households, and only while the price spike lasts. To help fund instrument switching, perhaps such changes could be included in the proposed aid-for-trade part of the WTO's Doha agenda.

This standard answer has far greater power now than just a few years ago, thanks to the digital information and communication technology revolution. In the past it has often been claimed that such payments are unaffordable in poor countries because of the fiscal outlay involved and the high cost of administering such handouts. However, recall that in roughly half the cases considered above, governments reduce their trade tax rates, so even that intervention may require a drain on the budget of many finance ministries—as does replacing a non-prohibitive export tax with a ban. In any case, the option of using value-added taxes in place of trade taxes to raise government revenue has become common practice in even low-income countries over the past decade or two. Moreover, the digital revolution has made it possible for conditional cash transfers to be provided electronically as direct assistance to even remote and small households, and even to the most vulnerable members of those households (typically women and their young children).

What if countries are still unsatisfied with the contribution of their farmers to national food security—as reflected, for example, in food self-sufficiency ratios—or feel their farmers are missing out on the benefits of rapid economic growth and industrialization? Again, agricultural import protection measures are far from the best way of dealing with these sociopolitical concerns. Alternative measures include subsidizing investments in agricultural research and development, rural education and health, and roads and other rural infrastructure improvements. If the social rates of return from these investments are currently high and above private rates of returns, as is typically the case in developing countries, expanding such public investments will be nationally beneficial. So, too, could be improvements in land and water institutions that determine property rights and prices for those key farm inputs. Should those changes at the same time reduce rural underemployment and poverty, and slow out-migration of workers to urban areas, they may also lower the risk of sociopolitical instability.

Encouraging countries to switch from trade to domestic policy instruments for addressing non-trade domestic concerns is obviously a considerable challenge, and not least because governments' stated policy objectives are not always the real motives for market intervention. Yet some reform has been possible during the past two decades. Attempts to understand the political economy forces behind different countries' experiences at reform has begun, but much political econometric work remains to be undertaken. More research is needed also on the political economy of biofuel policies, so as to understand how greater energy security and greenhouse gas emission reductions can be achieved in the large fuel-importing countries without having to resort to biofuel policies that are compromising global food security. A better understanding of public and private stockholding behavior is required before optimal food stockholding policies can be devised. Would a coordinated regional rice storage agreement among Asian governments reduce their reluctance to engage more in the international rice market, for example?

Greater transparency of the extent of government interventions in markets, more analysis of their welfare, distributional, environmental, and transfer efficiency effects, and much broader dissemination of these findings are needed to counter the political influence of powerful, narrow vested interests. Key indicators to monitor are not only daily or at least monthly prices of key staple foods in international markets, as are now available on FAO and IFPRI websites, but also timely data on global stocks of key staple foods. As more domestic food prices are added to international price data portals, movements in the ratio of domestic to international food prices can also be monitored. Movements in that ratio—especially if price data more frequent than annual can be assembled—may provide an early warning of the extent to which a country is altering its trade barriers in an attempt to insulate its domestic market from international price fluctuations.

ACKNOWLEDGMENTS

The author is grateful for discussions with Will Martin and Signe Nelgen, for comments from Chris Barrett and other workshop participants, and for funding support from Cornell University, the Australian Research Council, and Australia's Rural Industries Research and Development Corporation. Views expressed are the author's alone.

NOTES

1. Since the 1980s, some high-income governments have also provided decoupled assistance to farmers. Because that support, in principle, does not distort resource allocation, its NRA has been computed separately and is not included for comparison with the NRAs for developing countries when the focus is price distortions. Not included are estimates of the subsidy equivalent of irrigation water or biofuel

policies. On the difficulties of measuring these excluded items, see Dinar (2012) and de Gorter et al. (2011), respectively.

2. The relative rate of farmer assistance is defined in percentage terms as:

$$RRA = 100^{*}[(100 + NRAag^{t})/(100 + NRAnonag^{t})-1]$$

where $NRAag^{t}$ and $NRAnonag^{t}$ are the percentage NRAs for the tradable parts of the agricultural (including noncovered) and non-agricultural sectors, respectively. See Anderson et al. (2008).

3. This is of course not to deny that high costs of trading across borders of low-income countries reduce their ability to trade profitably; but the above findings also relate to higher-income countries, and refer to the most tradable of farm products.
4. This is consistent with a global model back-casting exercise to see how much different global markets would have been in 2004 if the trade policies of 1980-84 had remained in place. The results suggest the share of farm production traded globally would have been no more than one percentage point different as a result of that quarter-century of reform (Valenzuela et al. 2009, Table 13.8).
5. There is yet another way in which the sporadic imposition of food export restrictions generates an international public "bad," namely in reducing food-deficit countries' willingness to remain dependent on food imports. The rise of agricultural protection in East Asia from the early 1970s may have been partly driven by such concerns following food-exporting countries' reactions to the 1972–74 food price spike.

REFERENCES

ADB. 2011. *Asian development outlook 2011.* Manila: Asian Development Bank.

Aksoy, M. A., and B. Hoekman. 2010. *Food prices and rural poverty.* London: Centre for Economic Policy Research. Published for the World Bank.

Anderson, K., ed. 2009. *Distortions to agricultural incentives: A global perspective, 1955–2007.* London: Palgrave Macmillan and Washington DC: World Bank.

——— Forthcoming. Trade barriers and subsidies: Multilateral and regional reform opportunities. In *Global crises, Global Solutions,* 3rd edition, ed. B. Lomborg. Cambridge and New York: Cambridge University Press.

Anderson, K., J. Cockburn, and W. Martin. 2011. Would freeing up world trade reduce poverty and inequality? The vexed role of agricultural distortions. *The World Economy 34*(4): 487–515.

Anderson, K., Y. Hayami, and A. G. Mulgen. 1986. *The political economy of agricultural protection: East Asia in international perspective.* London: Allen and Unwin.

Anderson, K., M. Ivanic, and W. Martin. 2012 Food price spikes, price insulation and poverty. Paper presented at the NBER Conference on the Economics of Food Price Volatility, Seattle, August 15–16.

Anderson, K., M. Kurzweil, W. Martin, D. Sandri, and E. Valenzuela. 2008. Measuring distortions to agricultural incentives, revisited. *World Trade Review 7*(4): 1–30.

Anderson, K., and S. Nelgen. 2011. What's the appropriate agricultural protection counterfactual for trade analysis? In *The Doha development agenda: An assessment,* ed. W. Martin and A. Mattoo, 325–54. London: Centre for Economic Policy Research and the World Bank.

——— 2012a. Agricultural trade distortions during the global financial crisis. *Oxford Review of Economic Policy 28*(2): 235–60.

———. 2012b. Updated national and global estimates of distortions to agricultural incentives, 1955 to 2010. www.worldbank.org/agdistortions (accessed March 30, 2013).

Anderson K., and A. Strutt. 2012. Global food markets in 2030: What roles for farm TFP growth and trade policies? Paper presented at the AARES Pre-Conference Workshop on Global Food Security, Fremantle, February 7.

Anderson K., and E. Valenzuela. 2008. *Estimates of global distortions to agricultural incentives, 1955 to 2007.* Washington DC: World Bank. www.worldbank.org/agdistortions (accessed March 30, 2013).

Briones, R.M. 2011. Regional cooperation for food security: The case of emergency rice reserves in the ASEAN Plus Three. *ADB Sustainable Development Working Paper Series.* No. 18 Manila: Asian Development Bank.

de Gorter, H., D. Drebik, and E.M. Kliauga. 2011. Understanding the economics of biofuel policies and their implications for WTO rules. Paper presented at the IATRC annual meetings, St Petersburg FL, December 11–13.

Dinar, A. 2012. Economy-wide implications of direct and indirect policy interventions in the water sector: Lessons from recent work and future research needs. Policy Research Working Paper 6068. Washington DC: World Bank.

Do, Q.-T., and M. Ravallion. 2012. Coping with food price shocks: Trade versus social protection policies. Paper presented at the NBER Conference on the Economics of Food Price Volatility, Seattle, August 15–16.

Fan, S., M. Torero, and D. Healey. 2011. *Urgent action needed to prevent recurring food crises. IFPRI Policy Brief 16.* Washington DC: International Food Policy Research Institute.

FAO, IFAD, IMF, OECD, UNCTAD, WFP, World Bank, WTO, IFPRI, and UN HLTF. 2011. *Price volatility in food and agricultural market: Policy responses.* Rome: FAO. In collaboration with IFAD, IFPRI, IMF, OECD, UNCTAD, WFP, World Bank and WTO.

Hertel, T. W., and J. Beckman. 2011. Commodity price volatility in the biofuel era: An examination of the linkage between energy and agricultural markets. In *The intended and unintended effects of U.S. agricultural and biotechnology policies,* ed. J. G. Zivin and J. Perloff, 189–221. Chicago: University of Chicago Press. Published for NBER.

IEA. 2011. *World energy outlook 2011.* Paris: International Energy Agency.

Ivanic, M., and W. Martin. 2008. Implications of higher global food prices for poverty in low-income countries. *Agricultural Economics 39*:405–16.

Ivanic, M., W. Martin, and H. Zaman. 2012. Estimating the short-run poverty impacts of the 2010–11 surge in food prices. *World Development 40*(11): 2302–2317 (November).

Johnson, D.G. 1973. *World agriculture in disarray.* London: Fontana.

Krueger, A.O., M. Schiff, and A. Valdés. 1988. Agricultural incentives in developing countries: Measuring the effect of sectoral and economy-wide policies. *World Bank Economic Review 2*(3): 255–72.

Martin, W., and K. Anderson. 2012. Export restrictions and price insulation during commodity price booms. *American Journal of Agricultural Economics 94*(2): 422–27.

OECD. Producer and Consumer Support Estimates, OECD Database 1986–2010. http://www.oecd.org (accessed September 26, 2011).

Ravallion, M., S. Chen, and P. Sangraula. 2007. New evidence on the urbanization of poverty. *Population and Development Review* 33(4): 667–702.

Rosegrant, M., S. Tokgoz, and P. Bhandary. 2013. The Future of the Global Food Economy: Scenarios for Supply, Demand, and Prices. In *Food Security and Sociopolitical Stability*, ed. C Barrett. Oxford: Oxford University Press.

Sandri, D., E. Valenzuela, and K. Anderson. 2007. Economic and trade indicators, 1960 to 2004. Agricultural Distortions Working Paper 02, Washington DC: World Bank. www.worldbank.org/agdistortions (accessed March 30, 2013).

Tyers, R., and K. Anderson. 1986. Distortions in world food markets: A quantitative assessment. Background Paper No. 22. World Development Report 1986. Washington DC: World Bank.

———. 1992. *Disarray in world food markets: A quantitative assessment.* Cambridge and New York: Cambridge University Press.

Valenzuela, E., D. van der Mensbrugghe, and K. Anderson. 2009. General equilibrium effects of price distortions on global markets, farm incomes and welfare. In *Distortions to agricultural incentives: A global perspective, 1955–2007*, ed. K Anderson. London: Palgrave Macmillan; Washington DC: World Bank.

World Bank. 2011. *Pink sheets.* http://econ.worldbank.org (accessed 30 March 2013).

11

Food Security and Political Stability: A Humanitarian Perspective

DANIEL MAXWELL

The 2011 famine in Somalia put humanitarian food security crises back on the agenda of the international community. With its causes linked not only to environmental, climatic, and food price factors, but also to internal conflict, competition among regional powers, and the global war on terror, the Somalia famine revived many long-standing concerns about the link between extreme food security crises and stability, security, and other political imperatives. The Somalia famine may have been an outlier in terms of the severity of the crisis and the extent of human suffering it engendered, but the famine was in many ways characteristic of the kinds of crises likely to be experienced in the medium-term future. The causes and consequences of these crises—as they relate to political stability—are important to understand.

This chapter explores the links between food security and political stability from a humanitarian perspective. Humanitarian agencies and actors have traditionally been called upon to respond to crises of various kinds, to protect human life and dignity, and to contain the human costs of crises. Food insecurity has been both a cause and consequence of those crises, but for much of recent history, the humanitarian establishment has focused primarily on the latter: food insecurity following emergencies. This chapter will trace recent food security crises and the international humanitarian response in order to draw conclusions about the relationship between food security and political stability, describe the major challenges facing policymakers as they address these two concerns, and suggest effective policies that work within the constraints on humanitarian action.

TYPES OF CRISES, DRIVERS OF CRISES

Acute food insecurity is frequently the result of humanitarian disasters of different types. Classically, disasters were divided by their nominal causal factor, or predominant triggering factor. FAO/WFP (2008) depicts increasing numbers of food security crises over time, caused both by "natural" and "human-induced" causes. Rapid-onset natural disasters include earthquakes and tsunamis, as well as events that might be generally but not very specifically predictable, such as storms and floods. Slow-onset natural disasters are mainly droughts that take several months or even years to develop, thus enabling (at least in theory) some amount of mitigation and preparation for response. Human-made disasters often involve conflict, but are sometimes driven by political factors or government policy without militarized conflict. For example, the prolonged crisis in Zimbabwe and the famine in North Korea in the mid-1990s were both major food security crises that were caused almost entirely by political factors.

Several observations about humanitarian crises should be made. First, the number of disasters has grown. EMDAT-CRED (2012) notes that the number of natural disasters since the turn of the nineteenth century, and the numbers of people affected, have increased dramatically. However the number of people killed in disasters has declined—better response and mitigation have reduced the number of fatalities.

Second, while the causes of crises are both natural and human-made, the consequences of crises that are more human-made (particularly those that are conflict driven) are more difficult to manage. The relative budgets of contemporary humanitarian response to different kinds of disasters makes clear the extent to which international response is heavily focused on conflict emergencies. Many—if indeed not most—humanitarian disasters today are triggered by some combination of factors, both natural and human-made (FAO/WFP 2010). Political instability can be the consequence of either kind of disaster. Sometimes, natural disasters, such as the 2010 earthquake in Haiti, can be a source of significant political instability. Past famines in Ethiopia (1972 to 1973 and 1984 to 1985) are widely believed to have been associated with regime change, but were triggered mainly by droughts that were poorly managed (Lautze and Maxwell 2006). On the other hand, internal conflicts are self-evidently about political instability, and often the drivers of food security crises. Both types of emergencies (which are, in fact, rarely easily distinguished in practice) are taken here to be potential sources—and manifestations—of political instability. The difference is that in primarily natural disasters, the state may be in a better position to mitigate and respond to the crisis (or at least in a better position to be assisted). In internal conflicts, the state may well be a party to the conflict—hence one of the causes of the crisis, and therefore not interested in external assistance to address the crisis. Of the 20 countries receiving the most humanitarian aid (food assistance and otherwise) since 2000, 18 of them have been involved in some kind of internal conflict (Development Initiatives 2011).

The third major point to be noted is that causal factors have evolved. It is no longer particularly helpful to break down crises between natural and human-made even if donors use these categories to break down resource allocation. Virtually every contemporary crisis is caused by a combination of factors. And even if the primary shock may be completely natural, such as an earthquake, the nature of the humanitarian emergency engendered by the shock is shaped almost entirely by political considerations.

A 2010 report by the Feinstein International Center summarized several additional emergent causal factors or drivers of crises of the future. Demographic changes are probably the most predictable of these drivers. Virtually all the projected net global population growth between 2020 and 2050 will take place in the developing world, and almost all will be in urban and peri-urban areas. Governments will have little chance to adapt to this growth; infrastructure and services will not be able to keep pace in many countries. This portends an increasing urbanization of poverty, vulnerability, hunger—and disasters. Food security crises remain predominantly rural, but there is increasing evidence that the locus of crisis is slowly shifting towards urban areas (Pantuliano, Buchanan-Smith et al. 2011). Population sizes are expected to double in much of sub-Saharan Africa. Less is certain about HIV/AIDS, but ever-higher populations of AIDS orphans are generally expected (Feinstein International Center 2010).

The impact of climate change will also have a major impact on humanitarian disasters and the response. The Feinstein Center predicts so-called first order impacts—changes in agricultural production systems, global disease vectors and epidemiological factors, and the resulting disruptions of livelihood systems that have been well documented by many studies. It also foresees second order impacts—the tendency for political systems to become more repressive as the rate of change outpaces the ability of systems to adapt (Feinstein International Center 2010). The implication is that climate change will increase the intensity and frequency of "natural" disasters (floods, droughts, and so on) and lead to more political crises, as well.

The impact of globalization on disasters is another factor. Globalization portends more rapid economic growth, but also greater inequality; the growth of unsustainable economic activity in pursuit of short-term gains in unregulated markets; and higher levels of human migration, both voluntary and forced. More marginalized groups mean more people at risk of disaster. The impact of globalization on the state appears unclear and perhaps contradictory—perhaps a more curtailed role, or perhaps a greater assertion of sovereignty. With regard to humanitarian action, both trends are visible. Already evident is the greater penetration of global markets into local economies. In some cases, this can result in greater vulnerability of localized (and previously much more isolated) markets to global price volatility and in more rapid transfer of economic shocks. On the other hand, the globalization of markets likely serves to reduce the impact of localized production shocks—at least for net food purchasers. And more integrated markets have opened up the possibilities of cash and voucher-based programming—a more rapid and in many cases, more efficient response to localized food access crises, as well as one frequently preferred by recipient communities (Lentz et al. 2012; Violette et al. 2012).

Fourth, the nature of conflicts driving food security crises has also changed. In the1990s, much of the increase in internal conflict was driven by local grievances that had been masked by the bipolar nature of international relations between East and West between 1948 and 1989. Some of these post-Cold War conflicts were political, but many were resource based, and many took the form of conflict between ethnic or other identity groups. In recent years, while resource-based conflict remains prevalent, some conflicts have become again more ideological and polarized—and in many cases "asymmetric," particularly in the context of the global war on terror. While this no doubt shapes foreign policy and security considerations, it also profoundly affects humanitarian policy: nine of the top ten recipient countries of humanitarian assistance in 2011 have conflicts that involve at least one group labeled as a foreign terrorist organization by the US government (Development Initiatives 2011).

Fifth, the very term "crisis" or "emergency" once implied a short, acute episode with an identifiable cause, a beginning, and an end. In the twenty-first century, there are very few crises that fit such a description. Long-lasting crises, which may have many causes but no clear ending and limited potential for recovery—so called "protracted crises"—are much more the norm today.

A 2010 report by the UN Food and Agriculture Organization (FAO) and the World Food Programme (WFP) outlined the nature of protracted crises and their implications for food security. Protracted crises are defined first and foremost by longevity or duration. The FAO Global Information Early Warning System (GIEWS) tracks information provided by most governments and shows states of humanitarian emergencies (so defined because they require external assistance) lasting for up to 30 years in some countries. Conflict, weak governance, and the breakdown of services and institutions are also common characteristics, although not all countries in protracted crisis have necessarily experienced overt conflict.

The consequences for livelihood systems and food security are stark. Assistance that could have been used for developmental purposes is increasingly limited to life-protecting humanitarian aid. And the prevalence of food insecurity—as measured by the FAO undernourishment indicator and the International Food Policy Research Institute Global Hunger Index (GHI)—is significantly higher in protracted crises (FAO/WFP 2010). Countries in protracted crisis on average have a prevalence of food insecurity that is nearly three times higher (37 percent) than all other developing countries.

More alarmingly, in 2010, 24 countries in Africa alone reported crises to GIEWS, and 19 of these had been in crisis for at least eight of the previous ten years (the FAO/WFP criterion for longevity to define a protracted crisis). By contrast, two decades earlier, only 12 countries in Africa reported crises, and only five of those fit the criterion for protracted crises. This strongly suggests that protracted crises—not the acute short-lived emergencies for which the international humanitarian apparatus is designed—are the "new norm." This is certainly the case in Africa, but also in many parts of South Asia and a handful of countries elsewhere in the world.

The implications are profound. Not only is food insecurity significantly worse in these countries, the very nature of these crises means that the government of the affected country is often not in a position to address either the causes or consequences of the crisis through concerted public policy action, and in fact may be party to conflicts that drive the crisis (Sudan being the classic example). Donor funding going to these crises now consumes the vast majority of humanitarian budgets. In 2010, roughly 70 percent of humanitarian assistance was going to countries that had been in crisis eight years or more, 20 percent went to countries reporting crises lasting four to seven years, and only 10 percent went to crises of three years or less duration. Yet much of the rhetoric and most of the response mechanism is still built on the presumption of defined, acute emergencies (Development Initiatives 2011).

Perhaps the most troubling finding of the FAO/WFP report is that, of the five African countries that were in protracted crisis in 1990 (Sudan, Somalia, Ethiopia, Angola, and Mozambique), only Mozambique was no longer in protracted crisis by 2010. So the difficulties of emerging from this kind of crisis are very different from the notions of "recovery" that are often bandied about—even by the organizations and national governments that have to deal with these crises (FAO/WFP 2010). All of this points to the need for greater emphasis on the gray area between humanitarian approaches and the standard development approaches to improving food security of the recent past. This gray area emphasizes risk reduction, recovery, asset protection, and livelihood diversification in chronically vulnerable areas (Maxwell et al. 2010).

To summarize this picture, several key points should be noted about the changing nature of crises. First, the number of disasters has grown over time and causal factors have evolved. It is no longer particularly helpful to try to distinguish between natural and human-made crises, although the terms remain in widespread usage. Virtually every contemporary crisis is caused by a combination of factors.

Second, many factors are expected to shape the crises of the future, including demographic changes and an increasing urbanization of poverty, vulnerability, and hunger. Climate change is expected to have a major impact on future disasters. And the impact of globalization will make already at-risk populations much more vulnerable to the kinds of food price spikes seen in

2008 and again in 2011. Third, the nature of conflicts driving crises has also changed from primarily local resource-based, ethnic conflict to more ideological and polarized and asymmetric conflicts. Fourth, protracted crises—which have multiple causes but no clear ending and limited potential for recovery—are now the norm. Food insecurity is demonstrably worse in these crises, and the nature of the crisis undermines the ability of the affected government to respond. These crises now consume the vast majority of international resources for humanitarian action—even though the problems are not what the international humanitarian system was designed to address. Political instability is both a cause and consequence of disasters. Indeed, containing the political consequences of crises—rather than acting to prevent or reduce human suffering—has become the major rationale for humanitarian action in the contemporary era.

CONTEMPORARY HUMANITARIAN ACTORS AND ACTIONS

Contemporary humanitarian actions have many antecedents, but most histories trace their origins to the work of Henri Dunant and the founding of the International Committee of the Red Cross (Walker and Maxwell 2008). Most contemporary humanitarian actors now pursue somewhat different objectives from the ICRC, or at least ones that are only partially related to the genre of humanitarian action that grows out of the Red Cross movement. But contemporary humanitarianism is in considerable turmoil—some even use the term "malaise" (Donini et al. 2008) to describe it—as several trends suggest.

First, humanitarian action has moved from relative obscurity during most of the Cold War to become an important—and highly contested—arena. Official budget allocations for humanitarian action have grown over the past two decades from around $3 billion in 1990 to more than $11 billion in 2010. This trend is unlikely to reverse anytime soon, particularly in light of the increasing number of disasters each year and the greater vulnerability of higher numbers of people.

Throughout this period, food assistance (traditionally food aid) has been the biggest single category of humanitarian response. Not only has food response to humanitarian emergencies been by far the biggest single category of humanitarian expenditure, the proportion of requirements that actually attract donor funding for food varies from 79 to 90 percent of the requested amount—whereas for most other sectors, the average has only been about 50 percent (OCHA, various years). With a few exceptions, most humanitarian appeals remain underfunded, and the longer-term recovery from crises woefully neglected (Maxwell et al. 2010). While acute emergencies receive the lion's share of funding, and food assistance the biggest share of that, there is little evidence that significant progress is being made to address the underlying causes of food insecurity in these emergencies, or that recovery from crises—much less, crisis prevention and risk reduction—is taken as seriously as crisis response (Development Initiatives 2012).

Second, there are different perspectives on the role or basis of humanitarian action. These have been characterized as "Dunantist"—after Henri Dunant, the founder of the Red Cross movement—on the nonpartisan end of the spectrum, to "Wilsonian"—after President Woodrow Wilson, who saw pragmatic international action both government-led and nongovernmental, as a key element of US policy, to "new humanitarian," a movement that began in the late 1990s that deliberately sought a much more politically engaged form of humanitarianism. A Dunantist approach to food security emergencies would imply basically only providing immediate food and nutrition assistance to people affected by a disaster, for as long as the effects of the disaster lasts; a Wilsonian approach would probably recognize the linkages between acute food insecurity in a crisis and a more chronic form of vulnerability to disasters, and emphasize economic growth, livelihoods diversification, and disaster risk reduction as the appropriate response. A new humanitarian approach would emphasize the right to food, the political blockages to the fulfillment of that right, and political contracts between rulers and populations that prevent the outbreak of famines or food crises. But there is no simple, linear continuum of types of humanitarian organizations.

The more general role of humanitarian action has also varied. During the Cold War, humanitarians were viewed, with a few exceptions such as during the Biafra war or the Ethiopian famine of 1984 to 1985, as "angels of mercy"—well intentioned and good-hearted, but fundamentally political lightweights on the international scene. With the outbreak of numerous localized conflicts that affected large numbers of people, but which commanded no particular geopolitical rationale for international engagement, humanitarian action in the 1990s increasingly became a substitute for foreign policy engagement to end conflicts—hence the rapidly growing humanitarian budgets. But in practice, this has meant that containing the humanitarian damage, rather than addressing the problem causing the conflict, was the modus operandi of the day—nowhere more obviously or tragically so than in the Bosnian war. Since 9/11, humanitarian action has become part and parcel of the foreign policy of donor countries and domestic policy of many affected countries—with profound implications for the ability of humanitarian action of whatever political stripe to access affected populations (Walker and Maxwell 2008).

Reflecting these diverse humanitarian perspectives, the number and type of humanitarian actors have expanded considerably. Not only are more agencies engaged, there are many new actors. Once largely the domain of governments and a handful of large UN agencies or international NGOs, now many entirely local (often partisan) solidarity-based organizations or organizations linked to diaspora groups, operate as NGOs. New donors and organizations in the Middle East and Southeast Asia have emerged. And increasingly, military and private commercial interests are taking on humanitarian roles.

Third, the capacity to respond to food security crises has improved substantially. Not that long ago (2004), Levine and Chastre found responses to food security crises in Central Africa were not based on assessment or analysis, tended to rely on a limited handful of interventions, and were not having very much impact. Since then, the ability to analyze trends and causes of food insecurity has improved, and the options for mitigating and responding to food security crises have expanded considerably. Whereas in-kind food aid—typically tied to source markets in the donor country—was the dominant response a decade ago, there are now a multiplicity of forms of response, including cash and vouchers, different modalities of in-kind food; various forms of support to livelihood so as to both improve food production and bolster the purchasing power of crisis-affected populations; and a wide range of much more effective nutritional responses, particularly ready-to-use foods; and community-managed forms of outpatient care (Maxwell and Sadler 2011). As a result—at least until the Somalia famine—mortality rates in acute food security crises had been declining, yet there is little doubt that the number of people pushed closer to the edge of survival has grown—particularly in the most food-insecure areas of the globe.

Fourth, funding for humanitarian response is generally not allocated on the basis of needs or impartiality (Darcy and Hofmann 2003). This trend has only gotten worse over time, leading to some crisis situations being called "hidden" (Checchi and Roberts 2008) or even "normal" (Bradbury 1998). The basis for allocation of funding has more to do with geopolitical or strategic interests than with an impartial assessment of need, or even with the so-called "CNN effect" of media publicity (Olsen et al. 2003).

Fifth, while much of the emphasis in humanitarian action is on the international response, international law is quite clear that national governments have primary responsibility for the protection and well-being of their citizenry (ICISS 2001). Nevertheless, humanitarian action and state sovereignty have a somewhat fraught relationship. For much of their existence, international humanitarian actors have had a "state-avoiding" tendency, reinforced perhaps by the frequent absence (or in some cases, the predatory nature) of state actors in the arenas where humanitarian assistance was needed most. In the immediate post-Cold War era, international donors funded the humanitarian agencies, but national governments were either ignoring the problems humanitarian agencies were called on to address—or were actively causing them.

Lastly, in the aftermath of numerous crises in chronically at-risk countries, a different kind of programmatic response has arisen that addresses these crises as predictable—and long-term—manifestations of poverty and vulnerability, not as humanitarian emergencies per se. Safety nets or social protection programs, led and at least partially funded by national governments—not just international humanitarian agencies—are now a major feature of the response to chronically high levels of food insecurity throughout much of Eastern and Southern Africa, with the Ethiopian Productive Safety Net Programme (PSNP) as the archetypal example. The intent of this program is to provide adequate access to food in the short term, while building household and community assets in the medium term, and hopefully eventually reducing the need for such a comprehensive program in the long term. To date, programs have been more successful at providing a safety net than at encouraging graduation and self-reliance. For the most part, donors and national governments view these as development programs, not humanitarian programs, although many of them have flexible funding and allocation mechanisms that can be ramped up in bad years.

Over the past several decades, humanitarian action has grown rapidly from a relatively obscure form of international response to a major—and highly contested—arena. Throughout this period, the primary form of response has been related to food security—mostly in-kind food aid. The ability to analyze and respond to food security crises has improved dramatically in the past five to ten years, and a much-expanded range of response options now provides not only immediate access to adequate food, but also forms of livelihood support that may bolster both access and production, and a new range of nutritional interventions that have been much more successful at controlling malnutrition problems (Maxwell and Sadler 2011). Safety nets and social protection programs have emerged to take the place of humanitarian response in the case of predictable, chronic food insecurity, but which may nevertheless require short-term (and often longer-term) assistance.

While the obligations of the state are clearly specified in international law, the actual role of the state in a given crisis is highly variable—ranging from absent to predatory, to facilitative to protective, and with the question of state sovereignty and international obligations not always clear. That said, humanitarian aid is clearly not allocated impartially on the basis of need. There is some evidence of both improved capacity to respond to crises and decreased mortality in crises, but particularly since 9/11, there have been more ominous trends, as well.

THE SECURITIZATION OF AID AND AID MANIPULATION

Humanitarian assistance is increasingly linked to security or political objectives, undermining its traditional—although somewhat mythical—neutral and independent status. Some have called this trend the "securitization of aid," or even "humanity as a weapon of war" (Brigety 2008; Howell and Lind 2009).

As a consequence, countries that have insurgent movements that are anti-Western in orientation and linked to terrorist organizations, or movements that are widely believed to have such links, have seen comparatively greater levels of assistance—both military assistance and development/humanitarian assistance. Significant humanitarian assistance has been funneled through the military, rather than through traditional aid agencies. Further, there has been an increasing "securitization" of the objectives of assistance, regardless of the type of agency or actor through which the funding is channeled. This assistance has been guided by a widely held belief that poverty and underdevelopment are key drivers of insurgency and terrorist movements—hence the notion of "winning hearts and minds" has become central to both counterinsurgency strategies and the aid allocation strategies of some donors, particularly in certain countries.

This trend is broader than just US foreign assistance, but has strongly influenced US assistance. In the past decade, US Official Development Assistance funding made through the traditional mechanism of USAID has decreased markedly, while funding channeled through the Department of State and the Department of Defense has risen. Since 2005, these trends have reversed somewhat, with declining percentages of assistance passing through the Department of Defense. However, the overall trend towards more securitized aid seems to be continuing as USAID objectives are increasingly tied very explicitly to promoting security objectives.

This change has resulted in an obvious blurring of the lines between humanitarian, developmental, and security assistance and has forced agencies to accept explicit security or counterinsurgency objectives. This in turn has led to increased worries about the security of humanitarian workers, who—despite the evidence—still regard themselves as neutral and impartial. Indeed, kidnappings and killings of aid workers have increased over time, particularly in a few countries. While it is uncertain whether this deterioration in the safety of aid workers is explicitly linked to the securitization of aid (Fast 2010), the trend highlights questions about whether humanitarian principles and international humanitarian law still govern humanitarian responses in situations of conflict, the extent to which these principles and laws are recognized by all actors, and the shrinking of humanitarian space.

Another recent strand of analysis has focused less on principles and more on the question of the effectiveness of aid provided in promoting security and stabilization objectives. The "winning hearts and minds" hypothesis posits that poverty, underdevelopment, and unmet human needs are important drivers of insecurity, and conversely, that meeting humanitarian need, providing economic opportunity, and promoting recovery will contribute to stabilization, conflict reduction, and counterinsurgency.

While this view is widely held, there is little evidence to support it. Recent research has shown that providing aid according to security objectives in Afghanistan has shown "very little evidence of aid projects winning hearts and minds or promoting stability." Instead, "one of the main reasons given by the Afghans...interviewed for the growing insurgency was their corrupt and unjust government" (Wilder and Gordon 2009, 2). The study found that allocating large amounts of money through funding mechanisms such as the Commanders Emergency Response Program in highly insecure regions often ended up exacerbating problems like corruption, and may even be fueling local conflict.

The Commander's Emergency Response Program budget has grown rapidly—from about $400 million in 2007 to $1.2 billion in 2010 in Afghanistan, for example. In the meantime, targeting assistance by security criteria—rather than by poverty or humanitarian criteria—allocates the most resources to the most conflict-affected countries or areas of countries, at the expense of other places with equally pressing needs and potentially higher probability of developmental or humanitarian impact from assistance.

The securitization of aid is a subset of the more general question of aid manipulation. Humanitarian aid has often been a resource that parties in conflict situations have sought to control, divert, or capture in some way in order to support their own (non-humanitarian) objectives. The evidence of this is as strong now as ever. Food aid in crises has been particularly subject to manipulation, in part because it is viewed as a powerful tool, and in part because it is available and difficult to conceal in what is often an otherwise resource-scarce environment. Local actors of course sometimes divert food aid for sheer economic gain, but more often these interventions are an attempt to influence the behavior of states, nonstate actors, and affected populations toward some particular end. Robert Paarlberg noted 25 years ago that the manipulation of food power rarely actually achieved those objectives, but my recent review of this topic found numerous attempts to utilize food assistance in humanitarian emergencies to achieve a variety of outcomes (Maxwell 2012). Table 11.1 depicts the ways in which food aid has been manipulated.

As Table 11.1 makes clear, nearly all parties involved in conflicts or response to humanitarian emergencies have attempted to manipulate food aid toward these ends at some point or another, including donors, national governments of recipient countries, nonstate actors and insurgent movements, humanitarian agencies themselves, and even recipient communities. Actions included attempting to use aid to favor one group over another, to address insecurity, to induce belligerents into peace talks, to influence the movement and registration of internally displaced persons, to sell for the purchase of weapons, to protect aid convoys, or to buy protection for recipient communities.

In almost all of the cases reported, the policy objective of the manipulation of aid was not achieved, but the attempt came at significant humanitarian cost. Nevertheless, unless and until there is a stronger consensus that humanitarian assistance is only for humanitarian purposes and stronger sanctions against its misuse are implemented, the manipulation of aid will almost certainly continue. It is simply too tempting to all parties.

Table 11.1. Actors, actions and outcomes: The instrumentalization of food assistance

Actors	Country	Action	Outcome
Donor agencies	Somalia	Aid withdrawal	• Significant factor in causation of 2011 famine • Little political or security improvement
	Afghanistan	"Securitization" of aid objectives	• Not clear—similar securitization of aid efforts have been counter-productive in security terms
	North Korea	Aid as a carrot—"food for talks"	• Famine was not prevented • No major change in regime policy or behavior
Recipient country governments	North Korea	"Triaging" access to food for large groups	• Perhaps a million people starved • Regime controlled access to food and survived • Underground private markets emerged—contrary to government policy
	Sudan	Manipulating registration of IDPs	• IDPs threatened with aid cut-off, but agencies—not government—blamed • Agencies reversed cut-off policy
Rebel movements and non-state actors	Ethiopia	Diverting funds to buy arms	• Not clear—evidence is so limited that the accusation was formally withdrawn
	Biafra	Landing/currency exchange fees financed the war	• Biafra was ultimately defeated • 180,000 people starved • Aid manipulation likely prolonged the war—and hence the suffering and loss of life
Humanitarian agencies	Darfur	"Food for access" and "Food for protection"	• Food diverted (undermined impartiality and neutrality) • Probably contributed to general deterioration of security
Recipient communities	Southern Sudan	Food taxation	• Food diverted to army • Humanitarian impact not clear
	Somalia	Food redistribution	• Likely mitigated local conflict • Likely reduced food security impact of food aid

Source: Maxwell (2012)

EARLY WARNING AND THE RESPONSIBILITY TO PROTECT

Research conducted in the early 1990s attempted to come to grips with an increasingly obvious dilemma in slow-onset (often drought-triggered) food security crises: the missing link between early warning and response (Buchanan-Smith and Davies 1995). Nearly 20 years later, this dilemma remains: why is it that the international community can see a crisis coming, yet fail to take actions to prevent, mitigate, or respond to it until it is a full-blown crisis? After the famine in the Sahel in the 1970s, the US government and several international organizations invested heavily in famine early warning systems, including the US-funded Famine Early Warning Network (FEWS NET) and several UN-led systems. Most chronically risk-prone countries now also have national systems, as well. The ability to monitor and predict food security crises has never been better, and yet, time and again, governments, donors, and humanitarian agencies continue to fail to prevent or mitigate crises (Humanitarian Policy Group 2006).

Several explanations posited by Buchanan-Smith and Davies remain relevant, including lack of trust between donors and recipient governments or donors and agencies about the actual dimensions of the problem. Political failures, similar to those accounting for the manipulation of aid, have been compounded in recent crises by counterterrorism legislation in many donor countries (Pantuliano, Mackintosh et al. 2011). Risk aversion on the part of both donors and agencies appears to have increased under these circumstances (Bailey 2012). There is little doubt in policy circles that early, protective responses are both less expensive financially and protect affected populations more effectively, yet a variety of factors continue to prevent this obviously optimal means of addressing food security crises.

After a brief interlude in which there seemed to be an emerging global consensus about the responsibility of the international community to intervene to protect the rights and dignity of people caught in crises—the "responsibility to protect" (R2P)—recent trends have seen the reassertion of nationalistic sentiments or sovereignty, with substantial constraints on the independence of humanitarian actors to respond to crises. While this mainly impinges on crises of human rights' violations, the overlap with humanitarian food security crises is substantial. The doctrine of responsibility to protect essentially posits that when a country is unwilling or unable to protect the rights of its citizens, the international community has an obligation to intervene. Growing out of the report of the International Commission on Intervention and State Sovereignty (ICISS) in 2001, R2P rapidly gained popularity by the mid-2000s. Overwhelmingly affirmed by the UN as its guiding principle in 2005, it was, from the start, on a collision course with other powerful interests and positions.

After the failure of attempts to intervene to protect citizens in recent crises at the end of the civil war in Sri Lanka, in Myanmar after Cyclone Nargis, and especially in Darfur, R2P has fallen on hard times and many observers now speak of an era of R2P pessimism. The expulsion of humanitarian agencies from Sudan in 2009—allegedly for providing information on war crimes to the International Criminal Court—or new legislation in many countries treating humanitarian agencies, particularly international NGOs, as suspect organizations, are examples of this reassertion of sovereignty at the expense of humanitarian actors' ability to intervene in crises. Even non-state actors are asserting much more "sovereign" control over activities conducted in areas they control: witness for example the virtual shutdown of humanitarian activity in south central Somalia by al-Shabaab, in the middle of the worst food security crisis in a decade. How far the pendulum will swing back toward sovereignty and away from international accountability and responsibility remains to be seen, but the trend is both clear and worrying.

THE COLLAPSE OF HUMANITARIAN SPACE

The combination of these factors has resulted in a substantially more constrained and difficult operating environment for humanitarian actors, both national and international—generally labeled the "collapse of humanitarian space" (Hammond and Vaughn-Lee 2012). Beyond the decline of the traditional principles of humanitarian impartiality and independence, the manipulation of aid by other parties, and failed attempts at R2P, new factors have recently emerged.

Security conditions for humanitarian agencies and staff have deteriorated badly (Stoddard et al. 2006), leading to increasing reliance on remote management (Egeland et al. 2011) or the increased "bunkerization" of aid agencies in situ (WFP 2010). Further, in the context of the global war on terror, legislation has been introduced that severely limits humanitarian response, to ensure that assistance does not end up benefitting terrorists or other proscribed groups. Most famously represented by so-called Office of Foreign Asset Control laws in the United States, such legislation is now common in many Western countries (Pantuliano, Mackintosh et al. 2011). These laws criminalize the transfer of resources to proscribed groups, whether accidental or intentional—whatever the motive or objective of the agency providing the assistance.

Terrorist groups themselves have also severely curtailed the actions and independence of humanitarian actors. Although it is not popular to talk about, deteriorating security, the attempts of local authorities to manipulate aid, and counterterrorism laws each played a major role in ensuring the absence of major humanitarian actors from areas of Somalia affected by famine in 2011 and in significantly slowing the response even after the famine was declared (Maxwell et al. 2012).

Despite significant challenges to humanitarian aid efforts, recent developments have included some bright spots. Improved technologies have allowed us to detect crises sooner—and to respond more quickly where the political will to do so exists. The range of response options has also widened, enabling more tailored responses. The use of cell phone technology, for example, allowed a cash response to the famine in Somalia to ramp up quickly in 2011 after the famine was declared, and it became clear that food aid was simply not an option.

New, non-Western/Northern actors have emerged and the role of local organizations in responding to humanitarian crises has expanded significantly, so that traditional humanitarian agencies have increasingly focused as much on partnership as on implementation. The existence of experienced and locally embedded organizations in Somalia in 2011 was the only means of addressing the famine after virtually all international agencies were denied access. And there were a number of nontraditional donors engaged in the Somalia crisis—many from Islamic countries, with their own implementing partners and their own coordination mechanisms.

Marginal areas have seen increased integration of local markets into national, regional, and global marketing systems, which has opened up new options—particularly cash or voucher-based transfers—for providing assistance in emergencies. This has particularly been applied in food security crises, but may be used to provide for most any kind of commodity markets can supply. This method is cheaper and faster than managing a logistical humanitarian supply chain.

While there are some bright spots, the constraints to effective humanitarian action to prevent, mitigate, and respond to food security crises have probably not been greater since World War II. Humanitarian assistance—particularly from major Western donors—is increasingly linked to political or security objectives. Counterinsurgency or counterterrorism measures put substantial constraint on operational activities, even in circumstances where the same concerns may make funding more available. But this trend certainly makes it difficult to argue that humanitarian aid is independent or impartial—regardless of its source. The lack of independence or impartiality is not lost on other belligerents in conflict situations.

Many of these factors have come together to substantially reduce the space for humanitarian action in crises, limiting the ability of the international community to intervene in crises, even when they deem it important. And the conundrum of early warning but delayed response remains as much an obstacle today as it was 20 years ago. One need look no farther for evidence of this than Somalia in 2011.

THE SOMALIA FAMINE

The United Nations declared a famine in south central Somalia on 20 July, 2011—the first time that actual famine conditions had been reported anywhere for nearly a decade, and the first time in history that a famine was declared in real time using empirical data and an agreed-upon set of thresholds.[1] There were early warnings of a crisis—as early as October, 2010—but a number of factors came together to cause the famine and to seriously compromise efforts to respond to it. The drought was certainly a cause, but the drought was merely one factor. It led to a steep drop in crop production and increased livestock mortality that substantially cut people's direct access to food and means of income at a time when food costs were increasing steeply. Somalia is dependent on imported food even in good years, and the combined impact of a production shock, falling incomes, and steeply rising prices for food from both domestic and international sources combined to turn the impact of a long running crisis into a catastrophe.

The conflict between the fledgling Transitional Federal Government (TFG) and its allies on one side and the Islamist insurgent movement al-Shabaab on the other had been ongoing since 2006, when the TFG was reinserted into Mogadishu by the Ethiopian army. Much of the affected area of the country had been in a series of low-grade, localized conflicts ever since the civil war that overthrew the Siad Barre government in 1991. Since the early 2000s, however, the conflict had taken on elements of regional power competition—with Ethiopia backing the TFG and Eritrea backing various insurgent movements, including al-Shabaab—and the global war on terror, as links between al-Shabaab and al-Qaeda became more evident and more pronounced. US drone strikes against al-Shabaab leaders and other insurgents became more frequent. Some US-based humanitarian organizations were forced out because of suspicion of complicity in the strikes and direct threats against their staff.[2]

There was no new spike in the fighting immediately prior to the famine, but the conflict, the external engagement (Ethiopia, Eritrea, the African Union, and the United States), and the collapse of humanitarian space as a result of the war were all implicated in the famine. The impact of counterterrorism laws in the United States and other donor countries complicated efforts to prevent, mitigate, and respond to the famine. The war and the lack of a central state apparatus, the drought, and the food price spike added to a long-standing crisis of livelihoods in Somalia, as well as an underlying environmental crisis. The Somalia famine was a textbook case of an acute emergency superimposed on a protracted crisis.

Despite ample early warning, there was scarcely any response beyond the usual level accorded to a protracted crisis, until the middle of July 2011, when the famine was declared. This was to some degree influenced by a long-observed tendency to simply tolerate higher indications of humanitarian crisis in

Somalia than elsewhere—a phenomenon first noted by Mark Bradbury (1998) but reconfirmed since by many others. Normally, a crisis involving a major production shock (the drought) and a price shock would have elicited a massive food aid program. The World Food Programme had withdrawn from areas controlled by al-Shabaab in 2010, however, and was not allowed back into the area during the run-up to or during the famine. WFP had been accused—both by the UN monitoring group and the BBC—of lax oversight arrangements with its transporters, so that as much as half the food aid being sent into south central Somalia was going missing (UN Monitoring Group 2010). Since much of this aid could have ended up in the hands of al-Shabaab, the United States cut off its funding. At the time, this action was explained simply and purely as compliance with Office of Foreign Assets Control (OFAC) regulations, but some observers suspected it was an attempt to undermine al-Shabaab and bolster the credibility of the TFG by making humanitarian assistance available only in TFG-controlled areas (Gettleman 2010).

By 2011, the only organization still able to negotiate a food response with al-Shabaab was the International Committee of the Red Cross. Al-Shabaab opposed the importation of food aid and not only refused WFP's return, but also eventually forced the ICRC to shut down its food pipeline. The humanitarian community did not have an adequate contingency plan for this situation, despite early warning (Darcy 2012; WFP 2012), and was forced to ramp up a major cash response quickly, amid some uncertainty about whether it would work or simply lead to more food price inflation (FEWS NET 2011; WFP 2011).

As humanitarian conditions worsened in the first half of 2011, large-scale population movement began, both internally within Somalia and across borders to Kenya and Ethiopia. Al-Shabaab restricted some of this movement and in some cases forced displaced people to return to their places of origin prior to the short rains in November, in order to increase farming activity and agricultural recovery.

Due to the restrictions of al-Shabaab, almost all of the response was managed remotely from Nairobi, with only a limited number of Somali staff of international organizations and the staff of a number of Somali NGOs implementing actual interventions on the ground. The limited access that some international agencies did have declined steadily throughout late 2011 and early 2012, as al-Shabaab closed down their operations. But other factors, specifically the Hawala banking system and a well-functioning cell phone network, enabled the cash transfer program to function, and indeed to scale up to assist over a million households. Global prices fortunately declined after mid-2011, and the cash transfer program did not appear to cause further food price inflation (Cash/Voucher Monitoring Group 2012).

OFAC and other counterterrorism laws were at least partially behind the withdrawal of WFP in 2010 and made it increasingly difficult to engage in areas controlled by al-Shabaab throughout 2011and 2012. The withdrawal of WFP in 2010 did not have an immediate impact on food security, because El Niño rains that year resulted in one of the most bounteous harvests in south central Somalia in recent times. But the predictable La Niña effect followed, indicating drought for the Horn of Africa. Late 2010 and early 2011 were no exception. After the famine was declared, The United States Agency for International Development (USAID) obtained a license from OFAC for some humanitarian activities, but the activities covered by the license remained vaguely defined, and in any case, did not cover prevention and mitigation activities, nor did the license apply to the period prior to the declaration. This partly explained the delayed response, but other factors were important as well, most notably a reluctance to commit major additional resources until there was incontrovertible evidence of the humanitarian crisis—in other words, until there had been a spike in mortality. That evidence was graphically provided by the declaration, but waiting until the declaration was, once again, waiting until it was too late.

Once the famine was declared, a major response ramped up, and ramped up quite quickly given the constraints. But for the Somali population caught in the famine, it was not quickly enough. Although a retrospective assessment of total mortality in the famine has not yet been conducted, the loss of life from the famine is certainly going be in the multiple tens of thousands—perhaps as many as one hundred thousand. The Somalia famine touched on all four pillars of food security: a major production shock resulting from the drought; a food access shock from deteriorating income possibilities and steep food price inflation; a malnutrition crisis from access and utilization constraints; and overall stability of food sources undermined by the conflict, the underlying livelihoods crisis, the collapse of the state, the extreme restrictions on humanitarian access, and by political actors—both internal and external—with priorities other than preventing humanitarian catastrophes.

IMPROVING FOOD SECURITY AND POLITICAL STABILITY IN HUMANITARIAN CRISES

Several conclusions emerge from this discussion. First, the causal relationship between food security and political stability in humanitarian emergencies is complex and difficult to generalize: food insecurity can be caused by conflict or political instability, and political instability can be caused by food insecurity. Militarized conflict—an extreme form of political instability—has clearly been a major driver of humanitarian emergencies, particularly since the end of the Cold War, with high levels of food insecurity a common consequence of these emergencies. At the same time, many of the local drivers of conflict have been related to control over land and other natural resources, which are ultimately linked to people's livelihoods—and therefore to their food security (Alinovi et al. 2008). The relationship between the independent and dependent variables can be understood in any given context, but it is circular and iterative, not linear. At times, there is no demonstrable relationship between the two. Substantial levels of food insecurity can exist without there being any driver related to political instability and without necessarily causing major political instability—so a certain amount of caution is justified regarding any general theory of the link between the two.

Second, there are some common drivers of both political instability and food insecurity. Climate change is at least partly implicated for both in the Darfur conflict, for example as rainfall patterns changed, nomadic camel herders had to migrate farther and farther southward to find dry season grazing and water, which brought them increasingly into conflict with other ethnic and livelihoods groups and made the lack of a designated homeland or "Dar" for the nomadic groups more evident (Young et al. 2005). Needless to say, the Darfur conflict was quickly politicized by other actors—predominantly the ruling party in Khartoum—for their own purposes, so it would be wrong to blame the Darfur crisis predominantly on climate change. Nevertheless, it likely played a crucial underlying role. Increasing frequency of drought and climate variability is equally implicated in food security crises elsewhere.

It seems unlikely that there will be any change in the foreseeable future in a number of drivers of food insecurity: short-term weather impacts and medium-term climate change impacts, the volatility of global and local food prices, or the number of localized conflicts ripe for manipulation. We can conclude that the number of localized food security crises is unlikely to decrease. This reality has major implications for both humanitarian preparedness and response, and for policymakers worried more broadly about the implications for political stability. The social protection responses rolled out on a national scale in Ethiopia and piloted in a number of other countries have certainly made progress in providing a safety net, but the jury is still out on whether such programs actually offer a broadly accessible "ladder" out of poverty and chronic food insecurity. Substantially more resources will probably be required to achieve the latter objective at scale.

Third, the relationship between humanitarian actors and policymakers or actors more primarily focused on political stability is problematic. I have argued that that causation can run both ways, so presumably the policy-making logic can, too: that is, achieving food security can be a means to the end of political stability, but the evidence suggests that political stability is also important for achieving food security. I have also argued that nearly all parties in conflicts, protracted crises, and food security emergencies are all too happy to manipulate humanitarian assistance to achieve other goals—whether economic, political, territorial, or security goals—and that these mixed motives have severely undermined humanitarian objectives and access. This has long been the rationale for separating humanitarian response from partisan political objectives, as suggested by numerous international agreements, both formal (international humanitarian law) and informal (humanitarian principles).

The far more serious question is not whether improved food security is generally good for political stability—most any reasonable person, humanitarian or otherwise, could agree to such a proposition—but rather whether, under certain circumstances, a judicious amount of short-term food insecurity is a useful tool for promoting longer-term stability or other political or security objectives. There is no doubt that humanitarian assistance has been restricted, channeled, and in other ways manipulated to favor one side in conflicts. Indeed doing so now clearly seems to be part and parcel of US policy, at least in certain countries. There is little doubt that the response to the Pakistan displacement crisis in 2009—which included major food security elements—was deliberately channeled to strengthen the Pakistani government and weaken the Taliban, without evidence of any concern for the humanitarian consequences of doing so (Péchayre 2012).

There were earlier indications that US policy in Somalia was devoted to using humanitarian assistance, at least in part in an attempt to bolster the political credibility of the TFG, without regard to the fact that needs were demonstrably higher outside of the tiny areas controlled by the TFG. Given the tragic delay in responding to the 2011 crisis in Somalia as it worsened in the spring and early summer of 2011, one cannot help but wonder if, at some level, policymakers thought that some amount of a humanitarian crisis in al-Shabaab controlled areas might actually be a good strategy to undermine al-Shabaab—and hence a good strategy toward longer-term political stability in the Horn of Africa. Although the evidence is circumstantial, this interpretation is consistent with earlier policy, and one of few coherent explanations for the poor response until a full-blown famine was declared.

In the context of the kinds of crisis likely to be experienced over the coming decade, some kind of policy decision has to be made that allows for policies and interventions to promote stability and food security, while not victimizing the people most vulnerable to these crises. Reconciling these two priorities will not be an easy task. First, a top priority is ensuring some amount of space for needs-based humanitarian action—certainly response and presumably prevention—that is distinct from political interventions in situations of extreme political instability. Second, all parties in a crisis situation must be held accountable for the consequences of their actions, including slow or inadequate responses. Finally, nations must seek greater consensus to prevent the instrumental manipulation of humanitarian assistance for non-humanitarian purposes. The hard evidence is that the schemers rarely achieve their purposes in such circumstances, yet the consequences for people caught in crisis are dramatically worsened, and the dampening of humanitarian objectives only legitimizes the theft, diversion, and other abuses. These observations alone should convince US policymakers to prioritize the reduction of risk and protection of human life and dignity in crises over fleeting or illusive short-term advantages in crises of both food insecurity and political instability.

ACKNOWLEDGMENTS

The author would like to acknowledge feedback on earlier drafts of this paper by Peter Walker and Antonio Donini, and also feedback from participants at the workshop on "Food Security and Political Instability" held at Cornell University on June 18–19, 2012. All errors or omissions are solely the responsibility of the author

NOTES

1. The materials for this section are mostly drawn from an upcoming special edition of the journal *Global Food Security* (Maxwell et al. 2012).
2. The agencies denied any involvement in military strikes, and no evidence has ever emerged to suggest that they were. But the suspicions about US-based agencies were so high that continued presence was impossible.

REFERENCES

Alinovi, L., G. Hemrich, and L. Russo. 2008. *Beyond relief: Food security in protracted crises.* Rugby, UK: Practical Action Publishing.

Bailey, R. 2012. *Deciding to delay? Famine early warning and early action.* London: Chatham House.

Bradbury, M. 1998. Normalising the crisis in Africa. *Disasters.* 22(4): 328–338. Brigety II, R. 2008. *Humanity as a weapon of war: Sustainable security and the role of the US military.* Washington, DC: Center for American Progress.

Buchanan-Smith, M., and S. Davies. 1995. *Famine early warning systems and response: The missing link.* London: IT Publications.

Cash/Voucher Monitoring Group. 2012. *Phase I: Final report.* Nairobi: CVMG.

Checchi, F., and L. Roberts. 2008. Documenting mortality in crises: What keeps us from doing better? *PLoS Medicine* 5(7 e146): 1025–32.

Darcy, J. 2012. *IASC Real time evaluation: Somalia drought crisis response 2010–12.* New York: Inter-Agency Standing Committee.

——, and C. A. Hofmann. 2003. *According to need? Needs assessment and decision-making in the humanitarian sector (Vol. HPG Report 15).* London: Overseas Development Institute, Humanitarian Policy Group.

Development Initiatives. 2011. *Global humanitarian assistance report 2011.* Wells UK: Development Initiatives.

Donini, A., L. Fast, G. Hansen, S. Harris, L. Minear, T. Mowjee, and A. Wilder. 2008. *The state of the humanitarian enterprise. Humanitarian agenda 2015: Final report.* Medford, MA: Feinstein International Center, Tufts University.

Egeland, J., A. Harmer, and A. Stoddard. 2011. *To stay and deliver: Good practice for humanitarians in complex security environments.* New York: OCHA.

EMDAT-CRED. 2012. The international disaster database. Retrieved May 2012.

FAO/WFP. 2008. *The state of food insecurity in the world 2008: High food prices and food security-threats and opportunities.* Rome: FAO/WFP.

———. 2010. *The state of food insecurity 2010: Food insecurity in protracted crises.* Rome: FAO.

Fast, L. A. 2010. Mind the gap: Documenting and explaining violence against aid workers. *European Journal of International Relations 16*(3): 365–89.

Feinstein International Center. 2010. *Humanitarian horizons: A practitioner's guide to the future.* Medford: Tufts University.

FEWS NET. 2011. *Markets functioning in Southern Somalia.* Nairobi: FEWS NET.

Gettleman, J. 2010. U.N. to End Some Deals for Food to Somalia. *New York Times*, March 12.

Hammond, L., and H. Vaughn-Lee. 2012. Humanitarian space in Somalia: A scarce commodity. HPG Working Paper. London: Overseas Development Institute.

Howell, J., and J. Lind. 2009. *Civil society under strain: Counter-terrorism policy, civil society and aid post-9/11.* Sterling, UK: Kumarian Press.

Humanitarian Policy Group. 2006. *Saving lives through livelihoods: Critical gaps in the response to the drought in the Greater Horn of Africa. HPG Briefing Note.* London: Overseas Development Institute.

ICISS. 2001. *The responsibility to protect: Report of the International Commission on Intervention and State Sovereignty.* New York: International Commission on Intervention and State Sovereignty.

Lautze, S., and D. Maxwell. 2006. Why do famines persist in the Horn of Africa? Ethiopia, 1999–2003. In *The "New Famines": Why famines persist in an era of globalization*, ed. S. Devereux, 222–44. London: Routledge.

Lentz. E, S. Passarelli, and C. Barrett. Forthcoming. The timeliness and cost effectiveness of the local and regional procurement of food aid. World Development Special Edition on Impacts of Innovative Food Assistance Instruments.

Levine, S., and C. Chastre. 2004. Missing the point: An analysis of food security interventions in the Great Lakes. HPG Network Paper 47. London: Overseas Development Institute, Humanitarian Policy Group.

Maxwell, D. 2012. Those with guns never go hungry: The instrumental manipulation of food assistance in conflict. In *The Golden Fleece: Manipulation and Independence in Humanitarian Action,* ed. A. Donini, 197–218. Sterling, VA: Kumarian Press.

———, N. Haan, K. Gelsdorf, and D. Dawe. 2012. The Somalia famine of 2011. *Global Food Security*. Special edition.

———, and K. Sadler. 2011. Responding to food insecurity and malnutrition in crises. In *World Disaster Report 2011*, 124–45. Geneva: International Federation of Red Cross and Red Crescent Societies.

———, P. Webb, J. Coates, and J. Wirth. 2010. Fit for purpose? Rethinking food security responses in protracted humanitarian crises. *Food Policy, 35*(2): 91–7.

Office for the Coordination of Humanitarian Affairs (OCHA). Various Years. Overview of the Consolidated Appeal Process. New York: OCHA.

Olsen, G. R., N. Carstensen, K. Høyen. 2003. Humanitarian crises: What determines the level of emergency assistance? Media coverage, donor interests and the aid business. *Disasters 27*(2): 109–26.

Pantuliano, S., M. Buchanan-Smith, V. Metcalfe, S. Pavanello, and E. Martin. 2011. City limits: Urbanisation and vulnerability in Sudan: Synthesis report.

Pantuliano, S., K. Mackintosh, S. Elhawary, and V. Metcalfe. 2011. *Counter-terrorism and humanitarian action HPG Policy Briefs 43.* London: Overseas Development Institute.

Péchayre, M. 2012. Politics, rhetoric and practice of humanitarian action in Pakistan. In *The Golden fleece: Manipulation and independence in humanitarian action,* ed. A. Donini, 149–170. Sterling, UK: Kumarian Press.

Stoddard, A., A. Harmer, and K. Haver. 2006. Providing aid in insecure environments: Trends in policy and operations. HPG Briefing Paper No. 24. London: Overseas Development Institute, Humanitarian Policy Group.

UN Monitoring Group. 2010. Report of the Monitoring Group on Somalia pursuant to Security Council resolution 1853 (2008). UN Security Council (S/2010/91). New York: United Nations.

Violette, W, A. Harou, J. Upton, S. Bell, C. Barrett, M. Gómez, and E. Lentz. Forthcoming. Recipient satisfaction with locally procured food aid rations: Comparative evidence from a three country matched survey. *World Development* Special Edition on Impacts of innovative food assistance instruments.

Walker, P., and D. G. Maxwell. 2008. *Shaping the humanitarian world (Global Institutions).* London: Routledge.

WFP. 2010. Humanitarian assistance in conflict and complex emergencies: June 2009. Conference report and background papers. Rome: WFP.

———. 2011. *Food supply and market situation in Southern Somalia.* Rome: World Food Programme.

———. 2012. *Somalia: An evaluation of WFP's portfolio.* Rome: WFP Office of Evaluation.

Wilder, A. and S. Gordon. 2009. Money can't buy America love. *Foreign Policy,* December 1.

Young, H., A. M. Osman, Y. Aklilu, R. Dale, B. Badri, and A. J. A. Fuddle. 2005. *Darfur: Livelihoods under siege.* Medford, MA: Feinstein International Famine Center, Tufts University.

12

Moral Economies of Food Security and Protest in Latin America

WENDY WOLFORD AND RYAN NEHRING

Protests over food insecurity—often referred to as food riots—are not new, nor are they usually primarily about food. Protests over access to food happen when traditional institutions that govern food consumption and production fail or are broken. Such norms, rules, and regulations have been referred to in the literature as the "moral economies" produced through customary land and labor relationships. First outlined in 1971 by the British Marxist historian E. P. Thompson, the term "moral economy" refers to the collective reasoning particular groups of people appeal to as a way of justifying, defining, and defending the correct organization of society, including how society's productive resources like food and land ought to be distributed (Scott 1976; Watts 1983). These moral reasonings are produced through and embedded in historically and culturally specific social relations. As such, they are always in formation, but they are cast into sharp relief when a social group's economic or political position is challenged or when the productive resource is seen as dangerously scarce (Thompson 1993). Indeed, as Thompson pointed out, riots over high food prices in England during the 1700s were not "a 'natural' or 'obvious' response to hunger but a sophisticated pattern of collective behavior, a collective alternative to individualistic and familial strategies of survival" (266).

Across history, people have reacted with food riots to new marketing practices they saw as unfair, or a violation of traditional expectations. Likewise, protests around high food prices in Latin America are not necessarily about the prices themselves, which remain historically low when viewed over the past 40 years (World Bank 2012), nor are they about scarce resources, as Latin America has relatively abundant arable land and is a net food exporter. Rather, protests are about expectations, reflecting the abrupt nature of change in prices and a collective sense of social injustice. Millions of people in Latin America are hungry at any given moment, but only a small fraction protest; the way in which they do so is shaped by historically situated norms of justice and, increasingly, rights. The concept of a moral economy is thus a useful one for understanding mobilization around food as it moves analysis beyond simple calculations of supply and demand to include political histories, relationships between the state and society, and the crucial role of expectations.

Unfortunately, attention to moral economies has not factored into the prominent analyses of food insecurity by top international organizations. Since 2003, the UN World Food Programme Vulnerability Analysis and Mapping Unit has conducted more than 80 comprehensive food security and vulnerability assessments worldwide.[1] Vulnerability in this context is "determined by [people's] ability to cope with their exposure to the risk posed by shocks such as droughts, floods, crop blight or infestation, economic fluctuations, and conflict. This ability is determined largely by household and community characteristics, most notably a household's or community's asset base and the livelihood and food security strategies it pursues" (WFP 2009, 27–28). While this is a useful measure of livelihood vulnerability, it describes only the objective conditions of insecurity and does not help to explain protests over access to food, which turn on perceptions, expectations, norms, and collective organization—in other words, on moral economies.

Because protests are embedded in localized moral economies, it is important to understand the social dynamics—particularly regarding production, consumption, and distribution of food—within particular communities and in relation to particular "protest events." In this chapter, we distinguish between two types of sociopolitical protest: spontaneous mobilization and longer-term sustained mobilization. These two types of conflict may manifest in similar ways as public demonstrations, but they often derive from different conditions and have different effects. We suggest that in Latin America today, spontaneous food protests are more likely to occur on the consumption end (due to high or rapidly changing prices), while sustained, organized protests are more likely to occur on the production end in relation to access to resources, particularly among the landless rural poor or historically marginalized groups such as indigenous peoples. Small farmers in Latin America are net food buyers, so while producers are also consumers, production-related protests throughout Latin America focus on the nature of markets for agricultural goods, inequalities in land ownership, and the weak presence or capacity of the state.

Spontaneous mobilization and sustained mobilization are clearly linked and in some cases become indistinguishable or overlap completely. Both occur when moral economies of access and use are violated, but spontaneous mobilizations involve communities that may be only loosely linked—individuals who may share a common sense of social justice, but do not have other connections. Almost by definition, markets are relatively anonymous places—coordinated by the so-called invisible hand—that bring together a broad range of individuals and interests. When high food prices, combined with extreme levels of poverty and food insecurity, threatened rural and urban consumers in Haiti, Honduras, and Bolivia in 2008, widespread spontaneous protests ensued.

Sustained mobilization, on the other hand, requires organized actors who build on collective ideals and for whom mobilization constitutes a ready piece of political culture. This explains the high level of sustained mobilization around access to food in Latin America, as there is perhaps no other region of the world as politically organized and motivated as rural Latin America. The so-called Pink Tide of left-wing governments voted in over the past ten years may have been a disappointment to people hoping for more radical structural change, but it has brought a widespread move to incorporate the demands of civil society within national policy and visions. The result might be characterized as a hybrid of neoliberalism and neo-developmentalism: strong-hand populism promoting rapid, extractive development with inclusion and a social dialogue. Across the region, there has been a push to dramatically rethink development, "order" national territories, reallocate factors of production (including people) to the most efficient sites, and develop better governance institutions, such as property registries and the rule of law. Ensuing tensions between inclusion and extraction and between populism and authoritarianism are shaping new political dynamics across the region. The promise of inclusion combined with a rejection of the more radical hoped-for changes is a dynamic that has fueled social and national revolutions throughout history (Skocpol 1979), and protests of this sort are most likely in countries with organized civil society actors and middle peasantries such as Brazil, Argentina, Mexico, Peru, Ecuador, and Bolivia.

Protest can have destabilizing effects, but can also be highly productive. One of the key reasons for Latin America's progress in reducing hunger over the past 10 to 15 years has been the strength of civil society and the high level of social mobilization. Across the region, social movements have demonstrated, marched, occupied public and private land, and forced political officials to focus on issues of inequality, distribution, and social justice. Social mobilization—which may look a lot like political instability—can be effective and positive for reducing hunger, poverty, and inequality. The organization of nonstate political actors—such as trade unions, social movements, indigenous confederations, and nongovernmental organizations—may traditionally have been seen as politically threatening (O'Donnell and Schmitter 1986), but these social actors have been increasingly incorporated into mainstream politics in Latin America and today have an instrumental voice in policy choice and implementation. Recent regionwide studies have found that the most effective governance derives from collaboration between state and society actors where both demonstrate technical competence and civil society retains the ability to step outside the partnership and protest irregularities or inequities

(Dagnino 2003). Civil society actors provide marginalized or insecure populations with a political voice. They can also serve a positive governance function by giving the government and international community ears on the ground for information and communication, and legs on the ground to implement political changes or policies (Keck and Abers 2006). On the whole, civil society is stronger with less inequality and more opportunities for participation in governance.

In this chapter, we describe the factors currently contributing to or shaping food insecurity in Latin America, highlighting important new mobilizations around access to land and food with a specific focus on Brazil. Brazil is unique in having abundant access to land, water, and cheap labor, as well as both a highly organized rural social movement (the Movement of Rural Landless Workers) that has won considerable political space over the past 15 years, and a highly organized rural elite that represents a profitable agroindustrial export sector. These two different groups are represented by two separate ministries for rural issues—the Ministry of Agrarian Development and the Ministry of Agriculture. The steps Brazil has taken to address food insecurity suggest three key lessons: the importance of bringing together production, consumption, and distribution; the importance of government legislation and spending in areas of social assistance, minimum wage policy, and agricultural research; and the value of a highly mobilized civil society that can push for policy change and social awareness around access issues.

LATIN AMERICA: THE CONTEMPORARY CONJUNCTURE

Throughout Latin America, significant progress has been made in decreasing the prevalence of hunger. With the exception of Haiti, Bolivia, the Dominican Republic, Guatemala, and Antigua and Barbuda, which are classified as "serious" in hunger severity, the International Food Policy Research Institute (IFPRI) classifies all of Latin America and the Caribbean as "low" or "moderate" in the severity of hunger (IFPRI et al. 2011, 18). Brazil has reduced hunger significantly over the past 15 years, cutting their Global Hunger Index score in half from 1990 to 2011 (IFPRI et al. 2011, 13).

In spite of the progress made in expanding access to food, food insecurity in Latin America is still an issue. One third of Latin Americans live below the poverty line and approximately 13 percent of the regional population lacks adequate income to meet their nutritional needs (Martínez et al. 2009, 10). These trends are particularly significant in light of the increase in agricultural productivity over the past 50 years (World Bank 2011). Between 2000 and 2008, Latin America's agricultural sector expanded by 39 percent (World Bank

2011), so that the region now has the most favorable food trade balance in the world. During this same period, however, rural poverty across the region fell by only 12 percent, leaving more than 50 percent of the rural population still in poverty (FAO 2010, 3; see Figure 1).[2]

Due to a combination of factors from speculative commodity investments to increased energy costs, expanding biofuels production, and reduced agricultural productivity due to erratic weather events (IFPRI et al. 2011; UNDESA 2008), countries across Latin America experienced spikes in food prices, food shortages, and increased poverty during the global food and economic crises of 2008 (Heidrich and Williams 2011). Researchers from the Latin American Trade Network estimate a 30 percent increase in the food price index for Latin American consumers between 2006 and 2008 (Heidrich and Williams 2011, 3), resulting in a 14 percent increase in food insecure people (Shapouri et al. 2009, 15). While the short-term causes attributed to increasing and volatile food prices are critical in understanding and responding to food insecurity in Latin America, it is also important to consider the historical events and processes that have conditioned changes in agricultural production, access to land, and the economic vitality of rural regions (IAASTD and McIntyre 2009; Rocha 2009, 53).

Since the 1950s, Latin American agricultural development has been shaped by demands for foreign exchange and rapid intensification, with rural land and labor seen as underemployed resources for industrial, urban modernization. Policymakers argued that agriculture required technological and operational scaling up with a shift from subsistence production to commodities and integration into the international economy (Byerlee et al. n.d.; Staatz and Eicher 2006). While the 1970s saw economic growth in several major economies in the region, the 1980s was a decade of economic instability and collapse. Structural adjustment policies imposed as conditions of international aid reworked and reduced state support for agriculture. The last two decades have seen an increase in competitive pressures, rising land prices, and increased inequality in land ownership (de Janvry and Sadoulet 2002).

Thus, the Latin American paradox of plentiful food production and food insecurity is a product of uneven development. Growth in agricultural production in Latin America is concentrated in specific regions where medium- to large-scale producers with access to export markets specialize in export crops (da Silva et al. 2010, 10); elsewhere, rural populations that lack access to public and private services and infrastructure continue to suffer from high rates of poverty and food insecurity (Trivelli et al. 2009).

One of the key manifestations of this uneven development is rapid rural–urban migration. Latin America as a whole has a relatively low percentage of rural residents, around 23 percent, but more than 60 percent of those considered poor live in rural regions (Robles and Torero 2010, 117). Even in Brazil, where the proportion of rural inhabitants is lower than the regional average, half of the rural population experiences food insecurity, with 20 percent of that group suffering from severe food insecurity (da Silva 2009, 367).

The poverty in rural areas is linked to the rise of urban poverty and food insecurity. The decline of cheap production for local markets (the purview of small farmers) hurts the poorest urban consumers. Recent spikes in food prices revealed the vulnerability of the urban poor, particularly in countries that depend on food imports (IFAD 2010). Food price increases have a disproportionately larger effect on household incomes of the urban poor. While rural households can utilize subsistence production to at least partially mitigate food price volatility, urban populations are at the whim of the commercial market. For example, in Lima, Peru, the urban population purchases over 90 percent of its food from the market (Cohen and Garrett 2009, 4), and a 2007 IFPRI study found that the urban poor in Nicaragua spend on average 52.4 percent of their household income to purchase food, which increasingly comes from large supermarkets integrated with the global food supply chain. In some of Latin America's biggest economies (Argentina, Colombia, Brazil, Mexico), supermarkets constitute 45 to 75 percent of the market share in food retailing (Reardon and Berdegué 2002). The combination of poverty and reliance on commercial food markets makes it difficult for the urban poor to mitigate food price shocks. In 2008 as prices climbed for staple goods, urban riots occurred in Honduras, Bolivia, and Haiti. In Haiti, the pressure was severe enough that the government fell in April of that year.

Regional and local inequalities, sparse access to social assistance networks, and informal labor can also challenge families' food security. Historically marginalized groups such as women and indigenous peoples face particular barriers in fulfilling their nutritional needs. In Latin America, poverty rates among indigenous populations are higher and fall more slowly. Access to social services such as health care and education is traditionally lower among women and indigenous populations. All of these factors can have detrimental impacts on the physical and cognitive development of children. Without active support for marginalized groups (and particularly their children) poverty and hunger traps can be expected to persist. It is in this context that we can view food insecurity among marginalized populations as more than simply lack of food and income.

ENVIRONMENTAL TRENDS SHAPING FOOD SECURITY IN LATIN AMERICA

The contemporary moment in Latin America is shaped by new environmental concerns that are reworking land and resource access throughout the region. Three specific environmental dynamics are important for agriculture in the region: temperature increases are projected to be higher in South America than the global average, with a concentration on the high Andes and the Amazon region; precipitation variability is expected to be erratic and could affect generational knowledge on seasonal crop rotating and spark unanticipated floods or droughts; and extreme weather events are more difficult to mitigate and have the potential to affect large tracts of agriculture (Verner 2011, 2). Each climatic event has the potential to reshape both economic and social forces that govern regional and global food security. Industrial agriculture produces high yields for export but is susceptible to whole-crop or generalized losses in the face of erratic climatic scenarios (Altieri et al. 2012). The expansion of soy and sugar cane for many Latin American countries thus represents an important source of external revenue, but not necessarily a sustainable or dependable one.

Scarce energy resources and a push to reduce greenhouse gas emissions have pressured governments into supporting biofuel production to meet domestic and international energy demands. There are indications that global biofuel production is causing abrupt increases in the demand for agricultural commodities and is restructuring agro-energy price linkages. Energy prices are now being transmitted into agricultural commodities markets resulting in a twofold effect: price increases in land and productive inputs such as fertilizers, transport, mechanization, and processing, paired with increased state intervention in agricultural subsidies and price controls to match ethanol prices with the crude oil market (Naylor et al. 2007, 33). Consequently, global commodity prices have skyrocketed for food crops that are also used for ethanol production, such as sugar cane, maize, cassava, palm oil, soy, and sorghum. Globally, those same staples comprise about 30 percent of mean caloric intake by people living in chronic food insecurity (Naylor et al. 2007, 41). This represents a new phenomenon increasingly linked to food security and potential conflict. New lands are being transformed into producing energy and food for export with little attention to the social relations based around agricultural production and consumption patterns.

The contradiction between food security and adaptation to environmental change is illustrated vividly by the Sustainable Rural Cities program, initiated in September 2009 as part of a governors' agreement between Arnold Schwarzenegger, then-governor of California, and the governors of the Mexican state of Chiapas and the Brazilian state of Acre. Part of a new public policy campaign to generate Reducing Emissions from Deforestation and Forest Degradation (REDD) credits and improve rural security and sustainability, the president of Mexico, Felipe Calderon, announced a plan to remove forest inhabitants from the Lacandon Forest in southern Mexico and create at least 25 concentrated settlements—the so-called sustainable rural cities. The governor of Chiapas, Juan Sabines, claims that these cities are a sustainable means of alleviating poverty, mitigating flood risk, and combating climate change, but conflict has mounted as local indigenous residents call on traditional moral economies to protest the forced removals that result in concomitant loss of access to forest resources and cultural cohesion (Conant 2012).

RESPONSES TO FOOD INSECURITY

Responses to the challenges of food insecurity have been shaped by localized and regional moral economies around access and distribution. Civil society groups that emerged in the wake of authoritarian governments in the 1980s have situated food justice issues within older and broader narratives of dispossession, inequality, and tradition. New frames of the "right to food" and "food sovereignty" resonate with strong peasant traditions and communities that now have political access at multiple scales. As a result, Latin American countries have implemented a host of economic and social policies to mitigate the effects of hunger and malnutrition, including reductions in consumption taxes, price controls, import tariffs, export controls, extension of microcredit to producers, food procurement, input subsidies, conditional cash transfers (CCTs), school feeding programs, food for work programs, and construction of regional trade blocs (SELA 2010; Martinez et al. 2009). The governments of Ecuador,[3] Bolivia,[4] Brazil, and Venezuela all have legislation defining food as a right. The 2008 Ecuadorian Constitution goes further with the concept of *buen vivir*—or the right to a good life within the laws of nature. It is perhaps no surprise, however, that the ambitious *buen vivir* project has proven difficult to implement. Despite incorporating indigenous moral economies into the Constitution, the government has been unable or unwilling to fulfill promises of resource distribution as defined by existing mobilized civil society actors (Churuchumbi 2012). Significant large-scale and well-organized riots in both countries have ensued, highlighting the difficulties, and perhaps the inadequacy, of forming an effective state–society relationship based on citizen rights to nature.

In general, social movements have played a considerable role in reframing the debates around food insecurity, land tenure systems, and economic organization (Wolford 2010; IAASTD and McIntyre 2009). La Via Campesina and the Movimento dos Trabalhadores Sem Terra (MST), or Landless Workers Movement, serve as important examples. Their responses have been crucial in addressing the impacts of the food crisis across the region: indeed, the countries that have experienced the greatest decline in the proportion of people suffering from hunger in the last decade have been those that have actively employed food security policies and worked with social movements. In this section, we highlight some of the key policies in place across the region and then take a closer look at Brazil's response to hunger and food insecurity. Brazil stands out as an exemplary case in Latin America, achieving one of the lowest rates of hunger in the region by decreasing hunger from 10 percent in 1990 to 6 percent in 2006 (SELA 2010, 11).

Economic Policies

Economic policies ranging from taxes and customs duties, price controls, release of stocks, production support through credit extensions, market management through price controls, import tariffs, and export controls have been implemented variously in Bolivia, El Salvador, Brazil, Colombia, Ecuador, Peru, Costa Rica, Guatemala, Honduras, Nicaragua, Argentina, Paraguay, Uruguay, Venezuela, Mexico, Chile, and the Dominican Republic (World Bank 2008). Some nations have used tariff reductions to increase imports, while others have imposed export bans and taxes to promote domestic food markets. In response to the 2008 food crisis, countries implemented a range of monetary and fiscal policies with supply-side stimulus, nutrition, and other food programs (Cuesta and Jaramillo 2010, 21–22). In the Andean region, Venezuela and Bolivia implemented contractionary monetary policies, while Venezuela, Bolivia, and Ecuador employed expansionary policies (Cuesta and Jaramillo 2010, 19). Bolivia, for instance, combined contractive monetary and expansionary fiscal policies with supply side stimulus including reducing tariffs, implementing export bans, and subsidizing inputs, while implementing a school meals program and a Zero Malnutrition (*Desnutrición Cero*) program to combat hunger. The most common policy used in Latin American countries to increase production, according to Piñeiro et al. (2010), was reducing producer taxes. Production subsidies, especially for grain, as well as input subsidies (for fertilizer and seed, for example) are commonly used to promote agricultural growth. Although costly, these policies boost production in the short to medium term.

SOCIAL PROTECTION

Among the most effective institutional responses to food insecurity have been direct and indirect assistance to meet food security needs. In many countries, social movement actors have fought to define state support as a right, not a privilege, and expanded noncontributory social assistance policies are part of a mobilized moral economy that includes CCT programs to give the poor access to the market, targeted consumer price subsidies, food distribution, school food programs, and nutritional education programs. Targeted subsidies such as CCTs have proven to be particularly effective for five reasons, according to Santiago Levy (2008): they "1) directly increase the purchasing power of the poor; 2) allow households to adapt to relative price changes; 3) do not reduce the income of poor food sellers; 4) diversify diet and prevent a decrease in food spending; and 5) limit the extent of the support because these policies have clear exit strategies and are clearly presented as such" (Levy 2008; cited in Piñeiro et al. 2010, 6). Numerous examples attest to the success of CCTs. CCTs have demonstrated success in increasing food consumption, nutrition, and education, and reducing child labor. While cash transfers have an impact on decreasing food insecurity, it is also important to recognize their limitations. Such policies have not contributed to food sovereignty due to their market-oriented approach. Cash transfers effectively supply the poor with access to food through the market, but they do not address issues of land access and distribution and the ability of small-scale farmers to subsist off the land.

In addition to cash transfer programs, nationalized grocery stores and soup kitchens have alleviated hunger among some of the poorest in Latin America. In Venezuela, the government established Mercal, a government-run chain of supermarkets created to offer low-cost food to consumers. School feeding programs have also buffered the negative effects of food insecurity. According to the World Food Programme, school feeding programs reach approximately 80 million students in Latin America each day (see the link at http://www.wfp.org/content/school-feeding-capacity-development-project-latin-america-and-caribbean-region). School feeding programs benefit children by "enhancing enrollment and reducing absenteeism" and contribute to students' learning by "avoiding hunger and enhancing cognitive abilities," as well as improving children's nutrition (Bundy et al. 2009, 20). Recent advances in school feeding programs have started to address the production side. In Brazil, school feeding programs must procure a minimum of 30 percent of supplies from local family farms (Ministério de Desenvolvimento Social e Combate à Fome 2012). School feeding programs can provide an ideal avenue for increasing market access for family farmers and supporting more diverse and organized production built around local mobilization and participation between producers and municipal officials.

In conjunction with national efforts, regional blocs have been established to tackle the challenges of hunger and food insecurity. Spawned by the 2008 food crisis, increased attention to questions of food security have emerged in intergovernmental initiatives in Latin America, including the construction of food sovereignty regional alliances and the development of a regional health strategy in Central America and an agreement to implement cooperative programs for food sovereignty and security among participant countries in the Bolivarian Alternative for the Peoples of Our Americas (Piñeiro et al. 2010). Groups such as the Bolivarian Alliance for the Americas (ALBA) and the Andean Community have constructed regional guidelines for food security and food sovereignty, where the former is defined as "sufficient and stable food availability and the timely and permanent access to it by our people", and the latter as "the right of the peoples, their countries, or state associations to define food policies with third-party nations, leaving behind disloyal international trade policies and avoiding the production of food for generating fuel."[5] These multilateral agreements are designed to work in conjunction with national strategies—focused on building institutional capacity in the agricultural and livestock sectors (SELA 2010, 14).

BRAZIL: A REGIONAL LEADER IN FOOD SECURITY

Brazil has become a leader in addressing food security due to the simultaneous strength and unevenness of its large-scale agriculture and the highly visible and effective mobilization of its grassroots social movements. In the last 20 years, Brazilian agricultural production has increased dramatically in area, volume, and factor efficiency, with gains made primarily in large-scale export crops such as soy and sugar cane. In comparison, family farm production (mainly for the domestic market) has grown steadily, but more modestly (IBGE 2010). This inequality arose through centuries of unequal access to the most basic production resource—land. Brazil had and continues to have one of the most unequal distributions of land ownership in the world. Even after 15 years of progressive land redistribution (where large public and private properties have been expropriated by the state or purchased through publicly-supported market mechanisms to benefit the rural poor), farms under 50 hectares constitute 82 percent of the total but only occupy 13 percent of the total arable land, while farms over 500 hectares make up only 2 percent of the total but cover 56 percent of arable land (IBGE 2006). Despite these inequities, Brazilian family farms comprise over 75 percent of agricultural employment and supply more than 70 percent of the domestic food consumption, while only receiving 25 percent of total agricultural credit (CAISAN 2011, 16). Economic growth and a rising middle class in Brazil is increasing the nation's overall food demand, as well as demand for a more diverse diet. Family farmers are tailored to fill such a demand as they produce more diverse, higher-value consumption crops, with a higher yield per acre than industrial agriculture (Boyce et al. 2005). With limited productive resources and a continual expansion of export-based and biofuel production, however, small-scale family farmers in Brazil have had difficulty competing.

Over the past twenty years, several social movements, most notably the MST, have formed to protest unequal land distribution. Grounded in the popular slogan "Land for those who work it," the MST formed in 1984, when activists organized what has become the largest and most visible social movement in Brazilian history, and perhaps in the world. The MST began its struggle with a focus specifically on access to land and quickly moved into production matters, with a policy of creating socialist collectives. Over the past 10 to 15 years, however, the movement has expanded its struggle to a broad set of systemic issues, including the general economic structure of Brazilian agriculture. Although the MST has had little success fighting large-scale agroindustry, movement activists have become a political fixture, and politicians from town mayors to the national president are compelled to take a position on both land reform and the movement itself. This visibility—along with the rise of other movements and what Evelina Dagnino (2007) refers to as the "perverse confluence" of neoliberalism and democracy—has shaped a significant role for social movements around food, hunger, and production issues. The MST's mobilization placed enormous political pressure on the government to enact agrarian reform and promote family farm production (Schneider et al. 2010). As a response to national and international pressure, the federal government passed the Agrarian Act in 1993 and shortly afterwards established the Special Secretariat for Agrarian Issues. After two massacres of MST members by off-duty policemen became public knowledge in the mid-1990s, international pressure forced the Brazilian government to increase the attention paid to rural issues, and in 1999 MEPF became a ministry of its own—the Ministry of Agrarian Development (MDA). For the past decade, the MDA has been the most important political supporter of small family farm policies and initiatives, from credit access to agroecological production (Schneider et al. 2010; Ferreira et al. 2001).

In 2002, Luiz Inácio "Lula" da Silva was elected president. Lula had been a supporter of the MST from the beginning, and the MST threw the full weight of its membership behind all four of Lula's campaigns for the presidency. Under Lula (2003–2010), Brazil implemented a host of social protection policies to address hunger, malnutrition, and poverty (IFPRI 2010). The rapid decline of poverty in Brazil has been well documented: the percentage of the population living below the poverty line has dropped from around 26 percent in 2002 to now under 14 percent (Soares 2012), and undernourishment has dropped from 10 percent to 6 percent between 1990 and 2006 (IFPRI 2010, 1). A central tool the government used to increase the spending power of the poor was to increase the minimum wage. After controlling inflation with the Real Plan under Cardoso, the government doubled the official minimum wage in ten years (53 percent increase in real terms). This directly benefited the 28 million Brazilians working for less than the official minimum wage and also indirectly pushed wages up in associated sectors around the country (Schneider et al. 2010). Corresponding increases and expansion also occurred in Brazil's rural pension scheme, which brought even larger incomes into rural households.

In 2003, Lula implemented a national poverty alleviation and food security strategy called *Fome Zero* (Zero Hunger), explicitly targeting the structural causes of poverty, including the combined food and farm crises, inequality and income concentration, unemployment, weak farm policies and falling farm prices, and discrepancies between food supply and demand.[6] Over the past 10 years, Zero Hunger has improved access to food, contributed to income generation and rural employment, supported family agriculture, and encouraged social inclusion and participatory democracy (Rocha 2009: 53). Between 2003 and 2008, the number of people in poverty decreased by 27 percent (from 15.4 million to 11.3 million) and Brazil's extremely poor fell by nearly half. According to The Latin American and Caribbean Economic System (SELA), "through the policies of its Zero Hunger programme and thanks to its strong economic growth, [Brazil] has reduced the proportion of hungry people from 9 to 6 percent and eradicated 73 percent of child malnutrition, in a period of four years" (SELA 2010, 3).

Zero Hunger is a comprehensive approach to hunger. It works at national, municipal, and rural levels through emergency and long-term policies ranging from emergency food support and basic health care provisions to education, school food programs, subsidized food in supermarkets and restaurants, support for family farming, and extension of credit to family farmers. Policies are participatory and are implemented through food and nutritional security councils comprising public officials as well as civil society organizations (da Silva 2009). Brazil's experience with Zero Hunger suggests that rural communities need to be strengthened and incorporated through land distribution, farm development, market access for both producers and consumers, rural–urban linkages, state–society (or public–private) participation, improved linkages between healthy food and medical services to boost nutrition, and social protection policies (CCTs, food baskets, community kitchens, and so on) that improve food access for the most vulnerable.

Perhaps the most visible example of such integrated rural–urban development is the contemporary case of the Brazilian city of Belo Horizonte. In 1993 a set of policies was implemented by Patrus Ananias, mayor of the city and a member of the Workers' Party, as part of a general move towards political transparency, dialogue, and accountability. Ananias created a new agency to oversee food security in the city, with a 20-member council of social movement activists, business leaders, educators, and administrators monitoring the agency's activities. These policies, now celebrated for having reduced hunger, malnutrition, infant mortality, and rural poverty, included three main components. First, mobile food markets were established in areas of the city previously lacking fresh, healthy food. This was accomplished in part by requiring vendors who were awarded permits to sell food in wealthy parts of the city to sell in the impoverished areas on the weekends. Second, food was established as a right, and several programs intended to serve the undernourished were launched, including free clinics, school lunches, and a "popular restaurant" where lunches are served five days a week for one *real* (less than one US dollar). Finally, the city worked with small farmers—many of whom were recipients of land under the federal government's ongoing agrarian reform program—in surrounding areas to establish farm-to-table links that reduced the overall cost of food provisioning and provided farmers with secure access to markets. These citywide programs were integrated with a broader set of programs nationwide that included the largest land distribution and CCT programs *(Bolsa Família)* in the country's history.

The important structural lessons of Ananias' successful food security program, which became the basis of the Zero Hunger program, include the importance of citizen oversight, which kept profits from being siphoned off by an elite minority (or the more well-to-do poor) and forced government transactions to be as transparent as possible; the integration of production, consumption, and distribution into the same set of policies; and the focus on rights, instead of handouts. Indeed, most of the programs, such as the popular restaurant, are freely available rather than targeted at a small group, so these benefits avoided the social stigma of being tarred with the label of "the poor." Perhaps the biggest success of the program has been to keep costs low, around $10 million a year, since its inception.

Building on Belo Horizonte's experience, in 2003 Lula reestablished Brazil's National Food Security Council (CONSEA), which has been an instrumental national forum for advocacy and design of progressive policies oriented around food access and origin and the importance of family farms. The council is composed of governmental actors, such as the National Secretariat for Food Security, as well as numerous civil society actors, such as indigenous organizations and social movements, including the strong presence of the MST. CONSEA has played a large role in bringing together advocates for agrarian reform, the right-to-food, agroecology, and other environmental concerns, and communicating those concerns to the president of Brazil.

The government also created a National Food and Nutritional Security Plan (PNSAN), specifically designed to strengthen sustainable agroecological systems and the role of family farm production in feeding the country, as the backbone of the Zero Hunger framework. Its structural role is to link the programs and initiatives of various sectors to promote the human right to adequate food. PNSAN has actively promoted respect for social movement goals such as food sovereignty and the guarantee of the human right to adequate food, including access to water, as a state policy and in international negotiations and cooperation (Chmielewska and Souza 2011, 6; CAISAN 2011).

Connected to PNSAN is the National Program for Strengthening Family Farming (PRONAF). Established in 1996 as a measure to support new small farmers created through land distribution, PRONAF marked a change in the government's attitude towards small farms (Guanziroli and Basco 2010, 47). In the past 15 years, PRONAF has grown dramatically, and the program now offers credit for a broad range of agricultural activities, including family fisheries, female agriculturalists, and environmentally sustainable farming projects. During Lula's administration, PRONAF increased its beneficiaries from 291,000 in 2003 to 1.6 million in 2010 (Ministério de Desenvolvimento Agrario 2011, 3). PRONAF has developed a highly decentralized approach, which places responsibility on local extension services to identify beneficiaries and distribute credit. PRONAF provides financial resources and incentives for purchasing and improving family farm infrastructure, such as machinery, buildings, renewable energy solutions, and aquifer improvement. To complement this policy, the government is strengthening the institutional capacity of local municipal extension services. PRONAF also sponsors lines of credit for on-farm diversification as the means of cushioning smallholders from sudden production or consumption changes. Small loans (up to about $20,000) are available based on the size, income, and makeup of the family farm, at interest rates between 1.5 percent and 5.5 percent. The credit program was the backbone of a broader program to support the family farm as a significant productive agricultural force in Brazil; however, the poorest farms receiving the lowest interest rate are experiencing the lowest repayment rate, with 28 percent being behind on payments (Guanziroli and Basco 2010, 57).

In addition, PRONAF has been working in conjunction with the government's rural marketing programs under Zero Hunger to increase domestic demand for agricultural products produced on family farms—increasing food security from both the production and consumption end. Started in 2003, Brazil's Food Acquisition Program (PAA) is a government-sponsored food procurement program that utilizes the productive capacity of family farms to supply the nutritional needs of local public schools, food banks, community kitchens, charitable associations, and community centers for the needy (CAISAN 2011). Municipalities participating in the PAA utilize and strengthen local institutions that can better suit the demands and expectations of producers and consumers. As a public intermediary, the program introduces more competition in a market that typically provides limited access for resource-constrained farmers. The program provides latitude for local adaptation to various social and governance structures, while expanding farmer organization by working with cooperatives and associations.

The government's focus on family farm production has revitalized public support for rural livelihoods and policies targeting their productive capacity. The election of Lula's protégé, Dilma Rousseff, in 2010 signified a continuation of the same social protection and productive inclusion policies, although Dilma developed a new strategy to replace Zero Hunger. *Brasil sem Miséria* (Brazil without Misery) was launched in 2011 and symbolized the government's pledge to eliminate extreme poverty in the country. The plan calls for a national and regional focus to lift more than 16 million Brazilians out of poverty with improved income, productive inclusion, and public services. As a part of this strategy, Brazil's flagship poverty alleviation policy, the *Bolsa Família*, aims to increase its beneficiaries by 800,000 families (Ministério de Desenvolvimento Social e Combate a Fome 2011).

The new plan outlines ambitious goals to amplify the PAA (MDS 2011). Since its inception in 2003, the program's budget for public food purchases has expanded from just over R$80 million to now over R$790 million.[7] In 2011, the PAA purchased over three million tons of produce directly from 200,000 family farms. It distributed the produce to more than 15 million people suffering from food insecurity in some 2,300 municipalities (Conab 2012; MDA 2012). By 2014, Brazil without Misery plans to double the number of family farmers selling produce to the PAA (MDA 2012).

EXPANDING SOCIAL PROTECTION POLICIES TO REDUCE HUNGER

The evidence from Latin America suggests that governments can—and must—build comprehensive, articulated policy webs, while working with civil society actors to blunt the effects of food insecurity. Brazil's innovative experience in fighting hunger owes some of its success to the country's economic growth and its endowment of natural resources. It is impossible to deny, however, the paramount role of highly organized civil society in lobbying for and developing decentralized and progressive policies that bring together food production, consumption, and distribution. The "Brazilian path" of agricultural development has a tumultuous history characterized by unprecedented inequities in land resources that gave birth to the country's largest social movement, the MST. Years of neglect and unfulfilled promises from the national government strengthened and broadened civil society movements—both urban and rural. The nation's relatively recent return to democracy has ushered in a new participatory process of governance and policy formation. Local innovations and moral economies helped the state devise methods to serve as a protest broker, while providing political space for a multi-stakeholder and multilevel network of civil society actors intentionally engaged in the dynamic process of policymaking.

As the Brazilian case makes clear, mobilizations around access to food are situated within and draw upon traditional moral economies, including collective notions of fair prices, just distribution, and the right conduct of government and community. Other countries will be hard pressed to follow the Brazilian model, however, as it requires a solvent state, strong oversight, and well-coordinated interaction among the state, society, and economy. In the rest of Latin America, there are multiple factors that provide challenges—or opportunities—for combating food insecurity and political instability. These challenges include a growing reliance on extractive resource sectors that pit the state and private sector against local communities, but potentially provide resources for combating poverty; increasingly active civil societies with expectations of left-leaning or popular governments that may be either unrealistic or unfulfilled; environmental degradation in regions with both the most vulnerable populations (the steep Andean slopes) and the least vulnerable (the grasslands of central Brazil and Argentina) that require intervention or risk generating food shortages; reliance on currently structured global markets that are susceptible to price volatility; and the development of state–society, public–private partnerships of various kinds that deliver services but devolve responsibility to nongovernmental or shadow groups.

Latin American countries that have the capacity to implement political safeguards such as CCTs (Argentina, Mexico, Peru, and Brazil) or countries with highly organized civil societies (Ecuador, Venezuela, Bolivia, and Brazil) may find themselves in a position to mitigate the effects of environmental change and economic instability. Conversely, those countries without access to what could be called human, political, and economic capital will not be able to shield the rural and urban poor, whether producers or consumers, from increasing insecurity—insecurity that may not necessarily be food insecurity, but that will become food insecurity if not addressed. These are not necessarily the countries where organized political mobilization will occur, but they are places where spontaneous protests could break out over high food prices or citizens' inability to access food.

ACKNOWLEDGMENTS

The authors would like to thank Chris Barrett and Cynthia Mathys for their encouragement, work, and patience. Sara Keene also provided invaluable research assistance.

NOTES

1. The FAO has also developed specific initiatives dealing with food security and vulnerability analysis and mapping. The FAO-FIVIMS (Food Insecurity and Vulnerability Mapping System) Program defines vulnerability as "the presence of factors that place people at risk of becoming food insecure. These factors can be external or internal. External factors include trends such as depletion of natural resources from which the population makes its living; environmental degradation or food price inflation; shocks such as natural disasters and conflict; and seasonality, or seasonal changes in food production and food prices. Internal factors are the characteristics of people, the general conditions in which they live and the dynamics of the household that restrict their ability to avoid becoming food insecure in the future." See: http://www.fivims.org/index2.php?option=com_content&do_pdf=1&id=54 (accessed March 31, 2013).
2. While poverty and food insecurity should not be conflated, there is certainly a strong correlation between the two (food insecurity, for instance, is often an indicator of extreme poverty; see Martinez et al. 2009).

3. See Asamblea Nacional del Ecuador (2008).
4. República de Bolivia, "Constitución de 2009" Political Database of the Americas, Georgetown University, http://pdba.georgetown.edu/Constitutions/Bolivia/bolivia09.html (accessed April 1, 2013).
5. See ALBA website at http://www.alba-tcp.org/en/contenido/food-security-and-sovereignty-agreement-between-member-countries-petrocaribe-and-alba-alba (accessed April 1, 2013).
6. Brazil's Food Security Policy can be accessed online at www.fomezero.gov.br (accessed April 1, 2013). Any additional information drawn from this document is cited as BFSP.
7. This budget represents a minuscule amount at under 0.00016% of GDP and pales in comparison to PRONAF's budget of over $7 billion (CAISAN 2011, 23).

REFERENCES

Altieri, M. A., C. Nichols, and F. Funes. 2012. Agroecología: Única esperanza para la soberanía alimetaria y la resilencia socioecológica. Sociedad Scientifica Latinoamericana de Agroecologia (SOCLA), SOCLA's Rio+20 position paper presented at the Rio+20 United Nations Conference on Sustainable Development, June, 2012.

Asemblea Nacional del Ecuador. 2008. La consitución del Ecuador. Asemblea Nacional http://www.asambleanacional.gov.ec/documentos/Constitucion-2008.pdf (accessed February 2012).

Boyce, J. K., P. Rosset, E. A. Stanton. 2005. Land reform and sustainable development. Working Paper Series No. 98, University of Massachusetts, Amherst, MA.

Bundy, D., C. Burbano, M. Grosh, A. Gelli, M. Jukes, and L. Drake. 2009. *Rethinking School Feeding: Social safety nets, child development, and the education sector.* Washington, DC: World Bank Publications.

Byerlee, D., A. de Janvry and E. Sadoulet, n.d. Agriculture for development: Toward a new paradigm. Available at http://are.berkeley.edu/~esadoulet/papers/Annual_Review_of_ResEcon7.pdf (accessed April 1, 2013).

CAISAN—Câmara Interministerial de Segurança Alimentar e Nutricional. 2011. Plano nacional de segurança alimentar e nutricional—2012/2015. Brasília, Brazil.

Chmielewska, D. and D. Souza. 2011. The food security policy context in Brazil. Brasilia, Brazil: UNDP-IPC, Country Study No. 22.

Churuchumbi, G. 2012. Personal Interview, Cayambe, Ecuador, January 30.

Cohen, M. J., and J. L. Garrett. 2009. The food price crisis and urban food (in)security. Human settlements Working Paper series: Urbanization and emerging population issues No. 2. London: International Institute for Environment and Development. http://pubs.iied.org/pdfs/10574IIED.pdf (accessed June 2012).

Conab—Companhia Nacional de Abastecimento. 2012. Programa de aquisição de alimentos: Resultados das ações da Conab em 2011. Brasília, Brazil.

Conant, J. 2012. Should Chiapas farmers suffer for California's carbon? *Yes! Magazine.* Bainbridge Island, WA: Positive Futures Network, November 13.

Cuesta, J. and F. Jaramillo. 2010. Taxonomy of causes, impacts and policy responses to the food price crisis in the Andean region. Working Paper No. 674. Washington, D.C.: Inter-American Development Bank.

da Silva, J. G. 2009. Zero Hunger and territories of citizenship: Promoting food security in Brazil's Rural Areas. *About IFPRI and the 2020 Vision Initiative*, 367–374.

———, J. G., S. Gómez, and R. Castañeda S. 2010. *Boom agrícola y persistencia del la pobreza rural: Estudio de ocho casos.* Santiago, Chile: FAO. http://www.fao.org/alc/file/media/pubs/2009/boomagri.pdf (accessed April 1, 2013).

Dagnino, E. 2003. Citizenship in Latin America: An introduction. *Latin American Perspectives* 30(2): 3–17.

——— 2007. Citizenship: A perverse confluence. *Development in Practice 17*: 549–556.

de Janvry, A., and E. Sadoulet. 2002. Land reforms in Latin America: Ten lessons toward a contemporary agenda. Paper prepared for the World Bank's Latin American Land Policy Workshop in Pachuca, Mexico: http://are.berkeley.edu/~esadoulet/papers/Land_Reform_in_LA_10_lesson.pdf (accessed March 2012).

FAO. 2010. Rural territorial development and its institutional implications in Latin America and the Caribbean. Presented at the Thirty-First Regional Conference for Latin America and the Caribbean. Panama City, Panama, April 26–30. http://www.fao.org/docrep/meeting/019/k7838e.pdf (accessed April 1, 2013).

Ferreira, B., Gasques, J. G. and J. C. Conceição. 2001. *Tranformações da agricultura e políticas públicas.* Brasilia: Instituto de Pesquisa Econômica Aplicada.

Guanziroli, C. E., and C. A. Basco. 2010. Construction of agrarian policies in Brazil: The case of the national program to strengthen family farming (PRONAF). Rural Development Series Comun January–July 2010, Inter-American Institute for Cooperation on Agriculture.

Heidrich, P., and Z. Williams. 2011. Trade and inclusive growth in Latin America: Compensatory policies for trade liberalization. In *Trade and inclusive growth series brief, #73*. Buenos Aires: Red Latinoamericana de Política Comercial (LATN), Facultad Latinoamericana de Ciencias Sociales (FLACSO), Buenos Aires, AR.

IAASTD, and B. D. McIntyre. 2009. *International assessment of agricultural Knowledge, science, and technology for development (IAASTD): Latin America and the Caribbean (LAC) Report.* Washington, DC: Island Press.

IBGE. 2010. *Censo demográfico 2010.* Rio de Janeiro, Brazil, Instituto Brasileiro de Geografia e Estatística.

———. 2006. *Censo agropecuário 2006: Resultados preliminaries.* Rio de Janeiro: Instituto Brasileiro de Geografia e Estatística

IFAD. 2010. *Rural poverty report 2011.* Rome: International Fund for Agricultural Development.

IFPRI, A. U. Ahmed, R. V. Hill, L. C. Smith, D. M. Wiesmann and T. Frankenberger. 2007. The world's most deprived: Characteristics and causes of extreme poverty and hunger. 2020. Vision for Food, Agriculture, and the Environment Discussion Paper No. 43, Washington, DC: International Food Policy Research Institute.

IFPRI, S. Fan, and J. Brzeska. 2010. The role of emerging countries in global food security. IFPRI Policy Brief 15, Washington DC: International Food Policy Research Institute. Available online at http://www.ifpri.org/sites/default/files/publications/bp015.pdf (accessed May 8, 2013).

IFPRI, K. Grebmer, M. Torero, T. Olofinbiyi, H. Fritschel, D. Wiesmann, Y. Yohannes, L. Schofield, and C. Oppeln. 2011. Global hunger index 2011. The challenge of hunger: Taming price spikes and excessive food price volatility. Washington, DC: International Food Policy Research Institute.

Keck, M. E. and R. N. Abers. 2006. Civil society and state-building in Latin America. *LASA Forum*, Winter *37*(1), 30Ame.

Martínez, R., A. Palma, E. Atalah, and A. C. Pinheiro. 2009. *Food and nutrition insecurity in Latin America and the Caribbean.* Santiago, Chile: United Nations/ECLAC/WFP.

Ministéro de Desenvolvimento Agrário. 2011. Key facts and figures: Agrarian development and food policy in Brazil. 37th Session of the UN Food and Agricultural Organization (FAO) Conference, June 24, 2011.

Ministéro de Desenvolvimento Agrário. 2012. PAA—Balanço 2011 e dados 2012. Brasília, Brazil, accessed from personal contact at MDA, March 20.

Ministério de Desenvolvimento Social e Combate à Fome. 2012 Programa nacional de alimentação escolar. Brasília, Brazil. www.mds.gov.br/ (accessed June 2012).

Ministério de Desenvolvimento Social e Combate à Fome. 2011. Plan Nacional de Brasil sem Miséria. Brasília, Brazil, www.brasilsemmiseria.gov.br/ (accessed April 1, 2013).

Naylor, R. L., A. Liska, M. B. Burke, W. P. Falcon, J. C. Gaskill. 2007. The ripple effect: Biofuels, food security, and the environment. *Environment 49*(9): 30–43.

O'Donnell, G. and P. Schmitter. 1986 *Transitions from authoritarian rule.* Baltimore, MD: Johns Hopkins University Press.

Piñeiro, M., E. Bianchi, L. Uzquiza, and M. Trucco. 2010. *Food security policies in Latin America: New trends with uncertain results.* Winnipeg, Manitoba: International Institute for Sustainable Development.

Reardon, T. and J. Berdegué. 2002. The rapid rise of supermarkets in Latin America: Challenges and opportunities for development. *Development Policy Review 20*: 371–88.

Robles, M., and M. Torero. 2010. Understanding the impact of high food prices in Latin America. *Economia 10*: 117–64.

Rocha, Cecilia. 2009. Developments in national policies for food and nutrition security in Brazil. *Development Policy Review 27*(1): 51–66.

Schneider, S., S. Shiki, and W. Belik. 2010. Rural development in Brazil: Overcoming inequalities and building new markets. *Revista di Economia Agraria LXV*(2): 225–79.

Scott, J. C. 1976. *The moral economy of the peasant: Rebellion and subsistence in Southeast Asia.* New Haven: Yale University Press.

SELA, Latin American and Caribbean Economic System. 2010. Food security and food prices in Latin America and the Caribbean: Current situation and prospects. In *XXXVI Regular Meeting of the Latin American Council.* Caracas, Venezuela: The Press and Publications Department of the Permanent Secretariat of SELA.

Shapouri, S., S. Rosen, B. Mead, and F. Gale. 2009. Food security assessment, 2008–09. Washington, DC: US Dept. of Agriculture, Economic Research Service.

Skocpol, T. 1979. *States and revolution: A comparative analysis of France, Russia and China.* New York: Cambridge University Press.

Soares, S. 2012. Bolsa Família, its design, its impacts and possibilities for the future. Working Paper No. 89. Brasilia: UNDP-IPC.

Staatz, J. M. and C. K. Eicher. 2006. Agricultural development: Ideas in historical perspective. In *Agricultural Development in the Third World,* 2nd edition, eds Staatz and Eicher, 3–28. Baltimore, MD: Johns Hopkins University Press.

Thompson, E. P. 1993. *Customs in common: Studies in traditional popular culture.* New York: New Press.

Trivelli, C., J. Yancari, and C. De los Ríos. 2009. Crisis and rural poverty in Latin America. In *Working Paper No. 37: Rural Territorial Dynamics Program.* Santiago, Chile: Rimisp Core Support for Rural Development Research.

UNDESA. 2008. *World urbanization prospects: The 2007 revision, executive summary,* New York: United Nations.

Verner, D. 2011. Social implications of climate change in Latin America and the Caribbean. Economic Premise No. 61, Washington DC: The World Bank, Accessed online at: https://openknowledge.worldbank.org/bitstream/handle/10986/10084/634000BRI0Box300http000vx.worldbank.pdf?sequence=1 (accessed April 1, 2013).

Watts, M. 1983. *Silent violence: Food, famine and peasantry in northern Nigeria.* Berkeley, CA: University of California Press.

Wolford, W. 2010. *This land is ours now: Social mobilization and the meanings of land in Brazil.* Durham, NC: Duke University Press.

World Bank. 2008. *Food price inflation and its effects on Latin America and the Caribbean.* Washington, DC: World Bank.

———. 2011. *Data bank.* http://databank.worldbank.org (accessed April 2012).

———. 2012. What are the facts about rising food prices and their effects on the region. Available online at http://go.worldbank.org/CJYWKZPMX0 (accessed April 1 2013).

World Food Programme (WFP). 2009. *Comprehensive Food Security & Vulnerability Analysis Guidelines.* Rome, Italy: United Nations World Food Programme, available online at: http://documents.wfp.org/stellent/groups/public/documents/manual_guide_proced/wfp203202.pdf (accessed April 1, 2013).

13

Food Security and Sociopolitical Stability in Sub-Saharan Africa

CHRISTOPHER B. BARRETT AND
JOANNA B. UPTON

Sub-Saharan Africa suffers the unfortunate triple distinction among world regions of having in recent years the highest incidence of undernourishment (FAO 2011), ultra-poverty (Barrett forthcoming), and conflict-related deaths (UCDP/PRIO 2012). These three phenomena are inextricable and mutually reinforcing. The associations among food insecurity, poverty, and conflict are strong, and there are strong theoretical reasons to posit causal relations flowing in each direction among these three variables. In this chapter, we focus on recent and prospective drivers of food security with an eye to identifying stressors that might foster episodes of sociopolitical instability in the subcontinent over the coming decade. But it is important to recognize that this analysis necessarily abstracts somewhat from inextricable concerns about poverty and from the feedback effects from instability to both poverty and food insecurity, and falls well short of making formal causal claims.

As agreed at the 1996 World Food Summit, food security exists when "all people at all times have access to sufficient, safe, nutritious food to maintain a healthy and active life." Food security has three key components: *availability* of sufficient quantities of foods, *access* to those foods when and where they are needed, and proper *utilization* of accessed foods through appropriate preparation and consumption (FAO 1996; Barrett 2010). Food security is thus a question both of supply, of either producing or importing sufficient foods, and of demand, or the ability of the population to acquire and consume these foods given incomes, prices, market access, intrahousehold power relations, and so on. Assuring food security for any given population is a dynamic process that relies on a wide array of supply and demand factors, including the environment and ecology, human skills and capabilities, effective markets, good governance, and the avoidance of conflict. Furthermore, these factors interact dynamically. Stable food production engenders wealth that supports the ability and resources to manage land appropriately and protect the environment. Meeting food needs empowers people and societies to develop and expand their capabilities beyond food production. A lack of food, however, not only weakens society but can lead to unrest and challenges for governance.

Sub-Saharan Africa faces numerous challenges, as well as opportunities, in assuring its population's food security. While on the one hand it is resource-rich in aggregate, land degradation and limited access to available water resources, along with weak institutions, threaten to further degrade the natural resource base on which food production depends. Combined with the world's lowest rates of uptake of modern agricultural technologies, agricultural productivity is low enough that, unlike any other world region, food availability is limiting in much of sub-Saharan Africa.

As shown in Figure 13.1, two-thirds of sub-Saharan African countries lack adequate calorie or protein supplies—much less crucial micronutrients (minerals and vitamins)—to provide an adequate diet to every resident, even if available supplies were distributed equally within each country. This makes food productivity growth a first-order priority in sub-Saharan Africa, far more than in any other world region. Hence too the disproportionate importance of humanitarian food assistance in sub-Saharan Africa. In 2010, nearly 62 percent of global food aid (in metric tons) was destined for sub-Saharan Africa,

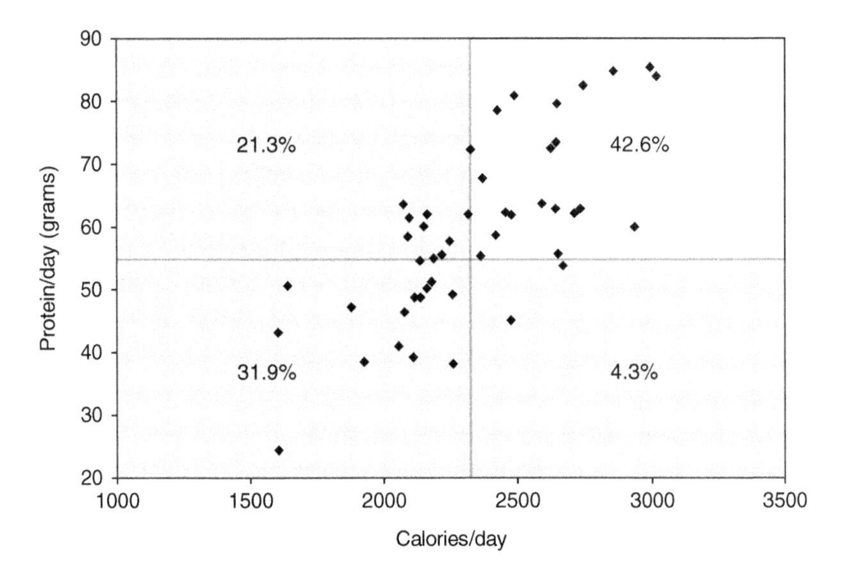

Fig. 13.1. Nutrient availability in 47 sub-Saharan African countries (2009, per capita, areas below minima are shaded)

Data Source: Food and Agriculture Organization 2012a (most recent data available from 2009).

including 61 percent of all emergency assistance and nearly 80 percent of all program aid provided to governments (WFP 2012).

As important as food production and availability issues are to sub-Saharan Africa's food security situation, the biggest driver of food insecurity is access. Throughout sub-Saharan Africa, widespread and often intense poverty impacts peoples' access to and utilization of foods, as well as their ability to finance innovations to enhance productivity. Unlike any other region of the world, the incidence of poverty—defined as the proportion of people living on $2.50 or less per day in inflation-adjusted 2005 terms—has scarcely budged in a generation in sub-Saharan Africa. The headcount rate of poverty in sub-Saharan Africa has oscillated between 80 and 84 percent since 1981, while the number of poor people has effectively doubled, from an estimated 320 million persons in 1981 to 610 million by 2005. More troubling, the ultra-poor—individuals living on no more than $0.62/day, one quarter of the global poverty line—are disproportionately concentrated in sub-Saharan Africa (Ahmed et al. 2007; Barrett forthcoming). While the global headcount rate of ultra-poverty has collapsed from 19 percent (704 million people) in 1981 to just 4 percent (221 million) in 2005, in sub-Saharan Africa, the ultra-poverty rate has remained about 20 percent, and the number of ultra-poor doubled from 1981 to 1999, from 84 to 165 million. The region is now home to 65 percent of the world's ultra-poor, up from just 12 percent in 1981. While poverty remains primarily an Asian phenomenon—because it is the world's most populous continent—ultra-poverty is primarily and increasingly an African condition and one closely related to food insecurity and conflict.

We explore the coevolution of food security and sociopolitical stability not only at the aggregate, subcontinental scale, but also at subregional and national levels for a few key countries so as to highlight the breadth of experiences within sub-Saharan Africa. Subregions examined include each of the economic and development communities that encompass different zones of the continent, as well as the land-locked states, and the Gulf of Guinea states that account for the overwhelming majority of Africa's crude oil production and exports.[1]

KEY CHARACTERISTICS OF SUB-SAHARAN AFRICA

Both supply- and demand-side factors combine to shape development and food security. Four such features have been especially key in sub-Saharan Africa. First, the region is land and water abundant, albeit with considerable variation within and among countries, and serious challenges to tapping and conserving those resources effectively. Second, its population is heavily rural, but growing and urbanizing rapidly. Third, agricultural productivity in the region is extremely low, and labor productivity growth has lagged in the face of rapid population growth and meager agricultural research and development investment. Fourth, the enormity of the subcontinent and the poor and sparse state of its physical infrastructure adds considerable cost and delay to trade, leaving sub-Saharan Africa unusually reliant on its own relatively unproductive agriculture. These four drivers reveal the most likely stressors for food security and sociopolitical stability in sub-Saharan Africa in the years ahead.

Land

Sub-Saharan Africa is land-abundant, averaging nearly three hectares of land, and about 0.2 hectares of arable land, per capita across 50 countries (FAO 2012a). In 2005, only about one-fifth of the potentially suitable cropland in sub-Saharan Africa was in cultivation (Bruinsma 2011); most of it was fallow or forest. While only about 14 percent of all land in the region is considered arable (FAO 2012a), it still holds nearly half of the world's uncultivated arable land (Deininger and Byerlee 2012), making it the world's primary remaining agricultural frontier. There is a fair amount of diversity, however, between sub-regions and countries. Some countries have too little land to feed their populations, while many of the most land-abundant nations are largely arid, with poor and degrading soil quality, or heavily forested with limited transport infrastructure. Soil quality throughout the region is poor and degrading; an estimated two-thirds of African land is already considered degraded, directly affecting about 65 percent of the population (UNECA 2007).

The distribution and security of land rights are major challenges. The relatively rich agricultural lands of many countries, particularly in southern and eastern Africa, have been subject to insecure land rights, including expropriation by the state, or have remained in the hands of an elite minority. In an extreme case, the government of Zimbabwe responded to the disproportionate ownership of lands by white settlers and their descendants by enacting laws to repossess and redistribute land (Swarns 2002). This largely arbitrary process not only led to a great deal of social unrest and violence, but undermined the productivity of both land and labor in a nation that otherwise had great agricultural potential (Coltart 2002). Insecure land tenure in Ethiopia, where the state owns all land and only grants extended leaseholds to farmers, has been shown to limit agricultural productivity and on-farm investment, even when such investments enhance the security of rights over land (Deininger and Jin 2006; Pender and Fafchamps 2006).

Proper land certification in some cases has positive impacts for smallholders, inducing them to maintain soils, make productive investments, and enhance land productivity (Holden et al. 2008). With some exceptions, however, state land reform efforts have commonly failed to deliver greater equality or security, as ostensibly intended, and have often instead advanced the interests of the wealthy and politically powerful (Lahiff and Cousins 2001). Whether or not the inequitable distribution of land undermines agricultural productivity, it creates the potential for rural unrest and challenges for meeting the food needs of the poorest (Moyo and Yeros 2005). The painful history of both colonial and neocolonial extractive investment makes both states and citizens wary of international investments in land.

Water

Water resources in sub-Saharan Africa are likewise abundant, but suffer from issues of distribution and management. The aggregate abundance of water in sub-Saharan Africa is concentrated in the equatorial region, from Sierra Leone to Uganda (FAO 2012b). These countries average between 20 and 100,000 cubic meters of renewable freshwater resource per capita per annum (World Bank 2012). Meanwhile, 70 percent of all countries in sub-Saharan Africa receive on average an order of magnitude less than this; 12 states average less than one thousand cubic meters per capita per year (World Bank 2012). A disproportionate 43 percent of the sub-Saharan Africa land mass ranges from semi-arid to hyper-arid (FAO 2012b). Even within the arid states, rain and water resources tend to be concentrated, leaving some subregions particularly arid and hence vulnerable to climate shocks and changes.

Desertification has been most stark in the northern tropic and Sahara, extending from the Sahel in West Africa to the Horn of Africa. Annual losses of rangeland and cropland to desertification are estimated to be on the order of about 200 km^2 in the northern region of Ghana, for example, and as much as about 3500 km^2 in Nigeria (UNECA 2007). Across sub-Saharan Africa it is estimated that 60 percent of all land is at risk of desertification (FAO 2012b), and 70 percent and 80 percent in Ethiopia and Kenya, respectively (UNECA 2007). Improved water and land management can help stem or even reverse those patterns, but climate change is probably the biggest determinant going forward.

Agriculture accounts for more than 80 percent of water use in sub-Saharan Africa, despite far less land suitable for irrigation or currently irrigated than any other world region (Bruinsma 2011). In 2005, the share of arable land in use that was irrigated ranged between less than 1 percent (including in Kenya) to a maximum of about 20 percent (in Mauritius and Sao Tome). The over-all average was close to 2 percent, whereas in the Horn of Africa and in the most populous states it remained less than 1 percent, and in the wealthiest ten states and in the SADC states it was approximately 2.5 percent (FAO 2012a). Just as with land suitable for cultivation, sub-Saharan Africa has far greater proportional capacity to expand the irrigated frontier than any other world region, which feeds its untapped agricultural potential.

Population and Labor

Population dynamics shape the supply of food, through influences on the availability and nature of the labor force, as well as the demand for food. From 1970 to 2000 sub-Saharan Africa's annual population growth rate averaged about 5 percent, the world's highest, leading to the population more than doubling, from about 296 to 669 million persons. Demographic patterns are also shifting in important ways. The population of young adults is growing; the current median age across sub-Saharan Africa is only 20, as opposed to roughly 40 in Europe and North America (UN DESA 2011). The end of the twentieth century also saw rapid urbanization in sub-Saharan Africa, as the urban population grew by 278 percent over that same period, from about 57.7 million to 218 million—more than three times faster than sub-Saharan Africa's rural population, which nevertheless nearly doubled, from about 238 to 451 million (FAO 2012a). Rapid urbanization has significantly increased the demands on food marketing chains, as urban populations depend almost entirely on markets to source foods.

In spite of trends toward urbanization, sub-Saharan Africa remains a heavily rural region. In 2000 approximately 64 percent of all sub-Saharan Africans lived in rural areas and 63 percent of those employed worked in the agricultural sector, an agricultural workforce of more than 400 million persons (FAO 2012a). The degree of movement from the agricultural sector varied by country and region. In West Africa the percentage of workers in agriculture fell by 13 percentage points, from 65 percent to 52 percent, between 1980 and 2000, while in the Horn of Africa by only 7 percentage points, from 84 percent to 77 percent. In the most developed country in the region, South Africa, only 9 percent worked in agriculture by 2000 (FAO 2012a).

Agricultural labor is limited less by numbers than by skills and resources. Education levels remain low across the continent; the expected average number of years of education increased by only two years, from six to eight, between 1980 and 2000. Education rates are generally lower and illiteracy rates higher in the rural sector (Sahn and Stifel 2003). The dearth of educated farmers limits the capacity for internally driven innovation in the agricultural sector and challenges uptake of innovations developed through research and development or imported from abroad. Agricultural skills are also limited at higher levels. Government support for agriculture has been fairly limited, and further challenged by national-level instability and conflict that has made it difficult to maintain research and development (R&D) programs in the field and to repatriate scientists sent abroad to train. The skilled human resources necessary for an effective agricultural R&D program have been scarce in sub-Saharan Africa, with high rates of turnover and difficulties maintaining skills after completing advanced degrees.

Agricultural Productivity and Poverty

Limited skills among farmers and agricultural researchers in sub-Saharan Africa have long hindered the development and uptake of modern inputs compared to other world regions. In the late 1990s, modern crop varieties accounted for only an estimated 20 to 25 percent of growing area of most primary food crops in sub-Saharan Africa (Evenson and Gollin 2003). Sub-Saharan Africa accounts for less than 2 percent of global fertilizer use, with most of that concentrated in South Africa. Scant production capacity leads to heavy reliance on expensive imports of inorganic fertilizer.

The comparatively low productivity of sub-Saharan Africa agriculture is arguably the biggest reason for the region's persistent and profound poverty. Global productivity, measured as the constant dollar value of food production per capita, increased nearly 2 percent per annum over the 1980–2000 period, leading to historic low real global food prices while in sub-Saharan Africa the value of food production per capita increased less than half as quickly, on average only from $550 per worker to $637 per worker from 1980 to 2000, a 16 percent improvement in 20 years (FAO 2012a). This helps explain persistent poverty in rural sub-Saharan Africa as slow progress in labor productivity characterizes the region's dominant sector of employment, which translates directly into low income growth. Land productivity increased somewhat faster in sub-Saharan Africa over the same period, from $60 per hectare to $102 per hectare on average, a 70 percent improvement. But even those partial productivity measures are still anemic by global standards.

Following Hayami and Ruttan's (1971) seminal approach, Figure 13.2 depicts the average land and labor productivity change by country from 1980 to 2000, for all countries in sub-Saharan Africa (including the United States and China as reference points reflecting the global agricultural productivity frontier). The dashed diagonal lines reflect constant labor/land ratios (arrayed from highest in the upper left corner to lowest in the lower right corner). The arrows begin at the 1980 value, with the arrowhead indicating the 2000 value for the labeled group of countries. Figures are expressed in logarithmic terms, so that a one unit increase represents approximately a doubling of factor productivity over the period. Rightward (upward) movements indicate increased food production per hectare (worker) with little or no growth in labor (land) productivity, while leftward or downward movements reflect (partial) productivity decline. The technology frontier is mapped by the arc of outer points from upper left to lower right.

When Timmer (1988) undertook similar analysis of sub-Saharan Africa, he found decline in both agricultural labor and land productivity, 1973–1984. While the divergence in productivity growth patterns at the country level is stark, the previous decline had by and large ended and unidimensional partial

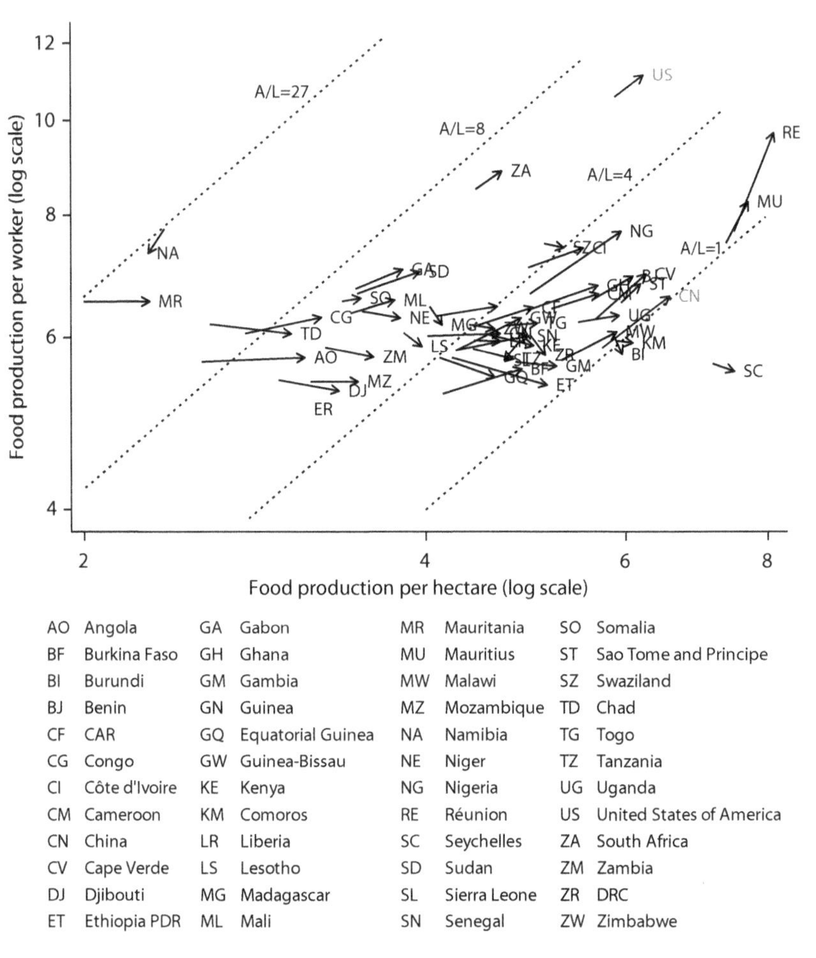

Fig. 13.2. International comparison of labor and land productivity: Sub-Saharan Africa, 1980–2000

Data Source: Food and Agriculture Organization, FAOSTAT 2012

productivity gains were achieved over the 1980–2000 generation. These trends are clearest if we take regional averages (not shown). The ECOWAS countries experienced on average the greatest productivity levels and increases in both dimensions of any other sub-region, of 78 percent in productivity per worker (from \$626 to \$1,112, in constant 2006 dollars), and of 117 percent in output per hectare (from \$94 to \$204). The IGAD states of the Horn of Africa, on the other hand, actually experienced a decline of 8 percent in productivity per capita (from \$455 to \$420), and an increase of 58 percent in productivity per hectare (from \$55 to \$87).

At the country level, we observe sharp improvements in both land and labor productivity in Mauritius, Nigeria, Reunion, and South Africa. Most sub-Saharan African countries stagnated in one direction or the other, however. For example, labor productivity did not change in Mozambique (remaining roughly $223 per worker per year) while land productivity increased by 38 percent (from just $24 to $33). Some countries even saw a decline in one or both factor productivity measures, including Burundi, the Democratic Republic of the Congo, Namibia, and Seychelles. But as is clear, only a minority of countries failed to show some progress in both dimensions, although gains were clearly concentrated overwhelmingly in improvements in land as opposed to labor productivity. The substantial gap between the United States and the sub-Saharan African countries also reflects the orders of magnitude difference in agricultural productivity between the regions.

These productivity results link directly to—indeed, are both cause and consequence of—the unusually high prevalence, depth, and persistence of poverty. As already noted, poverty rates throughout Africa are generally high, although there is considerable dispersion both within and among countries. Poverty is almost universally more prevalent and intense in rural areas in sub-Saharan Africa (Sahn and Stifel 2003). This leads to widespread, chronic problems of food access in food-growing regions of sub-Saharan Africa—a bitter irony indeed.

Market Access and Integration

Access to domestic and foreign markets and information are key components of food security. Farmers need access to markets both for inputs and to sell their goods. Consumers need reliable, inexpensive access in order to buy goods affordably. Unfortunately, poor roads limit market access across sub-Saharan Africa. The arid countries with low population density, such as Mauritania and Niger, have the severest dearth of roads, with approximately one kilometer (km) of road for each 100 km^2 of area, against the sub-Saharan Africa average of about 20 km. While a few states have extensive networks of paved roads, the average percentage of road distance paved across sub-Saharan Africa is only 25 percent. Even states that rely heavily on roads for trade have significant infrastructure challenges, such as Kenya, which has just 11 km of road per 100 km^2 area, only 13 percent of which is paved. Some (in particular, small) states, such as Mauritius and the Seychelles, rival the infrastructure of advanced countries with an impressive road network of about 100 km of paved road per 100 km^2, reflecting the considerable heterogeneity that typifies the region (World Bank 2012).

Poor transportation infrastructure poses particular problems for the 17 land-locked countries that face distances between 200 km and 1500 km from their capital cities to the nearest seaports. Train systems hold promise. However, in many cases railroad projects have struggled or failed, and the distances covered by active railways are limited and have declined in most countries over the past decades. The total distance of active railway lines (where data are available) actually declined by about 30 percent between 1986 and 2009 (World Bank 2012). Many initial projects attempting to link land-locked countries to coasts never succeeded. For example, while at least five railway lines were begun in coastal cities with the intent of reaching Niger, the construction of each of them stopped far short of that goal, and some are not even functional today where construction was completed. In other places, such as Madagascar, different sponsors laid rail of different gauge track, leading to breaks of gauge that add considerable cost and inconvenience to traffic.

Crossing borders to reach ports leads to high costs beyond the burden of poor physical infrastructure. Border crossings often entail fees—both official and unofficial—that increase transaction costs and delay deliveries. Duties at foreign ports can prove prohibitive, and political and social unrest also get in the way. For example, the conflict following Laurent Gbagbo's refusal to cede power following electoral loss in Côte d'Ivoire substantially upset the primary avenue for imports into Burkina Faso, leading to sharp price increases in imported foods in 2010 and 2011.

Across sub-Saharan Africa, insufficient transportation and communications infrastructure poses significant problems by impeding smooth market functioning. Weak market integration commonly renders government macroeconomic and sectoral policies ineffective by impeding market transmission of economic signals related to policy change (Moser et al. 2009). Similarly, without good access to distant markets that can absorb excess local supply, adoption of more productive agricultural technologies typically leads to a drop in farmgate product prices, erasing the gains from technological change and thereby dampening incentives for farmers to adopt new technologies that can stimulate economic growth. Markets also play a fundamental role in managing risk associated with demand and supply shocks: good market integration facilitates adjustment in net export flows across space and time, thereby reducing price variability faced by consumers and producers, while poor market integration leads directly to price volatility.

Trade

In part because sub-Saharan Africa remains heavily rural, and in part due to the high costs of commerce given poor infrastructure, sub-Saharan Africa relies less on food imports than does the rest of the world despite low agricultural productivity. This degree of self-reliance is all the more remarkable because many borders in sub-Saharan Africa were drawn arbitrarily by colonial interests, bisecting regions that are culturally similar and traditionally linked by production and marketing systems. The 1,000-mile long border between Niger and Nigeria, for example, cuts through the former Hausa empires that relied

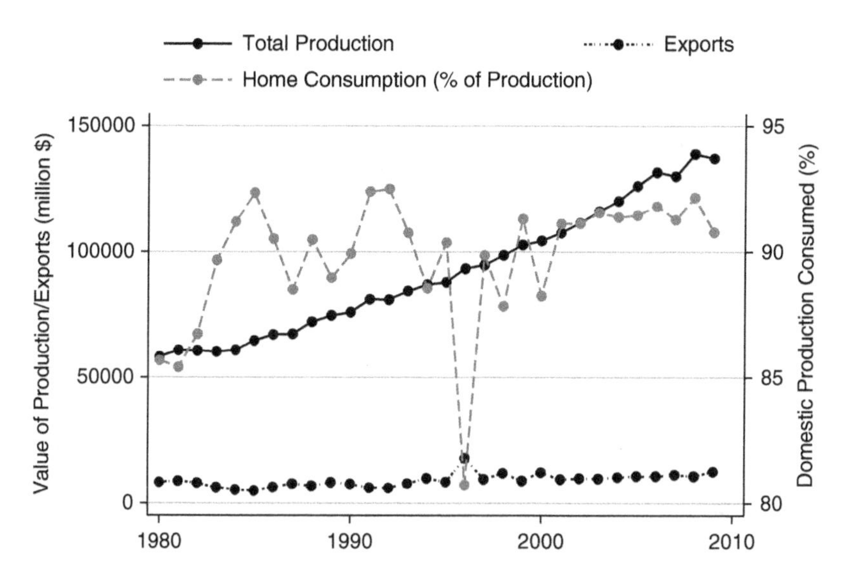

Fig. 13.3. Production, exports, and percentage consumed domestically: Sub-Saharan Africa averages, 1980 to 2009

Data Source: Food and Agriculture Organization, FAOSTAT 2012

on each other for food goods (mainly from Nigeria) and livestock (mainly from the North of Niger). A fair amount of interregional trade takes place unofficially across such borders, defying governments' attempts to control or even track it.

With that caveat in mind, it is striking how large a share of food consumed in sub-Saharan Africa is produced within the same country where it is eaten. Sub-Saharan Africa is overall a net food-importing region, meaning that less food is exported than is imported (Ng and Aksoy 2008). The bulk of production is consumed domestically and not exported (Figure 13.3). Between 1980 and 2009 the percentage of production consumed at home never fell below 80 percent, and was typically above 90 percent and increasing, on average, over the period. These figures are higher still among land-locked countries, as well as in the coastal, oil-producing states of the Gulf of Guinea. Export volumes have been noticeably flat, with virtually all expansion in output being absorbed by domestic consumers in sub-Saharan Africa.

THE AWAKENING LIONS, 2000–11

Much of sub-Saharan Africa consequently entered the twenty-first century seemingly ill-equipped to advance food security objectives by expanding supply and incomes at a sufficient pace to keep up with rapid population growth. Over the first decade of the new millennium, however, considerable divergence has taken place in the subcontinent, with a number of countries—mainly those that suffered violent conflict—continuing to languish, while some other sub-Saharan African economies began to take off. These "awakening lions" have been enjoying faster economic growth than the global average, and even faster than most of the East Asian "tigers." In eight of the last ten years, sub-Saharan African economic growth rates have exceeded those of East Asia (IMF 2012). At least seven sub-Saharan African economies (Ethiopia, Malawi, Mozambique, Rwanda, Tanzania, Uganda, and Zambia) posted annual average real GDP growth in excess of 6 percent for 2006 through 2011. Ghana has already met the Millennium Development Goal to halve hunger by 2015. Agriculture has begun to recover in the region, while population growth has slowed dramatically in a few countries. Overall, there has been notable progress in many quarters of sub-Saharan Africa over the past decade (Radelet 2010).

It is important to recognize, however, that many nations and subpopulations have been largely left behind. Overall income growth, which generally reduces the incidence of poverty, has commonly come with increased income inequality: the poorest even in the wealthiest countries still suffer from insufficient access to or availability of food. For example, the average Gini coefficient among the ten sub-Saharan African countries lowest on the United Nations Development Program (UNDP) Human Development Index (HDI) is 43, whereas for the top ten countries it is about 52, representing a more unequal distribution of income. Countries with the highest Gini coefficients include some of the "best off" by other indicators, such as South Africa, the Seychelles, the Comoros, and Botswana (UNDP 2011).

In addition, the natural resource base has been facing increased pressure, due to an apparent increase in extreme weather events such as droughts and floods and extensification of the agricultural frontier. Especially since the 2007–8 global food price spike, Africa's relatively abundant land and water resources have attracted considerable attention and investment from non-traditional sources, including native urban elites and foreign individuals, corporations, and sovereign wealth funds. These new resource deals have fueled stresses over resource access in a number of places, most notably in Madagascar, where a 2008 land deal with the Korean conglomerate Daewoo led directly to the overthrow of the Ravalomanana government.

Furthermore, while there have been some political moves to commit more resources to agriculture, supply-side developments in sub-Saharan African agriculture have remained anemic even in a period of accelerated overall economic growth. Meanwhile, the demand pressures unleashed by accelerated economic growth and increased urbanization, combined with sharply increased real food prices worldwide, have put considerable pressure on poor households to meet basic dietary needs and have, by most estimates, significantly increased food insecurity in the region. These food market pressures—a product of the confluence of both supply and demand factors—have been associated with sociopolitical unrest in quite a few sub-Saharan African states (Berazneva and Lee 2012).

Agricultural Productivity Growth and Food Availability

The first decade of the twenty-first century brought significant progress in agricultural productivity in some parts of sub-Saharan Africa, albeit with important divergence between countries (Figure 13.4). In spite of seemingly more frequent droughts and floods in the region, land productivity growth

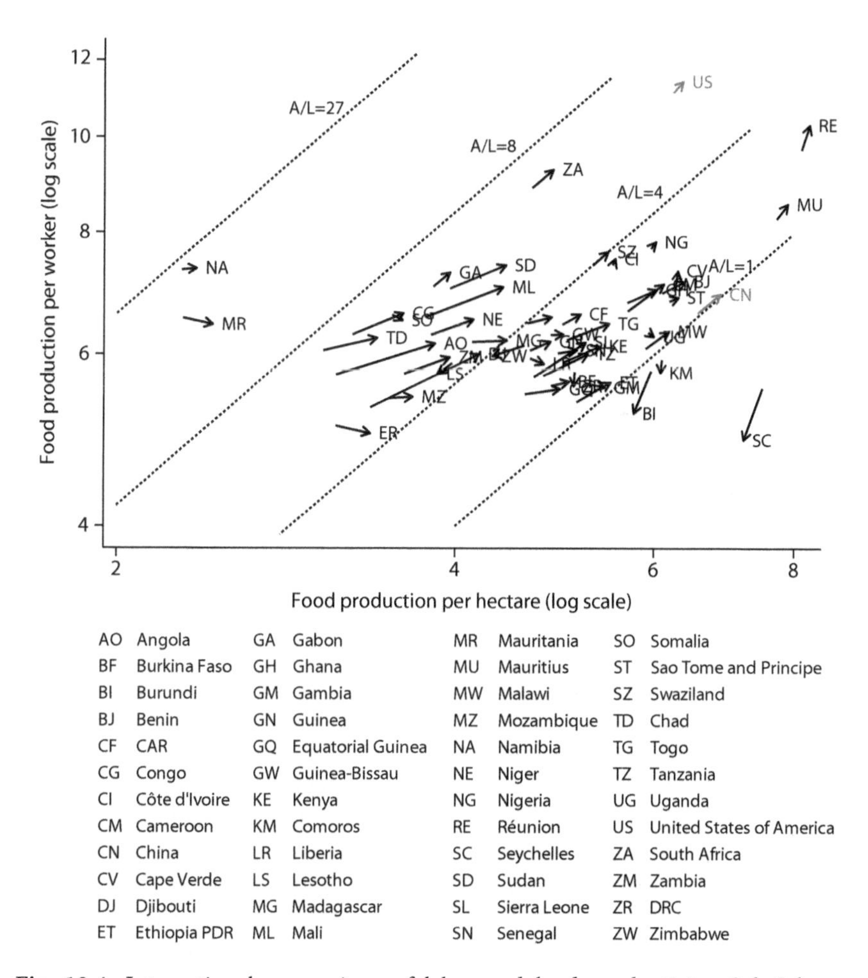

Fig. 13.4. International comparison of labor and land productivity: Sub-Saharan Africa, 2000–2009

Data Source: Food and Agriculture Organization, FAOSTAT 2012a

advanced significantly in each of sub-Saharan Africa's major regions between 2000 and 2009, with some increase in labor productivity, albeit not as rapid as yield growth. There was some notable catch-up on average by the central African states (ECCAS) and by the nations of the Horn of Africa (IGAD). The top ten countries in sub-Saharan Africa according to the UNDP HDI[2] improved faster on average than the subcontinent at large, on average by 16 percent in labor productivity (from $2,065 to $2,393 per worker, in constant 2006 dollars), and by 30 percent in land productivity (from $82 to $107 per hectare). On the other hand, the lowest 10 according to the HDI[3] improved on average by just 1 percent in labor productivity (from $336 to $341 per worker) and by 16 percent in land productivity (from $67 to $78 per hectare).

Some individual countries that improved to 2000 have stagnated, while others continue to progress, and in a few special (and small) cases—Mauritius and Reunion—have even surpassed the United States and China in average land productivity. Nigeria and Uganda, both of which made huge strides to 2000, stalled, at roughly $2,200/worker and $380/hectare, and $550/worker and $370/hectare, respectively. In some countries, food productivity declined sharply over the last decade, by 14 percent in labor productivity (from $458 to $395/worker) and 25 percent in land productivity ($99 to $75/hectare) in Zimbabwe, and similarly in the DRC, by 17 percent in labor productivity (from $316 to $261/worker) and less than 1 percent in land productivity (from $164 to $162/hectare). The declines in DRC and Zimbabwe reflect ongoing, bloody civil unrest in both countries around struggles for political succession, which is itself partly rooted in struggles to control natural resources in countries abundant in rich soils, water, and minerals. Burundi and the Seychelles similarly saw declines in both land and labor productivity over the decade. Meanwhile, countries such as Angola, Ethiopia, Malawi, Mozambique, and Tanzania showed significant progress, in the cases of Angola and Mozambique coinciding with the conclusion of drawn-out conflicts.

Part of the accelerated productivity growth in some countries is attributable to renewed government commitments to promoting and improving the agricultural sector, as reflected in the Comprehensive Africa Agriculture Development Programme (CAADP), an African Union program formed out of governments' shared commitment to agriculture-led development. CAADP sets goals related to promoting agricultural markets and regional integration, improving farmers' access to markets, combating inequality, and advancing agricultural technology. African governments agreed through CAADP to increase public investment in agriculture to a minimum of 10 percent of their budgets. This renewed commitment has sparked complementary efforts in cooperation with international donors, nonprofit organizations, and research institutions.

Great strides have been made in many specific areas. For example, the New Rice for Africa (NERICA) program, led by Africa Rice, an international research center based in Benin, introduced and promoted new, interspecific cultivars of rice developed through tissue culture techniques to cross African and Asian varieties that do not naturally interbreed. The resulting varieties have generated significantly greater yields and spread widely in West Africa over a short period of time. Meanwhile, drought-resistant maize seed that achieves more consistent yields in the face of water stress has been developed and has begun diffusing on the subcontinent.[4] A global initiative in turn successfully eradicated rinderpest, a disease that affects cattle and can be disastrous for sub-Saharan African pastoralists; the last outbreak was reported in 2001.[5] While private sector R&D has played a role, to date the vast majority of progress in developing improved plant and animal genetic material and natural resources management practices for sub-Saharan African agriculture have come from publicly funded national or international research efforts.

Fertilizer use has expanded substantially in sub-Saharan Africa over the past decade, fuelled in large measure by government subsidy programs and high-level attention afforded the subject by the 2006 Africa Fertilizer Summit in Abuja. Across sub-Saharan Africa, the tons of fertilizer nutrients applied increased by 40 percent, from 33 to 46 thousand metric tons between 2002 and 2009.[6] Fertilizer application rates remain very low in most countries in the region (with the exception of South Africa), at commonly 5 kg/hectare or less. Irrigation similarly remains extremely limited, and rapidly growing cities are increasingly competing with agriculture for access to fresh water. The release of new crop varieties and rates of farmer-level adoption of improved varieties increased noticeably from 1997–98 to 2009–10, signaling some progress in agricultural R&D and its impacts in sub-Saharan Africa over the past decade (Alene et al. 2011).

Meanwhile, CAADP rhetoric notwithstanding, the public agricultural R&D infrastructure of sub-Saharan Africa remains anemic. For example, where low- and middle-income countries in Asia and Latin America routinely have dozens, if not hundreds, of agricultural researchers per million of population economically engaged in agriculture, the comparable number is less than 10 in 24 of the 33 countries for which data are available (Alene et al. 2011). Extension services follow a similar pattern, with only one extension worker per thousand sub-Saharan African farmers, far less than the one per 200 hundred farmers ratio in developed countries.[7] Although budgets for agriculture are slowly increasing in many corners of the subcontinent, and there has been some expansion of scientific staffing for food crop improvement in particular countries and crops (Alene et al. 2011), the scientific human resource base on which to build remains woefully insufficient.

Overall, the past decade has brought spatially uneven acceleration in food productivity growth, leading to expanded food availability per capita in most countries in sub-Saharan Africa. FAO estimates of the numbers of undernourished people in sub-Saharan Africa nonetheless indicate an increase of more than 10 percent from 2000 to 2009, as the region topped 200 million hungry persons for the first time. Supply-side improvements clearly have not proved sufficient for ensuring food security for sub-Saharan Africans.

Population and Food Access

Sub-Saharan Africa remains the world's fastest growing region by population, rising from approximately 669 to 856 million people between 2000 and 2010. The FAO projects sub-Saharan Africa's population will reach 1.1 billion by 2020. This rate is, however, declining in some countries, with growth rates dropping to 2.5 percent in countries like Kenya and Nigeria as urbanization and economic growth begin to lower fertility rates (FAO 2012a). While sub-Saharan Africa will remain a majority-rural region for quite some time (Figure 13.5), urbanization rates are high and some subregions, in particular in West and Central Africa, will become majority urban by 2020.

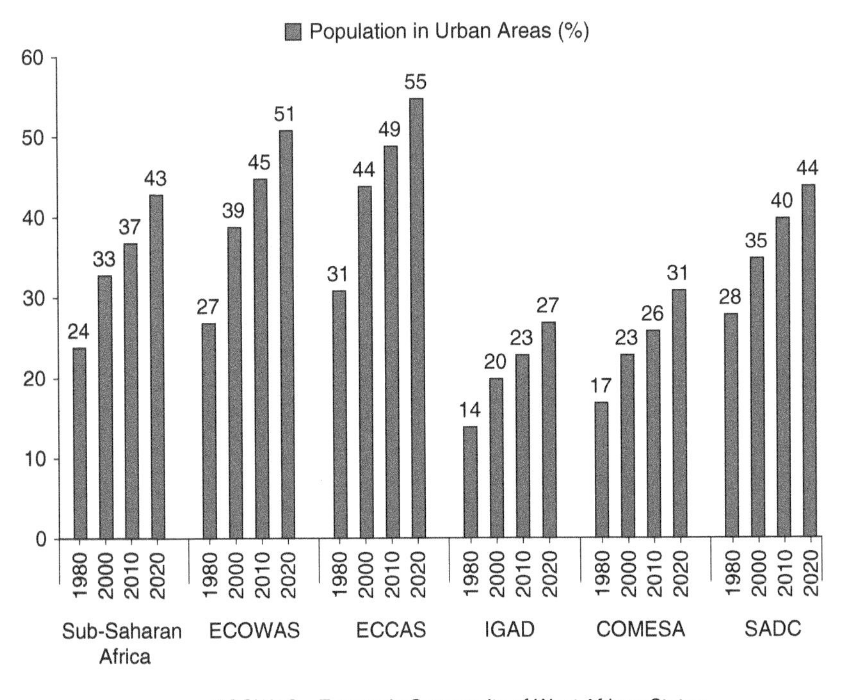

Fig. 13.5. Trends in urbanization, by subregion (percent)
Data Source: Food and Agriculture Organization, FAOSTAT 2012a

While poverty incidence and depth remain higher in sub-Saharan Africa than anywhere else in the world, they have nonetheless begun to decline in most countries. The headcount poverty indices, or percentages of the sub-Saharan African population living under $1.25 and $2.00 a day, declined from 58 and 77 percent in 1999 to 48 and 69 percent in 2008, respectively (World Bank 2012). While poverty reduction is not uniform across the sub-continent, with backsliding particularly in places that have suffered extended periods of civil unrest, rapid macroeconomic growth in countries such as South Africa, Rwanda, Uganda, Ghana, and Ethiopia is translating into fewer households living in abject poverty (World Bank 2012). Human progress is manifest as well in populations getting healthier, with increased life expectancies at birth (UNDP 2011).

The development of transportation and communication infrastructure, which is important for facilitating access to food and market integration, has been varied. Improvements in roads have been slow, although in some cases accelerated recently due to road building by Chinese investors. An information and communications technology revolution, on the other hand, has been sweeping large swathes—but not all—of sub-Saharan Africa over the past decade. Much of sub-Saharan Africa has skipped straight from reliance on dysfunctional postal systems and scarce, expensive, or nonexistent landline telephone services to cell phone and Internet connectivity that rivals that of many advanced economies. This again varies widely across the continent in availability, quality, and cost. But the percentage of inhabitants who use mobile phones exploded in the past decade, from virtually none in 1999 to approximately 60 percent of the population of sub-Saharan Africa by 2008 (Aker and Mbiti 2010). The intra-regional variation is still significant: in some countries (including South Africa and Nigeria) over 70 percent of the population owns a cell phone, whereas in others (such as Niger and the CAR) fewer than 20 percent do. Cell phones are also more prevalent in urban than in rural areas (Tortora and Rheault 2011). The spread of cell phone use has the potential to promote economic development, in particular by making agricultural markets more efficient and reducing price dispersion (Aker 2010; Jensen 2007). The region has become a pioneer in the development of mobile phone-based banking services and in crowd-sourcing of geospatially referenced information related to political unrest and natural disasters. Internet use is significantly less prevalent, but rising rapidly. The average penetration across sub-Saharan Africa is approximately 15 percent, with rates within countries ranging from near 1 percent (Ethiopia, Guinea, DRC) to 30 percent or more (Mauritius, Kenya, Nigeria; MMG 2012).

Population growth, urbanization, income growth, and increased connectivity have all fuelled growing demand for food across Africa, and increased agricultural market integration, as food marketing channels are showing early signs of value-chain transformations similar to those that occurred in Latin America and East and Southeast Asia over the past 20 years (Reardon and Timmer 2007; Gómez et al. 2011). These food-marketing channels are generally domestically oriented, however, in spite of the attention paid to export promotion. As global prices and sub-Saharan Africa food demand have increased, the region continues to rely heavily on its own output to feed its population. Net imports have increased slightly, but most of the region's added food demand has been met by expanded sub-Saharan Africa food production. The long-standing differences between coastal and land-locked countries have become more pronounced in the past decade, with the landlocked sub-Saharan Africa states now obtaining up to 95 percent of total food from domestic production.

The puzzle is that noticeable improvements in economic growth, reduced poverty, and somewhat accelerated food productivity growth have translated into more, rather than fewer, undernourished people in sub-Saharan Africa, for two reasons. First, the rate of undernourishment reflects the extraordinary heterogeneity of the subcontinent, as progress in some countries, especially those emerging from years of violence such as Mozambique, are more than offset by large, violence-wracked states such as DR Congo, Madagascar, and Sudan. Second, the global food price shocks of 2007–8 and 2010–11 outpaced economic growth in the region, leaving even previously food-secure individuals and households struggling to access adequate diets, especially as urbanizing populations became increasingly dependent on markets to obtain food.

Globalization and the Emergence of New Pressures

For decades, sub-Saharan Africa was a battleground for Cold War proxy fights and a member of various commercial and political networks that were legacies of the colonial era. That posture largely ended with the dawn of the twenty-first century. Sub-Saharan Africa's economic connections to other developing regions have been growing at an extraordinarily rapid pace. In the 1990s, the BRIIC countries (Brazil, Russia, India, Indonesia, and China) accounted for less than 1 percent of commercial trade with sub-Saharan Africa. Now they make up more than 20 percent (IMF 2012), with trade likely to continue to expand rapidly in the coming decade. Foreign direct investment, especially by Chinese firms in construction and in natural resources extractive industries, has likewise exploded throughout the subcontinent. But many outside companies now follow the example of communications giant Celtel or South African food retailer ShopRite and realize the commercial potential in growing African consumer demand: foreign direct investment in Africa is increasingly breaking from the neocolonial pattern of investment in resource extraction for export to emphasize more services provision and consumer goods delivery for domestic markets.

But alongside these generally positive changes have come new pressures and challenges, many of which exacerbate the region's food security situation. First, although sub-Saharan Africa is a net oil-exporting region, African agriculture remains highly vulnerable to higher oil prices, mainly because of the extraordinarily long and expensive supply chains between food surplus and deficit areas and between ports and many urban areas. The region's vulnerability to oil price shocks also arises from the increasingly tight connections between oil prices and those of key African food crops (maize and cassava) that are used for biofuel production elsewhere in the world and sub-Saharan Africa's nearly complete dependence on imported fertilizers made from hydrocarbons. Since few rural Africans are net sellers of staple food crops—and urban Africans are net buyers—the food price increases associated with oil price shocks are a growing threat as the region urbanizes and depends increasingly on market distribution of foods.

Second, sub-Saharan Africa has become the principal target for new investments in land and associated water resources in the wake of the 2007–8 and 2011 global food price spikes. Higher and more volatile commodity prices, combined with growing land scarcity (especially in Asia, the Middle East, and North Africa) have fuelled sharp increases in land investment. In 2009 alone, land acquisitions in sub-Saharan Africa were nearly 40 million hectares, greater than the aggregate agricultural lands of Belgium, Denmark, France, Germany, the Netherlands, and Switzerland (Deininger et al. 2011; Deininger and Byerlee 2012). While attention has focused on shady transfers to external investors, the true source of most land investment capital remains unknown. Some authoritative analysts claim that although the largest transactions have involved external investors, most of the acquisitions—by area as well as numbers of transactions—have at least officially been among nationals (Deininger and Byerlee 2012). But the admittedly incomplete Land Matrix database indicates that domestic investors accounted for just 9.7 percent of all land area acquired in sub-Saharan Africa through July 2012 (1.6 million of 16.9 million hectares), with almost 75 percent purchased by investors from other countries and 16 percent (2.4 million hectares) by investors from unknown countries.[8] In principle and under the right terms, these investments could be beneficial in providing much-needed capital, managerial expertise, and technical knowledge to help sustainably accelerate food productivity growth to feed sub-Saharan Africa's rapidly growing cities and bring down real food prices, thereby reducing poverty and hunger. The evidence indeed suggests that investors are targeting croplands with relatively high yield gaps and forestlands with good soils and ample water, all reasonably accessible to urban markets (Anseeuw et al. 2012).

But given the lack of *ex ante* clarity surrounding property rights in land and water in rural Africa and the meager—and largely unknown—net job creation and non-land complementary investment associated with these deals to date, land acquisitions have understandably acquired the pejorative label of "land grabs." Local perceptions and fears may be linked to the long history of extractive investment in resources, which have benefited a minority and in some cases spurred or helped finance conflict. Few people would rule out investment in sub-Saharan Africa's heavily undercapitalized agricultural lands, especially given their considerable and largely untapped natural potential based on soils and water availability. There are, however, competing uses for these lands—both agricultural and nonagricultural—and there are risks of displacing marginal rural populations and adding to preexisting agroecological stresses. Especially considering the lack of transparency around the terms of the deals to date, these factors have elicited appropriate concern from many quarters.

Governance and Sociopolitical Unrest

Sub-Saharan Africa's history in the late twentieth century was punctuated by difficult political transitions. The largely peaceful passing of the apartheid era in South Africa, culminating in the 1994 election that brought Nelson Mandela and the African National Congress to power, was a landmark event in a continent where state succession through open, competitive election had long been the exception. Ghana's peaceful, free, and transparent interparty electoral transfers of leadership with the 2000 and 2008 elections—after a checkered history of coups and sham elections—further reinforced an emerging, if fragile, acceptance of, and enthusiasm for, greater sociopolitical stability in the region. The 2012 peaceful transfer of power following the death of Prime Minister Meles Zenawi, the first nonviolent transition in modern Ethiopian history, was a further signal that perhaps governance is improving, at least among the best-performing economies in sub-Saharan Africa. Nation states' stability and administrative capacity to govern, develop infrastructure, implement policies, and enforce laws nonetheless remain variable, and largely weak, in sub-Saharan Africa.

There are a few attempts to measure the quality of governance, including the Transparency Index and the World Bank Worldwide Governance Indicators. The Ibrahim Index of Governance is a fairly comprehensive attempt to measure governments' success in achieving safety and the rule of law, participation and human rights, sustainable economic opportunity, and human development. According to the Ibrahim index, the lowest-ranking countries in sub-Saharan Africa are Somalia, Chad, and Zimbabwe, but DR Congo replaces Chad in the bottom three, with respect to sustainable economic opportunity, due to its dismal infrastructure and unwelcoming business environment. Mauritius, Cape Verde, and Botswana all consistently rank in the top three, followed closely by South Africa, which performs well in participation and human rights (Mo Ibrahim Foundation 2012). The generally poor state of institutions contributes to the dearth of infrastructure development and the still-limited implementation of policies that support agricultural and rural development.

Although sources of unrest are diverse and difficult to define precisely, several episodes over the past decade have been linked to the same pressures that challenge food security. The food price spikes of 2007–8 were associated with unrest in at least 14 different sub-Saharan Africa countries (Berazneva and Lee 2013; Bellemare 2011). Guinea, for example, suffered a year of riots and other unrest over high food prices. Food riots in Mozambique left at least ten people dead. The governments of both Mauritania and Cameroon were challenged during the same period, triggered initially by the rise in food prices. Similar pressures added fuel to political unrest sparked by electoral irregularities in Kenya and Zimbabwe. The generally peaceful nation of Burkina Faso experienced a few outbreaks of protest, in the spring of 2008 and again in the spring of 2011; both cases were dominated by urban consumers and merchants, who cited high food prices as a key complaint. In the 2011 case, prices were exacerbated due to conflict in Côte d'Ivoire that disrupted trade routes, and the government of Burkina Faso attempted to calm the population in part by price controls (IRIN 2012; Berazneva and Lee 2013). Berazneva and Lee (2013) note that food riots were disproportionately concentrated in urban areas of coastal countries more heavily reliant on imported foods and with lower per capita domestic food productivity.

Even with limited administrative capacity and financial resources, African governments have routinely tried to respond to exogenous price shocks with policy measures to buffer the impact on vulnerable populations. Governments able to administer such safety nets effectively have largely avoided social unrest related to food price spikes and related shocks. Demeke et al. (2009) report that by the end of 2008, at least 33 countries introduced or modified market or trade policies in an attempt to dampen food price increases. Several sub-Saharan African governments released grain from strategic reserves, others lowered duties on imported foodstuffs, while still others introduced emergency cash or food transfer programs. Government capacity and willingness to act so as to stem food price spikes that can drive vulnerable individuals and households into acute food insecurity seem key factors in peacefully relieving the sociopolitical pressure associated with price shocks when they occur.

The other main food security-related source of sociopolitical instability is competition for natural resources. Whereas food price spikes foster unrest disproportionately in urban areas, land and water disputes lead to conflict mainly in rural areas. These disputes arise from increasing population pressure, along with resource degradation, and are exacerbated by desertification, drought, flooding, and other disruptive natural phenomena. Ethnic tensions often align around competing interests in land and water. Sometimes this relates to competing uses, such as between crop cultivators and herders, which is the historical origin of many of the conflicts in Mali, Niger, Nigeria, Sudan, or Zimbabwe. In other places, tensions concern the disproportionate concentration of natural resources wealth in the hands of minority groups, as in Burundi, Kenya, Madagascar, Rwanda, South Africa, or Zimbabwe. Such places are perhaps especially prone to unrest related to land deals that seem to egregiously reinforce preexisting patterns of privilege and dispossession, as occurred in Madagascar in the wake of the 2008 Daewoo land deal. These land grabs follow on the heels of decades of government misuse of agricultural lands. Colonial powers and most independent governments have historically exploited agriculture, exhibiting a pronounced "urban bias" (Lipton 1977). Such distortions have undermined the potential of agriculture to contribute to economic growth, food security, and poverty reduction.

One needs to be careful, however, about leaping to the conclusion that food security is more than just one prospective driver of sociopolitical instability in sub-Saharan Africa. In particular, the areas where poverty and food insecurity are most intense and prevalent—the subcontinent's remote rural regions—are typically not the places where political unrest is most likely to occur in response to price shocks, land deals, or increasing land or water scarcity. This is especially true of disturbances likely to topple governments, which tend to be far more threatened by urban unrest, even if urban dissidents sometimes act in solidarity with rural kinsfolk. Hence central governments' ubiquitous efforts to try to assure adequate—and often subsidized—staple food supplies to assuage urban populations, above all in capital cities (Lipton 1977; Pinstrup-Andersen 1993).

Furthermore, instability and conflict directly impact food insecurity: the causality flows in both directions. Beyond the obvious adverse impacts of lives and property lost to violence, the mere risk of conflict commonly induces reallocation of labor and changes to crop and livestock portfolios as farmers try to reduce their vulnerability to attack (Rockmore 2012) and adversely affects health and education (Akresh and Verwimp 2006; Bundervoet et al. 2009). Farmers and traders both incur higher post-harvest losses and costs of storage and transport in riskier environments, impeding price transmission across space and time and helping foster localized price spikes (Moser et al. 2009). Moreover, by depressing incomes, sociopolitical unrest commonly exacerbates poverty and hunger (Collier 2003; Goodhand 2003).

LOOKING FORWARD: PROSPECTS FOR THE COMING DECADE

Major land acquisitions and rapid urbanization notwithstanding, sub-Saharan Africa will remain a heavily rural region disproportionately reliant on smallholder farming for the foreseeable future. Despite considerable recent progress, sub-Saharan Africa is likely to see increased populations at risk of hunger in 2025, in large measure due to the confluence of rapid population growth and higher real food prices (Rosegrant et al. 2012). So what sort of likely trajectories does this suggest for food security and sociopolitical stability over the coming decade?

Perhaps the most essential thing to keep in mind is that demand-side pressures will be considerable in the coming decade, and little can be done to slow them, given the region's demographic and economic momentum. Under the reasonable assumption that annual population and income growth will continue to average nearly 2.5 and 5.0 percent, respectively, the total income of sub-Saharan Africa will more than double by 2025; we should expect sub-Saharan Africa to demand at least half again as much food as it presently does, with market-mediated demand increasing even faster due to urbanization.

Since roughly 90 percent of sub-Saharan African food demand gets met by domestic production, the crucial implication is that African food production must increase sharply, along with imports, to forestall significant upward food price pressures. The very real prospect of higher global food prices, rapid population growth, and limited regional productivity growth leads to authoritative projections that the number of sub-Saharan Africans at risk of hunger will grow in the coming decade, even as the incidence drops slightly due to income growth (Rosegrant et al. 2012). Those higher global prices are most likely to impact the urban poor and those in rural areas that are well integrated with national and international markets.

Food production can increase through one of two pathways, both fraught with challenges. The first is increased use of inputs. The pressure to tap the significant potential for productivity growth in African agriculture and bring more uncultivated arable land into production will combine with the sharp increase in external demand for land and water to continue to fuel rapid, large-scale land acquisition. Shifting control over valuable natural resources—and popular local perceptions of land and water rights—is a major potential spark for social unrest, especially in rural areas. These inevitable land investments have the potential to generate income growth and improve food security, by boosting productivity through judicious investments that might close yawning yield gaps and generate environmentally and socially sustainable food supplies for domestic and regional markets. This potential may be realized if African states improve the terms and transparency of contracts and deals over land and other natural resources. The influence of China in competing for these resources may increase African bargaining power, and thereby open avenues for more advantageous relationships. The track record to date is, however, regrettably poor. If deals to acquire access to rural African land and water resources continue to be transacted in a less than transparent fashion and with little or no objective monitoring and evaluation, legitimate concerns will build around dispossession of the rural poor and despoliation of increasingly fragile natural resources—and the prospects for violent disputes over such transactions, or over deals gone bad, will be considerable.

The other method to substantially accelerate food output is by productivity growth through technological improvements. The prospects here are unfortunately limited, given scant agricultural R&D investments in the region. Many private companies have the capital and technical capacity to develop and disseminate new agricultural technologies but face limited potential profitability due to poor farmers' liquidity constraints and agroecological heterogeneity. Recent commitments to agriculture under the CAADP are promising, but the considerable lags involved in agricultural R&D and in disseminating new technologies to spatially dispersed farmers make the prospect of any substantial new breakthroughs unlikely in the coming decade.

Within these two fraught avenues for accelerated productivity growth in agriculture, there are four specific pathways that offer the brightest prospects, keeping in mind sustainable environmental management. Unlike the earlier Green Revolution in Asia, which developed and disseminated a few blockbuster improved seed varieties along with mass-produced inorganic fertilizers and standardized irrigation methods across vast homogeneous landscapes, the patchwork quilt of sub-Saharan Africa's heterogeneous agroecologies necessitates highly localized solutions. Development along any of the proposed pathways hence requires mobilizing local innovation by communities, farmers, and firms, and substantial change in conventional state-based agricultural R&D and extension services. These innovations could entail productive external investment in outgrower schemes or contract farming arrangements, which may preserve smallholder agriculture while increasing productivity due to improvements in more pro-poor value chains (Swinnen 2007; Reardon et al. 2009).

A first pathway is through the introduction of new cultivars, many of them genetically modified organisms (GMOs) or varieties developed in collaboration with international research laboratories. New GM varieties that are better able to withstand stresses like drought, pathogens, and pests could enable sub-Saharan Africa to begin its own belated Green Revolution. Expanded use of GMOs calls for adequate biosafety controls, yet considerable potential remains for political discord due to wildly varying views and policies with respect to GMOs. Western preconceptions have been the primary force driving controversy over GMO adoption in sub-Saharan Africa, arguably with consequences that are disadvantageous for the continent (Paarlberg 2002). Thus far, South Africa and Burkina Faso are the only two countries in sub-Saharan Africa with open cultivation of GM crops (James 2011), although many other sub-Saharan African countries have active research programs of on-station or even on-farm research with GM varieties. The pressure to expand GMO use will likely increase considerably in coming years for the same reasons pressure for land deals will increase: the confluence of eager investors and urgent need to meet growing domestic demand will compel action, often in the absence of adequate regulatory oversight. Where insufficient biosafety regulations are in place and enforced, or the contractual terms of release of GM varieties are perceived as exploitative, the almost-inevitable expansion of sub-Saharan Africa croplands cultivated with GM varieties could lead to conflict between agricultural and environmental interests, some of them external to sub-Saharan Africa.

Second, there is clear scope for expanded irrigation. Sub-Saharan Africa is by far the world's least irrigated agriculture region, largely because it remains too expensive for farmers to withdraw groundwater, given the cost of fuel and limited access to credit to invest in pumps and pipes. But the physical potential is considerable. Emergent technologies, such as solar-powered methods for withdrawing abundant groundwater supplies during periods of maximal evapotranspiration (Burney et al. 2009) or treadle pumps (Kay and Brabben 2000), could prove transformative if cost-effective means of financing uptake and maintenance emerge. When irrigation becomes affordable, attention then needs to turn to the complementary infrastructure that is necessary to facilitate adoption (Gollin et al. 2005). Disputes over access to water, however, have always been flashpoints in rural areas around the world.

Third, integrated soil fertility management methods to end or even reverse soil degradation could substantially increase yields without stimulating sharply increased reliance on unaffordable, imported fertilizers (Sanginga and Woomer 2009). Use of fallows in place of chemical fertilizers, for example, can be profitable for small farmers, if proper training and awareness programs are in place (Wanjiku et al. 2003). Proper use of livestock, such as by what has been termed "holistic management" involving short-term rotational grazing on marginal lands, has the potential to fix carbon, restore soils, and even combat desertification (Savory 1989). Efforts in this regard have been expanding, particularly in southern and eastern Africa, with support from USAID/OFDA (Bafana 2012).

Fourth, Africa's land abundance gives it a comparative advantage in livestock production. Both herd sizes and animal product output have been increasing rapidly. Crop–livestock integration not only offers valuable opportunities to sustain soils, but can provide regular financial liquidity to farmers who otherwise struggle to purchase essential inputs. Livestock also provide essential micronutrients—more than just calories and protein—that are especially crucial to women and children (Randolph et al. 2007). There is significant evidence that foods from animals are a better source of these essential micronutrients than supplements or plant-based foods (Allen 2003; Price 1939). A big challenge will be disease management, including of zoonotic diseases related to manure management that African governments have typically been unable to regulate effectively. There are opportunities to overcome these challenges. If the research community can contain poultry diseases and trypanosomiasis, thereby opening vast new areas for cattle, the impacts could be considerable on productivity growth, animal-source food supplies, and prices.

Beyond these productivity issues, several other trends merit attention. Sub-Saharan Africa's demography could prove a boon or a source of volatile tinder for political unrest. Sub-Saharan Africa's population is the world's youngest (UN DESA 2011); this implies more rapid labor productivity and income growth as sub-Saharan Africa's population moves into its most productive working years. These life cycle periods are typically characterized by great innovation, raising the tantalizing prospect that sub-Saharan Africa could quickly become the source of solutions to its and other regions' problems. Labor forces are aging and shrinking in most of the advanced economies, as well as China. With the rapid growth of the consumer population in sub-Saharan Africa, multinational executives already speak openly about the need to prepare for expansion into sub-Saharan Africa in the coming several years in order to tap a young, able workforce that is increasingly well connected to the rest of the world by information and communication technologies and markets that are ripe for growth.

The major downside of sub-Saharan Africa's demography is the rapid growth of the young working-age population in urban areas where employment prospects are currently dim. There are ambitious efforts under way in multiple sub-Saharan African countries to expand secondary and tertiary education. More education could lead to heightened expectations that could engender unrest, however, if not matched with commensurate job opportunities. Food price riots have been disproportionately concentrated in urban areas of coastal countries in recent years (Berazneva and Lee 2013). Sub-Saharan Africa is the region with the highest predicted urban population growth rates (FAO 2012a): by 2025, 16 of the 20 fastest growing cities in the world will be in Africa. Rapid urbanization, with population growth fastest in the region's economic laggards, increases the prospective volatility of young adults if they feel governments are failing to protect them from food price shocks or from the profiteering of outside or urban elite investors.

The threat of climate change, in particular desertification and increased climate variability, could contribute to or engender crisis. Some tropical grasslands and cultivated areas will become increasingly arid and unsuitable for cropping, and these risks will be exacerbated without irrigation.

Sea-level rise—which appears inevitable—could negatively impact coastal rice-producing areas (Chen et al. 2012), and also potentially contaminate freshwater supplies and agricultural lands. An additional risk is that increasing temperatures will expand the range of agricultural pests, and milder winters will increase the ability of pest populations to survive. Higher temperatures usually induce higher rainfall amounts, and some parts of sub-Saharan Africa do appear poised to enjoy resulting higher net productivity in pasture lands (Parry et al. 2007; Thornton et al. 2009). In many areas, however, rainfall distributions will become increasingly concentrated in extreme events, aggravating distributional problems with water availability. Greater inter-year volatility in rainfall and more extreme events, such as droughts and floods, put massive pressure on already tenuous agricultural systems (Barrett and Santos 2012). These varied threats posed by climate change, including desertification, all require strengthening the public sector's ability and readiness to moderate and respond to shocks.

A FRAUGHT PATH OF OPPORTUNITY

Sub-Saharan Africa is in the midst of a period of remarkable and under-recognized change. After a long period of generalized stagnation, even decline, a period of marked divergence has begun. The subcontinent's many "awakening lions" are experiencing rapid economic growth, peaceful political transitions, and significant progress in many well-being indicators, including falling poverty rates and improving measures of food security. But a substantial number of sub-Saharan African countries remain mired in civil unrest, political malaise, and weak economic performance. The coming decade is most likely to be characterized by highly varied experiences, which are impossible to summarize in a single subcontinent-wide indicator but can be best understood by assessing interconnected indicators and trends.

The great challenge for sub-Saharan Africa is that many of the subcontinent's major assets and opportunities for improving the food security of its rapidly growing population are also its biggest sources of prospective socio-political volatility. First, sub-Saharan Africa's considerable underutilized land and water resources offer an opportunity for significant expansion of food production that can generate increased employment for the rural poor, reduce real food prices in both the cities and the countryside, attract scarce foreign capital, and augment export earnings. Recent episodes of land grabbing, however, highlight the grave risks of unfettered, opportunistic investments that displace populations, undermine the sustainable management of sub-Saharan Africa's rich-but-fragile agricultural resource base, and sow social unrest. Land availability and productivity must simultaneously contend with the implications of climate change, especially desertification and the incidence and intensity of natural disasters.

Second, the yawning productivity gap between African smallholder agriculture and farmers elsewhere in the world offers the promise of productivity growth on the existing resource base. While price policies suffer from the food price dilemma—for example, raising food prices to help farmers hurts consumers, and vice versa (Timmer et al. 1983)—total factor productivity improvements can lead to all-around benefits. They typically improve farmer profitability while concentrating most welfare gains in the hands of poorer farm workers and consumers, who spend a large share of their meager incomes on food commodities (Minten and Barrett 2008). Productivity improvements may be enhanced by GMO adoption. Sub-Saharan Africa's scant agricultural research and extension capacity and extraordinary agroecological heterogeneity have slowed progress through conventional crop and livestock improvement methods, thereby increasing the latent demand for politically contentious genetically engineered crop varieties.

Third, the population bulge in sub-Saharan Africa promises an increased labor supply and potential innovation. Governments throughout the region are, somewhat belatedly, investing in the secondary and tertiary education necessary to channel this looming demographic dividend into productive nonagricultural employment. Yet education also creates a sociopolitical tinderbox of disaffected, underemployed urban youth if macroeconomic and sectoral policies do not foster rapid enough creation of attractive jobs to meet heightened expectations.

Finally, the region's sparse and low quality physical infrastructure—roads, ports, communications and electricity grids, and so on—is improving rapidly, thanks in part to modern information and communications technologies that let countries leapfrog more expensive twentieth-century systems, and in part to massive foreign direct investment, especially by China. These new technologies are rapidly enhancing market integration in the region (Aker 2010), but also raise new challenges to governments that have long suppressed dissent partly by controlling media and the flow of information. Foreign investment helps relax capital constraints and accelerate technology transfer, but can also raise a specter of neocolonialism and fan flames of nationalism in some countries.

It will be fascinating and sometimes nerve-racking to watch how these opportunities and their twinned tensions play out over the coming decade. Much will turn on how skillfully sub-Saharan African governments manage the opportunities provided by foreign investment and the growing working age population to accelerate productivity growth, while providing safety nets and social protection to ensure the food security of vulnerable subpopulations.

Widespread improvements in food security are within reach in sub-Saharan Africa, but will require concerted domestic and international efforts to reduce poverty and increase agricultural productivity for supplying domestic markets. Those two key factors will coincide with the pursuit of sociopolitical stability in the coming decade.

ACKNOWLEDGMENTS

We thank Cynthia Mathys, Rebecca Nelson, Marc Rockmore, and participants at the June 2012 authors' workshop for extremely helpful comments on an earlier draft. Any errors are our sole responsibility.

NOTES

1. The country members of each subregion are as follows: the Economic Community of West African States (ECOWAS) includes Benin, Burkina Faso, Cape Verde, Côte d'Ivoire, The Gambia, Ghana, Guinea, Guinea-Bissau, Liberia, Mali, Niger, Nigeria, Senegal, Sierra Leone, and Togo; the Economic Community of Central African States (ECCAS) includes Cameroon, the Central African Republic, Chad, Equatorial Guinea, Congo, and Gabon; the Common Market for Eastern and Southern Africa (COMESA) includes Burundi, Comoros, the Democratic Republic of the Congo, Djibouti, Egypt, Eritrea, Ethiopia, Kenya, Libya, the Seychelles, Madagascar, Malawi, Mauritius, Rwanda, Sudan, Uganda, Zambia, and Zimbabwe; the Intergovernmental Authority on Development (IGAD) includes Djibouti, Eritrea, Ethiopia, Kenya, Somalia, South Sudan, Sudan, and Uganda; the Southern African Development Community (SADC) includes Angola, Botswana, the Democratic Republic of the Congo, Lesotho, Madagascar, Malawi, Mauritius, Mozambique, Namibia, the Seychelles, South Africa, Swaziland, Tanzania, Zambia, and Zimbabwe; the landlocked states are Botswana, Burkina Faso, Burundi, the Central African Republic, Chad, Ethiopia, Lesotho, Malawi, Mali, Niger, Rwanda, South Sudan, Swaziland, Uganda, Zambia, and Zimbabwe; and lastly the Gulf of Guinea states are Angola, Benin, Cameroon, the Congo, the Democratic Republic of the Congo, Equatorial Guinea, Gabon, Ghana, Nigeria, Sao Tome and Principe, and Togo.
2. Seychelles, Mauritius, Gabon, Botswana, Namibia, South Africa, Cape Verde, Ghana, Equatorial Guinea, and the Congo.
3. Democratic Republic of the Congo, Niger, Burundi, Mozambique, Chad, Liberia, Burkina Faso, Sierra Leone, Central African Republic, and Guinea.
4. See for example the Water Efficient Maize for Africa Progress Report, Bill and Melinda Gates Foundation (http://www.aatf-africa.org/userfiles/WEMA-Progress-Report_2008-2011.pdf, accessed April 2, 2013).
5. The Global Rinderpest Eradication Programme report, http://www.fao.org/ag/againfo/resources/documents/AH/GREP_flyer.pdf (accessed April 2, 2013).
6. Based on FAO data on nitrogen and phosphate-based fertilizers.
7. This estimate appears reasonably widely, for example at http://allafrica.com/stories/201112221016.html (accessed April 2, 2013), but its original source is unclear.

8. Data drawn in late August 2012 from Land Matrix (2012), the online public database on land deals. Available online at http://landportal.info/landmatrix/get-the-detail?mode=map (accessed April 2, 2013).

REFERENCES

Ahmed, A. U., R. V. Hill, L. C. Smith, D. M. Wiesmann, and T. Frankenberger. 2007. *The world's most deprived: Characteristics and causes of extreme poverty and hunger.* Washington, DC: International Food Policy Research Institute.

Aker, J. C. 2010. Information from markets near and far: Mobile phones and agricultural markets in Niger. *American Economic Journal: Applied Economics 2*(3): 46–59.

———, and I. M. Mbiti. 2010. Mobile phones and economic development in Africa. *Journal of Economic Perspectives 24*(3): 207–32.

Akresh, R. and P. Verwimp. 2006. Civil war, crop failure, and the health status of young children. IZA Discussion Paper no. 2359.

Alene, A., Y. Yigezu, J. Ndjeunga, R. Labarta, R. Andrade, A. Diagne, R. Muthoni, F. Simtowe, and T. Walker. 2011. Measuring the effectiveness of agricultural R&D in sub-Saharan African from prespectives of varietal output and adoption: Initial results from the Diffusion of Improved Varieties in Africa Project. Conference Working Paper 7, Agricultural Science and Technology Indicators/International Food Policy Research Institute Conference, December 5–7, 2011.

Allen, L. 2003. Interventions for micronutrient deficiency control in developing countries: Past, present and future. *Journal of Nutrition 133*: 3875S.

Anseeuw, W., L. Alden Wily, L. Cotula, and M. Taylor. 2012. *Land rights and the rush for land: Findings of the global commercial pressures on land research project.* Rome: International Land Coalition (ILC).

Bafana, B. 2012. Brown revolution brings new hope. *Inter Press Service*, January 10. Available online at http://ipsnews.net/news.asp?idnews=106395 (accessed June 5, 2012).

Barrett, C. B. 2008. Smallholder market participation: Concepts and evidence from Eastern and Southern Africa. *Food Policy 33*(4): 299–317.

———. 2010. Measuring food insecurity. *Science 327*(5967): 825–8.

———. forthcoming. Assisting the escape from persistent ultra-poverty in rural Africa. In *Stanford synthesis volume on global food policy and food security in the 21st century*, ed. W. P. Falcon and R. L. Naylor. Palo Alto, CA: Stanford Center on Food Security and the Environment.

———, and P. Santos. 2012. The impact of changing rainfall variability on resource-dependent wealth dynamics. Cornell University Working Paper. Ithaca, NY: Cornell University.

Bellemare, M. F. 2011. Rising food prices, food price volatility, and political unrest. SSRN Working Paper.

Berazneva, J., and D. Lee. 2013. Explaining the African food riots of 2007–2008: An empirical analysis. *Food Policy 39*(April): 28–39.

Bruinsma, J. 2011. The resources outlook: By how much do land, water and crop yields need to increase by 2050? In *Looking ahead in world food and agriculture: Perspectives to 2050*, ed. P. Conforti, Chapter 6. Rome: FAO.

Bundervoet, T., P. Verwimp, and R. Akresh. 2009. Health and civil war in rural Burundi. *Journal of Human Resources 22*(2): 536–63.

Burney, J., L. Woltering, M. Burke, R. Naylor, and D. Pasternak. 2009. Solar-powered drip irrigation enhances food security in the Sudano-Sahel. *Proceedings of the National Academy of Sciences 107*(5): 1848–53.

Chen, C-C., B. McCarl, and C-C. Chang. 2012. Climate change, sea level rise and rice: Global market implications. *Climatic Change 110*(3–4): 543–60.

Collier, P. 2003. *Breaking the conflict trap: Civil war and development policy.* Washington, DC: The International Bank for Reconstruction and Development/ The World Bank.

Coltart, D. 2002. Zimbabwe's man-made famine. *New York Times,* August 7. Available online at: http://www.nytimes.com/2002/08/07/opinion/zimbabwe-s-man-made-famine.html?pagewanted=all&src=pm (accessed April 1, 2012).

Deininger, K., and D. Byerlee. 2012. The rise of large farms in land abundant countries: Do they have a future? *World Development 40*(4): 701–14.

———, J. Lindsay, A. Norton, H. Selod, and M. Stickler. 2011. Rising global interest in farmland: Can it yield sustainable and equitable benefits? Agriculture and World Development Series, World Bank.

———, and S. Jin. 2006. Tenure security and land-related investment: Evidence from Ethiopia. *European Economic Review 50*(5): 1245–77.

Demeke, M., G. Pangrazio, and M. Maetz. 2009. Country responses to the food security crisis: Nature and preliminary implications of the policies pursued. Rome: Food and Agriculture Organization.

Evenson, R. E., and D. Gollin. 2003. Assessing the impact of the Green Revolution, 1960 to 2000. *Science 300*(5620): 758–62.

Food and Agriculture Organization. 1996. Rome declaration on world food security and world food summit plan of action. World Food Summit November 13–17, 1996. Rome: FAO.

———. 2011. *State of food insecurity.* Rome: FAO.

———. 2012a. FAOSTAT, FAO Statistics Division. Available online at http://faostat.fao.org/ (accessed April 1, 2012).

———. 2012b. Terrastat: Natural resources and environment. Available online at http://www.fao.org/nr/aboutnr/nrl/en/ (accessed April 1, 2012).

Gollin, D., M. Morris, and D. Byerlee. 2005. Technology adoption in intensive post-green revolution systems. *American Journal of Agricultural Economics 87*(5): 1310–16.

Gómez, M. I., C. B. Barrett, L. E. Buck, H. De Groote, S. Ferris, H. O. Gao, E. McCullough et al. 2011. Research principles for developing country food value chains. *Science 332*(6032): 1154–5.

Goodhand, J. 2003. Enduring disorder and persistent poverty: A review of the linkages between war and chronic poverty. *World Development 31*(3): 629–646.

Hayami, Y., and V. W. Ruttan. 1971. *Agricultural development: An international perspective.* Baltimore, MD: Johns Hopkins University Press.

Holden, S. T., K. Deininger, and H. Ghebru. 2008. Impact of low-cost land certification on investment and productivity. *American Journal of Agricultural Economics 91*(2): 359–73.

Integrated Regional Information Network (IRIN). 2012. Analysis: Burkina Faso's uneasy peace. IRIN Humanitarian News and Analysis. Available online at http://www.irinnews.org/InDepthMain.aspx?indepthid=72&reportid=95060 (accessed March 12, 2012).

International Monetary Fund. 2012. IMF Data and Statistics. Available online at www.imf.org/external/data.htm (accessed May 2, 2013).

James, Clive. 2011. Global Status of Commercialized Biotech/GM Crops: 2011. *ISAAA Brief* No. 43. Ithaca, NY: ISAAA.

Jensen, R. 2007. The digital provide: Information (technology), market performance, and welfare in the South Indian fisheries sector. *Quarterly Journal of Economics 122*(3): 879–924.

Kay, M., and T. Brabben. 2000. Treadle pumps for irrigation in Africa. A report for IPTRID. Knowledge Synthesis Report No. 1—October 2000. Rome: International Program for Technology and Research in Irrigation and Drainage, Food and Agriculture Organization of the United Nations.

Lahiff, E., and B. Cousins. 2001. The land crisis in Zimbabwe viewed from south of the Limpopo. *Journal of Agrarian Change 1*(4): 652–66.

Lipton, M. 1977. *Why poor people stay poor: A study of urban bias in world development.* London: Temple Smith.

Miniwatts Marketing Group. 2012. Internet world stats: Usage and population statistics. Available online at http://www.internetworldstats.com/stats1.htm.

Minten, B., and C. B. Barrett. 2008. Agricultural technology, productivity and poverty in Madagascar. *World Development 36*(5): 797–822.

Mo Ibrahim Foundation. 2012. Ibrahim Index of African Governance: Data Report. London: Mo Ibrahim Foundation. Available online at www.moibrahimfoundation.org/download/2012-IIAG-data-report.pdf (accessed April 2, 2013).

Moser, C., C. B. Barrett, and B. Minten. 2009. Spatial integration at multiple scales: Rice markets in Madagascar. *Agricultural Economics 40*(3): 281–94.

Moyo, S., and P. Yeros. 2005. *Reclaiming the land: The resurgence of rural movements in Africa, Asia, and Latin America.* New York: Zed Books.

Ng, F., and M. Ataman Aksoy 2008. Who are the net food importing countries? Policy Research Working Paper 4457. Washington, DC: World Bank.

Paarlberg, R. L. 2002. The real threat to GM crops in poor countries: Consumer and policy resistance to GM foods in rich countries. *Food Policy 27*(2002): 247–50.

Parry, M. L., O. F. Canziani, J. P. Palutikof, P. J. van der Linden, and C.E. Hanson, eds. 2007. *Contribution of working group II to the fourth assessment report of the Intergovernmental Panel on Climate Chagne.* Cambridge and New York: Cambridge University Press.

Pender, J., and M. Fafchamps. 2006. Land lease markets and agricultural efficiency in Ethiopia. *Journal of African Economies 15*(2): 251–84.

Pinstrup-Andersen, P. 1993. *The political economy of food and nutrition policies.* Washington, DC: International Food Policy Research Institute.

Price, W. A. 1939. *Nutrition and physical degeneration.* San Diego: Price-Pottenger Nutrition Foundation.

Radelet, S. 2010. *Emerging Africa: How 17 countries are leading the way.* Washington, DC: Center for Global Development.

Randolph, T., E. Schelling, D. Grace, C. Nicholson, J. Leroy, D. Cole, M. Demment, A. Omore, J. Zinsstag, and M. Ruel. 2007. Role of livestock in human nutrition and health for poverty reduction in developing countries. *Journal of Animal Science 85*: 2788.

Reardon, T., and C. P. Timmer. 2007. Transformation of markets for agricultural output in developing countries since 1950: How has thinking changed? In *Handbook of Agricultural Economics Volume 3*, ed. Robert Evenson and Prabhu Pingali, 2807–55. Amsterdam: Elsevier.

——, C. B. Barrett, J. A. Berdegué, and J. F. M. Swinnen. 2009. Agrifood industry transformation and small farmers in developing countries. *World Development 37*(11): 1717–27.

Rockmore, M. 2012. Living within conflicts: Risk of violence and livelihood portfolios. Households in Conflict Network Working Paper Number 121.

Rosegrant, M. W., S. Tokgoz, and P. Bhandary. 2012. Future of the global food economy. Working Paper. Washington, DC: International Food Policy Research Institute.

Sahn, D. E., and D. C. Stifel. 2003. Progress toward the Millennium Development Goals in Africa. *World Development 31*(1): 23–52.

Sanginga, N., and P. L. Woomer, ed. 2009. *Integrated soil fertility management in Africa: Principles, practices, and developmental process.* Nairobi, Kenya: Tropical Soil Biology and Fertility Institute of the International Centre for Tropical Agriculture (TSBF-CIAT).

Savory, A. 1989. A solution to desertification: Holistic resource management. Paper 27a. Albuquerque, New Mexico: Center for Holistic Resource Management.

Swarns, R. L. 2002. For Zimbabwe white farmers, time to move on. *New York Times*, August 4, 2002. Available online at http://www.nytimes.com/2002/08/04/world/for-zimbabwe-white-farmers-time-to-move-on.html?pagewanted=all&src=pm (accessed April 2, 2013).

Swinnen, J. F. M. 2007. *Global supply chains, standards and the poor: How the globalization of food systems and standards affects rural development and poverty.* Wallingford, UK: CABI.

Thornton, P. K., P. G. Jones, G. Alagarswamy, and J. Andersen. 2009. Spatial variation of crop yield response to climate change in East Africa. *Global Environmental Change* (19)*1*: 54–65.

Timmer, P. 1988. The agricultural transformation. In *Handbook of Development Economics, Volume 1*, ed. Hollis Chenery and T. N. Srinivason, 275–331. Amsterdam, Oxford: North-Holland Press.

——, W. P. Falcon, and S. R. Pearson. 1983. *Food policy analysis.* Baltimore: Johns Hopkins University Press.

Tortora, B., and M. Rheault. 2011. Mobile phone access varies widely in sub-Saharan Africa. *GALLUP World*, September 16, 2011. Retrieved from http://www.gallup.com/poll/149519/mobile-phone-access-varies-widely-sub-saharan-africa.aspx (accessed April 2, 2013).

Uppsala Conflict Data Project, International Peace Research Institute (2012). Armed Conflict Dataset v.5-2012b, 1989–2011. Available online at http://www.pcr.uu.se/research/ucdp/datasets/ucdp_battle-related_deaths_dataset/ (accessed April 2, 2013).

United Nations, Department of Economic and Social Affairs (2011). Population Division, Population Estimates and Projections Section: *World population prospects: The 2010*

revision. Available online at http://esa.un.org/unpd/wpp/Excel-Data/population.htm (accessed April 1, 2012).
United Nations Development Program. 2011. *Human development index.*
United Nations Economic Commission for Africa. 2007. Africa review report on drought and desertification. Available online at http://www.uneca.org/csd/csd5/ACSD5-SummaryReportonDrought.pdf (accessed March 23, 2012).
Wanjiku, J., C. Ackello-Oguto, L. N. Kimenye, and F. Place. 2003. Socio-economic factors influencing use of improved fallows in crop production by small-scale farmers in western Kenya. *African Crop Science Conference Proceedings* 6: 597–601.
World Bank. 2012. The World Bank open data, by indicator. Available online at http://data.worldbank.org/ (accessed April 1, 2012).
World Food Programme. 2012. Food Aid Information System. Available online at http://www.wfp.org/fais/reports/ (accessed April 2, 2013).

14

Lessons from the Arab Spring: Food Security and Stability in the Middle East and North Africa

TRAVIS J. LYBBERT AND
HEATHER R. MORGAN

While lingering economic crises, recent food price spikes, and looming climate change have brought food security concerns to the fore in many policy arenas and in many countries, food insecurity and its broader social and political causes and consequences have been prominently displayed in the recent unrest across the Middle East and North Africa region. The complex links between food insecurity and political instability will profoundly shape the region in the coming decade—with important implications for global politics and markets. This chapter sizes up the current food security situation and other related concerns across the Middle East and North Africa, explores the political and policy responses to these concerns, and offers a framework of drivers that will shape the evolving interaction of food security and sociopolitical conditions in the coming decade.

Although extreme and chronic uncertainty regarding basic needs is typically suffered privately, the anxiety and outrage it produces can trigger public acts of protest. The oft-cited beginning of the Arab Spring—the self-immolation of a Tunisian street vendor named Mohamad Bouazizi—epitomizes such acts of desperation. That this single act is credited with catalyzing unrest across the Arab world indicates widespread latent frustration and outrage in the region. These popular sentiments are by no means homogeneous in the Middle East and North Africa, but nonetheless stem from some shared institutional and biophysical features and a confluence of political, economic, and demographic factors. Food security and sociopolitical stability will continue to play a central role in this unfolding transformation of the region.

Due to the relative scarcity of both arable land and water, the Middle East and North Africa region imports more than half its food, a higher import dependency than any other region of the world. With a hefty appetite for wheat, these countries are especially reliant on cereal imports and have seen cereal trade deficits grow steadily in the recent decade (Breisinger et al. 2010). In 2007 the Middle East and North Africa imported 58 million metric tons of cereals, making them particularly vulnerable to food price shocks (Harrigan 2011). Although the region has avoided widespread famine in recent decades, frequent droughts have triggered serious fiscal stress to many governments that are reliant on food subsidies to placate their populations. These droughts also directly threaten poor households in both rural and urban settings across much of the region. The 2008 food price spikes exposed this vulnerability, forcing countries to rethink their policy efforts to hedge this risk at the macroeconomic level and to reduce household vulnerability to such shocks.

Understanding the links between food insecurity and instability in the Middle East and North Africa requires an appreciation for differences among countries in the region. This chapter builds on a recent International Food Policy and Research Institute (IFPRI) classification of Middle East and North Africa countries according to their food security status and their mineral and oil resources, with discussion of recent policy responses to food security threats. The dimensions of heterogeneity captured by this classification provide an important framework for understanding food policies, past and future. Some of the region's countries have successfully leveraged general economic growth to reduce poverty in the past, but in others, this process has been far from complete and has stalled more recently. Undernutrition and hunger remain pronounced in many rural and urban settings, even as average poverty levels have fallen.

In the coming decade, land and water constraints will continue to tighten across the Middle East and North Africa, and may be exacerbated by lower precipitation and more severe drought as forecasted by climate change models. Additional critical stressors for shaping the food security landscape in the region include growing and young populations, high unemployment rates among young adults, state-dominated economies, and rapid and uncertain political changes in many countries. This chapter concludes with an exploration of these stressors and likely scenarios for the near future.

SIMILARITY AND HETEROGENEITY IN THE MIDDLE EAST AND NORTH AFRICA

Home to one of the world's richest countries, Qatar, and to desperately poor countries, including Sudan and Yemen, the Middle East and North Africa region is far from monolithic. Many of these stark differences stem from inequities in mineral resources: countries with a history of exploiting oil reserves differ politically, economically, and socially from those without oil. However, these nations are still profoundly similar in many ways.

Three primary types of similarities unite the Middle East and North Africa. First, the region shares several climatic features; even with some climatic diversity, these countries are on average dry and hot, which partially explains the region's relatively high reliance on food imports. Most climate models predict the region will also face a yet drier and hotter future.[1] Second, the Middle East and North Africa region shares important cultural and religious roots. As evidence of these similarities, the region is defined interchangeably by geography (Middle East and North Africa) or language and culture ("Arab world").[2] Third, the region shares some distinctive political and economic features. In a recent analysis of the economic underpinnings of the Arab Spring, Adeel Malik and Bassem Awadallah (2011) argue that several political and economic common denominators cut across the region:

> First, all across the Arab world both economic and political power is concentrated in the hands of a few. Second, the typical Arab state can be characterized as a security state; its coercive apparatus is both fierce and extensive. Third, the broad contours of demographic change and the resulting youth bulges are fairly common across the region. Fourth, Arab countries are mostly centralized states with a dominant public sector and, with a few exceptions, weak private enterprise. Fifth, external revenues—whether derived from oil, aid, or remittances—profoundly shape the region's political economy (4).

These interrelated features have important implications for both food security and sociopolitical stability. For example, the "youth bulges" that are common in the region (over a third of the Middle East and North Africa population is under age 15), combined with stagnant private enterprise, and the fact that "the limited economic opportunities that do exist are rationed by connection rather than competition," (Malik and Awadallah 2011, 3) distinguish the region as the worst place for a young person to find work. Every year since 1991, the region has had the highest youth unemployment rate in the world, ranging from 24 to 30 percent (International Labor Organization 2011). The tension created by expanding education among young adults without a proportionate increase in economic opportunity is a common concern in the region (Campante and Chor 2012).

Against this backdrop of broad similarities across the region, we now add dimensions of heterogeneity, shaping how Middle East and North Africa countries experience and respond to food security and instability. Table 14.1 uses an IFPRI food security typology to depict some of these differences. Food security status and mineral wealth are particularly relevant, given the region's heavy reliance on food subsidies. Across these categories, there are stark differences in income, food production, and percentage of the population in rural areas. While average annual income per capita in the relatively food insecure

Table 14.1. Food security typology for Middle East and North Africa countries (IFPRI 2010) with food security related indicators

	GDP per capita ($)	Food commodity production per capita* ($)	Rural Population (%)	Global Hunger Index	Macro Food Security (%)	Micro Food Security (%)	Food Security Status
Food security challenge countries							
Mineral resource rich							
Algeria	4,567	111	36	4	7.3	15.6	serious
Iraq	2,565	n.a.	33	n.a.	n.a.	27.5	serious
Libya	9,957	133	15	4	3.4	21	moderate
Sudan	1,425	148	58	21.5	8.4	37.9	alarming
Syria	2,893	237	49	10.5	9.7	28.6	serious
Yemen	1,300	44	72	11.2	15.4	59.6	ex. alarming
Mineral resource poor							
Djibouti	1,203	54	14	22.5	42.3	32.6	ex. alarming
Egypt	2,698	199	57	7.9	8.7	30.7	serious
Jordan	4,560	120	17	4	13.9	8.3	serious
Lebanon	9,228	258	13	4	16.5	15	serious
Morocco	2,796	163	41	5.9	8.2	21.6	serious
Tunisia	4,199	220	34	4	6.5	9	moderate
West Bank and Gaza	n.a.	135	28	11.2	31.9	11.8	serious

(Continued)

	GDP per capita ($)	Food commodity production per capita* ($)	Rural Population (%)	Global Hunger Index	Macro Food Security (%)	Micro Food Security (%)	Food Security Status
Food secure countries							
Mineral resource rich							
Iran	4,526	246	33	12.2	2.4	16.6	moderate
Bahrain	17,609	n.a.	3	n.a.	2.9	9	low
Kuwait	41,365	55	2	4	3.8	4	low
Saudi Arabia	15,836	104	19	4	4.0	9.1	low
United Arab Emirates	39,623	114	23	4	3.4	n.a.	low
Qatar	61,532	n.a.	28	n.a.	2.0	4	low
Oman	17,280	n.a.	5	n.a.	6.2	9.6	low
MENA average	12,903	146	29.0	8.4	11.2	19.1	n.a.
MENA—Food security challenge	3,949	152	35.9	9.0	14.0	25.4	n.a.
MENA—Food secure	28,253	130	16.1	6.1	3.5	8.7	n.a.
World average	9,174	233	49.4	12.1	n.a.	n.a.	n.a.

* Estimated as the value of edible commodities per capita not including value added in food processing, etc.
Source: Breisinger, et al. 2010, Breisinger, et al. 2012, World Bank 2012., Yu et al. 2010, Von Grebmer et al. 2010.

countries is less than $4,000, income in the food secure countries is more than $28,000, and twice as many people live in rural areas in the food insecure countries than in the food secure countries. On average, however, the Middle East and North Africa region has relatively high urbanization rates, compared to countries of similar income levels.

Differences in food security status in the region are also clear in Table 14.1. IFPRI's global hunger index varies widely in the region. The macro food security measure—food imports as a percentage of total exports and net remittances—captures the ability of a country to purchase food on international markets through export earnings (Diaz-Bonilla et al. 2000; Yu et al. 2010) and remittances (Breisinger et al. 2012).[3] By this measure, Djibouti and the West Bank and Gaza have serious macro food security challenges, with Lebanon, Yemen, and Jordan not far behind. A recent World Bank report offers an alternative macro perspective on food security differences in the region by considering cereal import dependency and fiscal position, indicating Lebanon, Jordan, Djibouti, Yemen, Tunisia, and Morocco as the most vulnerable countries. Syria, Egypt, and Sudan are less dependent on cereal imports, but fiscally strained (World Bank 2009). Every Middle East and North Africa country except Morocco is expected to become more dependent on cereal imports in the coming decade (Breisinger et al. 2010). Micro food security is measured as the percentage of children who are stunted—short for their age—a telling measure of chronic food stress (Breisinger et al. 2012). While this measure makes Yemen and Sudan very food insecure, even less poor Egypt, Morocco, and Syria have relatively high micro food insecurity. The final column in Table 14.1 provides a categorical assessment based on national macro and micro food security measure averages.[4]

Food security status differences reflect other dimensions of differences. Figure 14.1 depicts agricultural distinctions. Though the region's averages are the driest in the world, average annual rainfall by country varies from 50 mm to 350 mm. As a share of GDP, agriculture plays a more important role in food security-challenged countries than in food secure countries. This globally evident pattern has a distinctive flavor in the Middle East and North Africa due to the role of mineral resources in the economies' evolution and structure. The percentage of agricultural land that is irrigated also varies widely: less than 5 percent of agricultural land is irrigated in the Maghreb (Morocco, Tunisia, and Algeria), but more than 90 percent of Egyptian agricultural land is irrigated. Differences in irrigation infrastructure explain why some climate models predict improvements in agricultural productivity in Egypt and declines in the rest of the region (Cline 2007).

Although corruption and cronyism is a general problem in the region (Malilk and Awadallah 2011), countries with food security challenges tend to be more corrupt and less amenable to private business than their food secure neighbors. Similarly, while youth unemployment rates in the region are the highest in the world, there are some significant differences within the region, with the rate ranging from under 10 percent in some of the Gulf States, to

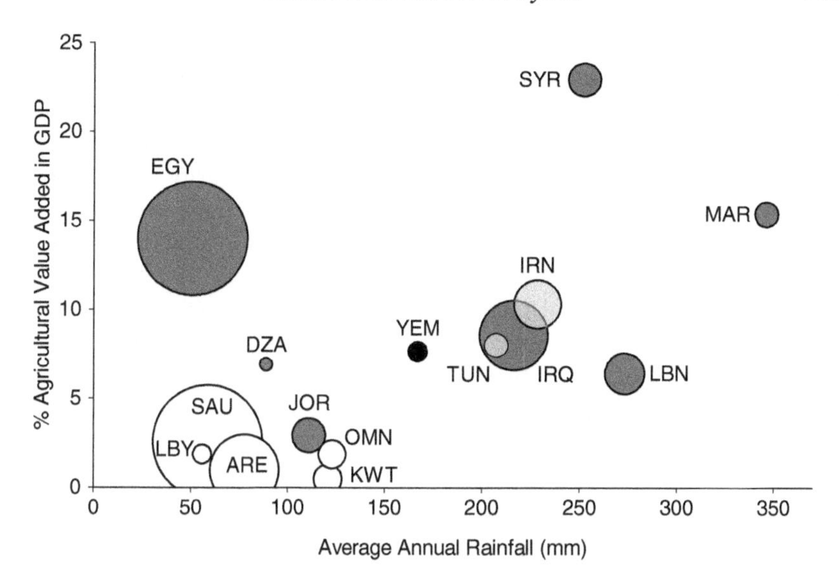

Fig. 14.1. Irrigated agricultural land, agriculture's contribution to GDP, and average rainfall in the Middle East and North Africa region

Source: World Bank 2012

Note: Bubble size indicates the percent of agricultural land that is irrigated. Bubble shading indicates food security status (see Table 14.1) with darker shading denoting more serious food security challenges.

ARE—United Arab Emirates, DZA—Algeria, EGY—Egypt, IRN—Iran, IRQ—Iraq, JOR—Jordan, KWT—Kuwait, LBN—Lebanon, LBY—Libya, MAR—Morocco, OMN—Oman, QAT—Qatar, SAU—Saudi Arabia, SYR—Syria, TUN—Tunisia, YEM—Yemen

above 35 percent in Algeria, Iraq, and Jordan. These rates are notoriously difficult to estimate in places with substantial informal economies (Egyptian estimates range from 20 to 40 percent, for example), but youth unemployment is clearly one of the region's major contemporary challenges.

RECENT FOOD SECURITY THREATS AND RESPONSES

Building on the prior section's broad characterization of the region, we now explore recent food security challenges and food policy responses in the Middle East and North Africa. After describing these experiences and responses, we assess how food insecurity and sociopolitical instability are interrelated in the region. The 2011 Arab Spring provides rich and recent regionwide evidence of potential interlinkages, which we exploit to make the discussion concrete. We then explore in greater detail four countries—Egypt, Morocco, Tunisia, and Syria—whose recent experiences of food insecurity each shed a different light on the discussion.

Food security implies that "all people, at all times, have physical, social, and economic access to sufficient, safe, and nutritious food which meets their dietary needs and food preferences for an active and healthy life"[5] and encompasses food availability, access, utilization, and stability. Given the region's longstanding food imports dependency, food security in the Middle East and North Africa has for decades hinged on macro trade and fiscal policies. Recent decades of population growth and more frequent drought have brought greater urgency to addressing macro food security threats. The region's population, now over 300 million, tripled from 1960 to 2006, maintaining 1.7 percent annual growth. Of the 12 driest winters in the region since 1902, ten occurred in the last two decades (Hoerling et al. 2011). These trends, combined with strained water resources and lagging yield increases, underlie widespread expectations of and increasing import dependence in the coming decades (Breisinger et al. 2012) and escalating food security urgency.

Since the 1970s, the common Middle East and North Africa policy response to similar episodes of urgency has aimed for greater food self-sufficiency by encouraging cereals production. Although many of the world's most valuable trade routes run through or near the region, longstanding intraregional and international trade barriers for nearly everything but oil have led to relatively anemic trade flows in the region (Malik and Awadallah 2011). In such a setting, the production-based approach that emphasizes food self-sufficiency may be preferred to a market-based approach that exploits comparative advantage and international trade. Still, it is important to acknowledge just how costly it is for the world's driest region to grow enough wheat to support 300 million people (Magnan et al. 2011). Some oil-rich nations have shifted their focus to accessing arable land in Africa through long-term contracts (see Deninger this volume), but this option is unlikely to be viable for Middle East and North Africa countries without mineral reserves (Wharton 2011).

Food security also requires food access, utilization, and stability at the household level. Many Middle East and North Africa countries have made important strides in alleviating poverty and improving primary and secondary education, generally improving household food security. However, micro-level food insecurity persists as a stubborn regional problem because of national disparities in assistance, resources, and infrastructure determined by who does and does not have favorable connections with the regime. This lopsided political power, combined with corruption, nepotism, and limited resources can produce pockets of severe food insecurity among disparaged minority groups. Indeed, those in power may have little genuine interest in widespread food security, potentially even benefitting politically from persistent food insecurity among some citizens.

Macro measures of economic growth in the region often mask fundamental weaknesses in the private sector. Only private enterprises with state connections seem to thrive in the Middle East and North Africa, and crony capitalism stifles much of the economic dynamism for incentives and opportunities in both rural and urban settings (Malik and Awadallah 2011). Although some countries—including Egypt, Tunisia, and Morocco—have recently attempted to integrate private agribusiness into agri-food chains, typically agriculture is either explicitly or implicitly state dominated. Productivity suffers since most agricultural input and output markets function poorly. Consequently, heavy food and energy subsidies are the most popular policy remedy in the region for household food security threats. While these subsidies do increase poor households' access to food, they tend to be poorly targeted, inefficient, subject to corruption, and fiscally expensive. Even in 2007—before food prices spiked—Syria, Egypt, and Morocco, respectively, spent 2.1 percent, 1.3 percent, and 0.7 percent of their GDP on food subsidies (World Bank 2009). By spurring consumption of cheap bread, flour, sugar, and oil, these subsidies may be contributing to rapid increases in obesity in the region (Asfaw 2007). Together, these regional food subsidy features suggest they effectively function as a quid pro quo with the broader populace to stave off feelings of disenfranchisement and to diffuse frustrations and dissent. While oil-rich countries can obviously afford greater generosity in this regard, versions of this "autocratic bargain" are apparent throughout the region (Diwan 2012).

In the past five years, much of the region has wrestled with increasing urgency with new twists on old food security concerns. In 2001, roughly 22 million people in the Middle East and North Africa were considered undernourished, but the 2007 to 2008 food price spike added an additional four million people to this tally (IMF 2008). Today, about 7 percent of the Middle East and North Africa population is considered undernourished. The rural poor—76 percent of all the poor in the region—spend 35 percent to 65 percent of the household budget on food. Since many Middle East and North Africa households hover just above the official $1.25/day poverty line, this price spike forced an additional 2.19 million people below the line. The urban poor and rural poor with small or no landholdings were especially hard hit, especially in Palestine, Yemen, Syria, Sudan, Egypt, and Jordan (ESCWA 2010).

In the wake of these food prices spikes, demonstrations and rioting occurred in Yemen, Jordan, Morocco, Lebanon, Bahrain, Saudi Arabia, Morocco, and Tunisia—often with protesters symbolically brandishing loaves of bread (Johnstone and Mazo 2011). Despite the fiscal distress that the price spikes imposed on the Middle East and North Africa countries without oil exports, many governments increased food and other subsidies in an attempt to calm the urban unrest (Harrigan 2011). Food subsidies were far and away the most popular policy response to price spikes, although many governments tried to provide added protection for poor consumers through a suite of policy responses and social protection programs, including reducing taxes, releasing stocks, and restricting exports for food grains, as well as cash transfers, food-for-work programs, food rations, and school feeding programs. While these policy responses and social protection programs aimed to protect poor consumers, they were often ineffective at targeting needy consumers, and likewise unsuccessful at defusing social unrest when enough other grievances were present.

LINKS BETWEEN FOOD INSECURITY AND INSTABILITY IN THE MIDDLE EAST AND NORTH AFRICA

The speed at which open dissent, protest, and revolution swept across much of the Middle East and North Africa in 2011 even surprised regional experts. Much of the region had been grappling with serious unemployment, inequality, food insecurity, and social unrest for years. While recent years have seen a general trend of growing dissatisfaction with life in much of the region (Breisinger et al. 2012), the young and the poor have felt particularly disenfranchised, both politically and economically. To maintain stability and placate their people, the authoritarian regimes of the region long relied on a mix of repression, coercion, and persuasion, including hefty food subsidies. With tensions brewing and budgets tightening, the 2008 food price spike pushed many countries to the brink and sparked food riots in many cities. When food prices spiked again in 2010, similar riots quickly escalated into massive and daring protests.[6] Although many factors set the stage for the Arab Spring, many claim food insecurity to be the smoking gun (Friedman 2012; Johnstone and Mazo 2011).

The region has several admittedly extreme examples of how severe unrest and conflict can undermine basic food security. The blockade of Gaza, resulting from the Israeli–Palestinian conflict, has directly induced severe food insecurity among many Palestinians living in Gaza. Iraq has also provided ample evidence of this causal impact, and the more recent uprisings, violence, and instability in Libya and Syria have shed additional light on this linkage. Syria powerfully demonstrates how sociopolitical instability can cause serious food security problems: the desperate and violent repression of civil unrest by the Assad regime has devastated the economy in general, reduced agricultural production substantially, induced rampant inflation in food prices, and triggered international sanctions that deepen food insecurity among an ever-expanding population of food insecure households (FAO 2012).

Conversely, food insecurity undermines stability in the region. There is strong evidence that food price spikes helped to precipitate the ensuing instability (Bellemare 2011; Johnstone and Mazo 2011; Lagi et al. 2011), but the causal role of food insecurity is more nuanced. Those who took to the streets and continued marching when even their lives were in danger were certainly not vulnerable and food insecure by standard measures of vulnerability, hunger, and undernutrition. Instead, they tended to be young, urban, and literate—many with university degrees. Given this reality, food insecurity could only have played a causal role in the ensuing instability if the protesters were motivated in part by concerns about the welfare of their poorer, less food-secure compatriots, or if the definition of food insecurity is expanded beyond traditional measures of vulnerability, hunger, and undernutrition. While the protesters may have been motivated by the former,[7] we focus here on the latter explanation.

Food security requires that people have access "at all times" to food "that meets their dietary needs and food preferences." Based on this definition, were the protesters who took to the streets food insecure even if they weren't hungry? There is little question that these protesters felt deeply insecure about many aspects of their future, including their careers, families, and participation in social and political processes. While there is no rigorous way to test the relative importance of broadly defined food insecurity, the set of lifestyle threats that motivated protesters in some countries may well have included food preferences. As we strive to understand how these broader food security concerns relate to general lifestyle threats, it is important to appreciate that the threat of food insecurity may be just as potent a trigger for desperate acts as food insecurity itself.[8] These early reactions to risk can be especially pronounced when prior experiences have created clear expectations about what a nutritionally, culturally, and socially acceptable diet looks like.

Neither the threat of future food insecurity nor insecurity in general is alone sufficient to trigger rioting, unrest, and instability. Of the other mediating factors, one appears to be a necessary condition: the conviction that these insecurity threats stem from a fundamental injustice—a sense of being cheated, deceived, betrayed, misled, or otherwise exploited (Fraser and Rimas 2011). When faced with the threat of future food insecurity and uncertainty in general, the moral outrage that wells up from such a conviction can spark a wildfire of protest. In sum, although food insecurity is neither necessary nor sufficient to trigger instability, the threat of food insecurity broadly defined, together with general lifestyle threats and insecurity and persistent perceptions of injustice, clearly contributed to the recent unrest in many Middle East and North Africa countries—and this threat may have been especially potent because many of these countries have growing masses of disenfranchised citizens whose sense of injustice and moral outrage has been fermenting for years. This confluence of factors in the Arab Spring and the attendant role of information and communication technologies is well deserving of the research attention it is now receiving.

To explore the complex interlinkages between food security and stability more concretely, we explore four country case studies: Egypt, Morocco, Tunisia, and Syria. While these descriptions are intended to be more concise than comprehensive, the cases highlight several salient dimensions to these interlinkages across substantively different contexts.

CASE STUDY: EGYPT

In the 1990s, economic reforms allowed the Egyptian government to reduce inflation and the government deficit and accelerate growth, but real wages barely changed, and unemployment and poverty increased (Korayem 1996). After 2000, Egypt experienced growth and increased economic efficiency due to various reforms under the guidance of the International Monetary Fund, but most ordinary Egyptians did not benefit much. In 2004, the Mubarak regime began to liberalize the economy through tariff and tax reductions, trade agreements, and facilitated foreign direct investment, which reduced bureaucracy and increased efficiency. Much of Egypt's growth came from the extraction of resources, primarily oil; rents, such as for use of the Suez Canal; and foreign aid from donors. Without much innovation, dynamism, or creative destruction, real wages stagnated, the vulnerability of the poor increased (Loewe 2012), and the economy left a growing pool of unemployed young workers on the sidelines (Coleman 2011). Rich Egyptians had become richer, but the position of those outside the elite had not improved, and even worsened; safety nets, public services, and subsidies were cut while crony capitalism and corruption flourished (Coleman 2011).

Crony capitalism, like other forms of corruption, is at the core of Egypt's various socioeconomic problems, both with and (now) without Hosni Mubarak. The entrenchment of economic and political power among insiders is deep, and connections, rather than hard work, education, or ability, determine one's economic opportunities. The Egyptian military commands extensive economic resources, ranging from manufacturing and retail to real estate and other services, and success in these arenas requires patronage more than entrepreneurship (Malik and Awadallah 2011). Plagued with private-sector stagnation, the Egyptian economy failed to create jobs for young workers, including university graduates (Coleman 2011), leaving job creation to the public sector. This is becoming increasingly unsustainable and, according to many observers, has "bred a colossal failure of expectations," with newcomers to the labor force having an "ingrained preference for high paid jobs in the public sector, where remuneration is usually de-linked from skills or productivity" (Malik and Awadallah 2011).

Against this backdrop, food security policy plays a pivotal role in contemporary Egypt. The average Egyptian family spends approximately 40 percent of its income on food (Credit Suisse 2011). When grain prices skyrocketed in 2008, bread riots broke out across Egypt, and Mubarak even ordered the army to bake loaves of bread (Wharton 2011). Since those in power had no incentive to deviate from the established policy, the government responded with subsidies that cost more than $15 billion annually, soaking up more than 10 percent of GDP (*Economist* 2012), but these subsidies are both inefficient (e.g., cheap bread is often used as animal feed) and subject to corruption. Approximately

40 million Egyptians, nearly half of Egypt's population, became reliant on ration cards (Johnstone and Mazo 2011). While these food frustrations helped fuel the 2011 riots, they were ignited by smoldering sentiments of broader disenfranchisement and frustration. Those protesting at Tahrir Square and clashing with police were mostly middle-class university students who were not immediately threatened with food insecurity. Senior FAO economist Abdolreza Abbassian argues, "To say that food is pushing them onto the streets is overstating the case," and notes that "justice, democracy, and equality" were major factors in the peoples' reaction in Egypt (Johnstone and Mazo 2011).

The June 2012 presidential election of the Muslim Brotherhood candidate, Muhammed Morsi, may have introduced more uncertainty than it resolved. Nevertheless, Morsi's election promises, which included improving the quality and availability of subsidized bread, showcased food insecurity as a primary concern. While Egypt's future political climate is quite uncertain, it is clear change will be as slow and bureaucratically cumbersome as ever. The government's lack of initiative and progress, combined with unreasonably high public expectations will sow seeds of ongoing frustration, making it difficult to tackle food insecurity, unemployment, and crony capitalism (*Economist* 2012).

CASE STUDY: MOROCCO

Relative to the rest of the Middle East and North Africa, Morocco tends to be more willing to compromise to maintain social and political order (Wright 2008). This political approach was clearly on display in 2011, when in response to demonstrations, King Mohammed VI quickly proposed modifications to the Moroccan constitution that devolved some of the powers of the throne to the parliament. Although several latent tensions in Moroccan politics and society remain, the country has largely succeeded in avoiding serious unrest—despite having a "serious" food security status. The fact that Morocco has been relatively stable compared to its Middle East and North Africa neighbors despite being at least as food insecure highlights the important role moral outrage plays in turning the threat of food insecurity into real instability. Some Moroccans waiver in their support for the throne and many complain about the government, but the majority stand behind the King as "commander of the faithful" and a descendant of the prophet Mohammed. Until now, this sentiment has protected the throne against eruptions of outrage over injustice and corruption, which continue to persist in the country, albeit less glaringly than in some neighboring countries.

King Mohammed VI has encouraged the development of the country's private sector, and Morocco has benefited from trade agreements with the EU and

United States. While Morocco relied heavily in the past on self-sufficiency policies that encourage costly cereal production, the past decade has seen a more concerted effort to exploit the comparative advantage of Moroccan agriculture and encourage the production of high-value horticultural exports (Magnan et al. 2011). In the ambitious Plan Maroc Vert[9] (PMV), the government seeks to coordinate the high-value modern agricultural sector with the work of traditional farmers. The plan aims to exploit productive resources more efficiently and increase smallholder farmers' income by linking them to higher value markets for fruit, vegetables, olives, and cereals. Although it is too early to tell whether the PMV will succeed in improving rural food security in Morocco, it is a welcome departure from cereal self-sufficiency policies that had little connection to market realities.

As in Egypt, Morocco's 2011 protests and demonstrations were the product of frustrations that included—but were not restricted to—food. While the protests were tame and inconsequential compared to protests in other Middle East and North African countries, much latent frustration remains, as evidenced by the unexpected victory of the moderate Islamist Justice and Development Party in the November 2011 parliamentary elections. The country experienced a serious drought in 2011 to 2012, pushing the cost of subsidies on staples to nearly 6 percent of GDP. With projections of increasing drought frequency, perhaps the biggest food security questions in Morocco relate to the fiscal sustainability of the country's food and fuel subsidies in light of the threat to national food security posed by limited rainfall and rising food prices.

CASE STUDY: TUNISIA

Tunisia is not immune to food insecurity, but has only "moderate" risk, according to recent IFPRI reports, and child stunting is not a serious problem (Breisinger et al. 2012). Tunisia relies heavily on cereal imports and is vulnerable to food price shocks, since it is fiscally constrained (Harrigan 2011). On June 13, 2008, protests against high food price inflation and unemployment occurred in the Tunisian city of Redeyef, with one demonstrator killed and dozens wounded. The protests signified Tunisia's susceptibility to food insecurity, while also reflecting protesters' general frustration and dissatisfaction with corruption, lack of freedom, suppression of political speech, and youth unemployment. Official youth unemployment rates in Tunisia are over 30 percent, with actual rates likely much higher (Arieff 2011). Like Egypt, Tunisia's youth unemployment problem is directly connected to corruption and crony capitalism.

Although observers have long warned that the youth unemployment problem made social unrest inevitable in countries like Tunisia (Wright 2008), the regime did little to create economic conditions to encourage job creation. To the contrary, the government phased out its National Employment Fund, which offered a safety net against unemployment and vocational training, in the early 2000s, making the lack of opportunities to those without regime connections even more intolerable (Goldstone 2011). In 2008, major demonstrations broke out in the formerly prosperous mining district of Gafsa, and the regime sent in the army to end the demonstrations. Adnan Hadji, a trade unionist imprisoned for the protests, attributed the demonstrations to "pollution, unemployment, disease, and maldistribution of wealth" (Arieff 2011). Ben Ali's last-ditch promise to provide "300,000 new jobs" is evidence of youth unemployment's central role in the revolution (Arieff 2011; Asfaw 2007; Malik and Awadallah 2011; Mude et al. 2009).

The fact that Mohammed Bouazizi—whose self-immolation in December 2010 triggered the massive protests that ultimately ended Ben Ali's regime—was a food vendor has been invoked as evidence that food security was a fundamental concern (Friedman 2012), but this seems like a stretch. His frustrations about the injustice of a system that denies opportunities to those without the proper permit, which in a corrupt system often requires bribes and connections, would have boiled over just as easily if he had been hawking sunglasses. While food insecurity alone was unlikely to have triggered his desperate response, the food price spike was nonetheless an "aggravating factor," according to World Bank President Robert Zoellick (NPR 2011). The eruptions of major demonstrations about the lack of jobs for young adults during both the recent food price spikes seems to provide evidence of such an aggravating relationship between youth unemployment and food insecurity—particularly in a context where moral outrage is never far from the surface.

CASE STUDY: SYRIA

The troubling violence and severe civil unrest Syria has experienced in 2011 to 2012 makes it an extreme case among its Middle East and North Africa peers. The underlying relationship between this profound instability and food insecurity is, however, relevant to much of the region. In Syria, severe droughts from 2006 to 2011 led to total crop failure and massive livestock mortality (Erian et al. 2011).[10] Within a few years, the Global Hunger Index for Syria doubled from 5.2 to 10.5 in 2010 (IFPRI 2011). Although Syria was previously comparable to Morocco, with a Global Hunger Index that barely changed during this period from 5.8 to 5.9, it is now more like Yemen with an index of over 10. Millions of farmers and herders faced a future in which their livelihoods were no longer viable. Migration appeared to be their only option—and by 2009 more than 250,000 headed for the city (Sands 2009).[11] The experience of a farmer on the Syria–Jordan border is telling:

> We used to have a good life, we lived like sheikhs. But things slowly got worse, the river dried up, the rains stopped. We sold one car, then the next, then the tractor. When there was nothing else worth selling, we loaded ourselves into our truck and left. There was no other choice. We drove for 14 hours to get to Dara'a. People in the other cars on the road were staring at us, as if we were homeless, or refugees, not Syrian citizens (Sands 2009).

Like thousands of other families, this family was not well received in the southern village of Dara'a, and their circumstances became dire. The dysfunctional Assad regime did little, apart from a lot of talk and some planning, to quell the social and economic pressures induced by these extraordinary climate stresses. It is no coincidence that the protests that provoked violent repression from the regime and quickly escalated to other cities began with disenfranchised youth and farmers in Dara'a. As the people of Dara'a were struggling to cope with severe drought and the social turmoil induced by these waves of migration, they were incensed by the fact that they could not buy or sell land without permission from corrupt officials (Friedman 2012). As the demonstrations spread, other forms of long-standing injustice channeled insecurity into open protest in Syrian cities. While the average youth unemployment rate in Syria of around 26 percent was not far from the Middle East and North Africa average in 2007, the disparity with the adult unemployment rate—around 4 percent—was the highest in the region, suggesting that young adults were excluded to a disturbing degree from meaningful participation in the Syrian economy (Kabbani and Kamel 2007).

In the Syrian case, food insecurity—always combined with moral outrage about the dysfunctional, repressive, and corrupt regime—was clearly a primary trigger for the instability that ensued. Many of the first protesters seemed to have experienced marked declines in their access to food sufficient to meet their expectations in the preceding few years. Similarly, the violence and severe instability directly worsened the food security situation in Syria. While weak rainfall has continued to suppress cereal production in important agricultural regions in Syria, civil instability has magnified this production problem by disrupting farmers' access to fertilizer and seeds, and in some locations, limiting their access to their land (FAO 2012). At the same time, food prices increased 50 percent from May to December 2011 (FAO 2012). Even before the civil unrest began in early 2011, micro food insecurity assessed by the Global Hunger Index had already doubled. The unrest has deepened the degree of insecurity among the food insecure, undoubtedly expanding the number of food insecure households dramatically.

PREDICTING FOOD SECURITY OR INSTABILITY IN THE COMING DECADE

Many countries in the Middle East and North Africa have experienced unprecedented change in the past few years. The immediate future in many of these countries remains extremely uncertain. Yet, the experiences described above suggest several dimensions that will importantly shape the region in the coming decade.

Table 14.2 lists drivers and stressors associated with five dimensions. For each of these dimensions, we indicate the likelihood of change for better or worse in the coming decade, using shaded rectangles to denote the *likely* range, and horizontal lines to denote the *possible* range of changes. We describe important aspects of these changes and, in the final column, implications for food security and instability.

Many of the Middle East and North Africa features already described will continue to shape the region in the near future. For example, populations in the region will continue to grow relatively fast, and youth and young adults will continue to dominate the demographics of the region. Constraints on arable land and water resources will continue to limit agricultural productivity and be aggravated by ongoing climate change. We take these persistent factors as given and, for each of the subsequent dimensions, explore aspects that are likely to be both influential and uncertain.

Climate: Climate change, including major droughts in this region and elsewhere, has been flagged as a "threat multiplier" for its apparent role in the uprisings of the Arab Spring (Johnstone and Mazo 2011). Drought frequency and receding water tables will continue to drive both food insecurity and instability in the region—particularly in resource-poor countries that are fiscally vulnerable to food price spikes. The two countries that appear most vulnerable to climate change threats are Morocco and Syria because of their relatively heavy reliance on rain-fed agriculture and forecasted changes in rainfall patterns in the Mediterranean region (Schilling et al. 2012). While many of these changes are projected to occur over several decades, the North Atlantic Oscillation may continue to have a pronounced effect in the shorter term (Hoerling et al. 2011). Serious water constraints are likely to worsen and will continue to trigger local conflicts over resources.

Politics: The only certainty about political institutions in the Middle East and North Africa region in the coming decade is how much they will matter. In most countries, the structural form and practice of institutions, along with their path is quite unclear. It seems likely, however, that the region's evolution of democracy will chart a different course than traditional Western liberalism (Richards and Waterbury 2008; Roy 2012). The military's political role will directly influence stability and food security in many countries. Widespread

Table 14.2. Dimensions and associated drivers that will shape the interrelationship between food security and sociopolitical stability in the coming decade

Dimension	*Drivers*	Likelihood of change in next decade [a] (Worse – Status Quo – Better)	*Implications*
CLIMATE	Drought Water	*Frequency/intensity of drought and average temperatures increase; Water for irrigation more limited; Groundwater recharge slows.*	Increasing burden on social safety nets; Greater environmental pressure; More volatile local food production; Increased rural vulnerability and local conflicts.
POLITICS	Representation Transparency Corruption Conflict Military	*Public/international pressure may induce meaningful reform, but vested interests and elites will resist real change; Unpredictable elections; Continued rise of Islamist parties; Unrealistic expectations; Foreign assistance remains important.*	Injustice and corruption fade slowly and revive periodically; Moral outrage persists in some countries; Minority–majority conflicts; Continued inefficiency in social programs with some subsidy reform.
PRIVATE SECTOR	Unemployment Crony capitalism Innovation Productivity Entrepreneurship	*Youth unemployment persists; Some reforms may improve private productivity; Uncertainty regarding political reform.*	Connections will continue to ration opportunities; Youth disenfranchisement remains source of tension; Improvements generally lag political change.

(Continued)

Dimension	*Drivers*	Likelihood of change in next decade [a]			*Implications*
		Worse	Status Quo	Better	
International Economy	Food prices Intraregional & international trade Oil demand	*Global economy recovers; Food and oil prices high and volatile; International trade to region increases, but intraregional trade continues lagging.*			Food import bill grows steadily; Oil exporters calm tensions and anxieties with generous subsidies; Oil importers forced to reform subsidies while facing rising tension; Trade stimulates some sectors.
Agriculture	Water/irrigation Trade Cereal production High-value crops Public research	*Cereal yields increase slowly; Some countries boost private sector exports of high-value crops; Public research/extension makes modest contribution, if any; Smallholder productivity remains low.*			The rift between modern and traditional agriculture grows; Rural food insecurity grows in oil-importing countries even as the value of agriculture expands; Rural-to-urban migration continues.

Note: [a] For each dimension, shaded boxes indicate likely changes and horizontal lines indicate possible changes relative to status quo.
Source: Authors

outrage over corruption and injustice played a central role in the recent unrest and will be a dominant force in the political landscape. Popular frustrations from unmet expectations and persistent problems will seriously challenge incoming regimes in the next decade, which may encourage them to adopt heavy-handed tactics reminiscent of the past.

Private sector: Where the politics go, the private sector is likely to follow. The region's state-dominant economies, crony capitalism, and ongoing minority–majority tensions make this particularly true. Uncertainty about improvements in the private sector is tied to political uncertainty. Real private sector improvements require changes in expectations, suggesting there will be an important lag between political and private-sector improvements—and that near-term prospects for the private sector in many Middle East and North Africa countries may be even worse than political prospects. While there will continue to be promising signs of investment and entrepreneurship in some sectors in some countries, these will likely remain isolated successes without much aggregate traction in the next decade.

International economy: Commodity prices—especially food and oil prices—will continue to be critical to the region. Predicted high and volatile global food prices will fiscally strain many countries. These governments will no longer be able to rely only on subsidy-based solutions to price spikes: generous subsidies may simply be financially infeasible. People are more likely to expect a remedy to root problems. The broader international economy will almost certainly recover before most of the Middle East and North Africa countries do. Isolated exports (for example, high-value agricultural products) and remittances will provide important links to the global and especially European economy for some sectors and countries, but these links will be too modest to bring much aggregate benefit.

Agriculture: In countries that rely importantly on agriculture, developments in their agricultural sectors will drive food security in the coming decade in important ways. For traditional rural producers, improvements in yield or irrigation could be important, but dramatic improvements are also unlikely in the current system. Water will continue to be a primary production constraint, limiting yield gains for most crops. Any changes in agriculture are more likely to occur in modern agricultural sectors aimed at high-value exports. Spillovers might benefit rural areas generally, but these changes might also add new pressures on water and land resources, creating local conflicts and tensions with traditional smallholders. For many Gulf countries, large-scale agricultural production developments on leased land elsewhere in Africa may be the dominant change in the sector. The effect of these dimensions and drivers on food security and sociopolitical stability will depend crucially on their interaction and on persistent factors, such as demographics and physical resources.

Any serious analysis will need to identify specific variables to represent key drivers in Table 14.2. Developing a comprehensive list of such predictors is beyond the scope of this chapter, but a few observations deserve attention. Given the intense scrutiny the region has recently attracted, there are several new and ongoing attempts to understand the near-term economic, political or agricultural prospects for the region.[12] These analyses explore several potentially interesting variables, including standard variables that capture fiscal trends, unemployment rates, demographic trends, political transparency, poverty, agricultural productivity, and novel measures of life happiness and satisfaction.

Building on these recent efforts, Table 14.2 and the broader discussion in this chapter suggest other variables that might be useful in monitoring the relationship between unrest and food security. Remotely sensed measures of rainfall and crop productivity and existing drought monitoring in the region are clearly important to short-term changes in these key relationships. Corruption measures and business climate surveys may track latent frustrations about politics and the economy. The youth unemployment rate relative to the general unemployment rate seems to be the single most important economic variable to track. For many of these potential predictors, recent changes are likely to be more important than current levels. This is particularly true in the Middle East and North Africa region, where citizens face great uncertainty, thus placing emphasis on their perceptions of changes in key features of daily life to reformulate expectations.

While lingering economic crises, recent food price spikes, and looming climate change have brought food security concerns to the fore in many policy arenas and in many countries, food insecurity, and its associated causes and consequences have been particularly potent and transformational in the Middle East and North Africa. The policy links between food insecurity and sociopolitical instability are forcefully on display in the contemporary Middle East and North Africa region in the wake of the Arab Spring of 2011, profoundly shaping the region's position in global politics and markets in the future. These complex food insecurity issues are an important indicator of sociopolitical tensions in the region, shaping policy responses to food security concerns.

Continued conflict, tension, and uncertainty will make food insecurity a critical concern in many Middle East and North Africa countries and a persistent threat to the viability of new political regimes and power structures. Although some glimmers of hope in the region continue to exist, broad structural impediments in economic and political realms, along with a hotter and drier climate and severe water constraints, will either prevent significant improvements or precipitate major ongoing changes and instability. Most regional households appear to be willing to give change a chance, but expectations for rapid improvements may be too high—and the broad frustration and moral outrage that sparked the Arab Spring of 2011 will be lurking close behind for years to come.

ACKNOWLEDGMENTS

We thank David Patel and participants at the "Food Security and its Implications for Global Stability" workshop held at Cornell University in June 2012 for their suggestions and contributions.

NOTES

1. Although a drier and hotter climate almost uniformly bodes very poorly for agricultural productivity in the region, Egypt stands out as one possible exception. Due to its access to the Nile River and widespread irrigation infrastructure, some projections of agricultural productivity show gains in Egypt due to the carbon fertilization effect (Cline 2007).
2. A few countries, such as Somalia, Sudan, Comoros, Mauritania, and Turkey, are included or excluded from the region depending on which definition is used.
3. Severe unrest in a country obviously affects both exports and remittances (e.g., Iraq), but note that the data used in Table 14.1 predates the Arab Spring.
4. Although useful, the standard caveat applies to these country-level indicators: they obviously mask substantial and important differences in food insecurity within countries, including major regional and ethnic differences.
5. Definition adopted at the 1996 World Food Summit.
6. Recent work confirms that this pattern stems from a causal link from increasing food prices to unrest (Bellemare 2011; Lagi et al. 2011).
7. Based on prevailing demographics and urban migration trends, a majority of these protesters would have had a direct family link to rural areas and some familiarity with the food security challenges of the rural poor.
8. Evidence from other settings suggests that the perception of risk and not just its realization can powerfully shape decisions (Dercon and Christiaensen 2011; Elbers et al. 2007).
9. Details for this “Green Morocco Plan” are available here: http://www.ada.gov.ma/en/Plan_Maroc_Vert/plan-maroc-vert.php (accessed December 18, 2012).
10. See also http://climateandsecurity.org/2012/02/29/syria-climate-change-drought-and-social-unrest/ (accessed 28 May, 2012).
11. See http://www.irinnews.org/Report/85963/SYRIA-Drought-driving-farmers-to-the-cities (accessed 28 May, 2012).
12. See, for example, Arieff 2011; Breisinger et al. 2011; Breisinger et al. 2012; Khouri et al. 2011; World Bank 2009.

REFERENCES

Arieff, A. 2011. *Tunisia: Recent developments and policy issues. Congressional Research Service Report for Congress.* Washington, DC: Congressional Research Service.

Asfaw, A. 2007. Do government food price policies affect the prevalence of obesity? Empirical evidence from Egypt. *World Development* 35(4): 687–701.

Bellemare, M. 2011. *Rising food prices, food price volatility, and political unrest.* Durham, NC: Duke University.

Breisinger, C., O. Ecker, and P. Al-Riffai. 2011. Economics of the Arab awakening: From revolution to transformation and food security. IFPRI Policy Brief 18:2.

———, and B. Yu. 2012. Beyond the Arab awakening: Policies and investments for poverty reduction and food security. IFPRI Food Policy Report February.

Breisinger, C., T. van Rheenen, C. Ringler, A. Nin Pratt, N. Minot, C. Aragon, B. Yu, O. Ecker, and T. Zhu. 2010. Food security and economic development in the Middle East and North Africa. IFPRI discussion papers.

Campante, F. R., and D. Chor. 2012. Why was the Arab world poised for revolution? Schooling, Economic Opportunities, and the Arab Spring. *Journal of Economic Perspectives 26*(2): 167–188.

Cline, W. R. 2007. *Global warming and agriculture: Impact estimates by country.* Washington, DC: Center for Global Development, Peterson Institute for International Economics.

Coleman, I. 2011. Egypt's uphill economic struggles. Council on Foreign Relations, February 2, 2011.

Credit Suisse. 2011. Emerging consumer survey 2011. Zurich: Credit Suisse Research Institute (January).

Dercon, S., and L. Christiaensen. 2011. Consumption risk, technology adoption and poverty traps: Evidence from Ethiopia. *Journal of Development Economics 96*(2): 159–73.

Diaz-Bonilla, E., M. Thomas, S. Robinson, and A. Cattaneo. 2000. Food security and trade negotiations in the world trade organization: A cluster analysis of country groups. IFPRI discussion papers.

Diwan, I. 2012. The role of governance in the Arab awakening. Paper presented at Food Secure Arab World. Beruit, Lebanon, February 6.

Economist. 2012. Egypt's government: The revolutionaries get a few morsels. *The Economist*, August 4.

Elbers, C., J-W. Gunning, and B. Kinsey. 2007. Growth and risk: Methodology and micro evidence. *World Bank Economic Review 21*(1): 1–20.

Erian, W., B. Katlan, and O. Babah. 2011. Drought vulnerability in the Arab region: Special Case Study: Syria. Background paper for the 2011 Global Assessment Report on Disaster Risk Reduction. Geneva: UNISDR.

ESCWA. 2010. Food security and conflict in the ESCWA region. Report prepared for United Nations Economic and Social Commision for Western Asia.

FAO. 2012. Civil unrest raises grave concern for food security. Global Information and Early Warning System on Food and Agriculture (GIEWS) No. 331.

Fraser, E., and A. Rimas. 2011. The psychology of food riots. *Foreign Affairs*, January 30.

Friedman, T. L. 2012 The other Arab Spring. *New York Times*, April 7.

Goldstone, J. A. 2011. Understanding the Revolutions of 2011: Weakness and resilience in Middle Eastern autocracies. *Foreign Affairs 90*: 8.

Harrigan, J. 2011. Food security in the Middle East and North Africa (MENA) and sub-Saharan Africa: A comparative analysis. Center for Economic Institutions Working Paper Series (No. 2011-5): 4–50.

Hoerling, M., J. Eischeid, J. Perlwitz, X. W. Quan, T. Zhang, and P. Pegion. 2011. On the increased frequency of Mediterranean drought. *Journal of Climate 25*: 2146–61

IFPRI. 2011. Global hunger index: The challenge of hunger: Taming price spikes and excessive food price volatility. Report prepared for IFPRI.

IMF. 2008. *World economic and financial survey.* Washington, DC: International Monetary Fund.

International Labor Organization. 2011. Global employment trends for youth: 2011 update. Geneva: ILO.

Johnstone, S., and J. Mazo. 2011. Global warming and the Arab Spring. *Survival 53*(2): 11–17.

Kabbani, N., and N. Kamel. 2007. Youth exclusion in Syria: Social, economic, and institutional dimensions. Middle East Youth Intiative Working Paper. Washington, DC: Wolfensohn Center for Development at the Brookings Institution.

Khouri, N., K. Shideed, and M. Kherallah. 2011. Food security: Perspectives from the Arab world. *Food Security 3*: 1–6.

Korayem, K. 1996. Egypt: comparing poverty measures. In *Poverty: A global review*, ed. Else Oyen, S. M. Miller, and Syed Abdus Samad, 189–209. Oslo, Scandinavian University Press.

Lagi, M., K. Bertrand, and Y. Bar-Yam. 2011. The food crises and political instability in North Africa and the Middle East. Working Paper.

Loewe, M. 2012. Social security in Egypt: An analysis and agenda for policy reform. ERF Working paper 2024.

Magnan, N., T. J. Lybbert, A. F. McCalla, and J. A. Lampietti. 2011. Modeling the limitations and implicit costs of cereal self-sufficiency: The case of Morocco. *Food Security 3*: 49–60.

Malik, A., and B. Awadallah. 2011. The economics of the Arab Spring. CSAE Working Paper Series.

Mude, A., S. Chantarat, C. B. Barrett, M. R. Carter, M. Ikegami, and J. McPeak. 2009. Insuring against drought-related livestock mortality: Piloting index based livestock insurance in northern Kenya.

NPR 2011. *The impact of rising food prices on Arab unrest.* Washington, DC: National Public Radio (February 18, 2011).

Richards, A., and J. Waterbury. 2008. *A political economy of the Middle East.* Boulder, CO: Westview Press.

Roy, O. 2012. The transformation of the Arab world. *Journal of Democracy 23*(3): 5–18.

Sands, P. 2009. Refugees because the rains never came. *The National*, September 4.

Schilling, J., K. P. Freier, E. Hertig, and J. Scheffran. 2012. Climate change, vulnerability and adaptation in North Africa with focus on Morocco. *Agriculture, Ecosystems & Environment 156*: 12–26.

Von Grebmer, K., M. T. Ruel, P. Menon, B. Nestorova, T. Olofinbiyi, H. Fritschel, Y. Yohannes, C. von Oppeln, O. Towey, and K. Golden. 2010. *2010 global hunger index: The challenge of hunger.* Washington, DC: IFPRI.

Wharton. 2011. Middle East's investments in African farmlands are rooted in food security fears. Arabic Knowledge@Wharton. Philadelphia: University of Pennsylvania (March 22, 2011).

World Bank. 2009. *Improving food security in Arab countries.* Washington, DC: World Bank.

———. 2012. Data catalog. http://data.worldbank.org. Accessed December 19, 2012.

Wright, R. B. 2008. *Dreams and shadows: The future of the Middle East.* New York: Penguin Press.

Yu, B., L. You, and S. Fan. 2010. Toward a typology of food security in developing countries. IFPRI discussion papers.

15

Food Security and Sociopolitical Stability in Eastern Europe and Central Asia

JOHAN SWINNEN AND KRISTINE VAN HERCK

The countries in the region from Eastern Europe to Central Asia can potentially play a significant role in global food production. The region is a major exporter of food, especially grains. It exports almost as much grain as the EU and the United States. Nevertheless, with 17 percent of global arable land, it produces only 11 percent of the world's crops, and productivity growth has been mixed over the past decades. In some parts, yields are not much higher than in the 1970s. Increasing food prices could stimulate production and productivity growth in the region.

In general, food security in Eastern Europe and Central Asia has improved significantly over the past years as a result of strong economic growth. While the countries have a common past and some common features in their institutional, political, and economic changes over the past decades, however, they vary in economic development, poverty, political institutions, and other important factors.

Throughout the region, food prices have been increasing, with price spikes in 2008 and 2010, and all governments have responded by intervening in food markets. Food-exporting countries in the region, such as Russia, Ukraine, and Kazakhstan, banned, taxed, or restricted the export of food, while importing countries mainly reduced their import tariffs in an attempt to reduce fluctuations in domestic food prices and ensure sufficient domestic food supply. The export restrictions taken by Russia, Ukraine, and Kazakhstan have affected both the food-importing countries in the region (since the Eastern Europe and Central Asia countries trade much food among themselves) as well as countries in North Africa and the Middle East, which heavily rely on imports from these Eastern European and Central Asian grain-exporting countries.

Some indicators suggest that the region might be susceptible to a "Central Asian Spring," similar to the civil protests and demonstrations that have characterized the Arab world in the spring of 2011. However, up to now it is unclear whether sociopolitical stability will radically change in the region and it is not clear that food security will play a major role in this. Economic growth prospects are relatively good for the region, including in the poorest countries, which are the ones most likely to experience adverse impacts of high food prices on political and social stability. Moreover, the region has important agricultural potential, especially in terms of wheat production. However, in order to realize the potential of the region, substantial investments in the agro-food industry are needed. In addition, the question remains to what extent an increase in the agricultural potential in the region will affect global food security, because in the domestic and foreign policy of the major grain-producing country, Russia, domestic food security and self-sufficiency have become increasingly important.

A HETEROGENEOUS REGION WITH A COMMON PAST

Until 1989 the Eastern Europe and Central Asia countries were united by their communist political and economic regimes in the Comecon system. After the fall of the Berlin wall in 1989, the system disintegrated. While the shift toward a market economy and democracy varied strongly among countries, there were several common features of transition.

In 1989, the former satellite countries in Eastern Europe established independent political control, and a few years later, the member states of the former Soviet Union declared sovereignty over their territories and became independent states. In some countries these political changes turned violent—the most extreme being conflicts in former Yugoslavia and the Caucasus. At the same time, most countries in Eastern Europe joined Western international institutions such as the Organization for Economic Co-operation and Development (OECD), the World Trade Organization (WTO), and the EU.

In almost all countries, the economic liberalization caused an initial decline of the economy, reflected by a decrease in GDP and an increase in poverty. However, from the mid-1990s in Eastern Europe and the 2000s (after the Russian financial crisis in 1998) in most other countries, economic growth resumed and contributed to poverty reduction and improved food security. In addition to the direct effects, inhabitants in the region have benefited indirectly from economic growth in the EU and the richer countries in the region, where many of them migrated to work. This resulted in a substantial increase in remittances (Swinnen and Van Herck 2009).

In recent years, the global financial crisis caused economic growth to slow down. Real GDP decreased in almost all countries in the region in 2009. In 2010, however, there was a strong recovery, and in 2010 and 2011 real GDP growth was already strongly positive in most countries. Exceptions to this recovery were richer countries in Eastern Europe, such as the Baltic states, Hungary, and Slovenia.

The agricultural sector in all countries—like the economy as a whole—has been affected by the transition to a more market-oriented economy. In the early 1990s, the liberalization of the sector caused a large decline in agricultural output and productivity. In the first years of transition, gross agricultural output and productivity strongly decreased in all subregions (Figures 15.1 and 15.2). In the poorer Central Asian countries, such as Kyrgyzstan, Uzbekistan, and Turkmenistan, the decline in agricultural output was limited, whereas in parts of the former Soviet Union such as the Baltics, Russia, and Ukraine, agricultural output declined by more than 40 percent. Agricultural output and productivity started to increase again after 2000.

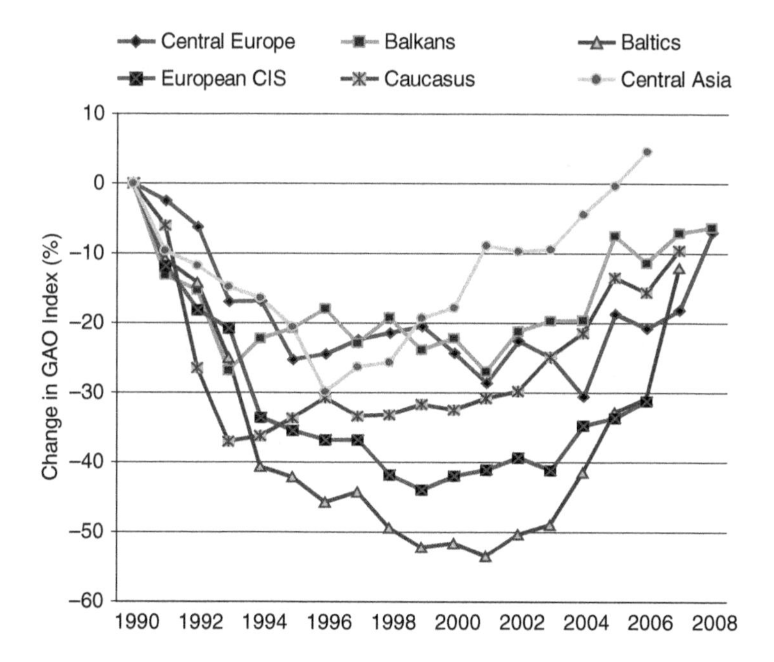

Fig.15.1. Evolution of gross agricultural output (GAO) (percent change)

Note: Central Europe includes Czech Republic, Hungary, Slovakia, Poland; Balkans includes Albania, Bulgaria, Romania, and Slovenia; Baltics includes Estonia, Latvia, and Lithuania; European CIS includes Belarus, Moldova, Russia, and Ukraine; Caucasus includes Armenia, Azerbaijan, and Georgia; Central Asia includes Kazakhstan, Kyrgyzstan, Tajikistan, and Uzbekistan.

Source: Authors' own calculations based on national statistics and FAOstat

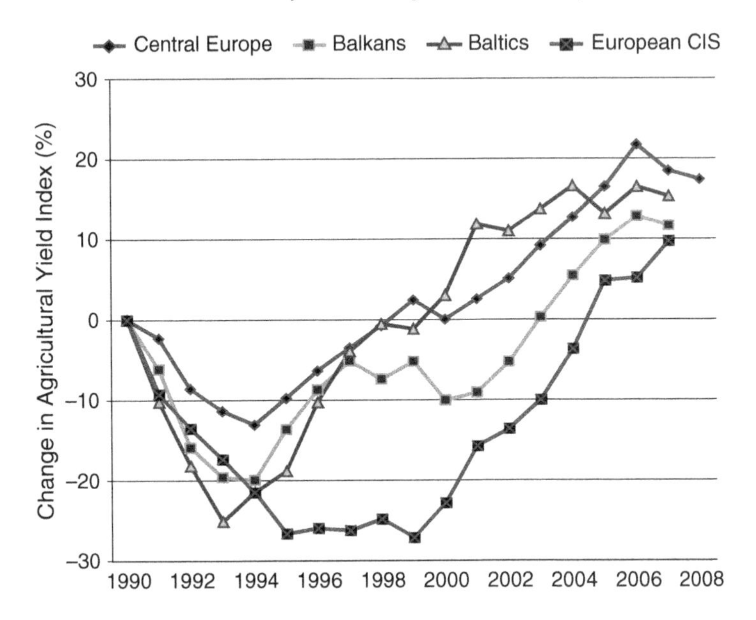

Fig. 15.2. Evolution of agricultural yields (percent change)

Note: Central Europe includes Czech Republic, Hungary, Slovakia, Poland; Balkans includes Albania, Bulgaria, Romania, and Slovenia; Baltics includes Estonia, Latvia, and Lithuania; European CIS includes Belarus, Moldova, Russia, and Ukraine.

The agricultural yield index is calculated as the average yield index of grains, sugar beet, and milk, where crop yield is output per hectare and milk yield is output per cow. Given the sensitivity of grain and sugar beet yields to weather conditions, these are calculated as three-year moving averages.

Source: Authors' own calculations based on national statistics and FAOstat

Today, the Eastern Europe and Central Asia region is very heterogeneous, both in terms of economic development and poverty, food security, food production, and trade, and in terms of political and social structures. In the rest of this section we review some key indicators of these factors, as well as differences within the region, which are essential to understanding the impact of food security on sociopolitical stability within and outside the region.

Political Regimes

There is major variation among the political regimes in the Eastern Europe and Central Asia region, ranging from full-fledged democracies (the EU new member states—Bulgaria, Czech Republic, Estonia, Hungary, Latvia, Lithuania, Poland, Romania, Slovakia, and Slovenia—and the Western Balkans) to strong authoritarian regimes (some of the Central Asian countries, such as Kazakhstan, Uzbekistan, Belarus, or Azerbaijan). There are also countries with characteristics of both democracy and autocracy, such as Russia, Armenia, and Tajikistan. One common feature of the political regimes in Central Asia is the presence of "semi-permanent presidents" (Gawrich et al. 2010). In all Central Asian countries, the first presidents after independence ruled at minimum about 15 years. In addition, in three out of the five countries, Kazakhstan, Tajikistan, and Turkmenistan, these presidents also established a form of personal cult.[1]

Political (In)Stability and Conflicts

Over the past 20 years, several conflicts and disturbances in the region have taken place, ranging from peaceful civil demonstrations to outright war. Wars in the Caucasus, the Western Balkans, and Central Asia were almost all the result of ethnic conflict. In the Caucasus, most conflicts took place in the beginning of the 1990s in Armenia and Azerbaijan, more specifically in the region Nagorno-Karabakh, between ethnic Armenians and Azeri. More recently there has been warfare in Georgia-South Ossetia (2008) and long-lasting conflict in Chechnya (1994 to 2010). In the Western Balkans the population experienced civil war in the mid-1990s in the ex-Yugoslav countries due to ethnic conflicts between Serbs, Croats, and Bosnians. Conflict also flared up in Tajikistan and Uzbekistan in the 1990s and more recently in South Kyrgyzstan.

Besides violent conflicts, the region has seen several large peaceful civil demonstrations, so-called "color revolutions." These movements use nonviolent civil resistance, such as demonstrations and strikes, to protest against their governments, which they consider corrupt or authoritarian. The Rose Revolution in Georgia (2003), the Orange Revolution in Ukraine (2004), and the Tulip Revolution in Kyrgyzstan (2005) all took place after disputed elections and led to the overthrow of the government (Table 15.1).

Incomes, Poverty, and Food Security

The Eastern Europe and Central Asia region is heterogeneous in incomes, poverty, and food security indicators. Certain Eastern European and Central Asian countries have high GDP per capita and low levels of poverty measured at the international poverty line—including, not surprisingly, the EU new members. In addition, GDP per capita is relatively high and poverty levels low in a group of countries that are rich in natural resources, including Russia, Kazakhstan, and Azerbaijan. In the Western Balkans, except for Albania and FYR Macedonia, GDP per capita is relatively high and poverty is low. In other countries, such as Armenia, Georgia, Moldova, Kyrgyzstan, Tajikistan, and

Table 15.1. Urban and social instability in Central Asia (since 2000)

Armed conflict situation	
Kazakhstan	2005; 2006; 2009
Kyrgyzstan	2001; 2005; 2009
Tajikistan	2001; 2010
Turkmenistan	2002; 2008
Uzbekistan	2004; 2009
Pro-government terrorism	
Kyrgyzstan	2009
Tajikistan	2001
Turkmenistan	2006
Anti-government terrorism	
Kazakhstan	2004
Kyrgyzstan	2002
Tajikistan	2000; 2001; 2007
Uzbekistan	2004
Violent riot	
Kyrgyzstan	2005; 2010
Demonstration	
Kazakhstan	2009
Kyrgyzstan	2000; 2001; 2002; 2003; 2005; 2006; 2008; 2010
Uzbekistan	2003; 2004

Note: Armed conflict is defined as "distinct, continuous, and coordinated interaction involving opposing, organized armed forces"; Pro-government terrorism is defined as "distinct event related to a persistent, directed violence campaign waged primarily by government authorities, or by groups acting in explicit support of government authority, targeting individual, or 'collective individual,' members of an alleged opposition group or movement"; Anti-government terrorism is defined as "distinct event related to a persistent, directed-violence campaign waged primarily by a non-state group against government authorities"; Violent riots are defined as "distinct, continuous, and (un)coordinated action staged by members of a singular political or identity group and directed toward members of a distinct 'other' group or government authorities"; Demonstrations are defined as "distinct, continuous, and (un)coordinated largely peaceful action directed toward members of a distinct 'other' group or government authorities."
Source: UCDP/PRIO Urban Social Disturbance in Africa and Asia (USDAA) database

Uzbekistan, there are relatively high levels of poverty and undernourishment, although substantial improvements have been made in the past years.

Over the past years, almost all countries in the region experienced strong economic growth. In addition, remittances increased as inhabitants of the region have benefited indirectly from economic growth in the EU and the richer countries in the region, where many of them migrated to work. As a result, since the 1998 Russian crisis, more than 50 million people have moved out of poverty in Eastern Europe and Central Asia (World Bank 2010).

Despite these recent positive evolutions, several countries still have a large proportion of poor in their population, especially in Central Asia. In Uzbekistan, 77 percent of the population is living below the $2 a day poverty line. In Tajikistan and Kyrgyzstan the figures are respectively 51 percent and 29 percent. Moreover, there are large disparities within countries, with a disproportionate share of poor households in rural areas, as in most of the rest of the world (Macours and Swinnen 2008). For example, while 40 percent of the rural population in Kyrgyzstan is living in poverty, poverty affects only 24 percent of the urban population (World Bank 2011a).

Undernourishment has also decreased substantially in the past decade. For example, in Azerbaijan and Georgia, which had a high prevalence of undernourishment in the mid-1990s, undernourishment had almost vanished by 2007. In countries such as Armenia and Tajikistan, the situation improved significantly, but undernourishment remained high (at more than 20 percent of the population) in 2007. In the region's poorest countries, diets are monotonous, and the majority of energy is obtained from starch and cereals, while animal and livestock products represent only a small proportion of the diet, especially for the lowest income groups (Musaev et al. 2010). Undernourishment and a poor diet may result in poor health as reflected in a commonly used health indicator: stunting (Figure 15.3).

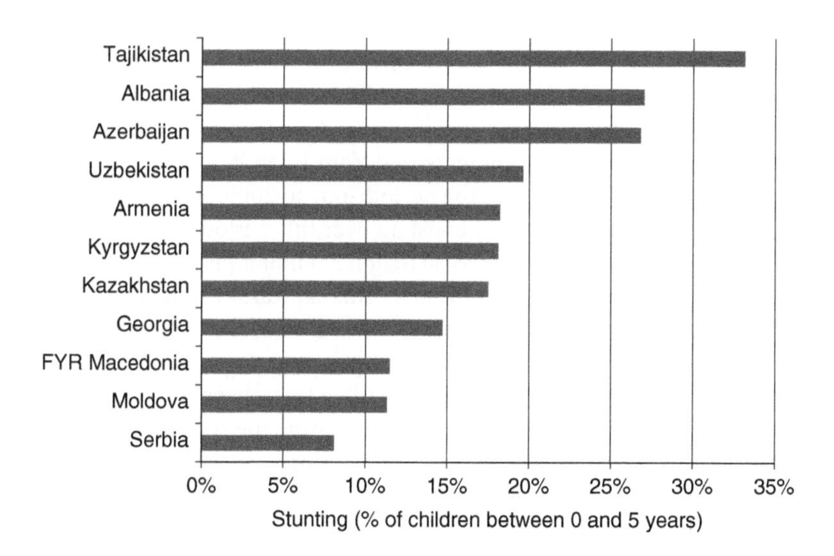

Fig. 15.3. Prevalence of stunting (percent of children between 0 and 5 years)

Note: Data are from the latest year available: 2005 for Albania, Armenia, Georgia, FYR Macedonia, Moldova, and Tajikistan; 2006 for Azerbaijan, Bosnia and Herzegovina, Kazakhstan, Kyrgyzstan, Serbia, and Uzbekistan

Source: World Bank (2012)

Food Production and Trade

The Eastern Europe and Central Asia region includes some important producers and exporters of cereals, such as Russia, Kazakhstan, and Ukraine. After an initial decline in wheat production in the first years after transition, wheat production started to increase again in the 2000s. Currently, the region is one of the most important wheat-producing regions in the world, producing 21 percent of the world's wheat between 2007 and 2009.

In contrast, almost all countries in the region, except for Hungary and Poland, are net importers of meat. The production of meat did not recover from a substantial decline in the first years after transition. However, in this context it is important to emphasize that the decline in meat production is not necessarily negative. The level and composition of agricultural output in the beginning of 1990s was reflecting the pre-reform situation, which was in general not based on the comparative advantage of the region, but on production allocations set by the central government. Meat and dairy production and consumption were heavily subsidized under the Communist regime (Liefert and Swinnen 2002). Liberalization of prices, subsidy cuts, and decentralization of production decisions induced a major shift toward crops, and in particular grain production (Anderson and Swinnen 2009). This evolution does not necessarily indicate that there are constraints for meat production in the region, but may mainly reflect a shift towards the comparative advantage of the region. In Russia, production has rapidly shifted from livestock to grain in the past decades (Liefert 2002; Liefert et al. 2010). Integration into the world market revealed that Russia was not able to compete with low-cost meat producers. As a result domestic meat production sharply decreased from the beginning of the 1990s, while meat imports started to increase, even during the economic crisis in the late 1990s. Overall, net import of meat increased from 1.2 million tons in the period 1992 to 1996 to 2.8 million tons in the period 2007 to 2009, although meat consumption remained approximately constant over this period.

The Eastern Europe and Central Asia region is a major player on the international grain market. The region (especially Russia, Kazakhstan, and Ukraine) produces 24 percent of global wheat exports, almost the same share as the EU15 (22 percent) or the United States (22 percent). Exports have increased substantially since the early 1990s, but are still extremely volatile compared to other major grain-exporting countries, which limits the region's contribution to global food security. Sedik's (2011) calculations suggest that the volatility in yields and output in Russia, Ukraine, and Kazakhstan were approximately the same as in Australia, and double of those of the United States and the EU. The volatility of exports was many times higher, in particular for Russia and Ukraine. An illustration of the volatility of exports from the region was in 2010, when drought and wildfires in Russia and Ukraine led to a number of export restrictions with significant effects on global markets.

Energy and Minerals

The Eastern Europe and Central Asia region includes some major producers of oil, gas, and minerals. Oil rents expressed as a percentage of GDP show that government incomes from oil are especially important in Azerbaijan (43 percent), Turkmenistan (20 percent), Kazakhstan (18 percent), Russia (14 percent), and to a limited extent in Uzbekistan (3.3 percent). Natural gas is also an important natural resource, especially in Uzbekistan (18.1 percent) and Turkmenistan (24 percent). In Russia and Azerbaijan, respectively, 3.6 percent and 3.9 percent of GDP are rents from natural gas production. Coal is important in Kazakhstan (5.5 percent), Ukraine (3.1 percent), Bosnia (2.4 percent), and Serbia (2.0 percent). Minerals are important in Kyrgyzstan (8.5 percent, mainly gold), Uzbekistan (8.1 percent: copper, tin, and gold), and FYR Macedonia (7.3 percent: copper, silver, and zinc).

These natural resources have major impacts on incomes, consumer expenditures, and government revenues. In addition, they matter in other ways for food security, since prices of food have fluctuated in parallel with prices of these commodities, and this correlation will become stronger in the future with growing use of biofuels.

Water

The mountainous regions of Kyrgyzstan and Tajikistan are the main sources of Central Asia's water, making Uzbekistan, Turkmenistan, and Kazakhstan dependent on their water supply. Historically this was not a problem, as water distribution was centrally organized within the Soviet Union.[2] However, since independence, water has been at the base of tensions among the Central Asian countries. In general, the downstream states (Uzbekistan, Turkmenistan, and Kazakhstan) are short on water supplies for agricultural production, but are rich in terms of energy supplies. In contrast, the upstream countries (Kyrgyzstan and Tajikistan) are rich in water supplies, but have few other natural resources, such as natural gas and oil, for which they are dependent on the downstream countries.

After independence, the upstream states invested in hydroelectric projects to become more energy independent. In winter, they release water from a number of large reservoirs to generate energy for heating. Unfortunately, this causes frequent flooding in the downstream countries. In summer, they keep water in their reservoirs in order to have enough capacity during the cold months, while farmers in the downstream countries rely on the upstream water supply for irrigation. These opposing demand patterns for energy and water resources have been at the base of continuous political tensions between upstream and downstream countries since their independence.

POLICY REACTIONS TO THE FOOD CRISIS

As a consequence of the 2008 food price spike, food-exporting countries in Eastern Europe and Central Asia banned, taxed, or restricted exports of food, while importing countries reduced import constraints, such as tariffs. An FAO survey found that one-third of the surveyed countries in the Eastern Europe and Central Asia region imposed export restrictions in some form, while one-third of the countries reduced import taxes after the first price spike (FAO 2009). Several countries also introduced price control mechanisms and intervention purchases in an attempt to ensure domestic food security.

All major grain exporters in the region implemented export restrictions to secure their domestic supply of grain and protect their local consumers from increasing food prices during the price spikes in 2008 and 2010. Ukraine was the first country to launch grain export restrictions. It introduced an export quota on wheat (and during certain periods, also on barley and corn) in 2007, 2008, 2009, and 2011. Russia, the largest exporter in the region, imposed substantial export taxes on grains in 2008 (barley and wheat) and an export ban on grains in 2011 (barley, corn, wheat, and flour). Kazakhstan, the third major exporter in the region, imposed an export ban on wheat from April to August 2008 and on oilseeds from September 2010 to June 2011. Even some net grain-importing countries in the region—Tajikistan, Turkmenistan, and Uzbekistan—introduced export restrictions (FAO 2011a).

The grain-importing countries in the region reduced import constraints to facilitate grain imports. In May 2008, for example, the Azerbaijan government removed the customs on grain and rice imports. In Moldova, the government removed the 5 percent import duty on wheat and the 20 percent VAT on imported grains (FAO 2011a). During the 2010 to 2011 price spike, Kyrgyzstan lowered its import duties by two-thirds (Al-Eyd et al. 2012).

Throughout the region, governments also intervened in other ways to minimize food price inflation. For example, in Ukraine, the State Agrarian Fund imposed limits on flour price mark-ups and retail bread prices (OECD 2009). The State Agrarian Fund also undertook intervention purchases and forward contracts, when grain is purchased at harvest time and sold later to smooth prices (FAO 2011b). In 2008, the Russian government imposed price controls for various food products, such as bread, milk, sunflower oil, and eggs, and attempted to stabilize domestic commodity prices through state grain purchases and sales by the State United Grain Company (OECD 2009; FAO 2011b). Kazakhstan has a state agency, the State Food Corporation, which is responsible for up to 29 percent of the domestic wheat purchases (FAO 2011b).

In Kyrgyzstan, the government increased social assistance payments, distributed wheat reserves to the poor, and increased the monitoring of processing and retail margins for primary products during the price spikes of 2008 and 2010 (Suiumbaeva 2009; FAO 2011b). In Georgia, the Tbilisi municipality opened groceries offering a 20 percent discount on basic products for vulnerable households (World Bank 2011b). In Uzbekistan, the government kept prices low by selling more flour from state resources (World Bank 2011b). In Tajikistan, the government reduced the VAT on wheat by half and implemented price controls on food in Dushanbe during the 2010 to 2011 price spike (Al-Eyd et al. 2012).

Impacts on Global Food Prices

Export restrictions by Kazakhstan, Russia, and Ukraine may have exacerbated global price spikes in recent years, since especially for wheat, these countries provide a substantial share of the world's exports (Headey and Fan 2008; Headey 2011). In fact, the timing of export restrictions and wheat price increases suggest some causality (Figure 15.4).

The first country to implement export restrictions on wheat was Ukraine. Dollive (2008) reports that when Ukraine imposed restrictive export taxes (essentially banning wheat exports) in 2007, the country's action triggered a cascade of effects. Coinciding with Ukraine's effective export ban, international wheat prices increased by 40 percent between July 2007 and the end of 2007. The export restrictions had a contagious effect on other countries imposing export restrictions, as it forced countries that were previously importing from Ukraine to switch to countries such as the United States, France, Australia, Argentina, and especially Russia and Kazakhstan. In the latter two countries the rapid increase in wheat exports halved the stock-to-use ratio, and in order to secure their domestic wheat supply, both Russia and Kazakhstan soon introduced export restrictions. Argentina also imposed export restrictions. As a result of these restrictions, demand for wheat from the remaining exporting countries surged, and wheat stocks further decreased. This led to an additional run-up in wheat prices, as wheat suppliers typically base their prices on existing stock level (Headey 2011).

When in the summer of 2010, destruction of the Russian and Ukrainian wheat harvest by drought and fire was forecast, the two countries imposed export restrictions on various grains, including wheat. Similar to the situation in 2008, this action coincided with a price spike on the international wheat market. In fact, the day that Russia announced its wheat ban, international wheat prices jumped immediately by around 10 percent. Overall, international wheat prices increased by 31 percent between September 2010 and the beginning of 2011 as a result of complex interactions among export bans, declining stocks, demand surges, and rising prices.

Fig. 15.4. Food prices and export restrictions by major grain-exporting countries

Note: The black lines indicate the start and end of the two periods in which Russia has imposed export restrictions on wheat and other cereals. The period A indicates a period during which there were restrictive export taxes on wheat exports, while period B indicates a period during which there was an export ban on wheat and other cereals. The grey lines indicate the start and end of the two periods in which Ukraine has imposed export restrictions on wheat. The period C indicates a period during which there were restrictive export taxes on wheat exports, while period D indicates a period during which there were restrictive export quotas on wheat exports.

Source: FAOstat

Impact in Food-Importing Countries

Overall, the poorer countries in the region rely heavily on grain imports (Table 15.2). More than 50 percent of the wheat supply in Armenia and up to 70 percent in Georgia was imported in the period 2006 to 2008, almost exclusively from Russia, Ukraine, and Kazakhstan. This might suggest that the export restrictions would have a strong impact on the food supply in these countries, but the overall impact seems to have been fairly limited.

One reason was that the policy responses reduced the effects. The impact was also limited because, after the introduction of export restrictions on wheat,

Table 15.2. Share of wheat imports from Kazakhstan, Russia, and Ukraine in total wheat imports for the countries in North Africa and the Middle East (averages 2006–2008)

	Proportion of imported wheat	Imports from Kazakhstan	Imports from Russia	Imports from Ukraine	Imports from the rest of the world
Armenia	57	9	86	3	2
Azerbaijan	40	46	51	3	0
Georgia	70	28	62	4	6
Kyrgyzstan	24	96	1	0	3

Source: UNcomtrade

most of the importing countries in the region rapidly switched to importing flour. In Kyrgyzstan, wheat imports decreased in 2008 and 2011 by 15 percent and 17 percent respectively, while flour imports increased by 70 percent and 431 percent. In fact, this shift from wheat to flour imports resulted in an overall net increase in wheat equivalents of respectively 8 tons in 2008 and 92 tons in 2011. Despite the fact that the price of flour is in general higher than the price of wheat, this indicates that overall food supply remained relatively stable in the importing countries.

Moreover, although food prices have increased substantially in several transition countries, the impact of such a price increase should not be overestimated. When we examine the evolution of real wages, food prices, and retail prices, wages increased substantially between the mid-2000s and 2009. The increase in real wages exceeded the increase in food prices and retail prices in all countries—even during 2008 and 2009. Hence, these data suggest—somewhat remarkably—that the slowdown and decline in GDP in 2008 and 2009 was not reflected in wages, and that any negative impact of the food price increase on food security may have been offset with wage increases. Rural households may have benefited from high food prices, while those employed in formal jobs may have been shielded by wage inflation. Possibly the most sensitive population is households without formal wage income, who strongly depend on falling remittances, and are net consumers of food.

Impact in Grain-exporting Countries

While export restrictions were introduced to ensure domestic food security during the food crises in 2007 and 2010, studies on Ukraine (Von Cramon and Raiser 2006) and Russia (Jones and Kwiecinski 2010; Welton 2011) found that the overall impact of the export restrictions on local food security was limited.

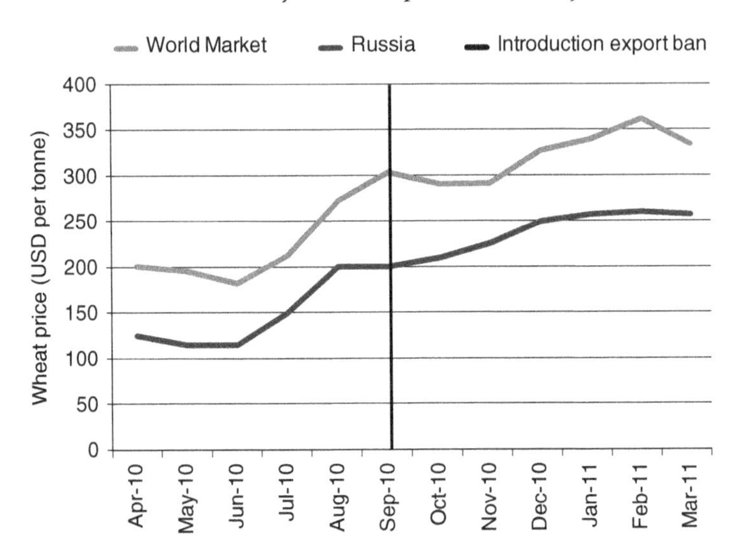

Fig. 15.5. Wheat prices in US and Russia in April 2010–March 2011

Source: Russian prices are obtained from the USDA FAS Gain Report; World market prices are proxied by US prices, which was the main exporter of wheat, and data are obtained from the World Bank

Consumers seem not to have benefited from a substantial decrease in food prices. In fact, interventions in the Russian wheat market in October 2007 at best delayed price increases by a couple of months. In January 2008, domestic prices for wheat increased rapidly despite intervention sales and the introduction of a prohibitive export tariff (FAO 2011a). When in September 2010, the Russian government introduced an export ban on wheat, Russian wheat prices continued to increase in parallel with world market prices (Figure 15.5). Similar findings hold for the price of bread, which evolved in Russia similarly to the price in the neighboring countries. A potential explanation is that large exporters hoarded grains until the export ban was lifted and they could benefit from high international prices (Welton 2011). Nonetheless, it is important to recognize that wheat prices contribute only a certain percentage of the final bread price, and the bread price represents only a marginal fraction of the consumer basket of an average Kazakh, Ukrainian, or Russian citizen.[3] The impact of lower feed prices on the price of meat or poultry is similarly small, as the livestock sector is highly concentrated, which does not provide an incentive for meat producers to pass lower costs on to their customers (Von Cramon and Raiser 2006).

Large losses for domestic grain producers also limited the overall food security impact of the export restrictions. In Ukraine, for example, the export quota introduced in October 2006 led to a loss in revenue of approximately $300 million by the end of 2006 (Von Cramon and Raiser 2006). In 2010, Ukrainian revenue losses from the export quota on grain were even estimated to be higher, approximately $1 billion (FAO 2011b). Since the poor, rural population is primarily involved in farm activities, the export restrictions may even increase poverty instead of decreasing poverty, as rural producers are not able to benefit from high output prices. Moreover, when farmers cannot profit from higher output prices, farmers have lower incentives to expand their production, even if the world market is giving exactly the opposite signal.

Exporters may also be hurt by the export restrictions imposed. In Russia, the loss of only the Novorossiysk port approached $40 million as a result of the 2010 export ban, according to Morgan Stanley (RIA Novosti 2010). Exporters are prohibited to export, but they also face substantial storage costs, financial costs, and reputational costs, since they may no longer be able to honor contracts signed before the export restriction. The uncertainty in the policy climate may also jeopardize future foreign direct investments in the agro-industry, which have been growing substantially in past years.[4]

In fact, the only ones benefiting from trade restrictions are feed and meat producers and flour millers, which benefit from lower input prices. In Russia, flour millers benefited during the first price spike in 2008, but not during the second price spike at the end of 2010, because flour exports were also banned.[5]

Sociopolitical Stability

Recent years have seen several cases of sociopolitical instability in Eastern Europe and Central Asia, ranging from war to mass demonstrations. Since 2008, two major armed conflicts, both cases of ethnic violence, have occurred. An armed conflict in August 2008 between Georgia (supported by Russia) and the separatist government of South Ossetia was fueled by a dragging ethnic conflict. In June 2010 in Kyrgyzstan, protests against government corruption and increased living expenses escalated in a violent clash between ethnic Uzbeks and Kyrgyz. In addition, several civil protests have arisen, such as the Russian demonstrations of 2011 to 2012 against the 2011 Russian election process that eventually restored President Vladimir Putin. Protests against political repression and government corruption in 2011 and 2012 also shook Romania, Croatia, Serbia, Belarus, Kazakhstan, Azerbaijan, and Armenia.

Overall, there is little evidence that this sociopolitical instability has been much influenced by the larger food crisis, possibly because of limited negative welfare effects. Nevertheless, those directly involved in conflicts in the region experienced important food security problems. In Georgia the 2008 war displaced around 130,000 people. An assessment of food security and nutrition of the conflict-affected population in March 2010 showed that 99 percent of the refugees were food insecure, depending heavily on external assistance to meet their basic food needs. A significant proportion of the households had a very monotonous diet (World Food Programme 2010a). More recently, the ethnic violence in southern Kyrgyzstan against the minority population of Uzbeks resulted in the displacement of more than 300,000 people. The majority of the refugees were food insecure, depending largely on humanitarian assistance and food gifts (World Food Programme 2010b).

In addition to the effects on the importing countries in the region, the export restrictions imposed by Kazakhstan, Russia, and Ukraine have had implications for other world regions. Some have argued that the export restrictions directly and indirectly affected protests in several countries in North Africa and the Middle East, and that the increase in food prices was a catalyst for what would later be the Arab Spring (Wright and Cafiero 2011).

The export restrictions imposed by Kazakhstan, Russia, and Ukraine affected food prices in the Arab world through two mechanisms: by impacting the wheat supply for some Arab countries and by raising global wheat prices. Several Arab countries are strongly dependent on wheat imports from Kazakhstan, Russia, and Ukraine (Figure 15.6). In the period 2005 to 2010, Syria was almost exclusively dependent on Russian and Ukrainian wheat (85 percent of all imports), while Algeria and Morocco were only importing a small amount of wheat from Kazakhstan, Russia, or Ukraine, respectively 5 percent and 9 percent. Tunisia, the country where the Arab Spring started at the end of 2010, got 47 percent of its wheat imports from Kazakhstan, Russia,

Fig. 15.6. Share of wheat imports from Kazakhstan, Russia, and Ukraine in total wheat imports for the importing countries in the region (averages 2005–2010)

Source: UNcomtrade

and Ukraine, while Egypt, where the protests started in the beginning of 2011, depended on the region for 44 percent of its wheat imports.

When Russia imposed an export ban in the marketing year 2010 to 2011, the importing Arab countries switched to other suppliers of wheat, such as the United States and France (Welton 2011; FAO 2011c). Most importing countries also had some grain stocks, so that in general there was no disruption of grain provision. Export restrictions also triggered excessive hoarding by several importing countries (Headey 2011). In Egypt, for example, the General Authority for Supply Commodities, the government body responsible for grain purchases, reported that wheat imports increased from 5.53 million tonnes in 2009 to 2010, to 5.66 million tonnes in 2010 to 2011, despite the Russian export ban.

Although the overall food supply remained relatively stable, the switch to different suppliers resulted in substantially higher wheat prices, and food prices in Egypt rose accordingly. Unlike in the importing Eastern Europe and Central Asia countries, however, wage inflation did not outpace food price inflation, so the negative impact of a food price increase on food security may not have been completely offset with wage increases. Moreover, since several of the Arab countries provided subsidies to keep local food prices low, the food crisis pushed up public spending, which may have been especially problematic for the oil-importing countries in the region (Wright and Cafiero 2011).

Some have suggested that the Central Asian regimes could be susceptible to a popular uprising similar to the Arab Spring. Many ingredients that characterized the Arab Spring are also there: repressive regimes and governments ruled by longtime dictators, widespread government corruption and nepotism, high unemployment, and inequality and—at least in some Central Asian countries—a rapid increase in the use of mobile phones, Internet, and social media, and a large, young population with limited prospects for the future (Huschke 2012; Zikibayeva 2011).

While there are some similarities with the Arab countries, however, there are also important differences. First, several of the countries in the region, such as Kazakhstan and Uzbekistan, benefited from a boom in the natural resources prices, which positively affected government budgets. Second, all countries in the region have experienced strong economic growth in the past and are forecast to also grow substantially in the future. High unemployment rates in the poorer countries are diluted by migration to the richer countries in the region. In addition to direct income effects of growth, social security programs have been scaled up in the past years (World Bank 2009).

Finally, the availability of social media is more limited than in the Arab world, where protestors could organize themselves through Internet, social media, and mobile phone networks. At the time of the protests, these technologies were already widespread in the Arab world. In some Eastern Europe and Central Asia countries, such as Kazakhstan and Kyrgyzstan, communication networks are well developed, but in other countries, such as Uzbekistan and Turkmenistan, this is not the case (Schmitz and Wolters 2012). Uzbekistan has witnessed a rapid increase in the spread of mobile phones and Internet use, but all data traffic—Internet, text messages, and emails—is controlled by the central government (Schmitz and Wolters 2012). In Turkmenistan, only 1.6 percent of the population is online, as private Internet access is very expensive and hence only accessible for the country's elite. Moreover, all data traffic in Turkmenistan runs through a central hub and is controlled by the government (Schmitz and Wolters 2012).

In addition, several protests and demonstrations in this region have had only a limited impact on governments. In Kyrgyzstan, rioters protesting against governmental corruption led in April 2010 to the installation of a new government that promised to have free elections. Russia, Belarus, Kazakhstan, Azerbaijan, and Armenia have had anti-government protests, but unlike in Kyrgyzstan, protests have led to no radical changes in government policy up to now.

It therefore remains unclear whether sociopolitical stability will radically change in the near future in the Eastern Europe and Central Asia region, and whether food security will play a major role. Also in question is the overall impact of climate change on the water–energy–food nexus in Central Asia and how it will affect sociopolitical stability. Most likely the impact will be heterogeneous. In some countries, such as Kazakhstan, water availability is projected to increase, while in Uzbekistan and Turkmenistan—two countries that already face serious constraints related to water availability—the average annual runoff of water is expected to decrease as a result of climate change (Nelson et al. 2010).

LOOKING FORWARD: INCOMES, ENERGY, AND FOOD SECURITY

The International Monetary Fund (IMF) estimates that between 2010 and 2017 incomes will steadily increase in most countries in Eastern Europe and Central Asia, with the strongest growth prospects in some of the resource-rich countries, including Russia (4 percent per year), Kazakhstan (6.5 percent per year), and Uzbekistan (6.8 percent per year). Several of these countries are highly dependent on oil and natural gas production. These two commodities represented approximately a quarter of Russian GDP and two-thirds of Russian exports in 2009. In the coming years, oil and energy prices are projected to increase substantially. The IMF estimates that, by 2020, oil prices may increase by approximately 80 percent (Benes et al. 2012). The region still has a substantial supply of oil and natural gas, especially in Siberia, and the Caspian, the Barents, and other Arctic seas look promising (International Energy Agency 2011). These prospects provide a huge potential for future economic growth in Russia, but such growth also makes the country vulnerable to potential pitfalls if the global economy plummets as it did in 2008.

The USDA foresees a substantial increase in grain production in the region by 2020, projecting an increase in grain production from 139 million tons in 2006 to 2010, to 180 million tons in 2012 (Liefert and Liefert 2012). Grain production in Kazakhstan, Russia, and Ukraine, the three major grain-producing countries in the region, could increase by up to 230 million tons (an increase of 80 percent compared to the 2004 to 2006 production level) (FAO 2008). Part of the projected gains comes from increased land use, and the remainder from increased land productivity.

After the transition from a centrally planned economy to a market-orientated economy, agricultural land use substantially decreased in most countries, so there may be scope to increase arable land use in the region, in particular if agricultural prices remain high.

The greatest potential increase in production, however, is likely to come from an increase in yields. In the first years of transition, agricultural yields of the major arable production in the region decreased strongly in all countries. For example, between 1990 and 1995, grain yields in Kazakhstan decreased by more than 10 percent per year. At the end of 1990s, grain yields reached a new low in the major grain-producing countries. Yields have rebounded in the past decade, but despite the recent increase, wheat yields in the main producing countries in the region are still substantially below yields in other major grain-producing countries in the world where there are similar climatological conditions.

Increased investment and better management and technology may further increase yields. Such yield increase will depend, of course, on incentives to invest, which in turn require favorable market, policy, infrastructure, and institutional conditions.

Climate change is also likely to affect yields in the future—but the effects may differ substantially within the region (Nelson et al. 2010). Grain yields may increase quite substantially with climate change in the major grain production areas. According to different climate change models used by IFPRI, yields of rainfed wheat and maize will increase by 1 to 2 percent per year in Ukraine and Kazakhstan between 2010 and 2050. In other countries in the region, productivity gains are expected to be positive as well, but more limited.

However, the question remains to what extent an increase in the agricultural potential in the region will affect global food security. The USDA projects that Russian average annual net grain exports will increase from 14 million tons to 24 million tons, and the three major grain exporters will increase their exports from 36 to 70 million tons (Liefert and Liefert 2012). If these projections are accurate, the region will become a dominant supplier of wheat to the world, with 30 percent of total wheat exports by 2020. Until now, however, exports have been volatile and largely affected by trade policy measures, such as the export bans, so prospects for future developments should be considered in the light of the Russian agricultural policy atmosphere.

Since 2009, food security and self-sufficiency have become increasingly important in Russian domestic and foreign policy (Liefert et al. 2010; Wegren 2011). This concern is reflected in the inclusion of food security motives in the National Security Strategy, which was adopted in May 2009 and later confirmed by President Dmitry Medvedev, who stated that the food security doctrine is crucial and "supplying food products is one of the cornerstones of security in general" (Wegren 2011, 142). Russia's efforts to upgrade livestock production with substantial subsidies to the livestock sector, as well as sanitary and phytosanitary import restrictions on live animals and animal products, suggest a similar interest in national food security (Liefert et al. 2010). The region's long-term grain export potential will strongly depend on policy-driven growth in the livestock sector and the trade-off between food and feed production (Liefert and Liefert 2012).

Russia's recent WTO accession reduces the possibility that Russia will put in place trade- and market-distorting policies in the future and will allow other WTO members to challenge the country's sanitary and phytosanitary import restrictions.

CONCLUSION

Increasing food prices in recent years have triggered a series of policy initiatives in Eastern Europe and Central Asia: food exporting countries banned, taxed, or restricted the exports of food, while importing countries reduced their import tariffs. The export restrictions by Russia, Ukraine, and Kazakhstan affected both the food-importing countries in the region and countries in North Africa and the Middle East, which heavily rely on imports from these major grain-exporting countries.

Rising food prices have been a catalyst for food riots, including the Arab Spring. Some indicators, such as the presence of repressive regimes ruled by long-time dictators, high inequality, and a young population with limited prospects for the future, suggest that the region could be susceptible to a "Central Asian Spring," similar to the civil protests and demonstrations that have led to the overthrown of several authoritarian regimes in the Arab world in the spring of 2011.

While several anti-government demonstrations have occurred, most have not been very successful. There are important differences with the Arab world, as several of the countries in the region benefited from a boom in the natural resources prices and high economic growth, which reduced unemployment. Moreover, the poorer countries' high unemployment rates are diluted by migration to the region's more prosperous countries. Prospects for economic growth are relatively good in the poorest countries, which might otherwise experience impacts of high food prices on political and social stability.

In terms of food production, the Eastern Europe and Central Asia region is in a good position to increase wheat production. However, in order to realize the agricultural production potential of the region, investments in the agri-food industry must increase. Agricultural investments in the more economically advanced transition countries, such as the new member states of the EU, have been one of the main engines behind productivity growth, quality improvements, and enhanced competitiveness through the introduction of vertical coordination mechanisms in the supply chain (Dries and Swinnen 2004; Dries et al. 2009; Van Herck et al. 2012).

These investments are overdue and will have to make up for decades of inattention. For example, in Russia, an important constraint is grain storage capacity (USDA 2011; Liefert et al. 2010). A large share is in bad condition, and the location of most elevators still reflects the patterns developed in the former planned economy, with more storage facilities in grain-consuming areas and less in grain-producing areas (USDA 2011).[6]

Poor institutions, infrastructure, and a lack of skilled workers are important constraints for investors. Rural infrastructure in most Eastern Europe and Central Asia countries is generally poor. Improvements in the public infrastructure may allow farmers to connect to markets by reducing transport costs. In addition, investments in rural infrastructure may improve access of rural laborers to urban areas and attract more off-farm employment, including opportunities offered by foreign investors. In short, investments efficiently reduce over-employment in the agricultural sector and stimulate pro-poor economic growth.

Finally, while the overall level of education is high in the region, the quality of the education and the level of practical and technical skills are low (Sondergaard and Murthi 2010). This not only affects agricultural productivity through reduced intersectorial labor mobility, but also constrains the adoption of new technologies in the agricultural sector. With respect to enhancing human capital within the agricultural sector, investment in agricultural R&D and extension services will be crucial.

ACKNOWLEDGMENTS

We thank Chris Barrett for encouraging our research on this project. We thank Chris, Steven Kyle and participants at the Cornell workshop for very useful comments on an earlier version of the paper. The research was financially supported by the Food Security and Sociopolitical Stability project and the KU Leuven's Research Council (Methusalem project). The authors are solely responsible for opinions in the paper.

NOTES

1. A personal cult arises when an individual uses mass media, propaganda, or other methods to create an idealized and heroic public image of himself (Gawrich et al. 2010).
2. The Soviet system of using this water for expensive irrigation of cotton and wheat farms caused major environmental problems—in particular salinization of soils and the shrinking of the Aral Sea.
3. While compared to the Western countries, expenditures on food were still relatively high in Russia, Ukraine, and Kazakhstan, respectively 28 percent, 31 percent, and 39 percent, flour and bread represents only a small share of these expenditures, even in poor households. In 2010, bread products represented 8.5 percent of the Kazakh consumer basket (Al-Eyd et al. 2012). In Ukraine, flour and bread represented respectively 0.54 percent and 3.87 percent of the consumer basket in 2006 (Von Cramon and Raiser, 2006). In Russia, the importance of bread and bakery products in the consumer was even lower as it only represented 2.07 percent of the consumer basket in 2008 (World Bank 2008).
4. For example, in Ukraine per capita investments in the agro-food industry increased by 23 percent, from 21 euro per capita in 2003 to 26 euro per capita in 2008. In Russia, the growth was even more remarkable as per capita investments in the food industry increased from 22 euro per capita to 31 euro per capita, or an increase of 40 percent (Hunya 2009).
5. However, it is important to note that the ban on flour exports was lifted in January 2011, while wheat exports remained banned until June 2011. This exception was made on request of Russian flour millers, especially Siberian millers whose capacity was not used to full extent (USDA 2010).
6. In 2011, grain storage capacity amounted to 118.2 million metric tons, but experts estimate 70 to 80 percent is in bad conditions (USDA 2011).

REFERENCES

Al Eyd, A. J., D. Amaglobeli, and B. Shukurov. 2012. Global food price inflation and policy responses in Central Asia. IMF Working Paper No. 12/86, Washington, DC.

Anderson, K., and J. F. M. Swinnen. 2009. Distortions to agricultural incentives in Eastern Europe and Central Asia. Agricultural distortions Working Paper 48624, World Bank, Washington, DC.

Benes, J., M. Chauvet, O. Kamenik, M. Kumhof, D. Laxton, S. Mursula, and J. Selody. 2012. The future of oil: Geology versus technology. IMF Working Paper, Washington, DC.

Dollive, K. 2008. The impact of export restraints on rising grain prices. Office of Economics Working Paper No. 2008-08-A, U.S. International Trade Commission, Washington, DC.

Dries, L., and J. F. M. Swinnen. 2004. Foreign direct investment, vertical integration, and local suppliers: Evidence from the Polish dairy sector. *World Development 32*(9): 1525–44.
Dries, L., E. Germenji, N. Noev, and J. F. M. Swinnen. 2009. Farmers, vertical coordination, and the restructuring of dairy supply chains in Central and Eastern Europe. *World Development 37*: 1742–58.
EBRD. 2009. Transition report 2009: Transition in crisis? London: European Bank for Reconstruction and Development.
FAO. 2009. The state of agricultural commodity markets: High food prices and the food crisis–experiences and lessons learned. Rome: FAO.
——— 2011a. Policy measures taken by governments to reduce the impact of soaring prices. Global Information and Early Warning System on Food and Agricultural (GIEWS). Rome: FAO.
——— 2011b. FAO Regional policy consultation on high food prices in Europe and Central Asian Region. Budapest: Summary of proceedings, FAO Regional Office for Europe and Central Asia.
———2011c. Near East food security update. Cairo: FAO Regional Office for the Near East.
Gawrich, A., I. Melnykovska, and R. Schweickert. 2010. More than oil and geography: Neopatrimonialism as an explanation of bad governance and autocratic stability in Central Asia. Paper presented at the workshop Neopatrimonialism in Various World Regions, August 23, 2010, GIGA German Institute of Global and Area Studies, Hamburg.
Headey, D. 2011. Rethinking the global food crisis: The role of trade shocks. *Food Policy 36*:136–46.
———, and S. Fan. 2008. Anatomy of a crisis: The causes and consequences of surging food. *Agricultural Economics 39*: 375–91.
Hunya, G. 2009. Wiiw Database on Foreign Direct Investment in Central, East and Southeast Europe: FDI in the CEECs under the Impact of the Global Crisis: Sharp Declines. Vienna: The Vienna Institute International Economic Studies.
Huschke, G. W. 2012. Central Asia: The new Arab Spring? Streit Council.
International Energy Agency. 2011. World energy outlook 2011. OECD/IEA, International Energy Agency annual projections, France.
Jones, D., and A. Kwiecinski. 2010. Policy responses in emerging economies to international agricultural commodity price surges. OECD Food, Agriculture and Fisheries Working Papers, No. 34. Paris: OECD Publishing.
Liefert, W. M. 2002. Comparative (dis?) advantage in Russian agriculture. *American Journal of Agricultural Economics 84*(3): 762–7.
———, and O. Liefert. 2012. Russian agriculture during transition. *Applied Economic Perspectives and Policy 34*(1): 37–75.
Liefert, W. M., Serova, E., and O. Liefert. 2010. The growing importance of the former USSR countries in world agricultural markets. *Agricultural Economics 41*(1): 65–71.
Liefert, W. M., and J. F. M. Swinnen. 2002. Changes in agricultural markets in transition economies. Agricultural Economics Reports 33945, Economic Research Service (ERS), United States Department of Agriculture (USDA).
Macours, K., and J. F. M. Swinnen. 2008. Rural–urban poverty differences in transition countries. *World Development 36*(11): 2170–87.

Musaev, D., Y. Yakhshilikov, and K. Yusupov. 2010. *Food security in Uzbekistan.* Tashkent: UNDP.

Nelson, G. C., M. W. Rosegrant, A. Palazzo, I. Gray, C. Ingersoll, R. Robertson, S. Tokgoz. 2010. *Food security, farming and climate change to 2050: Scenarios, results and policy options.* Washington DC: International Food Policy Research Institute.

OECD. 2009. *Agricultural policies in emerging economies: Monitoring and evaluation 2009.* Paris: Organisation for Economic Cooperation and Development.

RIA Novosti. 2010. Novorossiysk sea port may lose $40 mln over grain export ban. RIA Novosti, August 11. See http://en.rian.ru/business/20100811/160155627.html (accessed December 17, 2012).

Schmitz, A., and A. Wolters. 2012. Political protest in Central Asia: Potential and dynamics. SWP Research Paper 7, April 2012, Berlin.

Sedik, D. 2011. Food security and trade in the ECA region. Presentation at the *Course on Food and Agricultural Trade for ECA*, February 7–10, Vienna.

Sondergaard, L., and M. Murthi. 2010. Skills, not just diplomas: The path for education reforms in ECA. ECA Knowledge Brief, World Bank, Washington DC.

Swinnen, J. F. M., and K. Van Herck. 2009. The impact of the global economic and financial crisis on food security and the agricultural sector of Eastern Europe and Central Asia. Background paper for FAO, UN Ministerial Conference on the Social Impact of the Economic Crisis in Eastern Europe, Central Asia and Turkey, December 2009, Almaty, Kazakhstan.

United Nations Commodity Trade Statistics Database. http://comtrade.un.org/ (accessed April 5, 2013).

USDA. 2010. Grain export ban extended to July 1. GAIN Report, USDA, Washington, DC.

———. 2011. Russian Federation: Grain and feed annual 2011. Gain report RS1113, Global Agricultural Information Network, USDA Foreign Agricultural Service, Washington, DC.

Van Herck, K., N. Noev, and J. F. M. Swinnen. 2012. Institutions, exchange and firm growth: Evidence from Bulgarian agriculture. *European Review of Agricultural Economics 39*(1): 29–50.

Von Cramon, S., and M. Raiser. 2006. The quotas on grain exports in Ukraine: Ineffective, inefficient, and non-transparent. Institute for Economic Research and Policy Consulting in Ukraine; German Advisory Group on Economic Reform, Kiev.

Wegren, S. K. 2011. Food security and Russia's 2010 drought. *Eurasian Geography and Economics 52*(1): 140–56.

Welton, G. 2011. The impact of Russia's 2010 grain export ban. Oxfam Research Report, June 2011, Oxford.

World Bank. 2008. *Russian economic report.* Washington DC: World Bank.

———. 2009. *Turmoil at twenty: Recession, recovery and reform in Central and Eastern Europe and the former Soviet Union.* Washington DC: World Bank.

———. 2010. *The crisis hits home: Stress testing households in Europe and Central Asia.* Washington, DC: World Bank.

———. 2011a. International strategy note for the Kyrgyz Republic for the period FY12–FY13. Report No. 62777-KG. Washington, DC: World Bank.

———. 2011b. *Rising food and energy prices in Europe and Central Asia.* Washington, DC: World Bank.

———. 2012. *World Development Indicators 2012.* Washington, DC: World Bank.

World Food Programme. 2010a. *Emergency food security assessment Caucasus conflict follow-up study in Georgia.* Rome: World Food Programme.

———. 2010b. *Rapid emergency food security assessment in Osh and Jalalabad in Kyrgyz Republic.* Rome: World Food Programme.

Wright, B., and C. Cafiero. 2011. Grain reserves and food security in the Middle East and North Africa. *Food Security* 3(1): 61–76.

Zikibayeva, A. 2011. What does the Arab Spring mean for Russia, Central Asia, and the Caucasus. A Report of the CSIS Russia and Eurasia Program, Center for Strategic and International Studies, Washington, DC.

16

Food Security and Sociopolitical Stability in South Asia

ARUN AGRAWAL

Recent news reports on mountains of food grain stocks in India rotting because of inadequate storage and simultaneous widespread hunger may shock some. But to most they constitute a familiar story, and indeed are no more than a reprise of similar patterns of food losses in a country where a quarter of the population goes hungry or is undernourished (Bhattacharya and Subramanian 2012). Twenty percent or more of nearly all categories of food produced in South Asia is lost or wasted, most of it before it reaches consumers' tables (Gustavsson et al. 2011, 6–9). This lost food would be sufficient to feed nearly all who now go hungry. Disgust over food squandered while multitudes go hungry or starve led the People's Union for Civil Liberties in India, a voluntary human rights organization, to file a petition in 2001 for the right to food against the Indian government, six state governments, and the Food Corporation of India (Amrith 2008).

Such news reports highlight the extent to which distribution of and access to food, rather than shortfalls in aggregate food production, have been the key determinants of food security in the recent past in the Indian subcontinent. They also raise questions about the appropriate emphasis in policies aimed at increasing food security, if choices are necessary—and given scarce resources, they inevitably are. What combination of different measures should food security policies focus on, from among measures aimed at greater food production, reduction of food losses, improved access, better distribution, improved utilization, and reduction of vulnerability to disasters (Scrimshaw 1989)?

Since the 1950s, South Asian countries—Afghanistan, Bangladesh, Bhutan, India, Maldives, Nepal, Pakistan, and Sri Lanka—have made major strides in improving their agricultural output in an effort to protect and improve the food security of their populations (Bhalla and Singh 2010). Increased food production and improvements in policies for distribution of food to those with limited entitlements have contributed to improved health, life expectancy, and welfare in the region. They have done so despite two countervailing trends: a rapidly increasing population and a growing demand for food (Mittal and Sethi 2009; Ninno et al. 2007). Certainly, future estimates of food security in the region depend on the direction of these two trends, particularly as population growth and greater affluence raise the aggregate demand for food and also the demand for foods such as meat and fish that require more inputs, thereby imposing new strains on food security (Aggarwal et al. 2010; Bhalla and Hazel 1998; Lal 2011).

Rapid economic growth in the subcontinent should not allay concerns about food security and its impacts on human welfare and sociopolitical stability. Despite some success in addressing food production shortfalls, South Asian countries continue to have the largest numbers of people suffering poverty and malnutrition, with predictable consequences for infant mortality, life expectancy, gender equity, and other welfare indicators. Food production in the region has just about kept pace with increasing population. Food grain availability in India hovered near 170 kilograms per person in the early 1960s, for example, and has been near that threshold for most years of the twenty-first century (DES 2011). Similarly, other countries in South Asia, such as Nepal, Bangladesh, and Bhutan, have barely improved in terms of per capita food consumption over the past five decades. Particularly puzzling in this regard is the low caloric intake among poor households despite improvements in their incomes and increases in household expenditures (Deaton and Dreze 2009). These gaps suggest both the stubborn nature of food insecurity, particularly among poor households in South Asia, and the need for better data on nutrition and food intake in the region.

Future food security in South Asia confronts new concerns that parallel those related to global food security and the impacts of climate change on food production (Lobell et al. 2008; Rosegrant and Cline 2003). It also includes specific challenges associated with the region's rapidly changing demographic structure, changing composition of demand for food, inadequate food storage and irrigation infrastructure, and declining water supply and quality.

FOOD SECURITY AND SOCIOPOLITICAL STABILITY: SOME HYPOTHESES

To understand how climate change and sociopolitical stability are related, it is necessary to understand the causal pathways that connect them. A highly schematic representation of the relationship, the main focus of this chapter, may be as follows:

Climate change => Food insecurity => Social instability

Although this relationship is relatively straightforward at the abstract level, examining it more concretely is problematic because of both conceptual and data problems. Multiple factors and variables make up each of the above concepts. Assessing the relationships among these many different factors is difficult enough, but these relationships are themselves also influenced by additional factors.

Food security is a function of availability, access, utilization, and stability of food supplies, each a multidimensional concept in turn (Barrett 2010). Further, food availability may be influenced by climate impacts of different kinds, but it is also affected by soils, topography, technological changes, level of development of infrastructure, government policies in many different sectors, irrigation, and credit availability. The same logic holds for the other dimensions of food security.

Drawing from insights in the existing literature on environmental change and security, and also on the South Asian political context, Table 16.1 lists five potential hypotheses about the relationship between climate change impacts and food insecurity, and the intersection of food insecurity and social stability.

CLIMATE, DEMOGRAPHIC, AND AGRICULTURAL CHANGES IN SOUTH ASIA

How climate changes will influence future food production in South Asia, particularly in the next decade to two decades, is uncertain. This uncertainty is a corollary of the more general uncertainty about whether scientists can provide climate forecasts at spatial and temporal scales relevant for decision making (Cane 2010). Climate modeling for the Indian subcontinent indicates limited to no discernible effects on agricultural production by 2020 for temperature increases of less than 2 degrees Celsius, provided irrigated water remains available and agricultural pests do not increase much (Lal 2011). However, the rapid shift toward groundwater irrigation in the region and declining water tables and increased soil salinity suggest that, even without climate change induced uncertainties, water shortages will become far more acute in the region, particularly for India and Pakistan. Increased levels of glacier melt in Nepal and more erratic rainfall are likely to result in more aggravated instances of landslides and soil erosion. In Maldives, Bangladesh, and along the eastern coast of India, natural disasters associated with hurricanes and flooding are also likely to cause greater devastation than in the past (De et al. 2005; Mirza and Ahmad 2005).

Table 16.1. Major climate and food security hypotheses for South Asia

Hypotheses	Key relationships and data needs
1. Increasing temperatures in southern India together with more intense hurricanes will worsen food security along coastal areas	Climate-related disasters, cropping patterns along the coasts
2. Regional climate impacts on water availability in the northern plains in the subcontinent (Pakistan, India, Bangladesh), coupled with rising demand for ground water and deteriorating ecological conditions, will depress food production and availability within the next two decades	Climate impacts, crop production and water balance, regional hydrology, farmer adaptations to changing risks and costs of cultivating specific crops
3. Stagnant agricultural productivity and increasing demand for food will force governments in the region to increase their food import bills which will worsen existing fiscal pressures on the governments of India, Pakistan, Nepal, and Bangladesh, creating increased likelihood of need for food aid—particularly for Nepal and Bangladesh.	Agricultural production and trade, government revenues and expenditures
4a. Faced with budgetary pressures and social unrest owing to food insecurity, national and provincial governments in India will strengthen the public distribution system in terms of its targeting and coverage of recipients; 4b. Citizens in Nepal and Pakistan will face longer periods of political uncertainty and possible political violence as elite contenders compete for power and office	Agricultural production, social unrest, subsidized distribution of food, costs and benefits of food subsidies.
5. Increased food insecurity in more inaccessible parts of India, Nepal, and Pakistan will provide a greater supply of recruits for existing Left and Islamic social movements	Distribution of agricultural production and food access, strength and distribution of social movements of different kinds, number of incidents of violence and people involved in them

Source: Author

Observed recent climate changes include increased temperatures (Lal 2003; Siddiqui et al. 1999), greater uncertainty in the spatial and temporal distribution of rainfall (Giorgi 2006; Lal 2011), and some evidence of greater intensity and more frequent exposure to extreme events (De et al. 2005; Mirza and Ahmad 2005). Particularly important in this context is the timing and reliability of the South Asian monsoons because rainfed agriculture still accounts for more than 60 percent of the cropped area (IWMI 2010). Any adverse impacts on agricultural output in the short run will exacerbate the effects of declining availability of arable land (Rosegrant and Cline 2003), lower water availability for irrigation (FAO 2005), falling rates of productivity increases (Bhalla et al. 1999), and increasing demand for cereals and superior foods owing to both population and income increases.

Thus, general food security-related conclusions for the subcontinent include negative effects on food production in the southern part of the subcontinent and Sri Lanka, with particularly severe effects on rain-fed rice and wheat in the short run, and potential disruptions in food supplies owing to droughts, salinity in the drier parts of the Indo-Gangetic plain and the Indus Valley, and flooding and salt water intrusion along the coastal areas. Projections of impacts on overall food supplies are dire in the medium to longer run.

Aggregate food availability and numbers of people both have a strong impact on food security. The trends in these two factors have essentially balanced themselves over the past five decades in South Asia. Total population in the eight South Asian countries nearly tripled between 1960 and 2010: from 1.15 billion to 3.27 billion. Country growth rates have varied. Pakistan's population has grown the fastest, by almost four times; Sri Lanka the slowest, having doubled. The growth rate in India has been just below the average for the region.

Although growth in aggregate food production has about kept pace with population, there is much variation across the eight countries. Bhutan and Nepal show the slowest increase in aggregate production. Pakistan and Sri Lanka are at the other end of the scale. The data for Afghanistan show a fair degree of inter-annual variability and may be unreliable. India and Bangladesh have quadrupled their cereal production in the last half-century, more than keeping pace with their increased population. Overall, in most South Asian countries, food production per person has not changed much over the last five decades—and in the countries such as Maldives and Afghanistan where food production per person has changed, the change has been a major decline.

Just as food production per person has barely shifted in South Asia, the output per hectare has also stubbornly remained nearly the same since the gains during the Green Revolution years of the 1960s. Increases in the productivity of land have been extremely slow where they have occurred, and indeed, in some of the South Asian countries, productivity has declined. The only South Asian country with agricultural output comparable to other high-productivity countries is Bangladesh, with an average output more than 4,000 kilograms per hectare—mainly a result of higher cropping intensity. By comparison, China's agricultural productivity is more than twice as high as that of the two larger food producers in South Asia, India and Pakistan. This difference is attributable to much higher levels of application of inputs such as fertilizers and manure compared to all South Asian countries, as well as some differences in market access.

The effect of these general background patterns in the subcontinent is a picture of food security—in terms of food production and availability—that can be viewed either as stagnant or as stable depending on one's views about what the future might bring.

Between 1990 and 2008, the rate of increase in food consumption has varied between 0.2 percent per year in Pakistan and India to nearly one percent per year in Bangladesh. Nonetheless, the number of undernourished people in India is truly staggering; real numbers have risen, although the proportion has declined. The number of undernourished people has remained stationary in the other countries of South Asia as well. Life expectancy at birth has increased in all countries since the 1960s to about 65 to 68 years. The outliers on the positive side are Sri Lanka and Maldives, with a life expectancy of 75 years—nearly ten years higher than the mean for South Asia. On the negative side is Afghanistan, with a life expectancy nearly 20 years lower than the South Asian mean.

The numbers on demographic and food availability presented in this section simultaneously pose some reasons for both optimism and pessimism. They show how much South Asian nations have done to match the growing food needs of their increasing population. Setting aside the increase in the absolute number of undernourished persons in India, the overall pattern in South Asia is one of improving food security. But the evidence also suggests that all of the governments' and peoples' efforts have essentially allowed them to stay in the same place. The rising numbers of people and their increasing incomes and wealth have meant that the increases in aggregate production, in land and person productivity, and in a food basket with a better composition have done no more than prevent a striking rise in the number of undernourished people.

Particularly important in this context is the decline in productivity gains and stagnation in food output per person during the past decade. The causes of this decline can be traced to the slower growth rates of the major inputs that accelerated agricultural production during the 1960s to the 1980s. Relatively slow technological improvements, reduced public and private investment in agriculture, decline in the rate of growth of fertilizer use, as well as area under irrigation, have all contributed to the problem (Bhalla and Singh 2010).

The pressure of demographic change and economic growth on the demand for more and better-quality food is likely to continue to rise over the next decade to two decades. Existing projections, however, suggest that most South Asian countries will be able to meet the growing demand in the short run—until about 2020 to 2025 (Mittal 2006; Rosegrant et al. 1995), or that the shortfalls will be small (Dev and Sharma 2010). The possible exceptions may be Afghanistan, Bangladesh, and Pakistan.

But even if demand for food rises only slowly, no technological breakthroughs comparable to the Green Revolution technologies of the 1960s are on the horizon to enable South Asian farmers to meet even that slow growth in demand. Without a frontier that can be opened up for more food production—as in many African and Latin American countries—South Asian countries have only three options. First, they can substantially increase productivity per hectare of land, although they have not achieved this objective during the past two decades. Second, they can import substantial quantities of food from the three to four major food exporters in the world (World Bank 2010)—but they will need to compete effectively with China to do so. Or third, they can secure land in other regions of the world where it is still possible; they have done so to some extent, particularly in East and southern African countries, but there is a rising reaction against such "land grabs" (Borras et al. 2011; Cotula and Vermeulen 2009).

In short, although South Asian countries have been able to remain in place over the past 50 years by running relatively hard, they will need to run a lot harder to keep food insecurity at bay in light of current trends of population and economic growth, ecological change, and food production. And once the gaps in distribution of food to the neediest populations are taken into account, food insecurity seems almost certain to rise in the next two decades.

HISTORICAL EXAMPLES OF FOOD INSECURITY

Historical experiences of food insecurity in South Asia are for the most part qualitatively different from those that confront its nations today. But although the analysis deals with instances of acute, rapid, and disastrous declines in food availability and access, rather than chronic, low-grade shortages, the general conclusions about what it took to address the worst effects of these instances of food insecurity are still relevant for contemporary discussions of food insecurity.

Over the past century, South Asia has undergone a remarkable transition in the forms of food insecurity to which its people are exposed. Famines and gigantic losses of human lives characterized many of the most important instances of food insecurity until nearly the very end of British colonial rule. The disastrous Bengal famine of 1943 to 1944 constitutes a particularly prominent example of such losses in the twentieth century (Dyson 1991a, 1991b; Maharatna 1996). Although it has received much analytical attention, the 1943 famine was not the most disastrous in terms of lives lost. More than 150 years ago the famine of 1770 in Bengal killed nearly a third of the population—10 million people (Sharma 2001)—and the 1867 famine in Orissa killed a million people, a quarter of the province's population. The Bihar and Bengal famine of 1873 to 1874 did not lead to excess mortality, but three years later famines in Bombay and Madras provinces killed millions. A pan-Indian famine that occurred in 1896 and 1897 saw even more disastrous losses of life.

Many different factors account for these famines. A decline in food grain production is certainly relevant. Some historians have related such declines and years of lower production to climate patterns, including El Niño and the Southern Oscillation (ENSO) effects (Grove 1997). But as important to explaining the subsistence crises that only sometimes developed into famines causing massive death are shifts in cultivated area from food crops to cash crops, incentives for exports of grains and other agricultural products, price rises, a larger population, and indifferent government policies during famines in colonial India.

Although the British colonial government sought to develop a famine policy as early as the 1840s, its efforts flowered into the India Famine Codes of 1880 after experience with several famines during the late 1860s and 1870s in rapid succession (Ahuja 2002). The Famine Codes were an elaborate expression of the government's position and approach to addressing famines. As Hall-Matthews (1996, 216) points out, they have also been a source of borrowings for many later analysts of famines, including Amartya Sen, Alex DeWaal, and Jean Dreze. Their effectiveness in subsequent famines was compromised because they attempted to reconcile two conflicting principles: a laissez faire approach, according to which administration of relief would be tantamount to encouraging the poor to shirk work, and the need to provide such relief to reduce the risk of deaths on cataclysmic scales. On the one hand were the dictates of noninterference in the workings of the agricultural markets, economy in administration so as to remain within the constraints of available resources, and preventing dependency relations from developing between the state and the population. On the other hand, many administrators with direct experience of famines sought to heed the imperative of preventing the worst human suffering and loss of lives (Ahuja 2002; Klein 1984).

Indeed, official policy teetered between the desire to economize and the need to alleviate starvation and mass deaths. Its twists, reversals, tortuous justifications, and impacts are starkly visible in contrary government famine policy implementation during the 30 years between the Orissa famine of 1867 and the pan-Indian famine of 1899 to 1900. The basic relationship that emerges in this period and later is painfully obvious: when government mobilized sufficient resources and incurred the necessary expenditures to support effective famine relief, excess mortality because of famines was limited or absent. When the state sought to protect its rupees, catastrophic human suffering, and deaths on a holocaust scale ensued.

Nowhere is this inference more evident than in the contrasting cases of the Bengal and Bihar famine of 1873 to 1874 and the more sweeping famine in Madras and Bombay during 1876 to 1878. The comparison is instructive because the crop failures associated with the two famines occurred in such rapid succession, because official policy reversed itself in the period between the two events, and because the same administrator was in charge of implementing the different policies.

The implementation of famine relief during the 1873 to 1874 famine in Bihar might be one of the few examples of effective governance that unfolded in the state. The Bihar and Bengal famine was met with a highly interventionist and expensive official strategy of famine combat. Sir Richard Temple, charged with implementing the response to the famine, imported nearly a half million tons of food grains from Burma. To distribute the grain, the government undertook adequate preparations for transport, including plans to reach areas away from railway lines. At the site of relief works, people were paid even if tasks were not completed, and those deemed in need received sufficient grain and money. The scale of administration of famine relief was vast, with millions receiving employment, subsidized grain, and loans of money and food. The outcome was a near elimination of mortality in famine areas—but at the expense of 6.5 million pounds sterling. More than one hundred thousand tons of grain remained unused at the end of the famine campaign, and the profligacy of the effort came under widespread criticism as "spendthrift" (Klein 1984, 194).

The government switched course for the next famine two years later, when food production in larger areas was admittedly affected. The opprobrium heaped on the "excessively liberal" expenditures during the 1873 to 1874 famine, coupled with a decline in the government's finances, led to a mode of implementation that failed to procure sufficient food, although at a national level enough food was available to feed the country even at the height of the famine (Currie 1991). Nor did the government take the steps necessary to transport the food to the places it was most needed. Instead, it put in place a far more restrictive strategy of distributing the bare minimum of food necessary to avert starvation to those who could prove they were starving at the sites of food relief. And even those who proved they were starving could receive food only in exchange for labor at tasks deemed to be appropriate for development.

Temple was in charge of famine relief administration in Madras under instructions from Viceroy Lytton, and designed the "Temple rations" that reduced the payments to laborers in public works. He thereby saved the government two million rupees. The rationale for refusing to provide relief was the belief that "even in the worst conceivable emergency so long as trade is free to follow its normal course, we should do far more harm than good by attempting to interfere" (Currie 1991, 32). The principle provided a cover for the belief that preventing starvation deaths completely, as the government had done three years earlier in Bihar and Bengal, might be possible only at the risk of national bankruptcy. That such a belief was worse than wrong was clearly demonstrated. Without famine relief from the government, the famine killed an estimated 5.5 million people between 1876 and 1878. Similar sentiments of economy were in striking absence three years later during the Afghan war, when no expense was spared to provision the war.

Subsequent famines in the 1890s illustrate the same lessons. In the 1896 to 1897 famine, the government chose to implement a restrictive set of conditions under which the starving poor could receive famine aid. The famine is again estimated to have led to an excess mortality of nearly five million people. The effects of low rainfall were compounded by conversion of substantial areas of agricultural land to cash crops and promotion of crop exports at the same time. Three years later, when the government applied the administration of famine relief more liberally during the 1898 to 1899 famine, the number of people who died was far fewer even though the latter famine affected a larger area and more people in its scope (Damodaran 1995).

In his review of famines and their demographic impacts in colonial India, Ira Klein (1984, 186) asked what explains the great transition between the 50 years preceding the end of World War I and the half-century that followed. During the earlier period, population barely grew (from 255 million to 306 million), and famine, plague, disease, laissez faire colonial policy, and intolerable neglect destroyed tens of millions of lives. During the later period, population doubled (from 306 million to almost 650 million), despite the great Bengal famine of 1943. It is unlikely that the answer is merely the existence of a free press and democracy, as proposed by Amartya Sen (1981). Rather, the key seems to be government priorities: whether a government is committed to economizing and temporizing, or whether it is willing to mobilize the resources necessary to procure and move food to areas of scarcity and offer it at subsidized rates or provide adequate cash transfers for subsistence. This answer held for a time when per capita grain production in South Asia was far lower than it is today, when the infrastructure to distribute food was not well developed, when the main source of revenues for the government was agriculture, and when the capacity to import food quickly was limited.

Note that the answer is not so much about what caused the deaths as it is about what could have been done to prevent deaths. As many historians of famines have argued, the causes of famine mortality are diverse, and specific combinations of factors are pertinent to different instances of famines. But it is necessary to make a critical distinction between the factors that can lead to unimaginable suffering and loss of life—including the loss of those who never were born because of the decline in birthrates in post-famine years—from those that can prevent such suffering (Dyson and Maharatna 1992; Greenough 1992). This is perhaps the most relevant lesson from past famines for the analysis of contemporary food insecurity in South Asia. The government, whose economic and military capacities rested on the agricultural and forest revenues it collected from the people it ruled, could have consistently done far more to protect the lives it allowed to be wasted.

ADDRESSING CONTEMPORARY FOOD SECURITY CHALLENGES IN SOUTH ASIA

In contrast to the food crises of the eighteenth and nineteenth centuries that led to the loss of tens of millions of human lives, the primary manifestation of food insecurity in South Asia today is the grinding malnutrition and hunger that limits life chances without leading to immediate loss of life. The key contemporary question related to food security in South Asia, nonetheless, is analogous to the one that Klein asked about the shift away from famines: What is the set of measures that can enable a transition away from the malnutrition that is the fate of hundreds of millions in South Asia today to a more food secure condition? The question is all the more important for national governments in South Asia because their legitimacy and electoral fortunes rest in no small measure on their ability to prevent starvation deaths and deliver continued improvements in the welfare of their citizens.

And, in fact—with some exceptional periods notwithstanding—the experience of South Asian states in managing food security crises has for the most part been adequate. Their success in reducing famine-related mortality, especially in India, has led some observers to ask what lessons can be learned from the Indian experience for other regions (McAlpin 1987; DeWaal 1996). After considering Maharashtra, which experienced a drastic decline in food-grain production during 1971 to 1973 requiring imports of more than a third of its annual requirements of food, Michelle McAlpin (1987) remarked on the success of government policy in preventing mortality and limiting declines in birthrates, preventing disruption of agricultural production following the years of drought, and strengthening physical infrastructure. She identified six key lessons from the experience: intervene early, rely on local knowledge, focus on employment guarantees and improvements in monetary incomes, coordinate efforts of different actors, and learn from mistakes (McAlpin 1987, 398–400).

When it comes to quotidian food security challenges, the evidence is more complicated, even as it is clear that chronic food insecurity and undernutrition are more widely prevalent in South Asia than in any other world region except sub-Saharan Africa, and certainly the number of people suffering from food insecurity is greater in the region than anywhere else. The lack of progress in addressing adequate caloric intake despite the rise in incomes and rapid economic growth in the region is especially unfortunate. Table 16.2 identifies the essentials of food security at the conceptual level (Scrimshaw 1989, 2–12). These make sense in the abstract. But the lower caloric intake in India despite rising real incomes and expenditures and declining poverty has defied conventional wisdom (Deaton and Dreze 2009; Patnaik 2010). They make clear how much additional research is necessary to understand the relative contributions of these different strategies and their effectiveness in enabling improvements

Table 16.2. Addressing food security challenges

Food Availability (increasing production and reducing losses)	
Improve weather information and early warning systems	Agricultural research and extension
Develop agricultural insurance products	Rural credit and access to inputs
Secure participation and property rights for fisheries, forests, pastures, renewable resources	Research on heat/drought tolerant varieties and species
Manage water demand, and efficiency of use	Intensification of food production and use of inputs
Ensure minimum support prices	Enforcement of land use regulations and zoning
Hold buffer stocks	Diversification of national crop production portfolio
Ensure more efficient processing	Improved storage and handling
Provide protection from pests and rodents	Better prediction of food demand and use
Food Distribution and Access	
Create regional food security programs	National public food distribution systems
Improve efficiency of distribution/marketing	Minimum employment/wage guarantees
Establish equitable access to agricultural land	Entitlement programs including cash transfers
Food Utilization	
Specify nutritional standards	Meeting nutritional requirements
Provide immunization and vaccination programs	Addressing communicable diseases
Arrange public health education to reduce infectious diseases and protect against carrying agents	Specifying and enforcing fortification and processing standards

Source: Adapted from Scrimshaw (1989) and Dev and Sharma (2010)

in food availability, improving access to food for different population groups, and raising food consumption.

Further, the concrete steps governments take and how they translate these general recommendations into provisioning food and improving broad-based access to it are likely far more critical. Table 16.3 identifies the most important social protection strategies that five South Asian countries have used to address food security challenges for their populations.

Table 16.3 indicates that different South Asian countries have long sought to administer a variety of entitlement and distributive programs to enhance broad-based access to food for the most needy, yet the effectiveness of these programs has been in question for an equally long time. Concerns about costs,

Table 16.3. Major food security programs in India and Bangladesh

Country name	Programs and Type
Bangladesh	1. Micro-finance programs: One of the mainstays of poverty alleviation and social safety net programs in the country, widespread coverage, implemented by many NGOs and even as part of government agency programs. Effectiveness questioned by recent research. 2. Vulnerable Group Feeding Program: A food transfer program aimed at the extreme and chronic poor to address basic shortfalls in household food budget. Some evidence of long term impacts, particularly for the extreme poor (Matin and Hulme 2003). 3. Employment Generation Program for the Hard-Core Poor: Applies to rural areas of the country with priority in 81 high poverty sub-districts (Koehler 2009, 2011).
India	1. Public Distribution System: Provision of food through fair price shops (FPS) at subsidized prices to poor, eligible families. Close to 40 million tons of food-grains distributed during 2007-08. One of the oldest programs. Implemented through nearly half a million FPS. The scheme is marred by exclusion errors, leakage, and low quality of distributed food (Dev and Sharma 2010). 2. Nutrition programs include ICDS (Integrated Child Development Services, launched in 1975) targeting children and mothers; MDMS (Midday Meal Scheme, extended in 2004) targeting primary school children, expected to cover nearly 180 million children. Problems include inadequate coverage, and lack of impact (Dev and Sharma 2010). 3. National Rural Employment Guarantee Act has led to the implementation of the Mahatma Gandhi National Rural Employment Guarantee Scheme, and is an attempt to enforce legally the right to work. Largest ever public employment program in human history. Implementation has been more effective than predecessor schemes, in part because of the scale of implementation (Gopal 2009).
Nepal	1. Nepal's universal social pension scheme is one of the oldest social protection measures in the region. All Nepali citizens above 75 are entitled to a small monthly benefit distributed by district development councils (Koehler 2009). 2. Karnali Employment Guarantee Programme aims to provide 100 days of guaranteed employment to the 64,000 residents of the poorest district in Nepal (Koehler 2011).
Pakistan	1. Means tested cash transfers through the Benazir Income Support Program to approximately half a million beneficiaries; coverage expanded substantially to assist flooding victims in 2010 (Koehler 2011). 2. Employment generation scheme for rural unskilled workers to guarantee 100 days of employment at the daily minimum wage. 200,000 households to be covered in the first year (Koehler 2011).
Sri Lanka	Overall social protection arrangements modeled on Britain's post-war universal approach to welfare survived till the late 1970s, and accounts for high levels of human welfare indicators for Sri Lanka despite low per capita income (Samartunge and Nyland 2007). Extensive food subsidies following independence, expanded to include free education and health services for all citizens (funded by taxes on the commercialized plantation sector). From the early 1980s, Sri Lankan governments embraced the Washington Consensus, dismantled the universal approach to social protection, and focused instead on safety net services to low income groups (Samaratunge and Nyland 2007).

Sources: Indicated in text, Author

scale, targeting (Gopal 2009), additionality, rigor of program implementation (Jacob and Varughese 2006), corruption, leakage, coordination across different types of programs, and overall effectiveness (Hall 2008) have been the staple of studies examining their impacts (Barrett 2002).

Few South Asian countries have relied on cash transfer programs—conditional or otherwise—to address food insecurity or enhance social protection of the poorest. Such programs have been shown to be effective in a number of Latin American countries (Rawlings and Rubio 2005; Yap et al. 2000). They may well be worth emulating in South Asia as government revenues improve.

Despite some notable efforts at providing improved social protection to their populations, the overall support that households receive and the number of households receiving such support remains low in most South Asian countries. Existing efforts include employment and skill improvement programs, crop and health insurance, pension schemes, welfare services, cash and in-kind transfers, and food subsidies (Baulch et al. 2008). The limited coverage and small scale of overall expenditures (relative to Latin America or to northern countries) is at least in part responsible for the consistently low evaluations of the effectiveness of these efforts.

In their review of social protection programs in the developing world, there is a compelling argument (Barrientos and Hulme 2008) about how the scale of such programs—rather than the efficiency with which they are implemented—explains much of the variance in their effectiveness. Another important contribution to the subject strikes the same chord in arguing for the effectiveness of longer-term transfers that cover a substantial proportion of the population as part of a broader development strategy (Hanlon et al. 2010). The relatively small scale at which social protection programs in South Asia have been implemented for the most part—the Mahatma Gandhi National Rural Employment Guarantee Act (MGNREGA) in India being an important exception—certainly illustrates the point made by Barrientos and Hulme.

GOVERNING FOOD INSECURITY: INVESTIGATING THE LINKS BETWEEN FOOD AND MOBILIZATION

Few studies predict major declines in aggregate food availability for South Asia—particularly for India and Sri Lanka—over the next decade to two decades. But the continuing maldistribution of available food and lack of entitlements to it mean that hundreds of millions of people in South Asia are likely to continue to suffer from inadequate food and nutrition. Areas at particular risk include Afghanistan, Bangladesh, Nepal, and Pakistan at the Afghanistan border; Indian provinces such as Chhattisgarh, Jharkhand, Uttaranchal,

Bihar, and Orissa; and specific districts in states such as Maharashtra, Andhra Pradesh, and Rajasthan in India. Women and children and households classified as belonging to lower castes and indigenous groups will remain especially vulnerable.

In all of these cases, the level of food production is unlikely to be the key limiting factor, at least in the short run. Rather, governance deficits and state agencies' inability to respond effectively to food shortages experienced by specific social groups are far more likely to precipitate sociopolitical instability. One important mechanism that may activate such links between food and sociopolitical instability is outlined in Table 16.1: "Increased food insecurity in more inaccessible parts of India, Nepal, and Pakistan will provide a greater supply of recruits for existing social movements that can challenge the state."

Food insecurity, even when widespread, seldom poses a significant existential threat to the ruling elite, whether in a democracy or a nondemocratic political system. Writings on social mobilization provide important clues as to why deprivation and discontent by themselves are insufficient for political instability (Tarrow 2011; Tilly and Tarrow 2006). All too often, a simple decline in food security translates into amorphous welfare losses for those who are already poor or disadvantaged. The marginal groups that suffer such losses are by themselves seldom in a position to express their discontent in ways that are politically destabilizing to the elite. Social mobilization and organization are costly; there must be social and political entrepreneurs willing to bear these costs and possible political opportunities that can be exploited by such entrepreneurs (McAdam et al. 1996). Political elites who seek to recruit those disadvantaged under the status quo must exist and be willing to channel the dissatisfaction in non-electoral directions. Marginal groups are unlikely to pose politically effective challenges to the status quo when the state deters them by exacting substantial costs for expressing discontent and eliminating obvious avenues for expressing grievances without legal sanction.

But these deterrents are missing in many parts of South Asia. Political conditions in a number of locations in countries such as Afghanistan, Pakistan, India, and Nepal are conducive to the recruitment and arming of dissatisfied households and groups. Islamic social movements in Afghanistan, Pakistan, and Bangladesh and Communist challengers in parts of India and much of Nepal form a fertile social context through which to channel deprivation and dissatisfaction into substantial political challenges against the government and its enforcement agencies.

Some of these movements, such as the agrarian-based Marxist–Leninist Naxalite social movements challenging the nation state, date back to the 1960s in India, in states such as West Bengal, northwest Bihar, and eastern Uttar Pradesh. After being effectively repressed by the enforcement efforts of the Indian state, they have staged a comeback in several parts of India and are most active in poorer, forested, tribal areas (Borooah 2008). According to one official estimate, nearly 90 percent of the violence and deaths from violence in India are the result of Naxalite activities (GOI 2005, 39). A key factor in the increased influence of the Naxalites was the merger of two factions: the People's War Group (PWG) in tribal areas of Andhra Pradesh and the Maoist Communist Center (MCC) in Bihar. The merger has given both a new lease of life, with the PWG expanding its influence in Orissa and Chhattisgarh and the MCC in Jharkhand and West Bengal (Guha 2007). Coincidentally, some of these are also the parts of India where poorer households suffer from chronic food insecurity. Without suggesting that the Naxalites indeed seek to take over political power through violence (Gupta 2006), it is important to recognize the options they offer to poor disadvantaged groups for political mobilization: India's prime minister, Manmohan Singh, called them the greatest threat to the territorial integrity, prosperity, and well-being of the country (Ahuja and Ganguly 2007).

Other similar challenger movements include the Communist Party of Nepal–Maoist (CPN–M). Their rise in Nepal has been traced to the failure of half a century of development in that country and the presence of high levels of intergroup inequality (Murshed and Gates 2005; Sharma 2006). The intensity of their activities during the past decade has a strong association with districts characterized by greater poverty and inequality (Bohara et al. 2006; Do and Iyer 2010). Effective rebel recruitment strategies have included promises to address material deprivation, as well as political education and indoctrination (Eck 2010). CPN–M, after becoming part of the ruling coalition in Nepal, continues to play a major role in shaping political change in the country, supporting the formation of provinces based on ethnic and caste identity. The impact of a sudden deterioration in the food security situation in Nepal is likely to increase the support for the Maoist parties in Nepal and increase pressures on other parties for accepting demands for an ethnic identity-based political organization of the country. The political Islamist movements in Pakistan and Afghanistan provide yet other examples of existing elite contenders for power who can readily mobilize individual dissatisfaction with the status quo regime into politically effective challenges.

FUTURE PROSPECTS FOR SOUTH ASIA

Although South Asian countries have been able to just about maintain per capita production levels over the last five decades, their task in the immediate future is going to be much more difficult. Factors such as the loss of the agricultural frontier, threats to continued improvements in productivity due to constraints on irrigation, the lack of adequate investments in agricultural research that can deliver improved output, persistent infrastructure constraints that prevent widespread distribution of agricultural inputs, and politically expedient credit, irrigation, and input pricing policies combine to paint a dire picture of the state of South Asian agriculture. If South Asian governments are to meet the food needs of their citizens, they will need to run a lot harder, given the current trends of population and economic growth, ecological change, and continued reliance of food production on the monsoons. Once the gaps in distribution of food to the neediest populations are taken into account, the likelihood that food insecurity will increase is even greater over the next decade to two decades.

Government interventions to address massive subsistence crises have converted catastrophic food insecurity primarily into a crisis of chronic hunger and malnutrition, so that contemporary South Asia hosts the largest number of hungry and undernourished people in the world. Per capita food availability, per hectare food output, and average food consumption per person have remained stubbornly level during the past quarter century, hiding a vast chasm of inequality and lack of entitlements for the poorest citizens. But the success of some of the social protection programs—particularly in India and Bangladesh—in improving household expenditures for poor populations underscores the historical lessons offered by famines during the colonial period.

The key to addressing food insecurity is in large measure the commitment of national governments and decision makers to provide sufficient resources to meet the food needs of their citizens. When governments focus more on economizing and temporizing or on precision in targeting subsidized food or minimum wage employment opportunities, the effectiveness of food delivery is likely to be compromised. When they are instead willing to mobilize the necessary resources to procure and move food to areas of scarcity and offer it at subsidized rates or to provide adequate cash transfers to all who do not have enough for their subsistence, the problem of food insecurity is far less likely. To be clear, the point is not advocacy of inefficiency in subsidized food delivery; rather, it is about the relative benefits and costs of a focus on targeting, as opposed to broad coverage.

Despite the massive losses of human welfare and lives during the colonial period for which the colonial government can be held responsible, there is little empirical evidence that food deprivation directly translated into political mobilization and serious challenges to authority. But the failure of more than a half-century of development and poverty alleviation and persistent inequalities have provided useful fodder to left and Communist movements and may continue to be a source of recruits to these political movements. The growth of their influence during the last decade in particular points to the important relationship between lack of entitlements and access to food and the ability of elite challengers to gain new followers in inequality-ridden, marginal, and inaccessible environments. These locations are characterized by weak enforcement capacity of the ruling elite, lower costs associated with protests, and more opportunity to mobilize social deprivation into political organization. The examples of the Maoist movements in forested and mountainous areas of Nepal and the left groups in ill-governed areas of eastern and southern India suggest greater recruitment of followers in areas where higher levels of poverty and inequality overlap with ineffective governance. The clear lesson for governments in the region, even if their primary motivation is only to reduce the likelihood of violent protests and challenges to state authority, is that they improve substantially the performance of social protection programs in the regions where successful left political movements are active. Positive side effects of even instrumental government actions aimed at improved socio-political stability would be less hunger, better nutrition, and increased life-expectancy. More enlightened social protection policy would seek to improve implementation in all areas of need!

ACKNOWLEDGMENTS

My thanks are due to Elizabeth Renckens for research assistance with this paper. I am also grateful to the editorial team at Cornell University and the participants in the "Food Security and its Implications for Global Stability" workshop at Cornell University, June 18–19, 2012 for their comments and suggestions to improve the paper.

REFERENCES

Aggarwal, P. K., V. K. Mannava, and M. Sivakumar. 2010. Global climate change and food security in South Asia: An adaptation and mitigation framework. In *Climate change and food security in South Asia*, ed. R. Lal, M. Sivakumar, S. M. A. Faiz, and A. H. M. Mustafizur Rahman, 253–274. Kluwer Academic Publishers.

Ahuja, R. 2002. State formation and "famine policy" in early colonial south India. *Indian Economic and Social History Review 39*(4): 351–80.

Ahuja, P., and R. Ganguly. 2007. The fire within: Naxalite insurgency violence in India. *Small Wars and Insurgencies 18*(2): 249–274.

Amrith, S. S. 2008. Food and welfare in India, c. 1900–1950. *Comparative Studies in Society and History 50*(4): 1010–35.

Barrett, C. B. 2002. Food security and food assistance programs. In *Handbook of agricultural economics*, ed. B. L. Gardner and G. C. Rausser, 2103–2190. Amsterdam: Elsevier.

———. 2010. Measuring food insecurity. *Science 327*: 825–28.

Barrientos, A. and D. Hulme. 2008. Social protection for the poor and poorest in developing countries: Reflections on a quiet revolution. Brooks World Poverty Institute Working Paper #30. Manchester: University of Manchester, School of Environment and Development.

Baulch, B., A. Weber, and J. Wood. 2008. *Social protection index for committed poverty reduction: Volume 2, Asia.* Manila: Asian Development Bank.

Bhalla, G. S. and P. Hazel. 1998. Foodgrain demand in India to 2020: A preliminary exercise. *Economic and Political Weekly 32*(52): A150–54.

———, and J. Kerr. 1999. Prospects for India's cereal supply and demand for 2020. Food, Agriculture and the Environment Discussion Paper 29. Washington, DC: IFPRI.

Bhalla, G. S., and G. Singh. 2010. Final report on planning commission project Growth of Indian Agriculture: A district level study. New Delhi: Center for the Study of Regional Development, Jawahar Lal Nehru University. 239.

Bhattacharya, S., and S. Subramanian. 2012. A bumper crop but a bitter harvest. *The National*, June 4. Retrieved from http://www.thenational.ae (accessed April 10, 2013).

Bohara, A. K., N. J. Mitchell, and M. Nepal. 2006. Opportunity, democracy, and the exchange of political violence: A subnational analysis of conflict in Nepal. *The Journal of Conflict Resolution 50*(1): 108–28.

Borooah, V. 2008. Deprivation, violence, and conflict: An analysis of Naxalite activity in the districts of India. *International Journal of Conflict and Violence 2*(2): 317–33.

Borras Jr., S. M., R. Hall, I. Scoones, B. White, and W. Wolford. 2011. Towards a better understanding of global land-grabbing. *The Journal of Peasant Studies 38*(2): 209–16.

Cane, M. 2010. Climate science: Decadal predictions in demand. *Nature Geoscience 3*: 231–32.

Cotula, L., and S. Vermeulen. 2009. Deal or no deal: The outlook for agricultural land investment in Africa. *International Affairs 85*(6): 1233–47.

Currie, K. 1991. British colonial policy and famines: Some effects and implications of "free trade" in the Bombay, Bengal, and Madras Presidencies, 1860–1900. *South Asia 14*(2): 23–56.

Damodaran, V. 1995. Famine in a forest tract: Ecological change and the causes of the 1897 famine in Chotanagpur, Northern India. *Environment and History 1*: 129–58.

De, U. S., R. K. Dube, and G. S. Prakasa Rao. 2005. Extreme weather events over India in the last 100 years. *Journal of Indian Geophysical Union 9*(3): 173–87.

Deaton, A., and J. Dreze. 2009. Food and nutrition in India: Facts and interpretations. *Economic and Political Weekly 44*(7): 42–65.

DES (Directorate of Economics and Statistics). 2011. *Agricultural statistics at a glance, 2011*. Retrieved from http://eands.dacnet.nic.in/latest_2006.htm (accessed April 10, 2013).

Dev, S. M., and A. N. Sharma. 2010. Food security in India: Performance, challenges, and policies. Oxfam India Working Paper Series—VII. Delhi: Oxfam India.

DeWaal, A. 1996. Social contract and deterring famine: First thoughts. *Disasters 20*(3): 194–205.

Do, Q.-T., and L. Iyer. 2010. Geography, poverty, and conflict in Nepal. *Journal of Peace Research 47*(6): 735–48.

Dyson, T. 1991a. On the demography of South Asian famines: Part 1. *Population Studies 45*: 5–25.

——— 1991b. On the demography of South Asian famines: Part 2. *Population Studies 45*: 279–97.

———, and A. Maharatna. 1992. Bihar famine, 1966–67 and Maharashtra drought, 1970–73: The demographic consequences. *Economic and Political Weekly 27*(26): 1325–32.

Eck, K. 2010. Recruiting rebels: Indoctrination and political education in Nepal. In *The Maoist Insurgency in Nepal: Revolution in the Twenty-First Century,* ed. M. Lawoti and A. K. Pahari, 33–41. London: Routledge.

FAO. 2005. The state of food insecurity in the world 2005: Eradicating world hunger—key to achieving the millennium development goals. Rome: Food and Agriculture Organization of the United Nations.

Giorgi, F. 2006. Climate change hot-spots. *Geophysical Research Letters 33*: L08707. doi:10.1029/2006GL025734.

Gopal, 2009. NREGA social audit: Myth and reality. *Economic and Political Weekly 44*(3): 70–71.

Government of India (GOI). 2005. Annual report 2004–2005, Ministry of Home Affairs, New Delhi: Government of India.

Greenough, P. 1992. Inhibited conception and women's agency: A comment on one aspect of Dyson's "On the demography of South Asian famines." *Health Transition Review 2*(1): 101–05.

Grove, R. 1997. *Ecology, climate and empire: Colonialism and global environmental history.* Cambridge: White Horse Press.

Guha, R. 2007. Naxalites and Indian democracy. *Economic and Political Weekly 42*(32): 3305–12.

Gupta, T. D. 2006. Maoism in India: Ideology, programme, and armed struggle. *Economic and Political Weekly 41*(29): 3172–76.

Gustavsson, J., C. Cederberg, U. Senesson, R. Otterdijk, and A. Meybeck. 2011. *Global food losses and wood waste: Extent, causes, and prevention.* Rome: FAO.

Hall, A. 2008. Brazil's Bolsa Familia: A double-edged sword. *Development and Change 39*(5): 799–822.

Hall-Matthews, D. 1996. Historical roots of famine relief paradigms: Ideas on dependency and free trade in India in the 1870s. *Disasters 20*(3): 216–30.

Hanlon, J., D. Hulme, and A. Barrientos. 2010. *Just give money to the poor: The development revolution from the global south.* New York: Kumarian.

IWMI (International Water Management Institute). 2010. Managing water for rainfed agriculture. Water Issue Brief 10. Colombo: IWMI.

Jacob, A., and R. Varghese. 2006. NREGA implementation 1: Reasonable beginning in Palakkad, Kerala. *Economic and Political Weekly 2*: 4943–45.

Klein, I. 1984. When the rains failed: Famine, relief, and mortality in British India. *Indian Economic and Social History Review 21*: 185–214.

Koehler, G. 2009. Policies toward social inclusion: A South Asian perspective. *Global Social Policy 9*(1): 24–9.

———. 2011. Transformative social protection: Reflections on South Asian policy experiences. *IDS Bulletin 42*(6): 96–103.

Lal, M. 2003. Global climate change: India's monsoon and its variability. *Journal of Environmental Studies and Policy 6*(1): 1–34.

———. 2011. Implications of climate change in sustained agricultural productivity in South Asia. *Regional Environmental Change 11*(S1): S79–94.

Lobell, D. B., M. B. Burke, C. Tebaldi, M. D. Mastrandrea, W. P. Falcon, and R. L. Naylor. 2008. Prioritizing climate change adaptation needs for food security in 2030. *Science 319*: 607–10.

McAdam, D., J. D. McCarthy, M. N. Zald, eds. 1996. *Comparative perspectives on social movements: Political opportunities, mobilizing structures, and cultural framings.* Cambridge: Cambridge University Press.

McAlpin, M. B. 1987. Famine relief policy in India: Six lessons for Africa. In *Drought and hunger in Africa: Denying famine a future*, ed. M. Glantz, 393–414. Cambridge: Cambridge University Press.

Maharatna, A. 1996. *The demography of famines: An Indian historical perspective.* Delhi: Oxford University Press.

Matin, I., and D. Hulme. 2003. Programs for the poorest: Learning from the IGVGD. *World Development 31*(3): 647–665.

Mirza, M. M. Q., and Q. K. Ahmad, eds. 2005. *Climate change and water resources in South Asia.* Leiden: A. A. Balkema Publishers.

Mittal, S. 2006. *Structural shift in demand for food: India's prospects in 2020.* Working Paper 184. Mumbai: Indian Council for Research on International Economic Relations.

———, and D. Sethi. 2009. *Food security in South Asia: Issues and opportunities.* Working Paper #240. Delhi: Indian Council for Research in International Economic Relations.

Murshed, S. M., and S. Gates. 2005. Spatial-horizontal inequality and the Maoist insurgency in Nepal. *Review of Development Economics 9*(1): 121–34.

Ninno, C., P. A. Dorosh, and K. Subbarao. 2007. Food aid, domestic policy, and food security: Contrasting experiences from South Asia and sub-Saharan Africa. *Food Policy 32*: 413–35.

Patnaik, U. 2010. A critical look at some propositions on consumption and poverty. *Economic and Political Weekly 45*(6): 74–80.

Rawlings, L. B., and G. M. Rubio. 2005. Evaluating the impact of conditional cash transfer programs. *World Bank Research Observer 20*(1): 28–55.

Rosegrant, M., M. Agcaoili, and N. D. Perez. 1995. Global food projections to 2020: Implications for investment. Food, Agriculture, and the Environment. Discussion Paper #5. Washington, DC: International Food Policy Research Institute.

Rosegrant, M. W., and S. A. Cline. 2003. Global food security: Challenges and policies. *Science 302*(5652): 1917–19.

Samaratunge, R., and C. Nyland. 2007. The management of social protection in Sri Lanka. *Journal of Contemporary Asia 37*(3): 346–63.

Scrimshaw, N. S. 1989. Completing the food chain: From production to consumption. In *Completing the food chain: Strategies for combating hunger and malnutrition*, ed. P. M. Hirschoff and N. G. Kotler, 1–17. Washington: Smithsonian Institution Press.

Sen, A. 1981. *Poverty and famines: An essay on entitlement and deprivation.* Oxford: Clarendon Press.

Sharma, K. 2006. The political economy of civil war in Nepal. *World Development 34*(7): 1237–53.

Sharma, S. 2001. *Famine, philanthropy, and the colonial state: North India in the early nineteenth century.* Delhi: Oxford University Press.

Siddiqui, K.M., I. Mohammad, and M. Ayaz. 1999. Forest ecosystem climate change impact assessment and adaptation strategies for Pakistan. *Climate Research 12*: 195–203

Tarrow, S. 2011. *Power in movement: Social movements and contentious politics.* Cambridge: Cambridge University Press.

Tilly, C., and S. Tarrow. 2006. *Contentious politics*. Boulder: Paradigm Publishers.

World Bank. 2010. *Development in a changing climate: World Development Report, 2009.* Washington, DC: The World Bank.

Yap Y-T, G. Sedlacek, and P. F. Orazem. 2000. *Limiting child labor through behavior-based income transfers: An experimental evaluation of the PETI Program in rural Brazil.* Mimeo. Washington, DC: Inter American Development Bank.

17

When China Runs Out of Farmers

LUC CHRISTIAENSEN

China's success in feeding its population since economic reforms began in 1978 has been remarkable. The nation has more than doubled its grain output and virtually eliminated hunger. It is now feeding more than 1.3 billion people, or 20 percent of the world's population, on less than 11 percent of the world's agricultural land and less than 6 percent of its water. Yet, China is facing new agriculture and food security challenges that also bear on domestic and international sociopolitical stability. The defining question is no longer whether China will run out of food—with $3.2 trillion in foreign exchange reserves as of December 2011, that would be hard to imagine—but rather whether it will run out of farmers.

Institutional reforms and widespread adoption of land-saving technologies have driven China's agricultural growth, aided by increasingly open domestic, and also international markets, after World Trade Organization (WTO) accession in 2001. To the surprise of many, this dramatic expansion in agriculture happened even though small, fragmented farms continued to define the agrarian landscape and factor markets remained heavily distorted. Self-sufficiency was also largely retained as the model of choice to secure cereal supplies. But, both China's smallholder model and its self-sufficiency for rice, wheat, and maize are increasingly tested.

While agricultural incomes increased rapidly over the past three decades, incomes in the urban nonagricultural sectors grew even faster. This induced massive urban migration, substantially reducing the agricultural and rural labor force over time, to the point that rural wages have now also begun to rise. The aging of the rural demographic—both because it is especially the young who migrate and because of China's overall demographic evolution—is further compounding the pressure on rural wages, which in turn is inducing the substitution of capital for labor, or agricultural mechanization. This increases the pressure on the smallholder farm model. Smallholders must now not only expand to generate earnings comparable with urban off-farm employment, but also to capture the economies of scale from mechanization following rising labor costs.

But inertia in farm consolidation is substantial, given land tenure insecurity and the inherent transaction costs involved in exchanging the multitude of tiny, noncontiguous plots. Moreover, in mountainous areas, the scope for farm consolidation and mechanization is limited. Farmers' income growth in China is thus struggling to keep up with non-farm income growth, with the ratio of average urban to rural income reaching 3.53 in 2009 (compared with 2.1 in 1985). Farmers feel themselves increasingly left behind, even though their absolute living standards have improved substantially. This farm income problem poses a substantial threat to domestic sociopolitical stability.

How to best manage the farm income problem is undoubtedly one of China's prime policy conundrums and has been recognized as such in each of China's last three five-year plans. A dramatic increase in agricultural subsidization and protection has historically been common (de Gorter and Swinnen 2002; Hayami 2007) and avoiding such a politically expedient, but economically inefficient response is particularly exigent. Indeed, China has increased its farm subsidies already from a minuscule 100 million yuan in 2002 to 122.8 billion (2002) yuan in 2010 in response to the farm income problem.

Which agrarian structure to promote is one of the key questions in dealing with the farm income problem. Rapid consolidation in large-scale, corporate, mechanized farming units may provide the necessary economies of scale to enable mechanization. But, it is unlikely to provide the necessary rural jobs, and unmanageably large urban migration flows may follow. Moreover, even if feasible in practice, despite the reigning reality of millions of smallholdings on a multitude of scattered plots, radically reducing the number of smallholder farms may not be necessary to keep farming profitable. Factor market reforms, institutional innovations, investment in rural public goods, and secondary town development go a long way toward rendering smallholder agriculture (including part-time farming) commercially viable. Such reforms would enable China to use all available mechanisms to overcome the intricate farm income problem and dissipate its ensuing sociopolitical tensions.

The urgency of the farm income problem is compounded by China's lagging food supply. Two bouts of double-digit (real) food price inflation over the past three years indicate that aggregate food supply is struggling again to keep up with demand. As domestic agricultural labor costs rise and international imports of grain become more competitive following renminbi (RMB) appreciation, China is finding it increasingly taxing to secure all cereal supplies domestically and retain the cereal self-sufficiency model, especially since these more recent constraints are compounding China's longer-standing supply challenges of a limited land base and growing water scarcity. At the same time, demand for cereal feed (maize and soybeans)—not food (rice and wheat)—is propelling cereal demand as China's diets become more protein-rich. Changes in the agro-climatic conditions further add uncertainty to the production environment.

Greater reliance on cereal imports provides an alternative. More imports also reduce agriculture's pressure on the water supply, but this shift in policy may not be palatable politically, especially if it were to undermine China's food sovereignty. It may also perturb the international markets, especially if the shift is sudden. China's imports of soybeans have already increased dramatically from virtually zero in the mid-1990s to almost 60 million metric tons over the past couple of years (64 percent of world soybean imports). The country has also recently become a net importer of maize. Securing sufficient cereal supplies will remain a central concern of China's agricultural policy agenda in the years to come.

How China handles its farm income and food problems will have far-reaching consequences at home and abroad. Failure to make maximum use of smallholder agriculture to close the rural–urban income divide will undermine the nation's ability to overcome rural–urban tensions and maintain sociopolitical stability at home. On the other hand, unanticipated ripples in China's domestic food markets and a dramatic shift in reliance on the world cereal markets will reverberate internationally. This chapter will argue that many of China's food and farm challenges can be overcome by combining a moderate relaxation of China's current cereal self-sufficiency ambitions (in particular for maize) with institutional innovations in the land, labor, and capital markets to facilitate gradual farm consolidation and farm exit, while keeping smallholder farming profitable. This way agriculture will contribute maximally to reducing the rural–urban divide, especially when complemented with the ongoing investments in rural public goods and assertive secondary town development to spur rural off-farm employment. Such a strategy, further strengthened with the establishment of rural social security systems and ongoing rural–urban migration, would position China well to navigate its intricate farm income and food problems, while also contributing to sociopolitical stability domestically and internationally.

FOOD, FARMS, OR FIGHTS

Since China embarked on its reform agenda in 1978, introducing capitalist market principles and breaking up collective farms, the country's economic growth and poverty reduction have been nothing less than remarkable. Agriculture has been an important (and generally underappreciated) contributor to these developments, with agricultural GDP growing at an unparalleled 4.5 percent per year on average between 1978 and 2009. By providing abundant and cheap food, agriculture kept nominal wages low and paved the

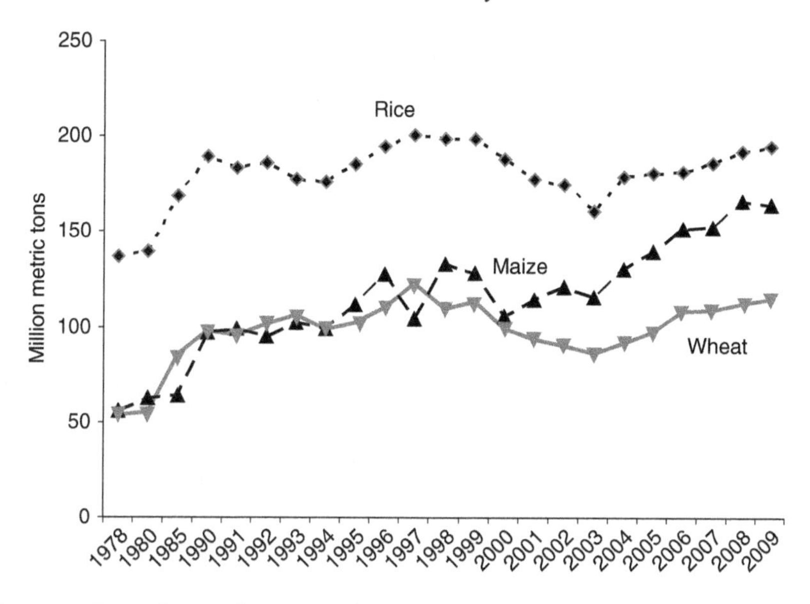

Fig. 17.1. Rice, wheat and maize production in China 1978–2009
Source: NSBC (2010)

way for China's export-led industrialization and its ensuing growth miracle. At the same time, by raising farm incomes, agriculture has been a driving force behind China's massive reduction in absolute poverty, aided by equal distribution of user rights to its farm land (Ravallion and Chen 2007).

Production of rice, wheat, and maize expanded rapidly, from 246 million metric tons in 1978 to 474 million metric tons in 2011 (Figure 17.1). With satiation of staple demand, rice and wheat production peaked in the late 1990s, while maize production continued to expand to feed the growing livestock sector. After a restructuring of China's stockholdings in the early 2000s and an accelerated decline in cultivated land area between 1999 and 2004 following the Grain-for-Green land restoration program, wheat and rice prices picked up again in 2003. Concerns about the country's ability to maintain its food self-sufficiency targets heightened and the production of rice and wheat have continuously increased since. China has traditionally maintained such targets to reduce its exposure to food supply disruptions in the international markets, which given the country's size, could easily translate into food shortages at home and sociopolitical instability.

China's agriculture has diversified beyond grains into fruits and vegetables, and also beyond crops, with livestock accounting for about a third of agricultural GDP in 2009 and fisheries for about 10 percent (up from 2 percent in 1980). These evolutions reflect ongoing dietary change and received an additional impetus from China's WTO accession in 2001, which induced China to align its trading patterns with its comparative advantage. This advantage is in labor-intensive products, such as horticultural or animal products (including aquaculture), whose exports rapidly increased, and not in land- and water-intensive bulk commodities such as grains, oilseeds, and sugar, whose exports fell. After WTO accession, China quickly became a net agricultural importer, with soybeans now making up the bulk of its agricultural imports (almost half, worth about $20 billion in 2009).

The introduction of the household responsibility system in 1978 and land-saving technological change thereafter—improved seeds and chemical inputs supported by the further expansion of irrigation—drove output growth. They were further aided by a liberalization in the domestic fertilizer, seed, and agricultural product markets during the 1990s as well as by the dramatic reduction in the nominal rate of assistance to nonagriculture, which further helped shift the domestic terms of trade in favor of agriculture (Anderson 2009). After a tremendous boost in total factor productivity (TFP) of 5 to 10 percent per year from 1978 to 1984, TFP continued to grow at 2 percent per year on average into the mid-2000s, on a par with the United States, many Western European countries, and Australia after World War II. At 340 kg per hectare in 2009, fertilizer use intensity in China is now among the highest in the world. Nearly 50 percent of China's cultivated land is supported by irrigation facilities. Groundwater irrigation is extensively used in the more arid northern parts of the country, while irrigation in the rest of China relies mostly on surface water.

Surprisingly, China's agricultural production has so far come almost entirely from small-scale and fragmented operations (0.6 hectares on average in 2008, spread across four to six plots) (Tan et al. 2006; Lohmar et al. 2010). Nonetheless, the number of producer and marketing cooperatives (called Farmer Professional Associations) are rising, especially since the adoption of the Farmers Professional Cooperative Law in 2007 (Deng et al. 2010). There are also an increasing number of anecdotes of companies cultivating large tracts of land. This foretells what is to come—mechanization and farm consolidation—as agricultural labor markets are tightening across the country.

To be sure, despite remarkable success in securing national supplies, a massive reduction in poverty, and successful domestic food market liberalization, a sizeable number of people remain poor and probably food insecure.[1] Many of them still live concentrated in the western provinces and the mountainous and ethnic minority areas. But a substantial share—more than half, according to World Bank estimates (2009)—now also live dispersed across villages in non-mountainous, nonminority areas, requiring targeted assistance. The growing rural–urban income divide (the farm income problem) and the growing tension between domestic food supply and demand (the food problem), however, are the most immediate concerns for domestic and international socioeconomic stability.

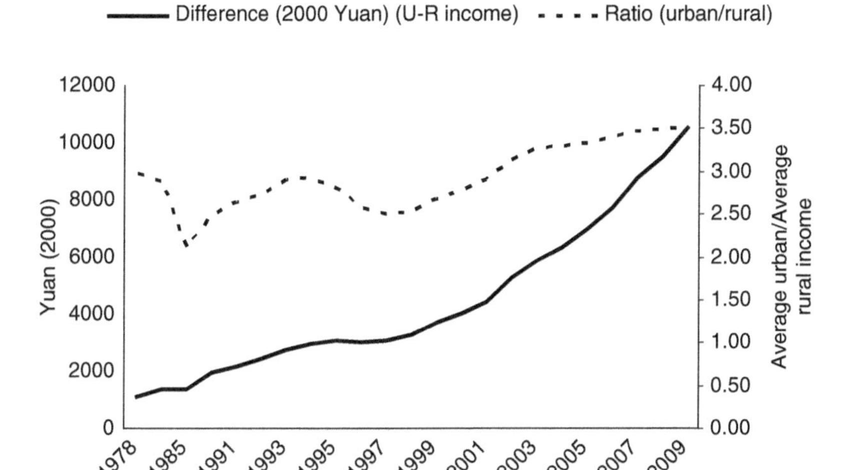

Fig. 17.2. Evolution of the difference between average urban and average rural incomes in China 1978–2009
Source: NSBC (2010)

China's Growing Farm Income Problem

After an initial decline in the rural–urban income gap between 1978 and 1984, when agriculture was especially buoyant, the ratio of average urban to average rural income increased from 2.1 in 1985 to 3.5 in 2009 (Figure 17.2). Rising discontent among China's rural populations has made addressing this rural–urban imbalance into a major policy objective since 2000, and following the "take-less-and-give-more" principle, the 10th and 11th five-year plans (2000 to 2005 and 2006 to 2010 respectively) focused on reversing the long-standing net fiscal flows from rural to urban areas to finance industrialization. After more than 2,000 years in existence, all agricultural and rural fees and taxes were abolished between 2003 and 2005. This was followed by an exponential increase in farm subsidies from a minuscule 100 million yuan in 2002 to 122.8 billion (2002) yuan in 2010.[2]

While the policies appear to have slowed down growth in the average urban to rural income ratio (Fan et al. 2011), in absolute terms, the difference in income has continued to rise rapidly, reaching 10,504 (2000) yuan in 2009.[3] Indeed, the pressure on China's fragmented smallholder farms to generate incomes and income growth commensurate with what could be earned outside agriculture continues unabated, now that not only urban wages, but also rural wages are rising rapidly. Very simple arithmetic shows for example that rice farmers should have expanded about 6 times between 2003 and 2008—from about 0.6 hectares on average to 3.6 hectares—to keep up with average rural income growth, when growing only rice. This is in an era of rising rice prices and after a sizeable increase in subsidization—without subsidization, rice farms would have had to grow 8.5 times.

Abundant agricultural labor has for the longest time been considered one of the defining features of China's agriculture, resulting in labor-intensive agriculture, with technological innovations focused on saving land. Urbanization and the aging of China's population are, however, rapidly reducing the agricultural and rural labor force. Until recently, about 20 million rural people were estimated to migrate to urban areas every year, while temporary migration was estimated at 100 million per year. These trends are reflected in the average total hours spent per household on the farm, which declined from 3,500 hours in 1991 to just over 2,000 hours in 2000 and only 1,399 hours in 2009 (De Brauw et al. 2011). Concentration of migration among the young further exacerbates the aging of the rural demographic, and with male migrants outnumbering females, agriculture in China is also feminizing.

The observed acceleration in rural wages around the mid-2000s (Zhang et al. 2011), even in the more remote areas and provinces, such as Gansu, suggests that these deep, slow-running demographic and economic forces have already started to bear themselves out in the labor markets (Figure 17.3). These developments test the viability of the current farm structure. Rising labor costs make capital, and thus mechanization of agriculture, more attractive, putting

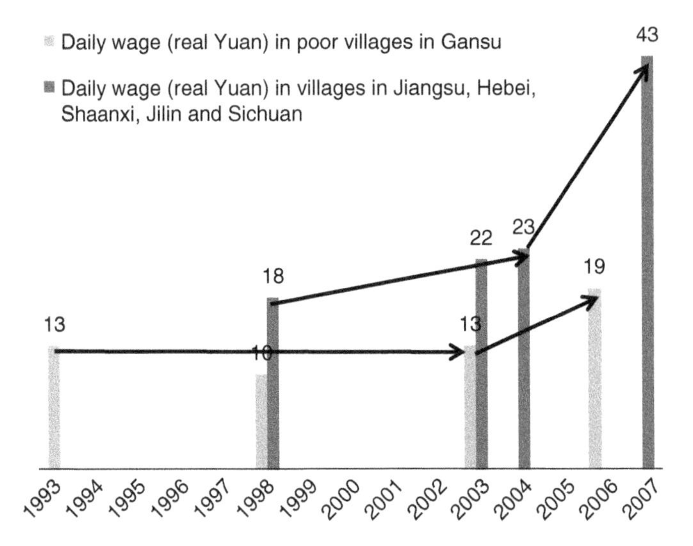

Fig. 17.3. Evolution of rural daily wages in selected rural villages across China since 2003

Note: Poor villages in Gansu concern 88 randomly selected villages from three nationally designated poor counties (Huining, Weiyuan, and Tianzhu Tibetan Autonomous Counties). Rural villages concern 101 randomly selected villages, 20 from Jiangsu, Hebei, Shaanxi, and Sichuan, and 21 from Jilin.

Source: Zhang et al. (2011)

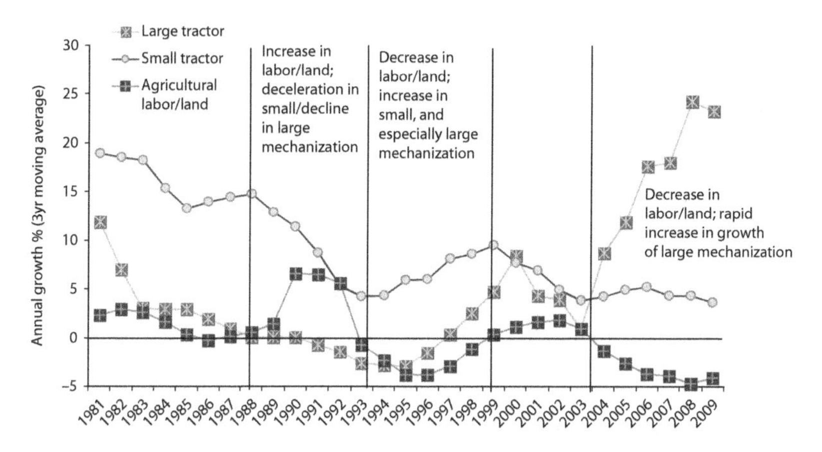

Fig. 17.4. Agricultural mechanization and the agricultural land/labor ratio in China during 1978–2009
Source: NSBC (2010)

upward pressures on the farm size, given the economies of scale that come with the indivisibility of machinery, as illustrated in Figure 17.4. Negative growth (i.e., a decline) in the agricultural labor–land ratio is associated with accelerating growth of tractor use, most recently especially large tractor use, a trend which is lately also actively supported by the government through machinery subsidies.

Enormous challenges of adjustment lie ahead. Aggregating land through land rental and sales to obtain more commercially viable farming entities entails significant coordination costs. These costs only multiply when the available plots for rent are small, scattered, and usually not adjacent to a farmer's existing plots. Moreover, while machinery rental services can help smallholders capture the economies of scale from mechanization and remain internationally competitive, farm size expansion is also necessary to remain competitive relative to what could be earned outside agriculture. This poses a particular challenge in mountainous areas, where smaller plots on more sloped land make mechanization and land consolidation even harder.

A dual farm structure may thus be necessary to make maximum use of agriculture's power to reduce the rural–urban divide, with part-time farming predominant in the mountainous areas and larger mechanized family farms characterizing agriculture in the plains. While large-scale, capital-intensive corporate farms might be feasible and effective at keeping food prices low, they generate little rural employment and increase pressure on urban migration. Yet not all farmers will be able to migrate, further increasing pressures to adopt agricultural policies of subsidization and protection as have historically been observed in Western countries and Japan (Hayami and Kawagoe 1989,

237–8). The increase in China's producer support estimate from 3 to 17 percent between the period from 1995 to 1997 and 2010, and the increase in the ratio of the producer to border price (the nominal protection coefficient) from 1.01 to 1.06 between the period from 1995 to 1997 and the period from 2008 to 2010 are striking reminders of the pertinence of these observations for China today (OECD 2011).

To be sure, agriculture cannot resolve the rural–urban divide alone. Achieving spatially equitable growth patterns will require a three-pronged approach focused on moving people off the farm (through urban migration), mitigating the circumstances of those left behind (through transfers), and modernizing agriculture. Failing to overcome the farm income problem and the associated rural–urban divide may ultimately motivate domestic socioeconomic instability, as it is between-group inequality that drives socioeconomic instability, much more than inequality per se (Milanovic 2011). Rising conflicts about rural–urban land conversions on the urban fringe in China, where the bulk of the benefit is captured by city governments and investors to promote growth, must also be seen in this context (Shouying 2012). High food prices may then provide the spark to set off the proverbial fire.

China's Re-emerging Food Problem

With growing farm income problems dominating the agenda since 2000, China's agricultural policies gradually shifted away from securing the aggregate food supply. Systematically higher food price inflation since 2003 compared with the consumer price index (CPI), punctuated by two bouts of double-digit food or grain price inflation over the past three years (Figure 17.5), has squarely put the "food problem" back on the agenda, as rising supply constraints are making it increasingly hard for supply to keep up with growing demand.

On the supply side, there is little scope for claiming new land for agriculture, the most straightforward way to expand production, while the competition for existing agricultural land is intensifying for alternative uses, including inhabitation, mining, and infrastructure. This problem is compounded by ongoing soil degradation that undermines the land's productive capacity, especially in the western parts of the country and where input use is limited (Ye and Van Ranst 2009). The decline in the quality and quantity of China's agricultural land base (from 130 million hectares in 1997 to 121.7 million hectares in 2008) led the government to adopt the Grain-for-Green program in 1999 and to institute the so-called Red Line in 2006. Under Grain-for-Green, sloped cropland is set aside to increase forest cover and prevent soil erosion, while the Red Line institutes a dynamic balance quota of 120 million hectares of land to be preserved for agricultural purposes. While both policies are contested—for threatening China's self-sufficiency goals in grain production by taking crop

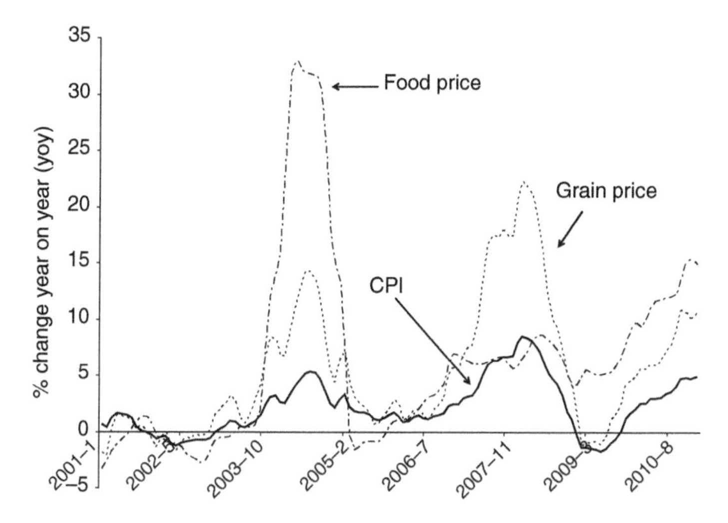

Fig. 17.5. Monthly food, grain and consumer price inflation in China during the 2000s
Source: World Bank (2011)

land out of production and for limiting the necessary conversion of land for economic growth, respectively—the debates in themselves illustrate that the scope for increasing agricultural production through (net) land expansion has essentially been exhausted.[4]

While multiple cropping through better water management can provide some relief, there remains limited scope for expanding irrigated agriculture. The competition for water from industry and human consumption is rising, and the available water base declining. In addition, water endowments are especially scarce in the north, where many water-intensive crops, such as soybeans, maize, wheat, and cotton, are grown. This has resulted in severe groundwater depletion.[5] China's low efficiency of current water use in irrigated agriculture provides some opportunities, but increased implicit trade of water through the import of more water-intensive agricultural products such as grains may well have to be part of the answer.[6]

There is also no more scope for increasing production by increasing chemical inputs. At more than 300 kilograms per hectare (340 kilograms per hectare in 2009), China is already among the highest fertilizer users in the world. Overuse of fertilizer has substantially contributed to nitrogen and phosphorus eutrophication of China's lakes and water systems. China is also one of the largest pesticide consumers in the world, much of it superfluous. Tests on thousands of fields across China show that fertilizer application rates could be cut in many cases by at least 30 percent through better agronomic practices, without loss of crop production or risk to the nation's aggregate food supply (SAIN 2010). This suggests important cost-saving opportunities, which can help resolve the food and farm income problem by keeping grain production competitive, domestically and internationally.

Appreciation of the renminbi is increasing the competitiveness of international agriculture. Land and water intensive cereals are especially vulnerable to renminbi appreciation (Gale and Tuan 2007), while more room for maneuver is left for Chinese fruits and vegetables, consistent with the global distribution of factor endowments and increased trade. The dramatic recent increase in soybean imports in China from about 28 million metric tons in 2006 to more than 56 million metric tons in 2011 (with most of the increase coming from the United States) to meet rising demand for animal feed may not be coincidental. In the summer of 2007, the renminbi passed the 7.5 RMB/US$ mark at which US soybeans became competitive on the Chinese market. Soybeans are now the largest export product of the United States to China. Similarly, after many years of net exports, China imported an estimated five million metric tons of maize in 2011, which coincided with the breach of the 6.5 RMB/$ mark in the spring of 2011, at which US maize was considered to become competitive on China's market (Gale and Tuan 2007).

Finally, the economic pressures of shifting from a labor-intensive to a more capital-intensive agrarian production system must be addressed in increasingly fragile ecosystems and in the face of rising climatic uncertainties. Some simulation models suggest that global warming is likely to be harmful to rainfed farms, but beneficial to irrigated ones—with the net impacts only mildly harmful at first, but growing over time (Wang et al. 2009). The impacts vary across regions, however, and the indirect effects of possible changes in the water flows, which are quite real, are not captured in these particular models. The most binding environmental constraint is likely water, especially in the northern plains, jeopardizing potential production gains from mechanization. These, and other simulations, underscore that at the local and regional scale, rising uncertainty about the production environment is the defining feature of climate change. This adds an important layer of uncertainty to a production system that is intrinsically uncertain to begin with.

On the demand side, feed demand is propelling overall domestic demand for grain, while demand for cereal food is halting. As households become richer and move to the cities, their diets shift away from grains to include more meats, eggs, and dairy. Urbanization itself is an important driver of this shift, with Chinese urbanites consuming on average only a third of the food grain eaten by their rural countrymen (Lohmar et al. 2010). These changing demand patterns provide some relief in China's pursuit of self-sufficiency for rice and wheat. The Chinagro II model predicts that consumption growth of rice and wheat between 2005 and 2030 will be virtually zero and easily met by domestic production (Keyzer and van Veen 2010). Annual meat consumption per person, on the other hand, is expected to increase by another 20 kilograms from about 53 kilograms per person per year in 2010 to about

73 kilograms in 2030, up from about 25 kilograms per person per year in 1990. Demand for feed grain (maize and soybean) should remain strong throughout 2030.

Dietary shifts also open up important new and remunerative employment opportunities for farmers, both directly through the more labor-intensive production of meat and dairy, fish, and fruits and vegetables, and indirectly through the increased demand for unskilled labor in the downward agro-processing industries and marketing. There are, however, important economies of scale in these downstream industries, promoting vertical integration of the supply chains to ensure a stable supply of high-quality agricultural products. These only grow as consumer demand for quality and food safety increases, and food standards tighten, putting the smallholder farm model under pressure. So far, they have not prevented smallholders from participating (Wang et al. 2009), yet there is an increasing contingent of large-scale animal farms and agri-businesses. As economies of scale grow and standards tighten, the pressures for consolidation and vertical integration will continue to grow (Reardon et al. 2010).

ADDRESSING THE FOOD PROBLEM: SECURING FOOD SUPPLIES

Following greater selectivity in agricultural land conversion, a rapid increase in agricultural subsidies, and an expansion in agricultural research and development (R&D) investment, overall rice and wheat production levels have caught up with their peaks of the late 1990s. At the same time, import dependence for feed grains has been growing, and worries about market reliance remain, not helped by restrictive trade practices by exporters in 2008 and 2011. Should China continue on its current path of strengthening its production capacity through investment in rural public goods and increasing its reliance on world markets for feed grains, including maize? And what are the implications for the world?

Continuing growth in TFP will be key to keeping smallholder farmers competitive and raising their incomes. During the past decade, China's investment in agricultural R&D has increased the most rapidly of any large nation. But much more will be needed.[7] In its agricultural R&D investment, China is increasingly betting on biotechnology. In 2008, it supplemented its ongoing research on agricultural biotechnology with a 26 billion yuan ($3.8 billion) "special program" focused on five staple crops (rice, wheat, maize, cotton, and soybeans) and three types of livestock (hog, cattle, and sheep). The focus is on crops for the domestic market in which China misses a natural comparative advantage.

But genetic engineering alone will not suffice and global concerns about limited germplasm exchange are rising. Domestically, potential shifts in consumers' attitudes toward genetically modified (GM) foods are lurking in the background. How broader awareness about GM crops would affect consumer acceptance is an important development to be watched. Against this background, GM maize for animal feed may have more potential in the short run. Internationally, concerns are rising about China's growing protectionism of its genetic resources. The increasing reluctance to exchange germplasm exacerbates the ongoing closure of the "global genetic commons," whose maintenance is of key strategic importance for world food security and sociopolitical stability (McCouch and Crowell 2012).

Since the late 1990s, agricultural R&D has also begun to attend to sustainable development through a series of environmental initiatives, including a national program on balanced fertilization and integrated pest management and a number of initiatives supporting research on the impacts of climate change. Given water scarcity, widespread non-point source pollution, and important economic gains from lower input use, these areas of investment will need to be ramped up. They will also require reforms in the extension system to put farmers and agricultural extension agents in closer contact (Hu et al. 2009).

Even agricultural TFP growth of 2 percent per year is unlikely to generate enough income to keep cereal farmers on the farm and self-sufficient. At the same time, renminbi appreciation is making international cereals more competitive, while WTO regulations limit space for inefficient protectionist support or subsidization. Bans on agricultural land conversion would hinder the necessary development of other sectors and economic growth. Sole focus on capital-intensive production through mechanization on large-scale farms would not generate the necessary rural jobs, exacerbating, instead of alleviating, the farm income problem. More erratic weather patterns following climate change are bound to make annual domestic production more volatile, increasing the need for occasional reliance on the world market (as in 2011), unless even larger and more costly buffer stocks are maintained. Limited *ex ante* engagement with the world market also leads to thinly traded world markets, which will subsequently be less able to absorb sudden demand shocks in case of domestic weather and crop failures. Greater reliance on imports, especially maize, for animal feed appears the rational response moving forward.

The massive increase in soybean imports observed over the past decade provides important lessons in this regard. It has been a rational response to rising resource constraints. Soybean production is among the most water-intensive crops. It also yields among the lowest value per cubic meter of water used (Figure 17.6). Importing soybeans thus equates with importing vast amounts of water, giving rise to an implicit trade in water. This relieves pressure on the water tables in the northern plains, where water scarcity is felt most. The move

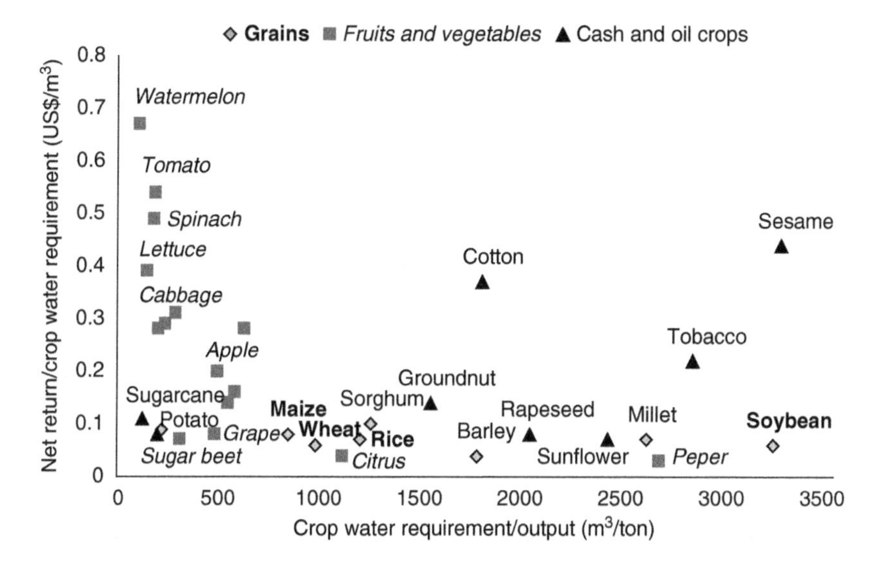

Fig. 17.6. Crop water requirements and net returns to crop water use for grains, fruits and vegetables, and cash crops and oil seeds in China

Note: Crop water requirement refers to crop evapotranspiration accumulated over the growing period of the respective crop. The water requirement for rice includes evapotranspiration and percolation. Net return refers to the difference between total revenue and material costs; labor cost is not included. Yield and price data for grains are based on 1999 to 2001 averages from the National Bureau of Statistics and FAOSTAT for other crops.

Source: Liu et al. (2007)

to relax self-sufficiency targets on soybeans first may thus have had important unintended positive effects.

Maize is also a large user of water, yielding an equally low return on water use. Most fruits (except citrus) and vegetables, including potatoes, are much less water intensive. They also yield much higher returns per cubic meter. With the exchange rate now at 6.3 renminbi per dollar and US maize competitive at 6.5 renminbi per dollar according to Gale and Tuan (2007), it makes economic, ecological, and social sense to extend the current strategy adopted for soybeans and relax self-sufficiency requirements for maize, the other main animal feed. China appears to be moving in this direction.

To be sure, rising maize imports by China would exert upward pressure on the world market price. However, if the increase is moderate, gradual, and predictable, it would provide the world's production system time to adapt, and disruptions in the world markets could be avoided. A limited effect on world prices (5 percent) is predicted by 2005 to 2030 simulations based on the computative general equilibrium Chinagro II model that enables endogenous reactions of world prices to changes in China's net trade flows (Keyzer and van Veen 2010). But these are preliminary estimates and the feed grain price rise may be higher if imports are higher.

But can the markets be relied upon to deliver the necessary maize supply at relatively stable prices? Given appropriate and timely engagement with the world market and exporting countries, the answer is yes. First, being primarily used as animal feed, the domestic political sensitivity of maize is less pronounced than that of rice or wheat, the core staples. World maize markets are also more liquid and diversified than those of rice, with financial instruments for hedging and securing supplies well established. And if feed grain shortages arise, domestic and imported nongrain feeds (for example, cassava and potatoes) remain a more water efficient alternative, as does the import of live animals or processed animal products itself (Zhou et al. 2008). But the latter would come at the expense of rural employment and income earning opportunities for Chinese livestock farmers, important for mitigating the farm income problem (Keyzer and van Veen 2010).

Second, risks could be diversified through bilateral partnerships and foreign land deals, as well as vertical integration or broader strategic investments in agriculture where there is still substantial potential for maize land expansion. The first two arrangements provide more direct control, but are also increasingly contested (Deininger and Byerlee 2011; Barrionuevo 2011). They could be complemented with investment in agriculture through bilateral development assistance and foreign diplomacy, especially in Africa, but also in neighboring countries with untapped potential (such as Laos, Cambodia, and Myanmar). Similarly, China could usefully increase its support to international agricultural research agencies such as those of the Consultative Group on International Agricultural Research. Support for raising agricultural productivity elsewhere has the added advantage that it indirectly fosters world food trade, which benefits all.

Third, from China's perspective, the reliability of world markets is largely endogenous, in that it results from coordination—or lack thereof—among its major players, of which China is one. The typical response to commodity price shocks has been for exporters to impose export restrictions, while importers lower import tariffs. These restrictive trade practices during a commodity price shock represent a classic collective action problem. One country refraining from this type of insulation faces greater shocks than would apply if countries collectively refrained from intervening. This should be enough to motivate WTO members to negotiate a collective agreement to limit the extent of price-insulating policies. As a ten-year member of the WTO, and the second-largest economy in the world, this presents an opportunity for China to exert a positive influence on global economic governance. It could build on its support for greater grain stock transparency agreed to during the June 2011 G20 meeting of the ministers of agriculture in Deauville, France. Increased transparency over global grain stocks reduces price volatility caused by speculative moves based on ill-informed perceptions about global supplies (Slayton 2009). Taking the lead on increased coordination and trust building represents the first and best plan of action for China—and the world—enabling greater reliance on feed grain imports.

ADDRESSING THE FARM INCOME PROBLEM: OVERCOMING INEQUALITY

To remain competitive intersectorally and internationally, Chinese farms must also grow in size. Rising TFP in agriculture alone is unlikely to close the growing income gap with nonagriculture, and the scope for a more rapid increase in food prices is also limited.[8] China's WTO commitments further limit the scope for raising farm incomes through increased protection at the border, although some room remains. Creating a commercially viable smallholder agriculture will require institutional innovations in the land, labor, and rural capital markets to facilitate farm consolidation and an orderly transition of labor out of agriculture, as well as an urbanization process that avoids urban overconcentration and facilitates secondary town development to absorb a significant part of this labor in a buoyant rural economy (Otsuka 2007; Christiaensen and Todo 2012).

The balancing act is to accommodate land consolidation and labor movements so as to remain self-sufficient in the core staples (rice and wheat) at internationally competitive prices, while at the same time maintaining commercially viable livelihoods for a sufficiently large group of smallholder farmers. Indeed, given China's current conditions, its landscape by 2030 will still be largely dominated by relatively small holdings, by any international standard.

Rationalize Factor Markets to Facilitate Farm Consolidation

As land cannot be sold under current regulations, land must be consolidated through land rental. Even if land sales were possible, sales would be slower and unlikely to function competitively for swift and steady land reallocation across households because of imperfections in credit markets. Land rental markets are the practical way forward in the near and medium term, while reforms in private land use rights and their marketability proceed. The trend in land rental, by aging farmers or farmers who have migrated to cities, is indeed strongly positive, from 7 percent of agricultural land in 2000, to 19 percent in 2008. In the more developed eastern provinces such as Zhejiang, rental reaches as high as 40 percent. But there are also increasing reports of complete village reorganizations into large, company-run farms, whereby remaining villagers give up their land in return for housing and lifetime compensation at the urban fringe. The scale of such consolidation initiatives remains unclear, as are the long-term welfare effects for the farmers involved.

Lingering concerns about tenure security uncertainties related to land registration (Kimura et al. 2011) and substantial coordination costs continue to hamper consolidation. These costs are even higher in mountainous areas, where plots are smaller and more scattered. Increasing tenure security could go a long way to deepen the rural land rental market, as could government-assisted voluntary exchange platforms. Enhancing rural social protection would also free up land for renting. Cultivated land continues to provide an informal safety net against unemployment, sickness, and old age. Important steps in this direction, including the expansion of low-income welfare payments, rural health insurance, and rural pensions, are anticipated under the 12th five-year plan. Further reforms of the *hukou* system[9] are desirable to free up land for farm consolidation and help farmers sort themselves according to their comparative advantage (Zhu and Luo 2010). When accompanied by small- and medium-town development, part-time farming could also be promoted, especially in mountainous areas where opportunities for land consolidation are more limited. Such a strategy can build on the "leaving the village without leaving the countryside" policy of the early 1990s, which proved hugely successful in generating rural nonfarm employment nearby, through the development of small towns in coastal areas (Gale and Dai 2002).

While these coastal towns often benefited from access to export markets or spillovers from growing cities, the conditions are much more favorable now to absorb the next wave of rural migrants by developing small and medium towns in China's interior provinces. Congestion increasingly plagues the coastal metropoles, coastal provinces are climbing up the skill and value chain, and labor-intensive industries are moving inward. Central support of such a strategy is consistent with the objective of reducing the regional and rural–urban divide. It requires increased investment in rural education and improved rural transportation to link towns with their rural hinterlands and to link farmers with their off-farm jobs in rural towns, often in related agro industries. Consistent with the gradual reform approach, the next reform phase of the *hukou* system could focus on facilitating migration to these small and medium towns.

Mechanization of grain cultivation will require greater access to bank capital in rural areas. According to the China Bank Regulatory Commission's finance service map database, only 31.4 percent of rural households had ever borrowed from banks at the end of 2009, and only 30.1 percent of micro- and small enterprises. Most agricultural loans (85 percent) come from Rural Credit Cooperatives. A comprehensive regulatory and supervisory framework for rural and microfinance institutions should be developed by the government, which should focus on creating the infrastructure and market environment for rural finance instead of direct public provision of financial services.

But credit-market imperfections can also be mitigated through institutional innovations such as machine rentals. The use of machine renting has historically been practiced in many Asian countries to capture economies of scale, and is also on the rise in China. If properly coordinated, machine rentals provide a practical and convenient way to substitute capital for labor, particularly for part-time farmers, who are often most in need of labor-saving technology. Together, machine rental and land consolidation could go a long way to help China's smallholders remain intersectorally and internationally competitive—which will be critical to help China simultaneously resolve its farm income and food problem.

Make Producer Cooperatives Work more Effectively for Smallholders

The need for volume, consistency, and quality in processing and marketing high-value agricultural products brings with it economies of scale and typically induces vertical integration of the supply chains. Producer cooperatives are a natural institutional route for smallholders to capture these economies of scale, providing important opportunities to address the farm income problem. Following the enactment of China's Farmers' Professional Cooperative Law in 2007, there has been a rapid expansion of farmers' cooperatives. Two types of cooperatives have emerged: farmer-led cooperatives and "company + household" cooperatives. In the latter model, shares are controlled by a leading agribusiness and other key shareholders, such as government officials, with farmers as participating members. Farmer-led cooperatives are usually controlled and owned by one or more large, wealthy households, with voting rights and profits proportional to the capital brought in, and smallholders participating as peripheral members. The company + household structures are often preferred by local government and supported through tax breaks, subsidized loans, and the provision of land.

While these new cooperative structures enable smallholders to tap into dynamic high-value markets, smallholder farmers in effect forego most of the benefits. They are largely excluded from the decision-making process in the cooperatives and miss out on the lion's share of the profits and the government's support, which is capitalized in the assets of the core members. The monopsonic power of the downstream agro-processing companies persists. In short, despite their important potential in mitigating the farm problem and reducing the rural–urban divide, cooperatives in China are not yet living up fully to that promise. The government should monitor these developments more closely and take steps to better protect the interests of minority shareholders while encouraging the formation of member-owned and farmer-controlled cooperatives.

MAKING MAXIMUM USE OF AGRICULTURE

China's success in feeding itself has defied the expectations of many. But, its smallholder agricultural model and focus on self-sufficiency are increasingly under pressure given rising agricultural labor costs, increasing competition from imports following renminbi appreciation, and growing agro-climatic uncertainty that compound the longer standing erosion of its land and water resource base. These growing supply constraints come in the face of rising cereal feed demand. This is causing tensions on two fronts—the farm income and food problems—and China's success in tackling both will be key for socio-economic stability both at home and abroad. Success will depend on the design of China's agricultural and urbanization policies moving forward.

Rapid urban income growth is forcing China's smallholder farmers to expand to earn a living commensurate with what could be earned outside agriculture, with rising agricultural labor costs now adding additional consolidation pressures to capture the economies of scale from mechanization. Large-scale, mechanized farming, which China increasingly appears to be embracing, could provide a way to keep production up and food prices down. But it would require draconian top-down measures to consolidate the multitude of fragmented smallholder farms, and it is unlikely to generate sufficient off-farm employment to absorb the agricultural labor force, which would instead be moved to the urban fringe and made largely dependent on transfers. Large-scale farming would also leave the income gap unresolved in the mountainous areas, where many of the poorer populations live and where consolidation is more challenging.

A better strategy is to make maximum use of agriculture to reduce the rural–urban income gap. Promoting part-time and commercially viable smallholder farms for maximum rural employment and a manageable urbanization process will require institutional reforms in the land markets to facilitate farm consolidation, and a more equitable distribution of the benefits from rural-to-urban land conversions. Continuing promotion of rental machinery services could further help small farmers capture the economies of scale from mechanization.

Reducing the rural–urban income gap will further require continued investment in rural public goods, in particular agricultural research and extension, as China is doing, but also rural infrastructure and education, including attention to the need to better prepare the next generation for employment opportunities outside agriculture. These public goods will need to be complemented with a greater emphasis on secondary town development, especially in the hinterlands, to enable part-time farming and a gradual transition out of agriculture. Better use could also be made of the rapid increase in farmer support by scaling back direct and indirect fertilizer subsidies and linking support to adoption of more sustainable and less nitrogen-rich land management practices or eco-services.

Important employment and income opportunities are also available in the production of high-value products for the domestic and international markets, in line with China's comparative advantage. To maximize their inequality-reducing potential, China's cooperatives need to be monitored more closely. Improvements in food safety will help safeguard domestic markets and maintain access to international ones. Together these measures could go a long way towards shaping an agrarian structure and urbanization pattern that mitigates the rural–urban income divide. Failure to address this divide could be especially destabilizing, especially when compounded by rising food price inflation, which has often marked key turning points in China's history.

Higher (real) food price inflation has indeed been China's reality over the past decade, reflecting how domestic supply is increasingly struggling to keep up with rising cereal (feed) demand. Better water and soil nutrient management, in addition to the institutional innovations highlighted above, could go a long way in relaxing supply constraints. Nonetheless, these strategies could be usefully complemented by greater reliance on the international market for cereal feeds—that is, by a downward revision of China's grain self-sufficiency objectives, particularly for maize, even if self-sufficiency targets for rice and wheat are maintained. This is economically and ecologically more efficient, and could help restore trust in world markets. If the shift is done gradually and transparently and paired with large-scale support to international agricultural research agencies and investments in agriculture in countries with untapped potential, such as in Africa, the world's agricultural system would have time to adapt and major disruptions could be avoided. Greater reliance on imports would also avoid potential tensions associated with bilateral land exploitation deals. The existing simulations suggest that such a strategy would have a limited impact on world market prices, although more research is needed to gauge the effects on future world price instability, as unanticipated ripples in China's food markets may reverberate like waves across the world.

ACKNOWLEDGMENTS

The author is grateful for insightful discussions and penetrating comments on earlier versions by Carter Brandon, Derek Byerlee, Wendao Cao, Shenggen Fan, Li Guo, Pei Guo, Jikun Huang, Patrick Labaste, Kei Otsuka, Alan Piazza, Scott Rozelle, Iain Shuker, Susanne Scheierling, Sari Soderstrom, Robert Townsend, Joergen Voegele, Xiaobo Zhang, and Jun Zhao, Chris Sall, Ulrich Schmitt, and seminar participants at the China World Bank Office in Beijing and World Bank Headquarters in Washington DC. The findings, interpretations, and conclusions expressed in this paper are entirely those of the author. They do not necessarily represent the views of the International Bank for

Reconstruction and Development/World Bank and its affiliated organizations, or those of the Executive Directors or the governments they represent.

NOTES

1. No direct estimates of the number of food insecure are available, but according to the official statistics, 36 million people were still living below 1196 yuan per year or $0.57/day in 2009. At such low levels of income, people are still struggling to meet their daily minimum needs, including of food. Even so, the percentage of food insecure people is small, given China's large population.
2. Subsidies include grain land based income transfers, grain land based input subsidies, new variety extension payments, and agricultural machinery subsidies.
3. The absolute and relative income differences reported here account for differences in the price evolution in rural and urban areas in China during 1978–2009, which were in effect minimal (NSBC 2010). However, they do not account for differences in cost of living between rural and urban areas in the base year (2000), and are thus likely to overstate the gap somewhat.
4. See Xu et al. (2006), Deng et al. (2006) and Chau and Zhang, (2011) for further views on each debate.
5. Water availability is especially a pressing concern in the North China Plain (which includes the provinces of Henan, Hebei, Shandong, and the northern parts of Jiangsu and Anhui), though Fischer et al. (2010) still see some scope for further expansion of irrigation in the northeast (provinces of Heilongjiang, Jilin, and Liaoning).
6. The implicit water content of a crop corresponds to its total accumulated evapotranspiration over the complete growth cycle. It is generally higher for grains and most cash crops, and lower for potatoes, vegetables, and most fruits (except citrus). If the amount of water lost through evapotranspiration is not replenished, for example, through local precipitation, crop water use will result in a net water loss locally and become unsustainable over time.
7. To maintain the yield growth experienced between 1995 and 2005, Chen and Zhang (2010) estimate that investments in agricultural R&D will have to increase by 15 percent annually between 2005 and 2020—by 43 billion yuan in 2005 prices, or $5.25 billion.
8. Between 2000 and 2010 the domestic grain food price index already rose by 73 percent. Increasing domestic food prices faster is unlikely and would not be politically palatable.
9. Under the *hukou* or household registration system, households cannot freely change the location of their registration, while the use of local public services such as education and health is confined to those registered in that locality. Consequently, many rural–urban migrants have difficulties accessing educational and health services.

REFERENCES

Anderson, K. 2009. Distorted agricultural incentives and economic development: Asia's experience. *The World Economy* *32*(3): 351–84.

Barrionuevo, A. 2011. China's farming pursuits make Brazil uneasy. *New York Times,* May 27.

Chau, N., and W. Zhang. 2011. Harnessing the forces of urban expansion: The public economics of farmland development allowances. *Land Economics 87*(3): 488–507.

Chen, K., and Y. Zhang. 2010. Agricultural R&D as an engine of productivity growth: China. Background paper to Foresight Project on Global Food and Farming Futures.

Christiaensen, L., and Y. Todo. 2012. Poverty reduction during the rural–urban transformation–The role of the missing middle. Paper presented at the Inaugural Conference of the Courant Center on Poverty, Equity, and Growth. July 1–3, 2009, Goettingen, Germany.

De Brauw, A., J. Huang, L. Zhang, and S. Rozelle. 2011. The feminization of agriculture with Chinese characteristics. World Development Report 2012 background paper. Washington, DC: World Bank.

De Gorter, H., and J. Swinnen. 2002. Political economy of agricultural policy. In *Handbook of Agricultural Economics,* ed. B. Gardner and R. Gordon, 1893–1943. Amsterdam: Elsevier.

Deininger, K., and D. Byerlee. 2011. *Rising global interest in farmland—Can it yield sustainable and equitable benefits?* Washington, DC: World Bank.

Deng, H., J. Huang, Z. Xu, and S. Rozelle. 2010. Policy support and emerging farmer professional cooperatives in rural China. *China Economic Review 21*: 495–507.

Deng, X., J. Huang, S. Rozelle, and E. Uchida. 2006. Cultivated land conversion and potential agricultural productivity in China. *Land Use Policy 23*: 372–84.

Fan, S., R. Kanbur, and X. Zhang. 2011. China's regional disparities: Experience and policy. *Review of Development Finance 1*: 47–56.

Fischer, G., T. Ermolieva, and L. Sun. 2010. Environmental pressure from intensification of livestock and crop production in China: Plausible trends towards 2030. CATSEI project report.

Gale, F., and H. Dai. 2002. Small town development in China—a 21st century challenge. *Rural America 17*(1): 12–19.

Gale, F., and F. Tuan. 2007. *China currency appreciation could boost U.S. agricultural exports. WRS-0703.* Washington, DC: United States Department of Agriculture.

Hayami, Y. 2007. An emerging agricultural problem in high-performing Asian economies. Policy Research Working Paper Series 4312. Washington, DC: World Bank.

Hayami, Y., and T. Kawagoe. 1989. Farm mechanization, scale economies and polarization. *Journal of Development Economics 31*: 221–39.

Hu, R., Z. Yang, P. Kelly, and J. Huang. 2009. Agricultural extension system reform and agent time allocation in China. *China Economic Review 20*(2): 303–15.

Keyzer, M., and W. van Veen. 2010. China's food demand, supply and trade in 2030: Simulations with Chinagro II Model. Presentation made at the Policy Forum of the CATSEI on Prospects of China's Agricultural Economy in 2030, November 19, Beijing.

Kimura, S., K. Otsuka, T. Sonobe, and S. Rozelle. 2011. Efficiency of land allocation through tenancy markets: Evidence from China. *Economic Development and Cultural Change 59*(3): 485–510.

Liu, J., A. Zehnder, and H. Yang. 2007. Historical trends in China's virtual water trade. *Water International 32*(1): 78–90.

Lohmar, B., F. Gale, F. Tuan, and J. Hansen. 2010. China's ongoing agricultural modernization: Challenges remain after 30 years of reform. In *China's Agricultural Modernization,* ed. R. Jeffries, 1–68. Nova Science Publishers.

McCouch, S., and S. Crowell. 2012 (Chapter 7 in this book).

Milanovic, B. 2011. *The haves and the have-nots: A brief and idiosyncratic history of global inequality.* New York: Basic Books.

NSBC. 2010. Statistical yearbook of China, 2010.

OECD. 2011. *Agricultural policies monitoring and evaluation 2011: OECD Countries and Emerging Economies.* Paris: Organization of Economic Cooperation and Development.

Otsuka, K. 2007. The rural industrial transition in East Asia: Influences and implications. In *Transforming the rural nonfarm economy: Opportunities and threats in the developing world,* ed. S. Haggblade, P. Hazell, and T. Reardon, 216–236. Baltimore: John Hopkins University Press.

Ravallion, M., and S., Chen. 2007. China's (Uneven) Progress Against Poverty. *Journal of Development Economics 82*(1): 1–42.

Reardon, T., C. P. Timmer, and B. Minten. 2010. Supermarket revolution in Asia and emerging development strategies to include small farmers. *Proceedings of the National Academy of Sciences 109*(31): 12332–12337.

SAIN. 2010. Improved nutrient management in agriculture: A neglected opportunity for China's low carbon growth path. Policy Brief No 1.

Shouying, L. 2012. *A third way for urbanization and land system reform in China.* Paper presented at the annual World Bank conference on land and poverty, April 23–26, 2012.

Slayton, T. 2009. Rice crisis forensics: How Asian governments carelessly set the world rice market on fire. Center for Global Development Working Paper 163. Washington, DC: Center for Global Development.

Tan, S., N. Heerink, and F. Qu, 2006. Land fragmentation and its driving forces in China. *Land Use Policy 23*(3): 272–85.

Wang, H., X. Dong, S. Rozelle, J. Huang, and T. Reardon. 2009. Producing and procuring horticultural crops with Chinese characteristics: A case study in Northern China. *World Development 37*(1): 1791–801.

World Bank. 2009. China: From poor areas to poor people: China's evolving poverty reduction agenda: An assessment of poverty and inequality in China. World Bank Report 47349-CN, East Asia and Pacific Region: Poverty Reduction and Economic Management Department. Washington, DC: World Bank.

——. 2011. *China Inflation Monitor.* Beijing: World Bank.

Xu, Z., J. Xu, X. Deng, J. Huang, E. Uchida, S. Rozelle. 2006. Grain for green versus grain: Conflict between food security and conservation set-aside in China. *World Development 34*(1): 130–48.

Ye, L., and E. Van Ranst. 2009. Production scenarios and the effect of soil degradation on long-term food security in China. *Global Environmental Change 19*: 464–81.

Zhang, X., J. Yang, and S. Wang. 2011. China has reached the Lewis turning point. *China Economic Review 22*(4): 542–54.

Zhou, Z., W. Tian, and B. Malcolm. 2008. Supply and demand estimates for feed grains in China. *Agricultural Economics 39*: 191–203.

Zhu, N., and X. Luo. 2010. The impact of migration on rural poverty and inequality: A case study in China. *Agricultural Economics 41*: 191–203.

18

Food Security and Sociopolitical Stability in East and Southeast Asia

C. PETER TIMMER

Since the end of the Vietnam War, known in Vietnam as the American War, the broad sweep of East and Southeast Asia (excluding China) has ceased to be the focus of concerns over security threats to the United States. To be sure, political instability remains a local threat in Myanmar, East Timor, even Thailand, the Philippines, and Indonesia. But the era of local conflicts pulling the United States into active military engagement in the region seems to be past.[1]

Still, local conflicts and political instability in East and Southeast Asia (excluding China) continue to carry dramatic implications for US economic and political interests. As the epicenter of the world's highly unstable world rice market, events in this region (along with China and South Asia) directly affect global food security and the nutritional welfare of the two-thirds of the world's poor who rely on rice for a large share of their calories. If there are any causal links between food security and political unrest, the probabilities are high that they will be triggered by events in East and Southeast Asia (excluding China) and will then be transmitted as rice price shocks to the global food economy.

The formation of rice prices in world markets has interested scholars and policymakers for decades.[2] Nearly half the world's population consumes rice as a staple food, and roughly half of these consumers are poor. Small farmers in Asia who use highly labor-intensive techniques typically produce this rice, which is mostly consumed where it is produced. International trade in rice is less than 30 million metric tons out of a global production of nearly 460 million metric tons. Only 7 to 8 percent of rice produced crosses an international border at an invoiced world price.[3]

Still, the world market for rice provides essential supplies to importing countries around the world and a market for surpluses in exporting countries. The prices set in this market provide signals to both exporting and importing countries about the opportunity cost of increasing production or consumption. It is disconcerting to exporters and importers alike if these market signals are highly volatile, because volatile prices for important commodities confuse investors and slow down the rate of economic growth.

Part of the long-standing interest in the world rice market has been precisely because of its volatility. The coefficient of variation of world rice prices has been much higher than that of wheat or corn for decades at a time.[4] Understanding this volatility has been difficult because much of it traces to the residual nature of the world rice market, as both importing and exporting countries stabilize rice prices internally by using the world rice market to dispose of surpluses or to meet deficits through imports. Thus, supply and demand in the world market for rice are a direct result of political decisions in a significant number of countries, especially in Asia.

As a rough generalization, there are three quite distinct food security environments in East and Southeast Asia (excluding China). First, there is "rich Asia," which contains Japan, South Korea, Taiwan, and Singapore. Despite heated political rhetoric, food security is not a serious issue in these countries as long as global food markets remain relatively open. But several countries in this category are hedging their bets on continued openness and are investing heavily in land acquisitions for food production in other countries.[5]

Second is "emerging Asia," which contains Malaysia, Thailand, Indonesia, the Philippines, and Vietnam. These countries are well advanced in their structural transformations, but the transition process is not completed in any of them, and regional food security issues remain on the welfare and political agenda. Indeed, policy actions to improve food security in these countries have the potential to radically destabilize the world rice market and consequently have an impact on global food security. Several of these countries are still relatively land-abundant and are entertaining both domestic and foreign investments in large-scale development of plantation-based food production, often in areas that are still forested.

Finally, there is "least developed Asia," which contains Myanmar, Cambodia, Laos, Papua New Guinea, and Timor Leste. In these countries, food security remains a daily concern for a significant fraction of the population, both urban and rural, although challenges to food security in any of these countries are unlikely to spill over to have regional or global consequences. On the other hand, successful economic modernization and rapidly growing productivity in rice production in these countries could easily destabilize the world rice market, with greater exportable surpluses than the market is able to handle at prices that will be remunerative to farmers in the rest of the region.[6]

This diverse array of countries is at the center of many global food security debates, mostly because the world rice economy is centered in the region. The two largest rice exporters, Thailand and Vietnam, are matched up with the two largest rice importers, Indonesia and the Philippines. The potential of

Cambodia and Myanmar to regain their historical status as major rice exporters will challenge political support to farmers throughout the region, as the level of protection provided in Indonesia, the Philippines, and even Thailand, will not be sustainable if an additional five to ten million metric tons of rice should come on the world market from these countries over the next decade. With rice consumption falling in nearly all of Asia, a major rural crisis cannot be ruled out.

At the same time, this region is buffeted by the extraordinarily rapid growth of their two bookend neighbors, China and India. The resource requirements of this rapid growth in both countries have spilled into all of these economies, but especially the resource-rich countries of Indonesia and Myanmar. Most of this resource demand has been for energy and minerals, and the development profession is well aware of the potential for such resource booms to set off intense political rivalries at both regional and national levels. In the current context, these conflicts will not necessarily be directly food related, but they will spill into the food sector because of competition for land, water, and labor. Demand from China might also directly affect food markets in East and Southeast Asia, as it is already the world's largest importer of cassava—much of it sourced from Southeast Asia—and soybeans, which makes it a competitor for supplies from the world market. China and India are also the two largest importers of palm oil from the region. China's sporadic imports of rice, mainly from Vietnam and Pakistan, have the potential to destabilize regional rice prices. Clearly, it is almost impossible to discuss food security and political stability in East and Southeast Asia without reference to China and India.

For quite different reasons, the Democratic People's Republic of Korea (DPRK), or North Korea, is also central to the region's food security and political stability. The DPRK has suffered several famines since the mid-1990s, is often the largest recipient of food aid in the region, consumes rice as its staple food (although maize, barley, and potatoes are also widely consumed), and yet has virtually no impact on the regional or global rice market. The "Hermit Kingdom" is almost impossible to analyze with the standard data and models used by economists, and its political economy remains an enigma even to specialists.

That said, the DPRK offers perhaps the scariest potential for food insecurity to spill into political instability—first internally, and then regionally—because the DPRK has nuclear weapons and a seeming willingness to use them in defense of its regime. The DPRK has used nuclear weapons as a bargaining chip in food aid negotiations, although most food aid donors see the hundreds of thousands of famine victims as justification enough for providing assistance.

Despite the secrecy and lack of access, the massive famine in the mid-1990s is reasonably well documented (Natsios 1999, 2001; Haggard and Noland 2007). Between one and three million people starved to death, many of them "triaged" because they were deemed "not essential" to the survival of the regime (Natsios, 1999). Signs of deep internal strife were reported, including rumors of a planned coup, but in the end the regime survived and tightened its control over food production and distribution.

The most recent challenges to food security in the DPRK in 2010 and 2011, caused in a narrow sense by bad weather, are better documented, as the Food and Agriculture Organization (FAO), World Food Programme (WFP), and United Nation Educational, Scientific, and Cultural Organization (UNESCO) have been invited by the government to observe the situation as a prelude to offering substantial food aid assistance (FAO–WFP 2011). But the major food donors are reluctant to assist the DPRK, as previous aid has clearly been siphoned off to feed government cadre and the military to a substantial degree, with little monitoring and evaluation possible by donor agencies. The most recent FAO–WFP report suggests that this attitude is changing, and the authors were surprised at how open their hosts were to visiting markets, interviewing households, and revealing official government statistics (FAO–WFP 2011). The new government of Kim Jong-un also seems more open to foreign input, and perhaps to a more market-oriented approach to the food economy. It is hard to see how to resolve the DPRK's food security problems without such an approach.

The combination of challenging events emanating from the world rice market, and intense resource demands on the economies of East and Southeast Asia coming from China and India, raises a host of difficult questions. For all countries in the region, reasonably stable rice prices in domestic markets are the political measure of food security, and this desire for stability drives much of the political economy of food security. Thus this chapter asks, what is the appropriate analytical framework for addressing these broad political economy issues? The standard model from the theory of international trade used to evaluate welfare losses from price changes in international and domestic markets is rejected in favor of a behavioral model of political economy (Timmer 2012). This behavioral model explicitly incorporates the role of empirical regularities of behavioral economics, especially loss aversion, time inconsistency, other-regarding preferences, herd behavior, and framing of decisions, which present significant challenges to traditional approaches to modeling international trade. The formation of price expectations, hoarding behavior, and the welfare losses from highly unstable food prices all depend on these behavioral regularities.

It is important, then, to understand actual responses by citizens to price changes, both economically and politically. Many of these responses—including a deep desire for stable food prices—cannot be expressed by citizens in markets they face, and thus spill directly into the political arena. Economists have long lamented that rice was more a "political" commodity than a "market" commodity. Behavioral economics, a relatively new subfield in the profession, has provided deep insights into why this is so, and what can and cannot be done about it.

SEEKING CAUSAL LINKS

Although not home to the two oldest Asian civilizations, those in India and China, other countries in East and Southeast Asia trace their historical and cultural identities back several millennia, including Japan, Vietnam, Thailand, and Burma. Most countries in East and Southeast Asia (excluding China), in fact, have extremely long histories of centralized government activity, political control, and social organization. The institutions that have grown out of these long histories vary radically from country to country, but all provide the context in which these societies approach the risks of poverty and famine. The political, social, and cultural dimensions of these institutions have historical roots that extend well beyond recent episodes of rapid economic growth. It is easy to forget that as recently as 1880, Indonesia, Thailand, and Japan had very similar levels of economic welfare (Maddison 1995).

Thus it is no surprise that many countries in this region seem stuck between outmoded political and cultural attitudes about food security, especially the role of rice in the nation's diet and culture, and modern economic risk management instruments, such as international trade and financial derivatives. Any understanding of the links between threats to food security and resulting political instability must be based on this historical reality, because political systems rooted in historical institutions may be unable to cope with sudden challenges that stem from global threats to food price stability. Alternatively, such political systems may simply design barriers that keep global instability from penetrating into domestic markets, with no concern for subsequent spillover effects into those global markets themselves.

This chapter aims to develop the rationale for a causal link from food insecurity to political instability, at least in the Asian context where rice is such an important factor in both food security and politics—for reasons that are often unrelated, or even antithetical, to food security, as with the use of guaranteed high crop prices to win the votes of rice farmers. There are clearly other causal links in the reverse direction, when political instability that is unrelated to food security threatens food security itself. Much of the political turmoil in East and Southeast Asia (excluding China) is of this latter type, originating in colonial conflicts or ethnic clashes over resources, political identity, and cultural values. In these circumstances there is often a virtuous circle that builds the trust of rural investors as political stability is established, and greater output and income in rural areas then contributes to political stability, partly through enhanced food security.

One reason for stressing the historical context of food security is the frequency and severity of external weather events on regional food production. These weather events, mostly El Niño and the Southern Oscillation (ENSO) dynamics, are particularly well studied in Indonesia, but they affect much of the region when the dynamics are severe. The external dynamics provide a strong case for a causal link from El Niño to rice production and local food security, and through trade changes, to global rice prices. Political instability has been associated with high food prices—a fact that has attracted considerable scholarly attention—although the direction of causation is not established (Lagi et al. 2011). ENSO dynamics offer a mechanism for attributing causation.

Scientists and their colleagues at Stanford University have extensively studied these ENSO effects on Indonesia's national and regional rice production and on world rice prices (Falcon et al. 2004). Using the August sea surface temperature anomaly (SSTA) to gauge climate variability, their work shows that each degree Celsius change in the August SSTA produces a 1.32 million metric ton effect on rice output and a $21 per metric ton change in the world price for lower quality rice. These relationships offer policymakers a forward-looking tool to prepare for threats to local food security (Naylor et al. 2009). These empirical relationships are a starting point for identifying causal links from exogenous weather events to rice production and trade impacts, which have the potential to be mediated by policy interventions that aim to stabilize domestic rice economies.

THE WORLD RICE MARKET

All commodity markets tend to track major macroeconomic developments, the volume of international trade, and currency values. The major food grain markets also have important economic and technological links, because of substitution possibilities in production and consumption. High corn prices will attract production resources away from soybeans, for example, and induce consumption of low-quality wheat as livestock feed. Still, the world rice market has several distinguishing features—especially the "thinness" of the international market for rice and the resulting price volatility—that make its performance quite distinct from the markets for wheat and corn.

Volatility in rice prices is driven by political decisions about rice imports and exports in most Asian countries, but also by the structure of rice production, marketing, and consumption in these same Asian countries—that is, by the industrial organization of the rice economy. Hundreds of millions of small farmers, millions of traders, processors, and retailers, and billions of individual consumers all handle a commodity that can be stored for over a year in a consumable form (although deterioration does set in after two years or more).

The price expectations of these market participants are critical to their decisions about how much to grow, to sell, to store, and to consume. There are virtually no data available about either these price expectations or their marketing consequences.[7]

As a result, the world rice market operates with highly incomplete and very imperfect information about short-run supply and demand factors. Because of this disorganized industrial structure and lack of information about the behavior of its participants, rice is a very different commodity from the other basic food staples, wheat and corn.[8]

Experience with world rice prices since the mid-2000s illustrates the importance of market structure to short-run price dynamics. The actual production–consumption balance for rice has been relatively favorable since 2005, with rice stocks-to-use ratios improving slightly. This stock build-up was a rational response to the very low stocks seen at the middle of the decade, and to gradually rising rice prices—exactly what the supply of storage model predicts (Working 1949). Short-run substitutions in both production and consumption between rice and other food commodities are limited. Until late 2007, it seemed that the rice market might dodge the bullet of price spikes seen in the markets for wheat, corn, and vegetable oils. The lack of a deeply traded futures market for rice also made financial speculation less attractive. Any speculative increase in rice prices would have to come from another source. Financial speculation simply cannot drive the world rice market.

The world rice market is very thin, trading just 7 to 8 percent of global production. While this trade share is a significant improvement over the 4 to 5 percent traded in the 1960s and 1970s, the global market is still subject to large price moves from relatively small quantity moves. In this, too, rice is distinguished from wheat and corn.

The global rice market is also concentrated. Thailand, Vietnam, India, the United States, and Pakistan routinely provide about 80 percent of available supplies. Only in the United States is rice not an important commodity from local consumers' perspectives. All Asian countries show understandable concern over their citizens' access to daily rice supplies. Both importing and exporting countries watch the world market carefully for signals about changing scarcity, while simultaneously trying to keep their domestic rice economy stable. These extensive policy concerns on the part of governments make rice a highly political commodity (Timmer and Falcon 1975).

As concerns grew in 2007 that world food supplies were limited and that prices for wheat, corn, and vegetable oils were rising, several Asian countries reconsidered the wisdom of maintaining low domestic stocks for rice. The Philippines, in particular, tried to build up stocks to protect against shortages going forward. Of course, if every country—or individual consumer—acts the same way, the hoarding causes a panic and extreme shortages in markets, leading to rapidly rising prices. Even consumers in the United States were not immune to this panic, as the run on bags of rice at Costco and Sam's Club in April 2008 indicated. Such price panics were fairly common in the twentieth century, but the hope was that deeper markets, more open trading regimes, and wealthier consumers able to adjust more flexibly to price changes had made rice markets more stable. It turns out this was wishful thinking, as the price record for rice shows.

Rice prices had been increasing steadily but gradually since 2002, but they began to accelerate in October 2007. Quickly, there was concern over the impact of higher rice prices in exporting countries, especially India, Vietnam, and Thailand. This concern translated into action as India and Vietnam moved to impose export controls.[9] Importing countries, especially the Philippines, started to scramble for supplies. Fears of shortages spread, and a cumulative price spiral started that fed on the fear itself.

The trigger for the panic came from high prices for wheat in world markets, an unexpected example of intercommodity price linkages. In India, the 2007 wheat harvest was damaged by drought and disease, problems also seen in many other parts of the world. The Food Corporation of India (FCI) had less wheat available for public distribution. For India to import as much wheat as it had in 2006, nearly seven million metric tons, would have been too expensive—politically, if not necessarily economically—because of the high world price. The FCI announced it needed to retain a larger share of rice from domestic production.

To bring about this larger role for rice in domestic distribution, India limited rice exports in October 2007 by imposing minimum export prices (MEP) that were higher than those prevailing in the world market. India is usually the second largest exporter of rice in the world, having shipped six million metric tons in 2007 (including over five million metric tons of non-basmati rice). An MEP higher than world prices should have stopped exports, but it was ineffective because exporters were able to evade the MEP. In April 2008, India announced a complete ban on exports of non-basmati rice, a policy the government could enforce. Other rice-exporting countries followed with their own controls, and rice prices started to spike.

The newly elected government in Thailand followed these events closely. It had a large political constituency among the poor and did not want consumer prices for rice to go up. The Thai commerce minister openly discussed export restrictions, and invited regional rice exporters to discuss an "OPEC" (Organization of the Petroleum Exporting Countries) for rice. Thailand was the world's largest rice exporter, shipping 9.5 million metric tons in 2007. Partly because of nervousness in the rice trade over Thai intentions, rice export prices in Thailand jumped by $75 per metric ton (metric tons) on March 28, 2008. Prices continued to skyrocket until, in April, rice for export cost over $1,100 per metric ton. This is the stuff of panics.

Price panics usually have their origins in the fundamentals of supply and demand. A number of market observers thought that low rice stocks accounted for the rising prices. Between 2000 and 2005, rice consumption had outpaced rice production. This was a mathematical inevitability if rice stocks were falling. Since the mid-1990s, China had reduced its rice stocks—a sensible response to growing reliance on trade as the buffer and to lower prices in world markets. Rice stocks were little changed in the rest of the world—indeed, the stocks-to-use ratio rose after 2005. Holding rice stocks in tropical conditions is extraordinarily expensive. A smoother flow of rice traded internationally offered the opportunity to reduce this wasteful stockholding.

But the sudden surge in rice prices in 2007 and 2008 demonstrated that something was happening beyond the fundamentals of supply and demand. Exporting countries were clearly willing to restrict exports of rice sharply to protect their own consumers. In responding, nearly all importing countries realized they were too dependent on foreign supplies for their own domestic food security. They quickly resorted to increasing domestic stockpiles, with a longer-run commitment to self-sufficiency in rice. Although larger stocks provide a greater degree of food security, they come at a very high financial cost, even when well managed to avoid deterioration in quality. In fact, excessive stockpiles of rice are a tragedy for poor consumers and for economic growth. Capital that is tied up in funding inventories does not contribute to stimulating growth in economic productivity (Timmer 2009a).

Sudden increases in demand for larger stocks, private and public, have a direct impact on demand in the world market. With a rapid and unexpected increase in short-run demand on the world market—the increase was roughly a quarter of normal annual rice trade—the world price would have to nearly double to get a new equilibrium. That is what happened. The fundamentals of rice supply and demand caused a gradual increase in rice prices between 2002 and 2007, but panicked hoarding caused the rice price spike in 2007 and 2008.

It would be nice if there were hard data to support this statement, but the data "vacuum" is precisely the point of the story here. Other than anecdotal evidence, mostly news stories, of grocery shelves being cleared of rice by frantic consumers, and farmers holding on to stocks expecting higher prices, data on changes in stocks by the multitude of small actors in the world rice economy are simply nonexistent. The main evidence is the price behavior itself, which was quite out of line with the "basics" of production, stocks, and consumption. Almost by their very nature, price expectations are unobservable, and hence hard to use when modeling decision making.

Fortunately, pricking the bubble and deflating expectations can end a speculative run based on herd psychology. This happened to the world rice economy. When the government of Japan announced in early June 2008, after considerable international urging, that it would sell at least 300,000 tons of its surplus World Trade Organization (WTO) rice stocks to the Philippines, prices in world rice markets started to fall immediately (Slayton and Timmer 2008). Once the price started to drop, the hoarding behavior by households, farmers, traders, and even governments also abated. By late August, medium-quality rice for export from Vietnam was available for half the price it had sold for in late April. Those millions of small farmers, traders, and consumers who had decided to hoard rice when prices were rising decided they could sell their supplies, or reduce the household inventory to normal levels. Demand for rice dried up, and the fall in prices gained momentum.[10]

There are three basic approaches to coping with the impact of high food prices once they hit world markets: domestic price stabilization, increasing supplies available in local markets, and providing safety nets to poor consumers. All three are directed at and must be managed by individual countries themselves, but donors and international agencies can play a substantial role as well in coordinating activities and providing resources—both financial and technical assistance.

The first approach is for individual countries to use market interventions to stabilize their domestic food prices. Such stabilization requires some capacity to isolate the domestic rice market from world markets and can only be implemented through government actions, although private traders can handle most of the actual logistics.[11] Such isolation runs directly against the spirit and, for many countries, the letter of WTO agreements. But it is a very widespread practice. Demeke et al. (2009) count 36 countries that used some form of border intervention to stabilize their domestic food prices during the 2007 and 2008 crisis.

Such policies can have a large impact globally. India, China, and Indonesia stabilized their domestic rice prices during the 2007 and 2008 food crisis by using export bans—or at least very tight controls—thus protecting well over two billion consumers from sharply higher prices. The policies pursued by these three countries demonstrate the importance of understanding local politics in policy formation, especially food policy. Although the end results were similar—food prices remained stable throughout the crisis—the actual policies pursued in each country were quite different (Slayton 2009b, Dawe 2010a).

India, Indonesia, and China are big players in the global rice market, even if their actual trade is limited. As David Dawe (2010b) emphasizes, stabilizing domestic rice prices in these large countries using border interventions might be both an effective and an efficient way to cope with food crises, even after considering the spillover effects on increased price volatility in the residual world market. Dawe emphasizes that unstable supply and demand must be accommodated somewhere, and passing the adjustment to the world market may be both equitable and efficient in a second-best world where fast-acting and well-targeted safety nets are not available.

The second basic approach to coping with a food crisis is to stimulate additional supplies through fast-acting programs. Nearly all countries tried to do something along these lines during the 2007 and 2008 crisis, whether by subsidizing fertilizer to get a quick production response or encouraging planting of short-season crops, even in urban gardens. If the high prices for food seen in the crisis actually get to farmers, they have strong incentives to search out these options themselves, but government assistance in gaining access to inputs or proper seed varieties can also help. In Asia, the short-run response of rice farmers to high prices was surprisingly vigorous, partly because of the availability of short-duration rice varieties and irrigated farming systems with multiple-cropping potential (Slayton 2009b). In Vietnam, which has three distinct cropping seasons for rice, production increased 6.3 percent in 2007 and 5.3 percent in 2008, compared with average annual increases of just 3.3 percent per year between 2005 and 2011. All of this increase in production, a total of 1.2 million metric tons, was put on the export market.

A variant of this second approach—stimulating a short-run supply response—is for countries to hold emergency food stocks as part of a broader strategy for providing food security to their citizens. Expectations of higher and more volatile food prices in the future should lead authorities to invest in larger food stocks than in the past. The "design rules" for adding to and disposing of these stocks, and their day-to-day management to avoid large storage losses, will be essential to making emergency food stocks a sustainable and cost-effective approach (Timmer 2009b).

One critical element of these rules will be to use international trade in the commodity as part of the provisioning mechanism, thus avoiding the extraordinarily high costs that can come from a strategy of total self-sufficiency. Even in countries as large as Indonesia, India, and China, where a high degree of food self-sufficiency is required simply because of the limited size of world grain markets, some interaction with these markets through a managed trade regime can lower the costs of food security. Managed trade regimes can be open and transparent, with clear rules on the nature of interventions, thus allowing the private sector to handle actual trade logistics.

The third approach to coping with a food crisis is to provide safety nets to poor consumers, either in cash or through the direct provision of food aid. This was the immediate—and almost only—response of the donor community to the recent food crisis. The safety net approach figures prominently in best practice recommendations from the World Bank, FAO, and WFP (World Bank 2005). The logic is clear: let high prices be reflected in local markets to signal the necessary changes in resource allocations to both producers and consumers, but protect the very poor from an irreversible deterioration in their food intake status. Efficiency is maintained, and the poor are protected. Barrett and Lentz (2009) cogently explain the behavioral foundations and research base on which this approach is based.

The difficulty is that food crises (as opposed to chronic poverty) are relatively short-lived events. Effective safety nets take a long time to design and implement, and they are very expensive if the targeted poor are a significant proportion of the population. Unless a well-targeted program with adequate fiscal support is already in place when the crisis hits, it is virtually impossible for a country to design and implement one in time to reach the poor before high food prices threaten their nutritional status. Even when a program is in place and can be scaled up quickly, as with the Raskin program of rice distribution to the poor in Indonesia, operational inefficiencies and simple corruption in deliveries may mean the poor are reached only at exceptionally high cost (Olken 2006).

After decades of rapid growth, the chronic poor in East and Southeast Asia (excluding China) have been reduced to less than 5 percent in the wealthier Asian countries, less than 20 percent in the emerging Asian countries, and perhaps 40 percent in the still poor countries of the region (Thapa and Gaiha 2011). But in the latter two categories, a substantial proportion of the population is "near poor," and thus vulnerable to shocks, such as spikes in food prices, that would push them below the poverty line. As a welfare issue—and one with immediate political consequences—spikes in rice prices have direct and immediate consequences for substantial numbers of people in this region. It is understandable that governments seek to prevent such spikes.

POLICY RESPONSES TO FOOD CRISES

Examples from three key countries in Asia help us to understand the political economy of actual responses to food crises (see Table 18.1). A brief summary of these responses by Thailand, Indonesia, and India during the food crises that began in 1972 and 2007, and during the collapse in commodity prices in 1985, reveals an underlying historical continuity as well as quite remarkable changes in policy approach over the period.

Thailand, usually the world's largest rice exporter since the 1960s, flipped its approach from stabilizing domestic rice prices in the early crisis to permitting full transmission of the price spike to producers and consumers in the most recent one. Export prices for rice from Thailand rose 138 percent between 1972 and 1973, whereas domestic retail prices rose just 13 percent. By contrast, export prices nearly doubled between 2007 and 2008, and so did domestic retail prices.

In the mid-1980s, Thailand passed the full brunt of price declines in world markets to its farmers, although consumers seemed to pay relatively higher prices during that period. In the mid-2000s, Thailand also initiated an expensive price support program for rice farmers, one reason the government was

Table 18.1. Comparing two world food crises for the rice economy

Country	1972/73	Price collapse in mid-1980s	2007/08
Thailand (exporter)	Banned exports and kept domestic prices relatively stable but destabilized world market	Passed low prices through to farmers, with increase in rural poverty	No control on exports and local prices followed world prices. Discussed forming a rice exporters cartel
Indonesia (importer)	Scrambled for imports but lost control of domestic prices. Led to new policies favoring agricultural development	Kept domestic prices above world prices but had surpluses and very high storage costs. Rural poverty declined	Already had high prices and did not import. Prices remained stable, but above world prices except at very peak
India (importer to exporter)	Sharply reduced imports with higher domestic prices and reduced food grain consumption. Stimulated more investment in raising rice productivity	Continued to expand rice investments, kept farm prices high and stable. Subsidized rice exports into a falling world market	Banned rice exports to stabilize domestic prices, with sharp impact on prices in world market. This policy was very popular for Congress Party

Source: Author

happy to pass through the higher prices in world markets in early 2008. This transition in price stabilization policy corresponds to the transition from authoritarian rule to popular democracy, with steps forward and backward along the way. Farmers remain a very large fraction of Thailand's electorate, and urban consumers have gotten used to relatively higher rice prices. Still, it is quite remarkable how radically Thailand's approach to rice price formation has changed.

Indonesia, as the world's largest rice importer over this period, shows a reverse transition in how to manage food security. During the food crisis in 1972 and 1973, the country lost control of its domestic prices—domestic retail prices increased 54 percent between 1972 and 1973—and the Suharto government almost lost political control as well, with widespread urban riots protesting high rice prices. In the most recent crisis, retail rice prices in Indonesia did not increase at all, and the democratic government took political credit for that fact. The contrast with Thailand goes even further. In the mid-1980s, Indonesia stabilized its domestic rice prices at levels well above the world price, to the point of generating substantial surpluses that needed to be exported at subsidized prices.

Although this policy was a very expensive undertaking, rural poverty in Indonesia continued to decline in the mid-1980s, in contrast to the rising rural poverty in Thailand (Ravallion and Huppi 1991). The commonality of policy experience for Thailand and Indonesia, however, is also striking, as both countries introduced price regimes that were much more favorable to rice farmers as democratic forces took increased political control.

India, of course, has been a democracy throughout this period. Historically, the country had been a regular rice importer, with supplies coming mostly from Burma and Thailand. India was slow to adopt Green Revolution rice technology despite being a leader in the wheat revolution. Even so, its rice imports in 1972 and 1973 were relatively small, and it got through the food crisis in those years with relatively modest increases in domestic rice prices—just 19 percent. Still, food grain consumption dropped sharply because of the El Niño-induced drought's impact on production of wheat and rice in 1973 and because of reduced imports of both food grains.

Indonesia's response to the rice shortages and high prices was paralleled in a similar response by India. Significantly greater attention to irrigation, research and extension, fertilizer availability and price, and maintenance of stable incentive prices led to a sharp increase in rice production over the following decades. As in Indonesia, these measures continued right through the price decline in world markets during the 1980s. Although the US dollar price of rice at the farm level was 10 percent lower in Thailand in 1985 than in 1973 (and these are nominal prices!), in Indonesia and India the farm level prices were 51 percent and 13 percent higher, respectively, in 1985 than in 1973. Clearly, the traditional importers felt threatened by the unreliability of the world rice market. Whether authoritarian or democratic state, food security required that far more resources be devoted to rice production.

This production initiative was much more successful in India than in Indonesia, at least in terms of import dependence. By the 1990s, India was a large and regular rice exporter, whereas Indonesia had reverted to substantial imports—over six million metric tons during the crisis year of 1998, a significant share of it from India. In 2007 India exported 6.3 million metric tons of rice. Even after the export ban, it still exported 3.3 million metric tons in 2008. Many of these shipments were basmati rice, which was not subject to the export ban. It is perhaps no coincidence that through most of the 1990s, poverty declined much more rapidly in Indonesia than in India. This remained the case until the Asian financial crisis in 1998 (Timmer 2004).

India faced a fundamentally different set of options during the 2007 and 2008 food crisis than it did in the 1972 and 1973 food crisis. As a large exporter, it had the opportunity to prevent domestic food prices from rising quickly by simply restricting trade. Of course, as a large exporter, such restrictions were likely to have an immediate impact on the world market—and they did (Slayton 2009a, 2009b). India took a lot of international political heat for its ban on rice exports, but the government argued that its first responsibility was food security for its own citizens. The subsequent national elections in May 2009 suggest that the electorate agreed with that position.

The underlying political economy of four decades of coping with rice price volatility, at least as seen through the lens of these three countries, is not hard to discern. In the short run, price stabilization is critical in the poorer countries (India and Indonesia, and Thailand in the early period). Both India and Indonesia learned that they could not stabilize rice prices at low prices because they needed their rice intensification programs to succeed. Millions of small rice farmers respond to incentives, whether in democratic or authoritarian regimes. With higher incentive prices domestically, despite low rice prices in world markets, rice production increased, and growth in consumption slowed.

In response to the impact on consumption of higher prices, both countries used physical distribution programs to alleviate the effect on poor households: the "below poverty line" (BPL) program in India and the Raskin program in Indonesia. Both safety net programs are very costly, with low efficacy. But the combination of price incentives to farmers and subsidies to consumers has proven politically popular in both countries. Prime Minister Manmohan Singh and President Susilo Bambang Yudhoyono were both reelected in 2009 with strong mandates. Part of their popularity stems from the price stability made possible by their common approach.

Most of what India and Indonesia did to cope with the world food crisis in 2007 and 2008 violates the guidelines provided by the World Bank and other donors for best practices in dealing with food price volatility (see World Bank 2005, 2009). Aggressive use of trade and stocks policy to stabilize domestic prices, combined with in-kind rice distribution programs to the poor, are all included in "policies to avoid." And yet both governments were rewarded with huge electoral victories in 2009, to the surprise of many outside observers. Do "bad" economic policies, at least with respect to food price volatility, make for "good" politics?

Surely the answer depends on how we define bad economic policies. The argument here is that government interventions to stabilize rice prices in domestic markets can be considered good economic policy if they are done right. Academics and donors have mostly denied this possibility in the past several decades, thus cutting government officials off from helpful dialogue, technical assistance, and funding to make these interventions more transparent, cost-effective, and supportive of market development. A different attitude is needed if the policy dialogue is going to be more fruitful, and if academics and policy analysts are going to have better training. We also need a new framework for thinking about how policy is adopted, the subject of the next section.

BEHAVIORAL DIMENSIONS OF FOOD SECURITY

Preventing food crises through better understanding of their fundamental causes, thus allowing implementation of better food policies, should be a high priority for food policy analysts. Once a food crisis hits, coping with its consequences becomes the main task at hand, with emergency food aid and other forms of safety nets hastily brought into play. But preventing food crises in the first place, *especially by preventing sharp spikes in food prices*, is obviously a superior alternative if a way can be found to do it. Understanding the behavioral dimensions of food security is an important first step. This section seeks to integrate new insights from behavioral economics into an understanding of why governments should stabilize basic food grain prices.[12] With a better understanding of "why," it is possible to suggest better approaches to "how."

The argument here is that highly volatile food prices—sharp spikes and price collapses—are undesirable for two separate reasons. First, it is increasingly recognized that volatile staple grain prices have serious consequences for economic welfare, especially for the poor (Timmer 1989; World Bank 2005).[13] Second, and the new argument here, spikes in food prices universally evoke a visceral, hostile response among producers and consumers alike. This response has deep behavioral foundations: the experimental and psychological literature shows clearly that individuals strongly prefer stable to unstable environments.[14]

Kahneman and Tversky (1979), for example, in their treatment of decision-making under risk, establish reference points for individual decisions as the basis for the widespread loss aversion that is the foundation of what they call "prospect theory." The pervasiveness of loss aversion among individual decision makers has immediate implications for how we should think about welfare losses from unstable food prices. Equal movements in prices up and down over time leave society worse off because the welfare losses from such price movements always outweigh the welfare gains. The asymmetry of welfare losses caused by loss aversion means that the "gains to trade" possible when prices are unstable will be less than the losses. This result alone explains much of the empirical political economy of food prices.[15]

Although this behavioral response is part of the reason that individuals tend to be risk averse, the implications are actually more profound. It is conceptually possible to hedge the risks from unstable food prices or to mitigate their welfare consequences for the poor using safety nets, but there are no markets in which to purchase stability in food prices directly. The message is clear. Citizens would willingly go to the market to buy food price stability, but such a market does not exist. Food price stability is a public good, not a market good. Understandably then, citizens turn to the political market instead. Only political action and public response from governments can provide stable food prices. Thus food becomes a political commodity, not just an economic commodity, and we will need a "behavioral political economy" to understand food policy.

Understanding the behavioral foundations of formation of price expectations will be critical to building this new political economy. In particular, the dynamics of herd behavior and the tendency of bad news—about terrorism, wild fires, or a sudden rise in rice prices in local markets—to serve as a focusing event in stimulating simultaneous, spontaneous behavior that results in panics provide robust insights into how individuals form price expectations and respond to them (Tversky and Kahneman 1986).

Governments that fail to stabilize food prices have failed in the provision of a quite basic human need that is rooted in behavioral psychology—the need for a stable environment. Governments that are successful in stabilizing food prices are usually rewarded politically: witness the landslide victories for Singh in India and Yudhoyono in Indonesia. Clearly, other factors contributed to the electoral success in both countries, but it is equally clear that the governments' abilities to provide stable domestic food prices when the rest of the world was experiencing a food crisis were politically popular.

The trick, of course, is to provide stability in domestic food prices at low cost to economic growth and participation by the poor. By and large, Asia has figured out how to do this as a domestic endeavor, but with large negative spillovers to world markets (Timmer 2009b). African countries do not have a viable strategy for stabilizing their domestic food prices, and the continent suffers even more from the instability in world markets transmitted from the Asian approach to food price stabilization (Jayne 2009). Indeed, the resource riches of Africa are attracting sizable investments from Asia (and elsewhere), but there seems to be little linkage between these investments, especially in land to produce food crops for export, and local food security.

The challenge to the development profession is thus twofold: to help Asia find more efficient ways to stabilize domestic food prices, especially for rice, with fewer spillovers to world markets, and to help Africa find a way to stabilize domestic food prices without introducing serious food economy distortions or retarding the development of an efficient private food-marketing sector.

INSIGHTS AND OUTCOMES

Three points stand out in summary. First, despite its declining economic importance in East and Southeast Asia (even including China and South Asia), rice remains the region's touchstone of food security in political terms. By and large, Asian governments define food security as a political concept by their ability to maintain reasonably stable rice prices in the main markets in their countries. The policies used to stabilize rice prices often include import and export controls on rice trade. Such controls can be effective, even for large countries, but they come with high external costs to the stability of the world rice market, which is also centered in this region. At the moment, there is no way to force these external costs to be internalized by the countries imposing them, and thus no effective way to prevent their use in the future.

Second, governments have good reason to fear the response of their citizens to a highly volatile rice economy. Behavioral economics provides helpful insights into why most people have a strong preference for stable prices: loss aversion is a powerful predictor of how people respond to changes in their circumstances. Accordingly, policy analysts need to help governments in the region design more efficient and effective rice stabilization programs, rather than denying the desirability or feasibility of doing so. Such programs will be put in place with or without our help. Perhaps, with better analytical input, these programs could be less costly and have fewer spillovers to the world market.

Third, there has been much discussion over the direction of causation in the relationship between food security and political instability. The place to test for causation is East and Southeast Asia. A formal analysis has not yet been conducted, but proposed research on the impact of ENSO events on regional rice production and trade, and from there to government interventions in defense of stable rice prices, offers an opportunity to identify causal drivers in a clear statistical fashion. The working hypothesis is that exogenous threats to food security—from a powerful El Niño, for example—directly threaten political stability in these countries, and government leaders understand that. Food security policies in the region are designed around this reality.

It seems appropriate to also ask what role the donor and academic communities in general, and the US government in particular, might play in improving food security, on one hand, and helping to maintain political stability, on the other. Are the links between the two sufficiently robust to provide guidance on interventions, investments, and policy advice?

This is treacherous territory. Interventions to improve food security in the short run can have negative long-run consequences. Political instability in corrupt and venal regimes, even if stimulated or ignited by food riots, may be the only way forward to a better-governed society. That said, historical experience in the East and Southeast Asia region (excluding China) does suggest that donors and policy analysts could play a much more helpful role in their assistance programs and analysis by recognizing the political imperative of a stable food economy. From that recognition would flow different priorities in foreign assistance: more efforts to build public-sector management capacity and less concern for improving the mechanics of markets for financial derivatives; more concern for stabilization mechanisms that work in a cost-effective manner and, perhaps via greater transparency in policy design, that would have smaller spillovers into world markets. Fewer resources might be devoted to documenting the benefits of free trade and to selling its merits to dubious policymakers.

For the United States, the single most important thing we can do to improve global food security is to place a high tax on the conversion of corn into ethanol. No other policy step has the potential for such far-reaching and positive effects on the global food economy. Of course, this action has very little directly to do with East and Southeast Asia (excluding China), except that it buys a lot of corn for its livestock sector. But if a signal went out that converting food into fuel is an environmentally destructive policy and the source of significant food insecurity, the lessons for cassava conversion to ethanol or palm oil conversion to biodiesel, might be understood and acted upon within the region.

ACKNOWLEDGMENTS

A lifetime of work incurs many debts, which I will not try to acknowledge here. The two key acknowledgments for this particular chapter are to Chris Barrett, who has pushed me pretty hard to make my arguments more compatible with mainstream economics (although we both recognize I am still far from the core), and to Erik Thorbecke, my commentator at the conference, who encouraged me to push further into the behavioral economics of political economy.

NOTES

1. This is, of course, a huge accomplishment, in view of World War II, the Korean War, and the Vietnam War. The obvious caveat to this sanguine view is the escalating tension in the South China Sea, where multiple claimants in the region, especially Vietnam and the Philippines, are challenging China's claim to the entire area. The defense treaty between the Philippines and the United States could draw the United States into a broader conflict over this issue.
2. The early standard works are Wickizer and Bennett (1941) and Barker et al. (1985).
3. This is a polite way of saying that there is a good deal of smuggling. Information on the world rice market is available from the "World Rice Statistics" section of the IRRI website (irri.org) (accessed April 11, 2013).
4. See the discussion of long-run price trends for rice, wheat and corn (maize) in Timmer (2009b) and Dawe (2010b).
5. The infamous Korean Daewoo deal for 1.3 million hectares of land in Madagascar, which precipitated a coup and the fall of the democratically elected government, is the best known example, but all of "rich Asia" is seeking some insurance against food export embargoes and boycotts as a way to preserve their import-dependent food security.

6. A forthcoming report from USDA comes to a similar conclusion. See *Southeast Asia's Rice Surplus* by Katherine Baldwin, John Dyck, Jim Hansen, and Nathan Childs, to be published in the fall of 2012 as an ERS/USDA (Economic Research Service, US Department of Agriculture) Electronic Outlook Report.
7. Indeed, even reliable price quotations for internationally traded rice are hard to obtain. The world rice market is quite opaque because most transactions are not reported publicly, and significant quality differences from lot to lot mean that "the price of rice" is impossible to define with the same precision as for publicly traded commodities such as wheat and corn.
8. This difference was pointed out clearly in Jasny's classic study of *Competition Among Grains* (Jasny 1940). He justifies his exclusion of rice from the study with the following observation: "The Orient is a world by itself, with its own climate, diet, and economic and social setup, and this makes it easy for us to omit it. The inclusion of rice would mean the discussion of two worlds. The writer would be satisfied to have mastered one" (p. 7).
9. To market observers, it was almost amusing that Indonesia announced a ban on rice exports early in 2008, before its main rice harvest started in March. Historically, Indonesia has been the world's largest rice *importer,* surpassed only recently by the Philippines, and no one in the world rice trade was looking to Indonesia for export supplies. But there was a rationale to the announcement by the Minister of Trade—it signaled that Indonesia would not be needing imports and was thus not vulnerable to the skyrocketing prices in world markets. The calming effect on domestic rice market participants meant that little of the hoarding behavior seen in Vietnam and the Philippines was evidenced in Indonesia (World Bank 2009).
10. As further evidence that psychology was driving prices in the world rice market rather than fundamentals, it was the *announcement* by the Prime Minister of Japan that rice supplies would be available to the Philippines, not their actual shipment, that pricked the price bubble and started the rapid decline in rice prices. By late 2009, Japan had actually not shipped any rice to the Philippines, and overall rice exports from Japan declined in 2008 rather than increased as promised in Rome by then Prime Minister Fukuda (Slayton, 2009b; 2010).
11. Isolation from the world market does not, of course, guarantee more stable prices. Indeed, for most countries, open borders to world markets lead to greater price stability, as local shortages and surpluses can be accommodated through trade.
12. A separate paper (Timmer forthcoming) attempts a more complete development of these themes.
13. I distinguish between "variance" in food prices, a standard statistical measure of price movements around an average or a trend, and which often has a substantial degree of predictability because of seasonal patterns and links to storage levels, and "volatility" in food prices, which emphasizes unforeseen spikes and crashes. Somewhat confusingly, "instability" is often used both ways.
14. Bernheim and Rangel (2005) stress the seriousness of the challenge from behavioral economics to mainstream welfare analysis, which is based on the principal of revealed preferences, a challenge first presented by Duesenberry (1949) and revived by Kahneman and Tversky (1979). If revealed preferences from choices about consumption, income generation, and time allocation, for example, are not "really" what individuals prefer, or they incorporate what others are doing, as the

experimental evidence from behavioral economics suggests, the normative foundations of consumer theory no longer hold. Without these foundations, such stalwarts of applied welfare analysis as consumer surplus no longer have a theoretical basis. The consequences are obvious for the arguments here: models that international economists use to prove the existence of "gains to trade" no longer hold, and theoretical arguments against stabilizing prices also disappear.

15. See Lindert (1991) for a summary of the empirical regularities in agricultural policy that cannot be explained by standard neoclassical economics. These include a bias against both imports and exports, an urban bias in poor countries when farmers are a majority of the population, and a rural bias when urban consumers are a majority of the population.

REFERENCES

Baldwin, K., J. Dyck, J. Hansen, and N. Childs. Forthcoming. Southeast Asia's rice surplus. ERS/USDA Electronic Outlook Report. Washington, DC.

Barker, R., R. W. Herdt, and B. Rose. 1985. *The rice economy of Asia*. Washington, DC: Resources for the Future..

Barrett, C. B., and E. Lentz. 2009. Food insecurity. In *International studies compendium project 2009*, ed. R. A. Denemark. Hoboken, NJ: Wiley-Blackwell.

Bernheim, B. D., and A. Rangel. 2005. Behavioral public economics: Welfare and policy analysis with non-standard decision-makers (August). NBER Working Paper No. W11518.

Dawe, David, ed. 2010a. *The rice crisis: Markets, policies and food security*. London and Washington, DC: The Food and Agriculture Organization of the United Nations and Earthscan.

——— 2010b. Can the next rice crisis be prevented? in *The rice crisis: Markets, policies and food security*. ed. Dawe, 345–356. London and Washington, DC: The Food and Agriculture Organization of the United Nations and Earthscan.

Demeke, M., G. Pangrazio, and M. Maetz. 2009. *Country responses to the food security crisis: Nature and preliminary implications of the policies pursued*. Rome: FAO.

Duesenberry, J. S. 1949. *Income, saving, and the theory of consumer behavior*. Cambridge, MA: Harvard University Press,.

Falcon, W. P., R. L. Naylor, W. L. Smith, M. B. Burke, and E. B. McCullough. 2004. Using climate models to improve Indonesian food security. *Bulletin of Indonesian Economic Studies 40*(3): 357–79.

Food and Agriculture Organization of the United Nations (FAO) and World Food Programme (WFP). 2011. Special Report: FAO/WFP Crop and Food Security Assessment Mission to the Democratic People's Republic of Korea. Rome: FAO and WFP.

Haggard, S., and M. Noland. 2007. *Famine in North Korea: Markets, aid and reform*. New York: Columbia University Press.

Jasny, N. 1940. *Competition among grains*. Stanford, CA: Food Research Institute, Stanford University.

Jayne, T. 2009. Market failures and food price spikes in Southern Africa. Paper presented at the Experts' Meeting on Institutions and Policies to Manage Global

Market Risks and Price Spikes in Basic Food Commodities, FAO Trade and Markets Division, October 26–27, 2009, Rome.

Kahneman, D., and A. Tversky. 1979. Prospect theory: An analysis of decision under risk. *Econometrica 47*: 269–91.

Lagi, M., K. Z. Bertrand, and Y. Bar-Yam. 2011. *The food crises and political instability in North Africa and the Middle East.* Cambridge, MA: New England Complex Systems Institute.

Lindert, P. H. 1991. Historical patterns of agricultural policy. In *Agriculture and the state: Growth, employment and poverty in developing countries*, ed. C. P. Timmer, 29–83. Ithaca, NY: Cornell University Press.

Maddison, A. 1995. *Monitoring the world economy: 1820–1992.* Paris: OECD Development Centre Studies.

Natsios, A. 1999. *The politics of famine in North Korea. Special report from the United States Institute for Peace.* Washington, DC: US Institute for Peace.

———. 2001. *The great North Korean famine.* Washington, DC: Institute of Peace Press.

Naylor, R., D. Battisti, D. Vimont, and W. Falcon. 2009. Agricultural decision-making in Indonesia with ENSO variability: Integrating climate science, risk assessment, and policy analysis. Final Report, NSF Award ID: 0433679, Stanford University.

Olken, B. 2006. Corruption and the costs of redistribution: Micro evidence from Indonesia. *Journal of Public Economics 90*: 853–70.

Ravallion, M., and M. Huppi. 1991. Measuring changes in poverty: A methodological case study of Indonesia during an adjustment period. *World Bank Economic Review 5*(1): 57–82.

Slayton, T. 2009a. Arson forensics: What set the world rice market on fire in 2008? Paper presented at the FAO Workshop on Rice Policies in Asia, February 9–12, Chiang Mai, Thailand.

———. 2009b. Rice crisis forensics: How Asian governments carelessly set the world rice market on fire. Working Paper No. 163. Washington, DC: Center for Global Development.

——— . 2010. The "Diplomatic Crop," or How the US Provided Critical Leadership in Ending the Rice Crisis, in Dawe, ed. *The rice crisis: Markets, policies and food security.*

———, and C. P. Timmer. 2008. *Japan, China and Thailand can solve the rice crisis—but U.S. leadership is needed. CGD Notes (May).* Washington, DC: Center for Global Development.

Thapa, G., and R. Gaiha. 2011. *Agriculture-pathways to prosperity in Asia and the Pacific.* Rome: International Fund for Agricultural Development (IFAD), Asia and the Pacific.

Timmer, C. P. 1989. Food price policy: The rationale for government intervention. *Food Policy 14*(1): 17–27.

———. 2004. The road to pro-poor growth: The Indonesian experience in regional perspective. *Bulletin of Indonesian Economic Studies 40*(2): 177–207

———. 2009a. Rice price formation in the short run and the long run: The role of market structure in explaining volatility. Center for Global Development Working Paper 172.

———. 2009b. Management of rice reserve stocks in Asia: Analytical issues and country experiences. Paper presented at the Experts' Meeting on Institutions and Policies to Manage Global Market Risks and Price Spikes in Basic Food Commodities, FAO Trade and Markets Division, October 26–27, 2009, Rome.

——. 2012. Behavioral dimensions of food security. *Proceedings of the National Academy of Sciences 109*(31): 12315–20.

——. Forthcoming. The political economy of food security: A behavioral perspective. In *Handbook on Food*, ed. R. Jha, R. Gaiha, and A. Deolalikar. London: Edward Elgar Publishers.

——, and W. P. Falcon. 1975. The political economy of rice production and trade in Asia. In *Agriculture in development theory*, ed. L. Reynolds, 373–408. New Haven: Yale University Press.

Tversky, A., and D. Kahneman. 1986. Rational choice and the framing of decisions. *Journal of Business 59*(4): 5251–78.

Wickizer, V. D., and M. K. Bennett. 1941. *The rice economy of monsoon Asia.* Stanford, CA: Food Research Institute, Stanford University, in cooperation with the Institute of Pacific Relations.

Working, H. 1949. The theory of the price of storage. *American Economic Review 31*(December): 1254–62.

World Bank. 2005. *Managing food price risks and instability in an environment of market liberalization.* Agriculture and Rural Development Department Report No. 32727-GLB. Washington, DC: World Bank.

——. 2009. Boom, bust and up again? Evolution, drivers and impact of commodity prices: Implications for Indonesia. Trade and Development Report, Poverty Reduction and Economic Management (PREM) Department, East Asia and Pacific Region (May 30 draft). Washington, DC.

Index